Elements of STOCHASTIC MODELLING

2nd Edition

Elements of STOCHASTIC MODELLING

2nd Edition

Konstantin Borovkov
The University of Melbourne, Australia

World Scientific

NEW JERSEY • LONDON • SINGAPORE • BEIJING • SHANGHAI • HONG KONG • TAIPEI • CHENNAI

Published by

World Scientific Publishing Co. Pte. Ltd.

5 Toh Tuck Link, Singapore 596224

USA office: 27 Warren Street, Suite 401-402, Hackensack, NJ 07601

UK office: 57 Shelton Street, Covent Garden, London WC2H 9HE

Library of Congress Cataloging-in-Publication Data

Borovkov, K. A. (Konstantin Aleksandrovich), author.

Elements of stochastic modelling / by Konstantin Borovkov (the University of Melbourne, Australia). -- 2nd edition.

pages cm

Includes bibliographical references and index.

ISBN 978-9814571159 (hardcover : alk. paper) -- ISBN 978-9814571166 (pbk. : alk. paper)

1. Stochastic processes. 2. Mathematical models. 3. Computer simulation. I. Title.

QA274.B655 2014

519.2'3--dc23

2014010103

British Library Cataloguing-in-Publication Data

A catalogue record for this book is available from the British Library.

Printed in Singapore

To all my teachers

Preface to the First Edition

The present text has been developed from lecture notes for a one-semester course on stochastic modelling. This is an introductory/intermediate level stochastic processes course taught to third year undergraduate students at the University of Melbourne by the author over the last five years. Most of the students doing the subject have already done a second year probability course, but otherwise have very different backgrounds. The reality is that some of them (in particular, actuarial students) are quite good in mathematics, while others may have problems with rather simple topics. Hence one has to try to include material suitable for almost all cohorts of students taking the subject as they should be able to follow. At the same time, the material has to be "challenging" (at least for most of them) as well as giving a reasonable overview of (at least, some part of) the huge area of stochastic modelling. On the other hand, it is mandatory to include certain topics in the subject which some cohorts of students have to cover in their courses. All this means that we need to have plenty of rather advanced material in the subject, but going into technical details and giving "normal" proofs would simply be impossible.

For this reason, in developing the course, we had to omit rigorous proofs almost completely, sometimes replacing them with sketches of arguments and trying to indicate why this or that particular assertion holds, and also how it is connected with other results in the area (or, even more generally, in mathematics and statistics). To compensate for this "rigour deficiency", we included, wherever possible, references to more specialised texts where students could find both the proofs of the cited/used results and more advanced material related to the topics covered in our text. So we hope that students who choose to more seriously study probability theory and stochastic modelling will be able to find all the missing proofs in the referenced literature. Also, we always attempt to explain rather than give ready recipes and to avoid making the overall exposition primitive. Our experience shows that this approach proved to be quite successful.

Why was the present text written? There exist quite a few very good texts on stochastic processes. The main problems with using them are that: (i) we would have to recommend students several texts in order to cover the material included in our course, and (ii) the level of exposition in most of the texts is either too low or too high for the group of students we have to teach. So eventually we had to choose to compile a new text and add references to those good books which may contain more detailed/better exposition covering various topics included in our text.

It should be stressed that we assume that the reader is already familiar with the elements of probability theory: this text is intended for students doing their *second* course in that discipline.

The book consists of ten chapters. Chapter 1 is a general introduction to the subject. Chapter 2 is devoted to (more or less rigorous) reviewing the basics of probability, introducing at the same time some elements of stochastic processes theory (and also such things as utility functions etc). We try to keep the exposition at a reasonably elementary level (in particular, we do not introduce the notion of martingale: it is dealt with in a companion subject taught at our department).

Chapter 3 is devoted to Markov chains. We give general definitions and basically concentrate on the case of denumerable state spaces. The main emphasis is made on examples. Chapter 4, in a sense, continues the previous one and deals with Markov decision processes. Again, we give the basic definitions and ideas and then attempt to illustrate them by examples.

Chapters 5 and 6 deal with continuous time jump Markov processes. The former is devoted to the exponential distribution and discusses in detail the Poisson process, whereas the latter covers the general case (including time-inhomogeneous processes) and then concentrates on birth-and-death processes. Chapter 7 applies the techniques developed in the two previous chapters to queuing systems.

Chapter 8 is devoted to renewal theory. We give the main facts from the area and also indicate how they are used in the theory of Markov chains.

Chapter 9 covers the basic elements of time series. We mostly discuss the structure of the classical time series models, and touch only very briefly upon the statistical aspects.

Chapter 10 deals with the basics of simulation of random numbers. We give an idea of how sequences of uniform (pseudo) random variables can be generated and then proceed to discussing in more detail methods for simulating non-uniform random variables. We also discuss elements of Monte Carlo techniques, including variance reduction methods, and introduce the reader to the Markov Chain Monte Carlo.

The passages set in small font are intended for the curious reader with a strong mathematical background. They contain elements of more advanced topics and often attempt to give a deeper insight into the course material.

At the end of each chapter, there is a separate section with problems. For readers' convenience, answers to problems and a few lists (including the list of abbreviations used in the text and the list of notations) are placed at the end of the book. (Talking about notation: throughout the text, $\log x$ stands for the natural logarithm of x. Perhaps there will be no better place to make this comment.)

Some of the examples and problems in the present text have been adapted—in most cases after substantial modification and/or extension—from the books and papers referenced both in the footnotes and at the end of the respective chapters. It is virtually impossible—as it is (almost) always the case with textbooks—to give credit to all the sources I used at all the instances. The only thing I can be quite sure about now, is that most of the mistakes in the text are mine.

Talking about mistakes: I would like to thank all the students who, having had a predecessor of this text as typed lecture notes, spotted dozens of typos and other bugs in it and kindly informed me about them. My special thanks go to my colleagues István Gyöngy and Anthony Brockwell who enthusiastically suffered the reading of the manuscript at the last stage of its preparation and helped me to improve the text. I also wish to thank the Department of Mathematics and Statistics of the University of Melbourne, which was a friendly and supportive environment for writing the present book.

Melbourne, Australia
October 2002

K. Borovkov

Preface to the Second Edition

The second edition is a revised and significantly expanded version of the original book. The newly added material comprised a substantial part of a companion course that followed the course the first book was based on, and that was taught basically to the same cohort of students. Therefore, when working on that companion course, the author encountered the same challenges: diverse backgrounds of students taking the course, the need to include a number of rather advanced mandatory topics *etc.* It was hence decided to follow the same philosophy as had been successfully used for the first course: a relatively low proportion of results is presented with full rigourous proofs, the main accent being on explaining rather than giving any ready recipes. It is hoped that the new version of the book is more "rounded-up" and would be suitable for a variety of courses, and so perhaps could be used for more than one course taught to a given group of students that may have different majors.

Belonging to the second part of a year-long combined course on stochastic modelling, the new material presented in the book is somewhat more advanced compared to the rest of the text. Accordingly, there are more proofs (although quite often not perfectly rigourous) in the newly added three chapters. Chapter 11 presents the first elements of the theory of martingales and stochastic Itô calculus. We prove neither any fundamental martingale inequalities nor any limit theorems for martingales, concentrating rather on discussing the nature of the processes, their use in stochastic modelling and presenting a few key technical tools. The same approach is followed when explaining the origin and nature of the Brownian motion process, Itô integration and stochastic differential equations. As in all the other chapters, we include a list of recommended literature where the interested reader can find more detail and advanced results.

Chapter 12 is devoted to diffusion processes. We explain the nature of diffusion processes, derive Kolmogorov differential equations for them and discuss how to use the latter to compute important characteristics of the processes such as their stationary distributions (when they exist), boundary hitting probabilities, the distributions of the first hitting times *etc.*

Chapter 13 presents elements of mathematical finance: we discuss there the ideas of arbitrage-free options pricing. The key concepts are first explained in the context of binomial and, more generally, finite discrete time markets, and then extended to the continuous time case, where the dynamics of the risky assets are described by the geometric Brownian motions (the situation often referred to as the Black–Scholes framework). The presentation often uses the material presented in the two previous chapters.

To make reading easier, we are using throughout the text the end of proof mark □ to denote the end of arguments demonstrating theorems and lemmata.

The first edition of the book received very positive reviews from specialists in the area. It is hoped that the second edition will be liked and enjoyed by both instructors and students at least as much as the first one.

I am very grateful to my colleagues István Gyöngy, whose lecture notes influenced part of the exposition in the newly added chapters and who provided me with important comments on the drafts of these chapters, Alexander Novikov, with whom I had the pleasure of discussing some aspects of the material included in those chapters, and Shaun McKinlay who proof-read the additional chapters. I also wish to thank the Department of Mathematics and Statistics of the University of Melbourne, for being a friendly and supportive environment for preparing the second edition of the present book.

Melbourne, Australia *K. Borovkov*
January 2014

Contents

Chapter 1

Introduction

> An engineer and a mathematician went to the races one Saturday and laid their money down. Commiserating in the bar after the race, the engineer says, "I don't understand why I lost all my money. I measured all the horses and calculated their strengths and mechanical advantages and figured out how fast they could run..."
>
> At that moment the mathematician pays for his beer and the engineer gets a glimpse of his well-fattened wallet. Obviously here was a man who knows something about horses. The engineer demanded to know his secret.
>
> "Well," says the mathematician, "first I assumed all the horses were identical and spherical..."

1. On the nature of mathematical modelling. Imagine a complex real-world system (say, the economy of a country or an ecological system) and suppose that one wants to know how the system will tend to evolve in the future. This is of particular importance when one is about to change something in the system (say, introduce a new tax or build a dam). Real-world experimentation is usually very risky and costly on the one hand, and far too slow, on the other, to think about trying it seriously. Instead, one develops a *model*—an "imitation" of the system of interest—and first studies and/or "runs" or "simulates" that model. It is very important to be able to study the evolution of the modelled system in "compressed time" and compare the consequences of different policies or possible actions for the system.

Any "non-physical" model constructed for analysis and simulation purposes will necessarily be mathematical. A *mathematical model* is a (rough) description of a class of real world phenomena expressed using mathematical symbolism. It is a powerful tool not only for understanding the underlying laws, but also—most importantly from the application point of view—for *predicting* and *controlling* the behaviour of the modelled systems.

The process of mathematical modelling could be divided into four stages.

The **first stage** is the statement of the laws relating the basic elements of the system to be modelled. One formulates and describes, in mathematical terms, connections between the objects of the system. An important aspect of that stage is the necessity of selecting the most important features of the system and omitting all irrelevant particulars (which would otherwise make the mathematical model completely intractable). It is amazing how crude a successful mathematical model can be. For example, when deriving the differential equation $x''(t) = -ax(t)$ for a clock's pendulum, one neglects (quite obvious) deviations of the assumptions made about the pendulum (a massless rod/thread, a point-size bob, small amplitude of the oscillations *etc.*) from the physical reality. The resulting *model* (often a system of differential equations or, more typical of our text, a random process given by its probability distribution) already belongs to the world of *mathematics*. Quite often one comes up with the same *mathematical model* for apparently completely different and unrelated real-life phenomena. Thus, the pendulum equation describes not only the movement of a clock's pendulum, but also oscillations of a weight attached to a spring, changes in characteristics of electromagnetic fields in various systems and so on.

It is common practice to involve experts on real-world systems of interest to get feedback on the validity of the model.

Quite often the derived mathematical model proves to be so complex that one cannot obtain an analytical description of its behaviour. Then one attempts to solve the problem numerically and/or to *simulate* the system by writing a computer program (using a special simulation package or a general purpose programming language). This approach has become extremely popular with the fast progress of computational hardware. But even when it is taken, it is still extremely important to know and understand the mathematical aspects of the model. It is needed for the correct formulation of the model itself and correct choice of approaches to solving the problem. In particular, sometimes one can analytically solve related mathematical problems, say, leading to analytical representations for some components of the model, and this can greatly reduce computation needed for simulation.

The **second stage** is the investigation of the mathematical problems to which the mathematical model leads. One of the main questions here is the solution of the so-called *direct problem*: given the "input data", obtain the "output data" as a result of analysis of the model. (For example, find the position of the pendulum at time $t = 10$ given that at time $t = 0$ it was at the stationary point and had a known velocity $x'(0) = -0.5$.) It is at this stage when the knowledge of the relevant parts of mathematics and computational skills become crucial for success. Mathematical problems arising when studying different models frequently turn out to be essentially identical. These typical mathematical problems can then be considered as independent objects abstracted from the phenomena. In the present course, we will mainly be dealing with that stage of mathematical modelling.

At the **third stage** we validate the model. We do this by verifying whether the results theoretically derived for the adopted model (reasonably) agree with the observed results. If the deviation lies outside appropriate limits (of which the determination and justification is another problem), the model cannot be accepted. When this is the case, one has to go back to stage one, try to determine if any essential factors had been overlooked, and formulate an alternative model.

Models often contain *a priori* undetermined characteristics (usually they are *parameters* of the models), and then one needs to solve the *inverse problem* which consists in finding the values of the characteristics such that the output of the model agrees with the empirical observations. Such a *fitting* of the models is closely related to problems of mathematical statistics.

The **fourth stage** is the subsequent analysis of the model. If one requires a more detailed or just better description of the phenomenon of interest, the existing model can prove to be unsatisfactory. This leads to the need of constructing a new, more precise, mathematical model. We return to stage one, but perhaps at another level of complexity.

2. Stochastics. Stochastic[1] models form a special class of mathematical models intended to describe a specific class of real-life phenomena that are characterised by the presence of *uncertainty*. That is, in an apparently random[2] fashion, in conducted experiments or empirically observed sequences of replications of the same *complex of conditions* (trails), cer-

[1]Greek *stokhastikos*, via *stokhazomai*, "aims at, guess" from *stokhos* "aim" (for archers).

[2]Middle English from Old French *randon* "great speed", from *randir*, "gallop" (cf. German *rennen* "to run, rush"). It is interesting to note that these two synonymic adjectives, *stochastic* and *random*, refer to two different aspects of the phenomena. "Stochastic" indicates the need of making guesses, uncertainty in results, whereas something made at "random" was made in a haste, not systematically, and hence tends to be irregular, chaotic.

tain events can either occur or not occur. There are numerous examples of such phenomena: biological populations, traffic systems, stock markets, complex computer networks and so on—one can find examples practically everywhere. What is common to all these examples, is that the phenomena are extremely complex, and there are far too many factors to be taken into account. Just imagine how complicated a detailed description of a single trial in the classical coin tossing experiment could be.

But how could one incorporate such uncertainty into a rigourous mathematical model?

To get a meaningful description, we require the phenomena to satisfy certain criteria. Namely,

(i) They should be of "mass character", that is, one could (at least theoretically) replicate unlimitedly many times the "same" complex of conditions that leads to our *random experiment* $\mathcal{E}$ (e.g., flipping a coin, births in a large population, phone calls *etc.*).

(ii) They should possess a property called the "statistical regularity": the relative frequency of events in a series of trials stabilises about a number between zero and one as the total number of trials increases. More precisely, if we have a series of "independent" identical *random experiments* $\mathcal{E}_i$ (i.e., replications of a certain complex set of conditions), $i = 1, 2, \dots,$ in which a certain event A can either occur or not occur, and if the event A occurs in[3]

$$n_A = \#\{\mathcal{E}_i,\, i = 1, \dots, n : A \text{ occurs in } \mathcal{E}_i\} \tag{1.1}$$

of the first n trials, then the *relative frequency* of the occurrence of A stabilises around a number between zero and one:

$$\frac{n_A}{n} \rightsquigarrow p \in [0, 1] \quad \text{as } n \to \infty. \tag{1.2}$$

Thus, tossing a fair coin leads to the limiting value $p = 1/2$ for the event $A = \{\textit{heads up}\}$, while, say, the proportion of 89 year old males who do not survive the next year is approx. 0.143.[4] This value p about which the relative frequency stabilises for large n is called the *probability* of the event A. For practical purposes, whenever one speaks about probabilities, the main meaningful interpretation is that, in a long series of independent replications of the random experiment, one would observe the tendency formulated as (1.2).

[3]Notation $\#S$ is used to denote the number of elements in the set S. In (1.1), the set S consists of all replications $\mathcal{E}_i$ of our random experiment with $i = 1, \dots, n$ such that the event A occurred in $\mathcal{E}_i$.

[4]According to the Australian 2008–2010 Life Tables (source: `http://www.abs.gov.au/`).

It is actually an empirical fact that once (i) is true, it is most likely that (ii) also takes place. *Why* this happens, is a very interesting and fundamental question. You may think of this as a consequence of the "chaotic" nature of the respective dynamical systems. Roughly speaking, this means that, as time passes, the system's state becomes—in two different senses—both very sensitive and insensitive to the initial conditions. Firstly, for any two (arbitrary) close initial states, the respective system's states will be quite far from each other after a long time. So such a sensitivity leads to apparently random outcomes when experimenting with the system. Secondly, for any small "volume" of the initial states in the state space, the set of the corresponding system's states at time t will tend to "fill" (or "spread over") the whole state space as $t \to \infty$, and this happens regardless of where the initial small volume was. Such insensitivity results in what we called above the "statistical regularity".

To illustrate this situation, consider a discrete time *dynamical system* given by the recurrent relation

$$x_{t+1} = f(x_t), \quad t = 0, 1, 2, \ldots, \quad x_0 \in \mathcal{X} = [0, 1), \tag{1.3}$$

with the right-hand side specified by the function

$$f(x) = \begin{cases} 2x, & x \in [0,\, 0.5), \\ 2(x - 0.5), & x \in [0.5,\, 1). \end{cases}$$

This is a sort of a continuous analogue to shuffling a deck of cards (when the cards from positions $1, 2, \ldots, 26$ go to places $2, 4, \ldots, 52$, while those at places $27, 28, \ldots, 52$ go to $1, 3, \ldots, 51$, respectively).

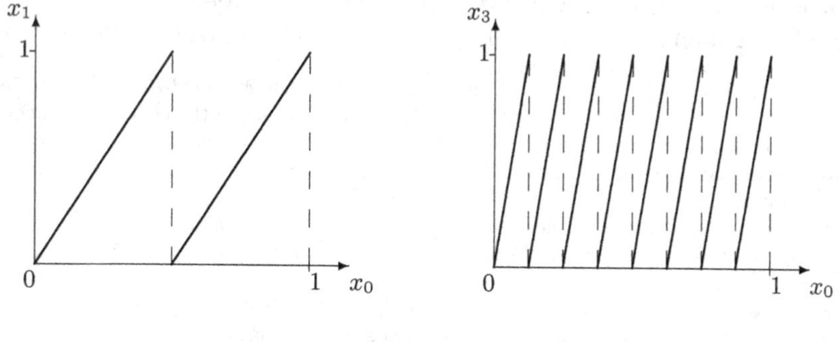

Fig. 1.1 Plots of x_1 and x_3 versus x_0 for system (1.3).

The plots of x_1 and x_3 as functions of the initial state x_0 are shown in Fig. 1.1. You can easily see what the general rule is and what the plot of x_t vs x_0 will look like for large t. For any two arbitrary close initial points x_0 and x'_0, the respective states x_t and x'_t will be quite far from each other for infinitely many values of t, while any small interval of the x_0's will eventually be mapped onto the whole $\mathcal{X}$. Moreover, for "almost all" x_0, the proportion of the time our x_t spends in a fixed volume A tends to a certain value $\mu(A)$ independent of the initial state x_0 (this is the assertion of the so-called *ergodic theorem*), which is actually our property (1.2). Thus, our completely deterministic system (1.3) displays behaviour similar to what we observe for stochastic systems.

On the other hand, for matter at the atomic and subatomic levels, where the classical physical theories fail and one needs to employ quantum mechanics, uncertainty is just an intrinsic fundamental property, and probability is the only language one can use to interpret the theory and describe what is happening.

One more important approach to the problem of randomness is from the point of view of *algorithmic complexity.* In two words, the idea is that once the algorithm[5] describing a system's evolution is long (i.e., complex) enough, the behaviour displayed by the system will appear random.[6]

For the time being, we simply observe that (1.2) is just an empirical law in a sense similar, say, to gas laws or Newton's second law: we know that it takes place for a certain class of phenomena (namely, when (i) and (ii) hold), and now the principal aim is to develop a mathematical theory that could be used to model such phenomena.

How can this be done?

The beginning of probability theory is often dated at 1654, the year when the famous correspondence on the so-called "Problem of Points"[7], an unsolved gaming problem of the time, took place between B. Pascal[8] and

---

[5]From Medieval Latin *algorismus*, after *Muḥammad ibn Mūsā al-Khwārizmī* (c.780, Baghdad – c.850), whose works introduced Arabic numerals and the concepts of algebra into European mathematics. His book *Kitab al-jabr we al-muqābalah* ("The Book of Integration and Equation") was translated into Latin in the 12th century and originated the term *algebra*.

[6]For more detail and references see, e.g., the article *Randomness and probability—complexity of description* in the *Encyclopaedia of Statistical Sciences*, Kotz, S., ed. V.7, Wiley, New York, 1986.

[7]Suppose two players stake equal money on being the first to win n plays in a game in which each play is a toss of a fair coin, heads for one player and tails for the other. Suppose the game is interrupted (by the wife of one of them, for example) when one player still lacks a plays to win, and the other b. How should the stakes be divided between the players?

[8]Blaise Pascal (19.06.1623–19.08.1662), a French mathematician, physicist and writer. Invented the first digital calculator, the syringe and the hydraulic press. About his life

P. Fermat[9]. But, of course, the computation of chances was not a completely novel exercise at the time of that correspondence. Anyway, during a period of almost three hundred years since that time, probability theory remained in a sense a sort of semi-heuristic science, with somewhat vague foundations and interpretation of its results (often quite strong and derived using very elaborate analytical techniques). One of its basic components was what is nowadays called the "classical probability". The term refers to situations where the random experiment of interest has only finitely many different possible outcomes, all of them being *equally likely*. The last word-collocation simply means that the situation is symmetric, and there is no reason to suspect that a particular outcome occurs more often than any other outcome of the experiment. Clearly, such an assumption is rather specific, and basing on it, one cannot go really far.

The importance of constructing modern axiomatic foundations of Probability Theory was recognised by D. Hilbert[10] who included that task in the sixth problem in his famous lecture on the key unsolved problem of mathematics delivered at the International Congress of Mathematicians in Paris in 1900. It appeared there as part of the more general problem of mathematical treatment of axioms of Physics:

> The investigations on the foundations of geometry suggest the problem: *To treat in the same manner, by means of axioms, those physical sciences in which mathematics plays an important part; in the first rank are* the theory of probabilities *and mechanics.*[11]

The first version of an axiomatic system to provide foundation for probability theory was suggested by S.N. Bernstein[12] in 1917. One can say that

see, e.g., Krailsheimer, A.J. *Pascal.* Oxford Univ. Press, New York, 1980.

[9]Pierre de Fermat (17.08.1601–12.01.1665), a French mathematician, the founder of the modern theory of numbers. About his life see, e.g., Mahoney, M.S. *The mathematical career of Pierre de Fermat.* Princeton Univ. Press, Princeton, 1973.

[10]David Hilbert (23.01.1862–14.02.1943), a German mathematician, one of the most universal and influential of his time, who made substantial contribution to establishing the formalistic foundations of mathematics.

[11]This is the standard (but somewhat loose) English translation of the original excerpt from Hilbert's talk: Durch die Untersuchungen über die Grundlagen der Geometrie wird uns die Aufgabe nahegelegt, *nach diesem Vorbilde diejenigen physikalischen Disciplinen axiomatisch zu behandeln, in denen schon heute die Mathematik eine hervorragende Rolle spielt; dies sind in erster Linie die* Wahrscheinlichkeitsrechnung *und die Mechanik.*

[12]Sergei Natanovich Bernstein (05.03.1880-26.10.1968), a Russian mathematician who contributed to various areas on mathematics and, in particular, laid the foundations of constructive function theory.

his approach was based on using Boolean algebras and eventually did not prove to be satisfactory.

The next attempt to construct a consistent formal probability theory was undertaken by R. von Mises[13] in the 1920s. The key element of the theory was the formal notion of a "collective"—an infinite sequence of experiments with definite numerical outcomes—having, along with (1.2), the following (also empirically observed) property: any subsequence chosen from a "collective" according to a rule specified in advance is again a "collective", with the same limiting value in (1.2). However, the theory proved to be rather cumbersome and very hard to work with.

A much simpler approach, reflecting not the "phenomenological" aspect of random experiments, but rather the natural properties of the limiting values of the relative frequencies proved to be much more successful. The approach basically combines *measure theory* (already quite well developed by the 1930s) with certain notions, specific to probability (first of all, with that of statistical independence), and was developed by another prominent mathematician of the 20th century, A.N. Kolmogorov[14]. In the next chapter we will review the basics of probability theory based on the Kolmogorov system of axioms. The review is rather brief, for we assume that the reader is already familiar with most of the material.

[13] Richard Martin Edler von Mises (19.04.1883–14.07.1953), an Austrian applied mathematician and philosopher. His approach to formalising probability theory was described in Mises, R. von. *Probability, Statistics and Truth*, Allen, 1957.

[14] Andrei Nikolaevich Kolmogorov (23.04.1903–20.10.1987), an outstanding Russian mathematician, one of the greatest personalities in the 20th century mathematics. It is hard to overestimate his contribution to various research areas in the field. The axiomatics of modern probability theory were published in: Kolmogorov, A.N. *Grundbegriffe der Wahrscheinlichkeitsrechnung*. Ergebn. d. Math. Heft 3, Berlin, 1933. [English translation: *Foundations of the theory of probability*, Chelsea, New York, 1950; 2nd edn. in 1956.]

Chapter 2

Basics of Probability Theory

To begin with, we would like to stress that Probability Theory is a rather special area of mathematics. Firstly, it is extremely broad in terms of its applicability: one can introduce probability on nearly any object belonging to any other area of mathematics, and this implies, in turn, its wide use in various applications. Secondly, unlike, say, Euclidean geometry and elementary calculus, probability theory is less supported by our everyday experience and hence is often counterintuitive. So usually it takes quite some time to develop "probabilistic intuition" enabling one to deal more confidently and feel more comfortable with probability, and successfully build and analyse meaningful stochastic models.

It may also be worth noting that, unlike some other areas of mathematics, one can hardly say that there is a small collection of standard methods or approaches one should learn to master probability theory. This is basically due to the above-mentioned diversity of probability-related problems. One can encounter situations where probabilities are defined on or related to practically any mathematical structure, and possible methods one can use to solve the respective problems will of course depend on both the nature of the structure and on how the probabilities were given. So approaches standard to different areas of probability theory use techniques from differential and integral calculus, linear algebra, functional analysis and many other fields.

The reader well familiar with the basics of probability theory may wish to skip most of this review chapter. It could later be used (if necessary) for reference when reading subsequent parts of the text. Note, however, that this chapter also introduces a few new important notions which the reader could encounter in her/his further studies. In particular, we do not recommend skipping Sections 2.9 and 2.10.

2.1 Probability Spaces

Any random experiment $\mathcal{E}$ is modelled by a **probability space** $(\Omega, \mathcal{F}, \mathbf{P})$, where

Ω is a set called the **sample space** (also referred to as the *space of elementary events/outcomes*),

$\mathcal{F}$ is a class of **events** in the experiment, and

$\mathbf{P}$ is a **probability** on $\mathcal{F}$ (probability is defined for events, not sample points!).

The correct choice of all the three components of the triple $(\Omega, \mathcal{F}, \mathbf{P})$ is the first crucial step in modelling.

Sample space. A sample space Ω is a set of "points" $\omega \in \Omega$ representing all possible outcomes of our random experiment $\mathcal{E}$ (sometimes ω is called the *chance*[1]). Recall that, as we have already said, in selecting an appropriate representation for the random experiment outcomes, we do not have to go into irrelevant particulars. For example, if we want to model the gaming situation of the "Problem of Points" mentioned in the Introduction, we begin with describing the outcome of a single toss of a coin. For this purpose, it suffices to designate two points, say, ω_1 for heads and ω_2 for tails. In fact, the points are often denoted by H and T, respectively. We are not interested in *where* the coin did land in the real-life experiment: it is irrelevant. The side that comes up is the only thing that matters.

To list all possible outcomes of the experiment consisting of tossing a coin twice (or, which is essentially the same, of tossing two identical coins), we observe that, to each of the two outcomes of the first trial, there correspond exactly two outcomes, H and T, for the second trial. That is, beginning with the sample space $\Omega_1 = \{H, T\}$ for a single trial, we now take our new Ω to be the set of all *ordered pairs* of the elements of Ω_1 (the first element standing for the outcome of the first trial, the second one representing the outcome of the second trial):

$$\Omega = \{HH, HT, TH, TT\}. \tag{2.1}$$

This is an elementary example, but even the greats used to make mistakes while dealing with such a simple—from our modern point of view!—problem. Thus, d'Alembert[2] wrote an article entitled *Heads and Tails* for

[1] From Latin *cadere*, to fall, befall.

[2] Jean Le Rond d'Alembert (17.11.1717–29.10.1783), a French mathematician, philosopher and writer. About his life see, e.g., Grimsley, R. *Jean d'Alembert, 1717-83*, Clarendon Press, Oxford, 1963.

the famous *L'Encyclopédie*[3] where he maintained that the probability of getting heads at least once in two tosses of a fair coin was 2/3. In other words, he did not distinguish between the outcomes HT and TH.

In the same way we obtain, for the experiment consisting of tossing n coins, a sample space Ω formed by all ordered n-tuples of the H's and T's. Note that the total number of points in that Ω is $|\Omega| = \underbrace{2 \times \cdots \times 2}_{n \text{ times}} = 2^n$.

Similarly, for rolling a single die we get $\Omega = \{1, 2, \ldots, 6\}$ ($|\Omega| = 6$), while for two dice one needs the sample space $\Omega = \{(1,1), (1,2), \ldots, (6,6)\}$ with $|\Omega| = 6 \times 6 = 36$ points.

The above examples lead to a very useful notion of *product spaces.* Suppose $\Omega_1, \ldots, \Omega_n$ are some sets (spaces), then their product[4] is

$$\Omega = \Omega_1 \times \cdots \times \Omega_n := \{\omega = (\omega_1, \ldots, \omega_n) : \omega_i \in \Omega_i,\ i = 1, \ldots, n\}.$$

The total number of points in Ω is easily seen to be

$$|\Omega| = |\Omega_1| \times \cdots \times |\Omega_n|,$$

so that when Ω_i are identical, we get $|\Omega| = |\Omega_1|^n$. Note that if all Ω_i are either finite or *countable*[5], and at least one of the "factors" Ω_j is countable, the product space Ω is also countable (can you show that?).

Product spaces are very often used to construct sample spaces for *compound* random experiments, which are combinations of simpler experiments (e.g., several tosses of a coin).

Of course, sample spaces do not need to be *discrete* (either finite or countably infinite). In most cases they are *continuums*, the basic case being the unit interval $[0, 1]$ serving as the sample space for choosing at random a point from the interval. Products of identical intervals are squares and cubes (sets of the form $[a, b]^n$ in the n-dimensional Euclidean space $\mathbf{R}^n$, which itself is the product of n copies of the real line).

[3] From Greek *enkyklios paideia* "general education" (just in case you didn't know).

[4] Also called the *Cartesian product*, after René Descartes (31.03.1596–01.02.1650), a French mathematician, scientist and philosopher, who has been called the father of modern philosophy by many renowned philosophers. The axiom *Cogito, ergo sum* (I think, therefore I am) is his most famous formulation. Giving tutorials proved to be fatal for this genius. In October 1649 he arrived in Stockholm to instruct in philosophy the 23-year-old ruling Queen Christina of Sweden, an ambitious patron of the arts and collector of learned men for her court. Three months later, he caught a chill that developed into pneumonia, and died.

[5] Countable (or *denumerable*) sets are the "smallest" infinite sets: they are sets of which the elements can be *enumerated.* That is, Ω is countable if there exists a one-to-one correspondence between it and the set $\mathbf{N} = \{1, 2, \ldots\}$ of all natural numbers (of course, if such a correspondence exists, it is not unique).

Once we have selected a sample space Ω for our random experiment, all the relevant *events* can always be expressed as collections of outcomes, i.e., *subsets* $A \subset \Omega$. One says that an event A occurs if the random experiment resulted in an outcome ω favourable to A, i.e., such that $\omega \in A$ (meaning "ω belongs to, or is an element of, A"). Thus, the event that there will be at least one heads in two tosses of a coin is the subset $\{HH, HT, TH\}$ of the sample space (2.1). The sample space Ω itself is called the *certain event* (for it always occurs!).

Denote by $\mathcal{F}$ the class of all events we want to consider in our random experiment. We need to have well-defined probability values for all elements of $\mathcal{F}$, and this imposes certain conditions on that class. When the sample space is discrete, we can include into $\mathcal{F}$ all the subsets of Ω and let $\mathcal{F} = 2^\Omega$, the class of all subsets of Ω (also called the **power set**).[6]

However, in the general case we *just cannot* take $\mathcal{F} = 2^\Omega$. The problem is that in non-trivial cases, it is simply impossible to consistently assign probabilities to all the elements of 2^Ω.

The most well-known example of such a situation shows that there is no way to define the length for *all subsets* of an interval (or a circle, which is basically the same) so that it would be invariant with respect to (w.r.t.) shifts.

One can argue as follows. Let $\Omega = \{(x,y) : x^2 + y^2 = 1\}$ be the unit circle on the (x,y)-plane. Take an arbitrary *irrational* number α (i.e., an α that cannot be represented as m/n for some integer m and n; for example, $\sqrt{2}$ can easily be shown to be irrational). For a point $s \in \Omega$, call the set $A(s)$ of all points obtained by rotating the (radius-vector) s by angles of the form $2\pi\alpha k$, $k = \dots, -1, 0, 1, \dots$, the *orbit* of s. In other words, if $s = (\cos\varphi, \sin\varphi)$ for some $\varphi \in \mathbf{R}$, then the orbit of s is $A(s) = \{(\cos(\varphi + 2\pi\alpha k), \sin(\varphi + 2\pi\alpha k)) : k \text{ is integer}\}$. If, for two points from Ω, their orbits have a common point, then the orbits will clearly coincide. Thus, for any $s_1, s_2 \in \Omega$, either $A(s_1) \cap A(s_2) = \emptyset$ or $A(s_1) = A(s_2)$. Denote by $\mathcal{A}$ the collection of all *distinct* orbits $A(s)$.

The next step is to *choose* a point from each of the orbits from $\mathcal{A}$. The set B of all thus chosen points is exactly what we are after. If we rotate B by a multiple of $2\pi\alpha$, the new set (which is a "copy" of B) will be disjoint with the original B! (Verify that!) Also, any $s \in \Omega$ belongs to one of such "rotated copies" of B. This means that we have got countably many disjoint "copies" of B covering our Ω. And this implies that the notion of "length" is meaningless for that set. Indeed, if the length of B is a positive number, we come to a contradiction by noting that the length of the circle Ω always exceeds the sum of the lengths of its disjoint subsets, and hence must be infinite. But if we assume that the length of B is zero, we again have a contradiction, for in that case the length of the circle Ω will be given by the sum of the lengths of countably many copies of B,

[6]The reason for such a notation is quite natural: in fact, 2^Ω is also a product space, all the factors being the *two point* sets $I = \{0, 1\}$, the total number of them being equal to the total number of points in Ω. Indeed, there is a one-to-one correspondence between subsets $A \subset \Omega$ and their *indicators* which are defined as functions $\mathbf{1}_A(\omega) = 1$ if $\omega \in A$ and 0 if $\omega \notin A$. The indicators could be thought of as "$|\Omega|$-tuples" of zeros and ones, the elements of I. Note that $|2^\Omega| = 2^{|\Omega|}$ (at least, for finite Ω's!).

which is equal to zero!

To avoid such unpleasant complications, one considers special smaller subclasses of 2^Ω which are assumed to satisfy several conditions. Before stating them, we recall some simple notions from set theory used in probability and some basic terminology of the latter.

For two subsets A and B, we write $A \subset B$ (A is a subset of B) if, for any $\omega \in A$, we also have $\omega \in B$. In probability, one says in that case that event A **implies** B (for whenever A occurs, B also occurs).

The **union** of events A and B is defined as the set

$$A \cup B := \{\omega \in \Omega : \omega \in A \text{ or } \omega \in B\}$$

(the event "A or B" which occurs when at least one of the events A and B occurs). If we have a family of events A_j, $j \in J$ (some index set), one uses notation $\bigcup_{j\in J} A_j$ for the union of all the events A_j (most typically, $J = \{1,\dots,n\}$ or $J = \{1,2,\dots\}$, and then one writes $\bigcup_{j=1}^n A_j$ and $\bigcup_{j=1}^\infty A_j = \bigcup_{j\ge 1} A_j$, respectively). Recall that $\cup$ is a *commutative* operation, i.e., the order does not matter: $A \cup B = B \cup A$.

The **intersection** of events A and B is the set

$$A \cap B \equiv AB := \{\omega \in \Omega : \omega \in A \text{ and } \omega \in B\}$$

(the event "A and B" that occurs when both events A and B occur). A convention similar to the one adopted for unions is used for denoting intersections of sets from families. The operation $\cap$ is also commutative.

Two events A and B are called **disjoint**, or **mutually exclusive**, if $A \cap B = \emptyset$ (the empty set, also called in probability the **impossible event**). If this is the case, the events cannot occur simultaneously (i.e., as a result of a single trial).

For an event A, the **complementary event** is defined as

$$A^c := \{\omega \in \Omega : \omega \notin A\} \equiv \Omega \backslash A.$$

More generally, the **set difference** is defined as $B\backslash A := \{\omega \in B : \omega \notin A\}$. The event A^c occurs iff[7] A does not, and $B\backslash A \equiv B \cap A^c$ occurs iff B occurs while A does not.

The binary operations $\cup$ and $\cap$ satisfy the **distributive laws**: for any three sets A, B and C,

$$A \cap (B \cup C) = (A \cap B) \cup (A \cap C), \quad A \cup (B \cap C) = (A \cup B) \cap (A \cup C). \quad (2.2)$$

[7] "Iff" stands for "if and only if".

Quite often it is convenient or even necessary to transform combinations of events into alternative forms. Along with the laws (2.2), one often uses the famous *De Morgan*[8] *laws* to get more convenient expressions:

$$(A \cup B)^c = A^c \cap B^c, \quad (A \cap B)^c = A^c \cup B^c. \tag{2.3}$$

You may wish to verify (2.2) and (2.3), which is a simple exercise in logic.

Note that if we interpret events as the corresponding *propositions*, the set operations of conjunction ($\cup$), disjunction ($\cap$), and complementation (^{c}) will become just the logical operations *or*, *and* and *not*, respectively. Using this interpretation is quite useful for understanding both the meaning of the combinations of events and their derivation and verification.

As we will see below, sometimes it is more convenient to work not with events, but with their **indicators** (or indicator functions) which are functions on Ω defined as

$$\mathbf{1}_A(\omega) := \begin{cases} 1 & \text{if } \omega \in A, \\ 0 & \text{if } \omega \notin A. \end{cases} \tag{2.4}$$

So an indicator function can take only two values, 0 and 1, the latter "indicating" that the respective event occurs for the given value of the argument ω of the indicator. To the standard set operations there correspond the following ones for the indicator functions:

$$\begin{aligned} \mathbf{1}_{A\cup B} &= \max\{\mathbf{1}_A, \mathbf{1}_B\} \equiv \mathbf{1}_A \vee \mathbf{1}_B, \\ \mathbf{1}_{A\cap B} &= \min\{\mathbf{1}_A, \mathbf{1}_B\} = \mathbf{1}_A \mathbf{1}_B, \\ \mathbf{1}_{A^c} &= 1 - \mathbf{1}_A. \end{aligned} \tag{2.5}$$

Having recalled all that, we will now state the requirements which must be satisfied by a collection $\mathcal{F}$ of subsets of Ω to make it a suitable candidate for the domain of a probability.

A1. $\Omega \in \mathcal{F}$, the whole sample space (the certain event) is an element of $\mathcal{F}$.

A2. If $A \in \mathcal{F}$, then $A^c \in \mathcal{F}$ as well.

A3. If $A_1, A_2, \ldots \in \mathcal{F}$, then $\bigcup_{n=1}^{\infty} A_n \in \mathcal{F}$.

Any class $\mathcal{F}$ having the above three properties is called a **σ-field** (or, equivalently, a **σ-algebra**), the Greek letter σ ("sigma") indicating that axiom A3 holds for *countably infinite* sequences of sets from $\mathcal{F}$.

If we required that A3 only holds for all *finite collections* of sets $A_i \in \mathcal{F}$, the class $\mathcal{F}$ would be called simply a *field*.

[8] Augustus De Morgan (27.06.1806–18.03.1871), English mathematician and logician. In particular, it was him who defined and introduced the term *mathematical induction*.

Note that:

(i) A1 and A2 imply that $\emptyset$ (the impossible event) is also in $\mathcal{F}$;

(ii) A3 and (i) imply that, for any *finite* collection of A_i's from $\mathcal{F}$, their union is also an element of $\mathcal{F}$ (or, equivalently, that for any $A_1, A_2 \in \mathcal{F}$ we also have $A_1 \cup A_2 \in \mathcal{F}$);

(iii) A2, A3 and De Morgan laws (2.3) imply that, for any finite or countably infinite collection of elements of $\mathcal{F}$, their intersection is also an element of $\mathcal{F}$;

(iv) for any two sets $A, B \in \mathcal{F}$, the set difference $A \backslash B \in \mathcal{F}$.

The couple $(\Omega, \mathcal{F})$ is called a **measurable space**, while sets from the class $\mathcal{F}$ are called **measurable subsets** of that space.

On one and the same sample space Ω, one can consider different σ-fields of events; the bigger the class of events, the more detailed description of the random experiment we get. Thus, in the two-coin example, we can select the following two σ-fields[9] of events on the sample space (2.1):

$$\mathcal{F}_1 := \{\emptyset, \{HH, HT\}, \{TH, TT\}, \Omega\}$$

and

$$\begin{aligned}\mathcal{F}_2 := 2^\Omega = \{&\emptyset, \{HH\}, \{HT\}, \{TH\}, \{TT\}, \{HH, HT\}, \{HH, TH\},\\ &\{HH, TT\}, \{HT, TH\}, \{HT, TT\}, \{TH, TT\}, \{HH, HT, TH\},\\ &\{HH, HT, TT\}, \{HH, TH, TT\}, \{HT, TH, TT\}, \Omega\}.\end{aligned}$$

Clearly, $\mathcal{F}_1 \subset \mathcal{F}_2$. Events from $\mathcal{F}_1$ "carry information" related to what happened to the first coin only: if we knew, for all the events from $\mathcal{F}_1$, whether they occurred or not in a particular trial, this would tell us how the first coin landed. And vice versa, if we just know whether the first coin showed heads or tails, the only events for which we can say whether they occurred or not, are those forming $\mathcal{F}_1$. On the other hand, the collection of events $\mathcal{F}_2$ specify completely (in the same sense) the outcome of the whole experiment with two coins.

More generally, for a sequence of n tosses of a coin (or, equivalently, for a single experiment with n coins), we can introduce a sequence of increasing σ-fields $\mathcal{F}_1 \subset \mathcal{F}_2 \subset \cdots \subset \mathcal{F}_n$ describing the "truncated subexperiments" with smaller numbers of trials (the events forming $\mathcal{F}_k$ are related to the first k trials only; one says that $\mathcal{F}_k$ represents all the information available by the

[9]Of course, when the sample space is finite, we actually do not need the prefix "σ"; we can just require that A3 holds for any collection of elements of $\mathcal{F}$ (the power set 2^Ω is finite in that case!).

time k). Such increasing families of σ-fields are called *flows of σ-algebras*, or *filtrations*, and are extensively used in modern branches of probability theory. We will meet them again in the last two chapters of the present book.

For a collection $\mathcal{G}$ of subsets of Ω, one says that a σ-field $\mathcal{F}$ is **generated** by $\mathcal{G}$ if $\mathcal{F}$ is the *smallest* σ-field containing $\mathcal{G}$. Such an object always exists since the intersection of an arbitrary collection of σ-fields on Ω is always a σ-field itself.

As we noted above, the real need for choosing a class of events smaller than just the power set 2^Ω arises when the sample space Ω is a continuum. When Ω is the real line $\mathbf{R}$ (or a subset thereof), one usually takes the class of events to be $\mathcal{F} = \mathcal{B}$, the so-called σ-field of *Borel sets*[10], which by definition is the σ-field generated by the class of all open intervals (a, b), $-\infty < a < b < \infty$ (or, when Ω is a subset of $\mathbf{R}$, the **trace** of $\mathcal{B}$ on Ω, i.e., the collection of all events of the form $\Omega \cap B$, $B \in \mathcal{B}$).

When using product spaces to model compound random experiments, one forms the so-called **product σ-fields**. If our sample space $\Omega = \Omega_1 \times \cdots \times \Omega_n$ is a product of spaces Ω_i endowed with σ-fields $\mathcal{F}_i$, $i = 1, \ldots, n$, respectively, then it is natural to take the σ-field denoted by $\mathcal{F}_1 \otimes \cdots \otimes \mathcal{F}_n$ and *generated* by all **measurable rectangles** $A_1 \times \cdots \times A_n$, $A_i \in \mathcal{F}_i$, $i = 1, \ldots, n$, to be the class of events we consider for the compound experiment.

The product

$$\mathcal{B}_n := \underbrace{\mathcal{B} \otimes \cdots \otimes \mathcal{B}}_{n \text{ times}}$$

is called the class of Borel sets in $\mathbf{R}^n$. One can show that, alternatively, $\mathcal{B}_n$ can be defined as the σ-field generated by all open balls in $\mathbf{R}^n$. This class (or its trace) is the standard choice when one uses the sample space coinciding with $\mathbf{R}^n$ (or its subset).

Once one has selected a class $\mathcal{F}$ of sets which one wants to consider as events in one's model, it remains to specify the probabilities of the events. But how is probability formally defined in our mathematical model? Recall the frequency interpretation (1.2). It implies that, for any event A, its probability $\mathbf{P}(A)$ is a real number between 0 and 1. The next observation, even without a formal model, was the main tool in all probability calculations from the very beginning of the theory.

[10]Named after (Félix-Édouard-Justin-) Émile Borel (07.01.1871–03.02.1953), French mathematician, creator of measure theory (of which modern probability theory is a sort of "offspring"). Also served in the French Chamber of Deputies and as minister of the navy (1925–40).

If events A and B cannot occur simultaneously in a random experiment $\mathcal{E}$, then the number of times the event $A \cup B$ occurs in the first n replications of $\mathcal{E}$ is $n_{A\cup B} = n_A + n_B$ (cf. (1.1)), and therefore the relative frequency of $A \cup B$ is

$$\frac{n_{A\cup B}}{n} = \frac{n_A}{n} + \frac{n_B}{n}.$$

From here we infer that, as $n \to \infty$, the limiting values of the frequencies (which we call the probabilities of the events) must satisfy the same relation. For a technical reason, when defining probabilities formally, one needs a somewhat stronger property. The mathematical definition of probability is as follows.

Let $(\Omega, \mathcal{F})$ be a measurable space. Any real-valued *set function*[11] $\mathbf{P}$ defined on the elements of $\mathcal{F}$ is called a **probability** (or **probability measure/law/distribution**, or simply **distribution**) on $(\Omega, \mathcal{F})$ if the following three conditions are met:

P1. For any $A \in \mathcal{F}$, $\mathbf{P}(A) \in [0, 1]$.

P2. $\mathbf{P}(\Omega) = 1$.

P3. If $A_1, A_2, \ldots$ is a sequence of disjoint events: all $A_i \in \mathcal{F}$ and $A_i \cap A_j = \emptyset$, $i \neq j$, then

$$\mathbf{P}\left(\bigcup_{i=1}^{\infty} A_i\right) = \sum_{i=1}^{\infty} \mathbf{P}(A_i). \tag{2.6}$$

In particular, taking $A_i := \emptyset$, $i > 2$, we obtain from P3 that, for any disjoint events A_1 and A_2,

$$\mathbf{P}(A_1 \cup A_2) = \mathbf{P}(A_1) + \mathbf{P}(A_2) + \sum_{i=3}^{\infty} \underbrace{\mathbf{P}(A_i)}_{=\mathbf{P}(\emptyset)},$$

where the right-hand side must be finite, which means that $\mathbf{P}(\emptyset) = 0$. We obtain from here the usual "addition theorem for probability":

$$\mathbf{P}(A_1 \cup A_2) = \mathbf{P}(A_1) + \mathbf{P}(A_2) \quad \text{provided that } A_1 \cap A_2 = \emptyset. \tag{2.7}$$

This property, in its turn, implies **monotonicity of probability**:

$$\mathbf{P}(B_1) \leq \mathbf{P}(B_2) \quad \text{provided that } B_1 \subset B_2.$$

Putting $A_1 := \Omega$ and $A_2 := \emptyset$, we see that (2.7) and P2 imply that $\mathbf{P}(\emptyset) = 0$.

[11] That is, a function whose argument is a set.

Property P3 is called **countable additivity** (or σ-additivity). Any non-negative set function $\mu(\cdot)$ with $\mu(\emptyset) = 0$, defined on a σ-field and having this property is called a **measure**. The two most widely used measures are:

(i) the *counting measure* on the set of all integers $\mathbf{Z}$ defined as $\mu(A) = |A|$ (the number of points in A), $A \subset \mathbf{Z}$ (i.e., $A \in 2^{\mathbf{Z}}$); it can also be thought of as given on $(\mathbf{R}, \mathcal{B})$ (or even on $(\mathbf{R}, 2^{\mathbf{R}})$, for there is no problem with defining such a measure for all subsets of $\mathbf{R}$), if we set

$$\mu(B) = \#\{\text{integer } n \in B\}, \quad B \subset \mathbf{R}; \tag{2.8}$$

(ii) the *Lebesgue*[12] measure on the real line (the "length"), which is defined as the (only) measure μ on $(\mathbf{R}, \mathcal{B})$ which is *invariant* with respect to shifts (that is, for any $B \in \mathcal{B}$ and $x \in \mathbf{R}$, one has $\mu(B + x) = \mu(B)$, where $B + x = \{y + x : y \in B\}$) and has the property that $\mu([0, 1]) = 1$. In this case $\mu([a, b]) = b - a$ for all $a < b$.

It is not hard to see that P3 can actually be replaced by a combination of **finite additivity**, which is a weaker version of P3 stating that (2.6) holds for *finite* collections of disjoint events (note that finite additivity is actually equivalent to (2.7)) of $\mathbf{P}$ with its "continuity" on monotonic sequences of events in the following sense:

$$\lim_{n\to\infty} \mathbf{P}(B_n) = \mathbf{P}\left(\bigcap_{n=1}^{\infty} B_n\right) \quad \text{if } B_n \supset B_{n+1},\ n = 1, 2, \ldots \tag{2.9}$$

Instead of (2.9) one can equivalently require that, for any *increasing* sequence of events (meaning that $B_n \subset B_{n+1}$, $n = 1, 2, \ldots$), the limit of the probabilities $\mathbf{P}(B_n)$ is equal to the probability of the union $\bigcup_{n=1}^{\infty} B_n$.

As we said before, there are many different σ-fields of subsets of a fixed sample space Ω. The nature of the experiment dictates which one should be used. But then there is again an enormous variety of different probabilities one can define on a fixed $\mathcal{F}$. Selecting the correct one is the most important part of the modelling process. Note that one often considers not a *single probability* on a given σ-field, but (sometimes rather rich) families of distributions depending, say, on certain parameters. Then, in the process of *fitting* the model, one has to estimate the values of the parameters from the data obtained in experimentation with (or from the observation of) the real-life system to be modelled. Such parameter estimation is one of the main classes of problems in *mathematical statistics*.

[12]Henry-Léon Lebesgue (28.06.1875–26.07.1941), a French mathematician who revolutionised the field of integration.

In non-trivial cases, the class $\mathcal{F}$ is so huge that it is impossible to explicitly specify probabilities for all the events. Even for finite Ω's, the number of subsets increases exponentially fast with the total number of different points in the sample space: $|2^{\Omega}| = 2^{|\Omega|}$, and listing the probabilities of *all events* is impractical. It is actually the basic task of probability theory itself to compute the probabilities of certain events from the known probabilities of some other, simpler events within the same model. The first step here is to see for what (as small as possible) subclass of events it suffices to specify probabilities to define the probability measure on the whole $\mathcal{F}$.

When the sample space Ω is discrete (finite or countably infinite), to specify a probability it suffices to say what the probabilities of all one-point sets are, i.e., list the values $p_\omega := \mathbf{P}(\{\omega\})$ for all $\omega \in \Omega$. Then, for any $A \in 2^{\Omega}$, we have from P3 that $\mathbf{P}(A) = \sum_{\omega \in A} p_\omega$.

When the sample space is a continuum, specifying the probabilities of all one-point events is not enough any more. For example, if one speaks of choosing a point at random from the interval $[a, b]$, $a < b$, it is natural to take $\Omega = [a, b]$, $\mathcal{F}$ = the class of Borel subsets of Ω, and to require that the probability of the point being in a subinterval $[c, d] \subset [a, b]$ depends only on the length $d - c$ of the subinterval, and hence is invariant with respect to shifts. Clearly, the probability of the interval $[c, d]$ is then given by $\mathbf{P}([c, d]) = (d - c)/(b - a)$. As for one-point events, the probability of each of them is then zero: $\mathbf{P}(\{\omega\}) = 0$, $\omega \in \Omega$. But the last relation holds for any of the huge variety of all continuous distributions, and therefore does not specify any particular distribution.

The key result allowing to specify probabilities in a constructive way is given by the Carathéodory[13] theorem. To explain what it is about, recall that any class $\mathcal{A}$ of subsets of Ω satisfying conditions A1, A2 and a relaxed version of A3 requiring that unions of *finitely many* elements of $\mathcal{A}$ should be elements of $\mathcal{A}$ as well, is called a **field** of sets. Fields are typically "much smaller" objects than σ-fields. For example, the field generated[14] by all semi-open intervals $(a, b]$ on the extended real line (including the points $\pm\infty$) is the class of all finite unions of such intervals, while the σ-field generated by the same $\mathcal{G}$ will be far bigger and contain a lot of "exotic" sets of very complicated structure.

The above-mentioned theorem states that if $\mathbf{P}$ is a set function defined on a field $\mathcal{A}$ of subsets of Ω and satisfying conditions P1–P3 on it (with

[13] Constantin Carathéodory (13.02.1873–02.02.1950), German/Greek mathematician.

[14] The meaning is the same as in the case of generated σ-fields: for a class $\mathcal{G}$ of subsets of Ω, the field generated by $\mathcal{G}$ is the *smallest* field containing $\mathcal{G}$.

the additional condition that the union appearing in P3 is an element of $\mathcal{A}$), then it can be extended in a unique way to a probability defined on the σ-field generated by $\mathcal{A}$. The proof of the theorem can be found, e.g., in Borovkov (2013).

Thus, to specify a probability distribution having desired properties, we can start by defining it for a relatively simple and small collection of events (say, semi-open intervals in the case of $\Omega = \mathbf{R}$), then proceed to the field generated by that collection of sets and try to verify if the set function meets the above criteria. Once this is done, we just apply the Carathéodory theorem to state that the desired distribution exists. Finding the *explicit values* of the probabilities of the events of interest is another story—this can be extremely difficult to do. But we will at least know that the probability distribution defining our stochastic model exists, and hence the model itself is consistent and can be worked with.

Before proceeding any further, we will have a closer look at the most important special case when the sample space Ω is the real line.

2.2 Distributions and Integrals

Very often one takes $\Omega = \mathbf{R}$ (or its subset, which is almost the same, or a product of real lines, a finite-dimensional Euclidean space). In this case, the class of events is usually the σ-field $\mathcal{B}$ of the Borel sets (which is generated by all open intervals, or by all semi-open intervals).[15] Since to specify a probability distribution on such a measurable space it suffices, according to what we said at the end of the previous section, to specify its values on all semi-open intervals only, it comes as no surprise that any probability on $(\mathbf{R}, \mathcal{B})$ is *uniquely determined* by its **distribution function** (abbreviated DF, sometimes also called the *cumulative distribution function* to distinguish it from probability densities)

$$F(t) = F_{\mathbf{P}}(t) := \mathbf{P}\left((-\infty, t]\right). \tag{2.10}$$

Indeed, for any semi-open interval $(s, t]$ we simply have

$$\mathbf{P}\left((s, t]\right) = F(t) - F(s), \quad s < t, \tag{2.11}$$

from the additivity of probability (2.7).

Any DF has the following three properties:

[15] Note that, despite its apparent simplicity, such a measurable space is very rich. Roughly speaking, any random process, however complex it may be, can be defined on $(\mathbf{R}, \mathcal{B})$ as the underlying sample space with the respective class of events.

F1. F is non-decreasing: $F(t) - F(s) \geq 0$, $s < t$.

F2. $F(t) \to 0$ as $t \to -\infty$, $F(t) \to 1$ as $t \to \infty$.

F3. F is *right-continuous*: $F(t) \searrow F(s)$ as $t \searrow s$.[16]

Property F1 follows from (2.11), F2 follows from the fact that $\mathbf{P}(\emptyset) = 1 - \mathbf{P}(\Omega) = 0$ and "continuity" of probability (2.9), while F3 is basically equivalent to that continuity.

Moreover, there is a one-to-one correspondence between distributions on $\mathbf{R}$ and functions satisfying F1–F3: for any such F, there exists one and only one probability distribution $\mathbf{P}$ on $\mathbf{R}$ having the DF F. And this is why probabilities on $\mathbf{R}$ are usually given by their DFs.

To analyse and better understand distributions on the real line, it is convenient to introduce the following classification.

1. Discrete distributions. A probability $\mathbf{P}$ on $(\mathbf{R}, \mathcal{B})$ (and also on a general measurable space) is called **discrete** if it is *concentrated* on a finite or countable set $\{x_1, x_2, \dots\}$: $p_j = \mathbf{P}(\{x_j\}) > 0$, $\sum_j p_j = 1$. The DF of a discrete distribution is a function increasing by jumps of the sizes p_j at the points x_j, respectively.

There is a temptation to add "and constant between the successive points x_j". However, there could be no "between". Suppose that the sequence $\{x_j\}$ contains all rational numbers (of which the set is countable). This sequence is then *everywhere dense* in $\mathbf{R}$, i.e., for any arbitrary small open interval (a, b), there is always at least one point x_j (and hence infinitely many of them) inside this interval. Therefore, for any x_j, there is no such thing as the "next" x_i—the next not in the order of the sequence, but in the sense that it would be the smallest member of our sequence exceeding x_j.

The simplest example here is the *degenerate distribution* I_a at the point $a \in \mathbf{R}$: its DF is $I_a(t) = \mathbf{1}_{[a,\infty)}(t)$. The distribution corresponds to the case when there is nothing random: with probability 1, one always gets the value a.

The second simplest example is the so-called *Bernoulli*[17] *distribution* B_p. Now two values, zero and one, are possible, and $1 - \mathbf{P}(\{0\}) = \mathbf{P}(\{1\}) = p \in [0, 1]$ (it will be convenient for us to keep using one and the same symbol $\mathbf{P}$ for different probability distributions, specifying each time what concrete law is assumed). The DF of this distribution has two jumps, at the points 0

---

[16] Notation $t \searrow s$ means that "t decreases to s".

[17] After Jacob Bernoulli (06.01.1654–27.12.1705), the first of the famous Bernoulli family of Swiss mathematicians, one of the founders of calculus (which he also applied in 1695 to the design of bridges) and probability theory. His pioneering work *Ars Conjectandi* ("The Art of Conjecturing") contained the mathematical statement and proof of the *law of large numbers*, a theorem showing that a simple mathematical model of the kind we are discussing here for a sequence of independent identical trials does display the fundamental property (1.2) of probability.

and 1, whose sizes are $1-p$ and p, respectively. Note that it can be written as

$$F(t) = (1-p)\mathbf{1}_{[0,\infty)}(t) + p\mathbf{1}_{[1,\infty)}(t),$$

i.e., as a weighed sum of degenerate DFs. This is a special case of the so-called **mixture distributions** which are combinations of the form

$$\sum_i a_i \mathbf{P}_i, \tag{2.12}$$

where the coefficients $a_i > 0$ add up to the unity, $\mathbf{P}_i$ are some distributions (on a common measurable space). Such distributions are often encountered in various stochastic models.

The meaning of mixture distributions can be best understood using the *total probability formula* to be discussed below. Namely, mixtures typically arise as the result of a "two-stage" sampling procedure. Suppose we have a "mixed" population consisting of individuals of different types encoded by different values of i, the probability of choosing an individual of type i being a_i, and the distribution of certain characteristics of the individuals within type i is described by $\mathbf{P}_i$. Then the resulting distribution for these characteristics describing the *whole mixed population* will be just our mixture (2.12).

Another important example to be mentioned here is the *Poisson*[18] *distribution* $Po(\lambda)$ with parameter $\lambda > 0$ on non-negative integers given by

$$\mathbf{P}(\{k\}) = e^{-\lambda}\frac{\lambda^k}{k!}, \quad k = 0, 1, 2, \ldots, \tag{2.13}$$

which is a good model for the total number of "successes" in situations where we have a large number of (almost) independent unlikely events.

The definitions of some other widespread discrete distributions will be given later.

2. Absolutely continuous distributions.[19] A DF F (or, which is the same, the respective distribution) is called **absolutely continuous** if there exists a **density** function $f(x) \geq 0$ such that

$$F(t) = \int_{-\infty}^{t} f(x)\,dx, \quad t \in \mathbf{R}. \tag{2.14}$$

[18] Siméon-Denis Poisson (21.06.1781–25.04.1840), a French mathematician famous for his work in electromagnetic theory and probability. His expression for the force of gravity in terms of the distribution of mass within a planet has been used in the late 20th century for deducing details of the shape of the Earth from measuring the paths of orbiting satellites.

[19] Sometimes also called just *continuous distributions*. However, that terminology is not quite correct, for a continuous distribution is just one with a continuous DF (thus having no "atoms", one-point sets of positive probability). But that property does not imply that the DF has a density!

Note that, for any density f, $\int_{-\infty}^{\infty} f(x)\,dx = 1$ and $f(x) = dF(x)/dx$ almost everywhere (i.e., on a set whose complement in $\mathbf{R}$ is of "total length" zero).

The integral here (and also in many other places below) is actually what is called the *Lebesgue integral*; this is a more general integral than the conventional, or *Riemann*[20] *integral*; with it, we can integrate over abstract measurable spaces (not only over finite-dimensional Euclidean spaces which is the case when one knows conventional integrals only, and this is very important in probability theory), and the class of functions one can integrate is much wider. Also, one integrates with respect to a measure, not just dx. Thus we get a powerful unified approach to both continuous and discrete cases, for sums are simple special cases of integrals (with respect to discrete measures, e.g., the counting measure (2.8)).

To illustrate the notion briefly, assume that $(\Omega, \mathcal{F})$ is a measurable space, and μ is a measure given on it. One starts defining the Lebesgue integral by considering *simple functions* on $(\Omega, \mathcal{F})$, i.e., functions $f : \Omega \to \mathbf{R}$ taking only finitely many different values: for a finite *partition* $A_1, \ldots, A_n$ of Ω (i.e., a collection of disjoint sets A_j whose union is the whole Ω) with all $A_j \in \mathcal{F}$, one has

$$f(\omega) = f_j, \qquad \omega \in A_j,$$

for some real numbers $f_1, \ldots, f_n$. For such an f, its integral with respect to μ is defined as

$$\int_\Omega f\,d\mu \equiv \int_\Omega f(\omega)\,d\mu(\omega) := \sum_{j=1}^{n} f_j\,\mu(A_j).$$

Then the notion can be extended to *arbitrary measurable* functions $f : \Omega \to \mathbf{R}$, that is, functions having the property that, for any Borel set $B \in \mathcal{B}$, the *preimage*

$$f^{-1}(B) = \{\omega :\ f(\omega) \in B\} \in \mathcal{F},$$

i.e., is *measurable* (note that a simple function is always measurable by definition). It is first done for non-negative measurable functions f by taking the limits of the integral values for sequences of simple functions *increasing* to f (it turns out that all the limiting values will coincide, and this common limit is called the Lebesgue integral $\int_\Omega f\,d\mu$ of f with respect to μ). For an arbitrary measurable function f, its integral is defined as the difference of the integrals of its positive and negative parts $f^+ := \max\{0, f\}$ and $f^- := -\min\{0, f\}$. And, for an $A \in \mathcal{F}$, we put $\int_A f\,d\mu := \int_\Omega f \mathbf{1}_A d\mu$.

For a detailed exposition of the Lebesgue integration theory, we refer the reader to any modern probability textbook (see, e.g., Section 2.11 of the present chapter). For the time being—and in most cases in what follows—you may think about integrals as just ordinary integrals from an elementary calculus course.

As examples, we mention here a few of the most popular absolutely continuous distributions.

The uniform distribution $U(a, b)$ on $[a, b]$, $-\infty < a < b < \infty$, has the density and the DF

$$f(x) = (b-a)^{-1}\mathbf{1}_{[a,b]}(x), \qquad F(x) = \begin{cases} 0, & x < 0, \\ \dfrac{x-a}{b-a}, & x \in [a, b], \\ 1, & x > 1, \end{cases} \tag{2.15}$$

[20](Georg Friedrich) Bernhard Riemann (17.09.1826–20.07.1866), a German mathematician. His ideas concerning geometry of space had a profound effect on modern theoretical physics, including relativity theory.

respectively. This is the standard model for *choosing a point at random* from the interval $[a, b]$. Due to the shape of its density, the distribution is sometimes referred to as the *rectangular distribution.*

The exponential distribution $Exp(\lambda)$ with parameter $\lambda > 0$ has the density and the DF

$$f(x) = \lambda e^{-\lambda x}\mathbf{1}_{[0,\infty)}(x), \qquad F(x) = \begin{cases} 0, & x < 0, \\ 1 - e^{-\lambda x}, & x \geq 0, \end{cases} \tag{2.16}$$

respectively. Clearly, λ is just a *scale parameter.* The distribution plays a distinguished role in stochastic modelling due to its certain unique properties to be discussed in detail in Section 5.1 below.

The gamma distribution $\gamma(\alpha, \lambda)$ with the shape parameter α and scale parameter $\lambda > 0$ has the density

$$f(x) = \frac{\lambda^\alpha}{\Gamma(\alpha)} x^{\alpha-1} e^{-\lambda x}, \quad x > 0,$$

where

$$\Gamma(z) = \int_0^\infty x^{z-1} e^{-x} dx, \quad z > 0,$$

is the *gamma function.*[21] Note that $\gamma(1, \lambda) = Exp(\lambda)$.

The normal distribution (a.k.a. the Gaussian[22] distribution, a.k.a. the second Laplace[23] error law) $N(\mu, \sigma^2)$, $\mu \in \mathbf{R}$, $\sigma > 0$, has the density

$$f(x) = \frac{1}{\sqrt{2\pi}\sigma} e^{-(x-\mu)^2/2\sigma^2}, \quad -\infty < x < \infty. \tag{2.17}$$

[21]This integral is actually well-defined for all complex a with $\operatorname{Re} z > 0$, and the function given by it is extended analytically to all complex numbers except $z = 0, -1, -2, \ldots$ (where $\Gamma(z)$ has simple poles). Its key properties include the identity $\Gamma(z+1) = z\Gamma(z)$ (integrate by parts!) leading to $\Gamma(n+1) = n!$, $n = 0, 1, 2, \ldots$ (historically, the gamma function was actually derived as an analytic extension of the factorial to non-integers).

[22]After Carl Friedrich Gauss (30.04.1777–23.02.1855), a German mathematician, called the "Prince of mathematicians". He created number theory, the method of least squares, developed complex analysis. Teaching was his only aversion. Gauss was at once an outstanding theoretician and remarkable practical mathematician, one of the greatest of all time. About his life see, e.g., Bühler, W.K. *Gauss: A Bibliographical Study*, Springer, New York, 1981; Hall, T. *Carl Friedrich Gauss: A Biography*, M.I.T. Press, Cambridge, 1970.

[23]After Pierre-Simon de Laplace (23.03.1749–05.03.1827), a French mathematician, astronomer and physicist who, in particular, established the stability of the solar system. He aided in the organisation of the metric system and served (for six weeks) as minister of the interior under Napoleon. About his life see, e.g., Gillispie, C.C. *Pierre-Simon Laplace, 1749-1827: a life in exact science,* Princeton Univ. Press, Princeton, 1997; Bell, E.T. *Men of Mathematics*, Simon and Schuster, New York, 1965.

There is no closed-form expression for the DF but the integral (2.14) of the density f given by (2.17).

Observe that μ and σ are the *location* and *scale* parameters, respectively, so that one actually only needs to know the so-called *standard normal distribution* $N(0, 1)$, whose DF is usually denoted by

$$\Phi(x) := \frac{1}{\sqrt{2\pi}} \int_{-\infty}^{x} e^{-t^2/2}\, dt. \tag{2.18}$$

The DF of $N(\mu, \sigma^2)$ is equal to $\Phi((x - \mu)/\sigma)$.

The normal distribution is extremely popular due to the fact that it well approximates probability laws when the resulting quantity is a sum of (almost) independent relatively small contributions. Mathematically this property is stated in the famous *central limit theorem*. Moreover, the normal distribution has several unique properties which make it, in particular, the most natural choice for modelling random errors.

3. Singular distributions. These are distributions whose DFs are continuous, and yet are constant almost everywhere. That is, the set of *growth points*[24] has zero total length for any singular DF. A famous example of such a function is given by the so-called *Cantor*[25] *ladder*; for more detail see e.g., Borovkov (2013) or Shiryaev (1984).

Of course, distributions do not need to be of *pure type*. Quite often one deals with mixtures of, say, discrete and absolutely continuous laws, a typical case here being the waiting time in a queueing system: with a positive probability, an arriving customer will not have to await service at all, but when he will, the respective (conditional) distribution is absolutely continuous. It turns out that various mixtures of the above three types is **all** that one can have. More precisely, the following *Lebesgue decomposition* holds true: any probability distribution $F(t)$ on the real line $\mathbf{R}$ has a unique representation of the mixture form: $F(t) = a_1 F_d(t) + a_2 F_{ac}(t) + a_3 F_s(t)$, where $a_i \geq 0$ and $\sum_i a_i = 1$, while F_d, F_{ac} and F_s are (uniquely defined by F) discrete, absolutely continuous and singular DFs, respectively.

In the case of a general measurable space $(\Omega, \mathcal{F})$, one says that two measures (in particular, probability distributions) μ and ν are *singular* (one writes $\mu \perp \nu$) if there exists a set $A \in \mathcal{F}$ such that $\mu(A) = 0$ and $\nu(A^c) = 0$. One says that μ is *absolutely continuous* with respect to ν (and writes $\mu \prec \nu$) if the relation $\nu(A) = 0$ for an $A \in \mathcal{F}$

[24]An $x \in \mathbf{R}$ is a *growth point* of the DF F if $F(x + \varepsilon) - F(x - \varepsilon) > 0$ for any $\varepsilon > 0$.

[25]Georg (Ferdinand Ludwig Philipp) Cantor (03.03.1845–06.01.1918), a German mathematician who founded set theory. See, e.g., Purkert, W. *Georg Cantor, 1845-1918*, Birkhäuser, Basel, 1987.

always implies that $\mu(A) = 0$ for the A. If this is the case, then, by the *Radon*[26]-*Nikodym*[27] *theorem*, there exists a *density* $f = \frac{d\mu}{d\nu} \geq 0$, i.e., a measurable function on $(\Omega, \mathcal{F})$ such that, for any $A \in \mathcal{F}$,

$$\mu(A) = \int_A f\, d\nu \equiv \int_\Omega f \mathbf{1}_A\, d\nu.$$

Thus, absolutely continuous distributions on the real line are actually absolutely continuous with respect to the Lebesgue measure, while both singular and discrete distributions are mutually singular with that measure.

2.3 Conditional Probability and Independence

Let A and B be two events on a probability space $(\Omega, \mathcal{F}, \mathbf{P})$, $\mathbf{P}(B) > 0$. Suppose we know that B occurred in our random experiment. Given that information, how likely is it that A also occurred in the same trial?

From the relative frequency interpretation point of view, we would like to know about what value the ratios n_{AB}/n_B will stabilise as the total number of trials $n \to \infty$ (indeed, we compute now the *relative frequency* of the trials in which A occurred among all trials in which B occurred). But

$$\frac{n_{AB}}{n_B} = \frac{n_{AB}}{n} \times \frac{n}{n_B} \rightsquigarrow \frac{\mathbf{P}(AB)}{\mathbf{P}(B)} \tag{2.19}$$

from (1.2). This is why the ratio on the right-hand side is taken as the formal definition of the **conditional probability** $\mathbf{P}(A|B)$ of A given B.

Note that, for a fixed B, the conditional probability $\mathbf{P}(\,\cdot\,|B)$ is, as a set function, a *probability* on $(\Omega, \mathcal{F})$ as well. It will also be a probability on the *truncated* sample space B (endowed with the "trace" σ-field $\mathcal{F} \cap B = \{A \cap B : A \in \mathcal{F}\}$; note that $\mathbf{P}(A|B) = 0$ for all $A \in \mathcal{F}$ such that $A \cap B = \emptyset$). So taking conditional probabilities given an event (of a positive probability) is basically a sort of reduction of the original sample space.

One of the most useful tools employing conditional probabilities is the so-called **total probability formula** (TPF). In its simplest form, this relation can be stated as follows. Suppose that $B_1, \ldots, B_n$ are mutually exclusive events: $B_i \cap B_j = \emptyset$, $i \neq j$, and consider an event A such that $A \subset \bigcup_{i=1}^n B_i$ (usually, $\{B_i\}$ forms a *partition* of Ω, i.e., the last union

[26]Johann (Karl August) Radon (16.12.1887–25.05.1956), an Austrian mathematician who, in particular, contributed to calculus of variations and differential geometry.

[27]Otton Marcin Nikodým (13.08.1887–04.05.1974), a Polish mathematician who made contributions to measure theory, functional analysis, differential equations, descriptive set theory and the foundations of quantum mechanics. One of the founders of the Polish Mathematical Society.

coincides with the whole Ω). Then

$$\mathbf{P}(A) = \sum_{i=1}^{n} \mathbf{P}(A|B_i)\,\mathbf{P}(B_i). \tag{2.20}$$

This relation is obvious from the probability property P3, since

$$\mathbf{P}(A|B_i)\,\mathbf{P}(B_i) = \mathbf{P}(AB_i)$$

and the union of the disjoint events ("sum of events") AB_i is clearly equal to A.

The family $\{B_i\}$ can be countably infinite; moreover, the relation holds in more general situations as well, when the sum is replaced with an integral. We will return to it later.

Thus, if one knows the probabilities of the "hypotheses" B_i (the so-called *prior probabilities*, before conducting the random experiment) and, for each of them, the conditional probability of the event A of interest given the hypothesis, then one can compute the probability of A (recall our illustration to mixture distributions (2.12) above). On the other hand, for any hypothesis B_j, one can also compute the conditional probability that it takes place given that A occurred (*posterior probabilities*, after the experiment). The probability is given by the *Bayes*[28] *formula*: under the same assumptions as for the TPF,

$$\mathbf{P}(B_j|A) = \frac{\mathbf{P}(AB_j)}{\mathbf{P}(A)} = \frac{\mathbf{P}(A|B_j)\,\mathbf{P}(B_j)}{\mathbf{P}(A)} = \frac{\mathbf{P}(A|B_j)\,\mathbf{P}(B_j)}{\sum_i \mathbf{P}(A|B_i)\,\mathbf{P}(B_i)}. \tag{2.21}$$

The formula is a simple consequence of the definition of conditional probability and (2.20).

When the conditional probability of an event given another event coincides with the unconditional probability of the former, one speaks of **independence**, which is one of the most important notions in probability theory. Equivalently, events A and B are said to be *independent* if

$$\mathbf{P}(AB) = \mathbf{P}(A)\,\mathbf{P}(B), \tag{2.22}$$

which is obviously the same as $\mathbf{P}(A|B) = \mathbf{P}(A)$ when $\mathbf{P}(B) > 0$. In terms of the frequency interpretation of probability, this means that the relative frequencies of the occurrence of A in a sequence of trials and in

[28] Thomas Bayes (1702–17.04.1761), an English theologian and mathematician who established a basis for probability inference (summarised in formula (2.21) named after him—apparently at the end of the 19th century); he was the first to explicitly introduce the notion of conditional probability. Bayes' only work published in his lifetime was *Divine Benevolence, or an Attempt to Prove That the Principal End of the Divine Providence and Government Is the Happiness of his Creators.*

its subsequence consisting of the trials in which B occurred, will tend to a common value.

Events $A_1, \ldots, A_n$ are said to be **mutually independent** if, for any subcollection of indices $\{i_1, \ldots, i_r\} \subseteq \{1, \ldots, n\}$,

$$\mathbf{P}(A_{i_1} \cap \cdots \cap A_{i_r}) = \mathbf{P}(A_{i_1}) \cdots \mathbf{P}(A_{i_r}). \tag{2.23}$$

If this is the case, then the knowledge that a fixed combination of events A_j occurred in a trial, would not change the chances for any other combination of the A_i's that did not participate in the first one, to occur.

It is not enough for independence to require that, for all pairs $i \neq j$, $\mathbf{P}(A_i A_j) = \mathbf{P}(A_i)\,\mathbf{P}(A_j)$; if this is the case, one says that A_j are *pairwise independent events.* This sort of independence is a weaker property, as can be illustrated by the famous *Bernstein tetrahedron example.* Suppose the faces of a symmetric tetrahedron are painted in three colours: one is red, the second is blue, the third is green and the fourth has all three colours on it. We roll the tetrahedron and see what face it stops on (all four are equally likely and hence have probabilities of 1/4 each). The three events $R = \{$there is *red* colour on the bottom$\}$, $B = \{\ldots blue \ldots\}$ and $G = \{\ldots green \ldots\}$ can easily be shown to be pairwise independent, but *not mutually* independent (indeed, if $R \cap G$ occurred, we know for sure that B occurred, too).

2.4 Random Variables and Their Distributions

Loosely speaking, *random variables* (RV) are quantities whose values depend on chance. Formally, the chance is now a point $\omega \in \Omega$, so that an RV $X = X(\omega)$ given on a probability space $(\Omega, \mathcal{F}, \mathbf{P})$ is a real-valued function of ω. For technical reasons, we require X to be $\mathcal{F}$*-measurable*: for any $x \in \mathbf{R}$, the set

$$\{X \leq x\} := \{\omega : X(\omega) \leq x\} \in \mathcal{F},$$

i.e., is an event. This requirement ensures that the probability of $\{X \leq x\}$ is defined!

If we just say that a function X is *measurable*, it is tacitly assumed that it is measurable with respect to the σ-field from the definition of the measurable space $(\Omega, \mathcal{F})$. In probability, one often deals with RVs measurable with respect to *smaller* σ-fields than $\mathcal{F}$ itself; note that if an RV X is $\mathcal{F}_0$-measurable for a σ-field $\mathcal{F}_0 \subset \mathcal{F}$, then it will automatically be $\mathcal{F}$-measurable as well.

Since the σ-field generated by the collection of half-lines $(-\infty, x]$ is clearly the class of Borel sets, it is not hard to show that the above definition of measurability is equivalent to the general one we have already mentioned before, in Section 2.2: for any $B \in \mathcal{B}$,

$$X^{-1}(B) = \{\omega : X(\omega) \in B\} \in \mathcal{F}.$$

In other words, it now makes sense to ask questions like "What is the probability that our random variable assumed a value from the interval $[a, b]$?" (for the set $\{X \in [a, b]\}$ is an event and its probability *is defined*). More generally, one speaks of a *measurable mapping* X of a measurable space $(G, \mathcal{G})$ to another measurable space $(H, \mathcal{H})$ if $X : G \to H$ and, for any $B \in \mathcal{H}$, the *preimage* $\{g \in G : X(g) \in B\} \in \mathcal{G}$. This notion can be used to define random vectors and processes as measurable mappings from the underlying probability spaces to the respective Euclidean or functional spaces. Also, one can define integrals of measurable functions, which makes it possible to compute the expected values of RVs.

Clearly, any constant is an RV. Further, it is not hard to show that any "reasonable" combination of RVs is again an RV More precisely, if $X_1, \ldots, X_n$ are RVs on a common probability space, and $f : \mathbf{R}^n \to \mathbf{R}$ is a measurable function, then $f(X_1, \ldots, X_n)$ is also an RV. Since any continuous function f can easily be seen to be measurable, all continuous functions of RVs (e.g., $X_1 + X_2$, e^X, $\max\{X_1, \ldots, X_n\}$ *etc.*) are RVs.

On discrete sample spaces Ω with $\mathcal{F} = 2^\Omega$, any function is evidently measurable. Therefore, in that case, any function $X(\omega)$ on Ω is an RV. When the sample space is more complicated, certain precautions are sometimes necessary.

If we restrict ourselves to observing the values of an RV X only, this amounts, in a sense, to reducing our random experiment to a "smaller one": we can switch to the real line $\mathbf{R}$ (the range of X) as our new sample space (with the natural choice of $\mathcal{B}$ as the σ-field of events on it), and to the probability distribution

$$P_X(B) := \mathbf{P}(\{\omega : X(\omega) \in B\}), \quad B \in \mathcal{B}, \tag{2.24}$$

induced on $\mathcal{B}$ by the mapping X (see Fig. 2.1). The probability P_X is called the **distribution of the RV** X. Note that the right-hand side of (2.24) is well-defined due to the measurability of X (the argument of $\mathbf{P}$ is an *event*). As long as only the characteristics of X are of interest, it does not matter on what probability space and how the RV was originally given. What matters, is just the distribution of X.

The DF

$$F_X(t) := P_X((-\infty, t]) = \mathbf{P}(X \le t), \quad t \in \mathbf{R}, \tag{2.25}$$

is called the **distribution function** of the RV X. As we already know, it uniquely defines the distribution of X.

According to the classification from Section 2.2, an RV X is called discrete, absolutely continuous or singular if its distribution is from the corresponding class.

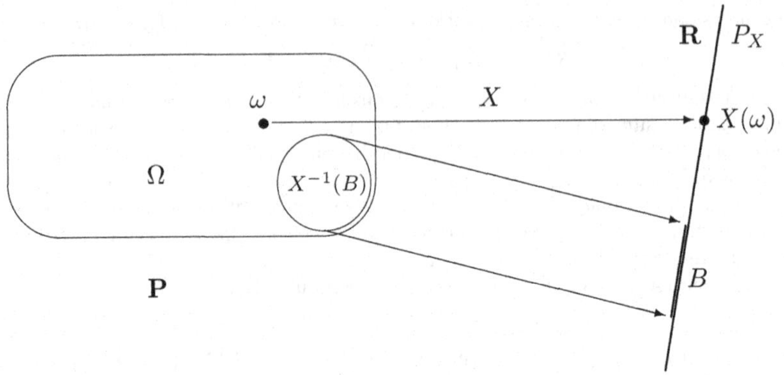

Fig. 2.1 The distribution P_X on $(\mathbf{R}, \mathcal{B})$ is induced by $X = X(\omega)$.

The distribution of a discrete RV X is given by the collection of the possible values $\{x_j\}$ of the RV and their probabilities $p_j = \mathbf{P}(X = x_j) \geq 0$, $\sum_j p_j = 1$. The distribution of an absolutely continuous RV X can be given by the density f_X of this distribution, a useful rough interpretation of which being the relation

$$P_X(\Delta) \approx f_X(x)\,|\Delta| \tag{2.26}$$

for a small interval $\Delta \ni x$, $|\Delta|$ being the length of Δ. From relation (2.26) one can easily derive the following useful formula: if $Y = g(X)$ for a strictly monotone and differentiable function g, then Y is also an absolutely continuous RV with density

$$f_Y(y) = \frac{f_X(g^{-1}(y))}{|g'(g^{-1}(y))|}. \tag{2.27}$$

A **random vector** $\boldsymbol{X} = (X_1, \ldots, X_d)$ on a probability space $(\Omega, \mathcal{F}, \mathbf{P})$ is a collection of RVs X_j given on that space. Equivalently, a random vector is a *measurable* mapping of $(\Omega, \mathcal{F})$ to $(\mathbf{R}^d, \mathcal{B}_d)$. The *distribution* of the random vector $\boldsymbol{X}$ (or, which is the same, the **joint distribution** of the RVs $X_1, \ldots, X_d$) is the probability on $\mathbf{R}^d$ given by

$$P_{\boldsymbol{X}}(B) = \mathbf{P}((X_1, \ldots, X_d) \in B), \quad B \in \mathcal{B}_d. \tag{2.28}$$

Note that one cannot specify the distribution of the random vector $\boldsymbol{X}$ by simply giving the "individual" distributions of the X_j's (called the **marginal distributions** of $\boldsymbol{X}$), for in that case we would lose all the information about the dependence of the components of each other. To illustrate this remark, consider a random experiment consisting of rolling two dice. The natural sample space Ω is the set of all pairs (i, j), $i, j = 1, \ldots, 6$ (this is clearly a product space), and, assuming symmetry, all 36 outcomes are equally likely. Let X_i be the number of points showing on the ith die, $i = 1, 2$, and $X_1' := X_2' := X_1$. Clearly, all X_i and X_i' have one and the same distribution (*uniform* on $\{1, \ldots, 6\}$), so that the marginal distributions of the vectors $\boldsymbol{X}$ and $\boldsymbol{X}' := (X_1', X_2')$ coincide. But the distributions of the vectors themselves will clearly be quite different: that of $\boldsymbol{X}$ is uniformly "spread" over the whole Ω, while the distribution of $\boldsymbol{X}'$ is concentrated on the "diagonal" $\{(i, i) : i = 1, \ldots, 6\}$.

The notion of the DF is meaningful and useful for distributions on $(\mathbf{R}^d, \mathcal{B}_d)$ as well. The DF of a random vector $\boldsymbol{X}$ (or its distribution) is the function

$$F_{\boldsymbol{X}}(\boldsymbol{t}) := \mathbf{P}(X_1 \le t_1, \ldots, X_d \le t_d), \quad \boldsymbol{t} = (t_1, \ldots, t_d) \in \mathbf{R}^d. \tag{2.29}$$

This is just the probability that the random vector $\boldsymbol{X}$ is in the "negative orthant" with the vertex at the point $\boldsymbol{t}$ (the left half-line in the one-dimensional case). Any DF on $\mathbf{R}^d$ satisfies somewhat complicated versions of the properties F1–F3. As it was the case in the one-dimensional case, there is also a one-to-one correspondence between distributions on $(\mathbf{R}^d, \mathcal{B}_d)$ and their DFs.

Indeed, knowing the DF, one can easily compute the probability of any "rectangle" $(s_1, t_1] \times \cdots \times (s_d, t_d]$, $-\infty \le s_i < t_i \le \infty$, by taking the differences of the values of the DF at the vertices of the rectangle. For example, in the two-dimensional case, as it is easy to see from the definition of the DF,

$$P_{\boldsymbol{X}}((s_1, t_1] \times (s_2, t_2]) = F_{\boldsymbol{X}}(t_1, t_2) - F_{\boldsymbol{X}}(s_1, t_2) - F_{\boldsymbol{X}}(t_1, s_2) + F_{\boldsymbol{X}}(s_1, s_2) \tag{2.30}$$

(an analogue of F1 for DFs on $\mathbf{R}^2$ states that all such differences are non-negative).

Now note that the field generated by rectangles simply consists of finite unions of rectangles. One can show that the multidimensional versions of F1–F3 imply that the set function defined on the field by the differences of the form (2.30) on rectangles will satisfy conditions P1–P3 (when the union is an element of the field). Therefore, by the Carathéodory theorem, it can be uniquely extended to a probability on $\mathcal{B}_d$, for that σ-field is generated by the above field.

The definition of discrete random vectors is almost identical to that of discrete RVs: they can assume at most countable many different values. Absolutely continuous random vectors $\boldsymbol{X} \in \mathbf{R}^d$ are defined as those whose distributions are absolutely continuous with respect to the Lebesgue measure ("volume") in $\mathbf{R}^d$. In terms of DFs, this means that there exists a (probability) density function $f_{\boldsymbol{X}}(\boldsymbol{x}) \ge 0$, $\boldsymbol{x} = (x_1, \ldots, x_d)$, such that the DF

$$F_{\boldsymbol{X}}(\boldsymbol{t}) = \int_{-\infty}^{t_1} \cdots \int_{-\infty}^{t_1} f_{\boldsymbol{X}}(\boldsymbol{x})\, dx_1 \cdots dx_d. \tag{2.31}$$

A rough interpretation for the density is similar to (2.26):

$$P_{\boldsymbol{X}}(\Delta) \approx f_{\boldsymbol{X}}(\boldsymbol{x})\,|\Delta| \tag{2.32}$$

for a small volume (say, a ball or cube) $\Delta \ni \boldsymbol{x}$, $|\Delta|$ being the volume (or area when $d = 2$) of Δ. Similarly to (2.27), one can see that if $\boldsymbol{g} = (g_1, \ldots, g_d) : B \to \mathbf{R}^d$ is a differentiable function given on a set $B \subseteq \mathbf{R}^d$ such that $\mathbf{P}(\boldsymbol{X} \in B) = 1$, having an inverse $\boldsymbol{g}^{-1}$ on its range and a non-zero Jacobian:

$$J(\boldsymbol{x}) := \det\left[\left(\frac{\partial g_j(\boldsymbol{x})}{\partial x_k}\right)_{j,k=1,\ldots,d}\right] \neq 0,$$

then the random vector $\boldsymbol{Y} := \boldsymbol{g}(\boldsymbol{X})$ also has a density given by

$$f_{\boldsymbol{Y}}(\boldsymbol{y}) = \frac{f_{\boldsymbol{X}}(\boldsymbol{g}^{-1}(\boldsymbol{y}))}{|J(\boldsymbol{g}^{-1}(\boldsymbol{y}))|}. \tag{2.33}$$

Random variables $X_1, \ldots, X_d$ are said to be **independent** if, for any $B_j \in \mathcal{B}$, $j = 1, \ldots, d$,

$$\mathbf{P}(X_1 \in B_1, \ldots, X_d \in B_d) = \mathbf{P}(X_1 \in B_1) \cdots \mathbf{P}(X_d \in B_d). \tag{2.34}$$

This is equivalent to saying that, for any B_j's, the events $\{X_j \in B_j\}$, $j = 1, \ldots, d$, are independent (cf. (2.23)).

Note that, for any (measurable) functions $g_1, \ldots, g_d$, if $X_1, \ldots, X_d$ are independent, so are the RVs $g_1(X_1), \ldots, g_d(X_d)$.

Independence of RVs takes place iff their joint DF can be factorised, i.e., has the form

$$F_{\boldsymbol{X}}(\boldsymbol{x}) = F_{X_1}(x_1) \cdots F_{X_d}(x_d). \tag{2.35}$$

Distributions having properties (2.34)–(2.35) are called (for obvious reasons) **product-distributions**; more generally, measures μ on $(\mathbf{R}^d, \mathcal{B}_d)$ having the property that

$$\mu(B_1 \times \cdots \times B_d) = \mu_1(B_1) \times \cdots \times \mu_d(B_d)$$

for some measures μ_j on $\mathbf{R}$, are called *product-measures* (μ is the product of the μ_j's; sometimes it is denoted by $\mu_1 \otimes \cdots \otimes \mu_d$). Area and volume in $\mathbf{R}^d$ are examples of product-measures.

If $\boldsymbol{X}$ is absolutely continuous, then (2.35) is equivalent to

$$f_{\boldsymbol{X}}(\boldsymbol{x}) = f_{X_1}(x_1) \cdots f_{X_d}(x_d). \tag{2.36}$$

Note that it can happen that all X_j are absolutely continuous, while $\boldsymbol{X}$ is not. For example, if $X_1 \equiv X_2 \sim N(0,1)$ (this means that the RV X_2 follows the distribution $N(0,1)$), then the distribution of $\boldsymbol{X}$ is concentrated on the diagonal $x_1 = x_2$. But the *area* of any (regular) line is zero, hence the

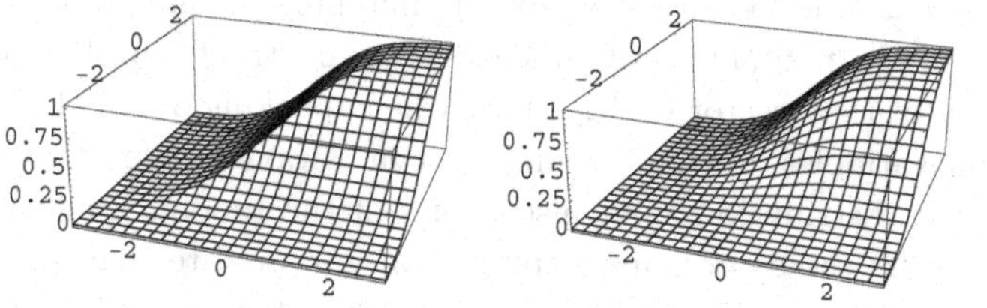

Fig. 2.2 The DFs of the random vectors (X_1, X_1) and (X_1, X_2), $X_i \sim N(0,1)$, $i = 1, 2$, being independent RVs.

distribution of $\boldsymbol{X}$ is singular to the Lebesgue measure and no joint density exists for (X_1, X_2).

But if $X_1, \ldots, X_d$ are *independent* and absolutely continuous, then $\boldsymbol{X}$ will also be absolutely continuous with the product-density (2.36). For example, if $X_j \sim N(0,1)$, $j = 1, \ldots, d$, are independent, then, from (2.17), $\boldsymbol{X} = (X_1, \ldots, X_d)$ has the density

$$f_{\boldsymbol{X}}(\boldsymbol{x}) = \prod_{j=1}^{d} f(x_j) = \frac{1}{(2\pi)^{d/2}} \exp\Big\{-\frac{1}{2}\sum_{j=1}^{d} x_j^2\Big\} = \frac{e^{-\boldsymbol{x}\boldsymbol{x}^T/2}}{(2\pi)^{d/2}}, \qquad (2.37)$$

where ^{T} denotes transposition, so that $\boldsymbol{x}^T$ is a column vector and hence the matrix product $\boldsymbol{x}\boldsymbol{x}^T = \sum_{j=1}^{d} x_j^2$ is a scalar.

A random vector $\boldsymbol{X} \in \mathbf{R}^d$ with density (2.37) is referred to as the *standard normal vector* in $\mathbf{R}^d$. Any random vector of the form $\boldsymbol{Y} = \boldsymbol{\mu} + \boldsymbol{X}A$, where $\boldsymbol{\mu} \in \mathbf{R}^m$ and A is a $d \times m$ matrix, is called an m-dimensional normal vector. When the rank of the matrix A is m, $\boldsymbol{Y}$ will also have a density which can be computed using (2.33). For the form of the normal density in the general case, see (2.61) below. Note that when, in particular, $m = 1$, so that $A = (a_1, \ldots, a_d)^T$ is a d-dimensional column vector, we just get a normal RV

$$\mu + \boldsymbol{X}\boldsymbol{a}^T = \mu + \sum_{j=1}^{d} a_j X_j$$

with parameters $(\mu, \sum_{j=1}^{d} a_j^2)$. This can be seen either from the convolution formula (2.64) for densities or using the machinery of moment generating or characteristic functions, see, e.g., Example 2.4 below.

Given the distribution P_X of an RV X, we can always construct a probability space and an RV on it having the distribution P_X. One can just take the "natural sample space" $\Omega = \mathbf{R}$ with $\mathcal{F} = \mathcal{B}$ and $\mathbf{P} = P_X$, and put $X(\omega) := \omega$ (the so-called "coordinate mapping"). Similarly, for a given distribution of a random vector, one can also construct a probability space and a copy of that vector on it. Thus, when speaking about distributions of random variables and/or vectors, we can always assume that we have not only the distributions, but also copies of the random elements themselves given by the coordinate mappings on appropriate spaces. However, if we have a complex stochastic system and are interested in some "global" characteristics of the system, we cannot separate particular random variables/vectors, which are parts of the model, from each other and study them isolated. In the general case, they will be dependent, and that aspect of the model should not be neglected.

Very often it is either very hard or impossible to find the distribution of a random characteristic of interest exactly. However, one can try to approximate it using simpler systems/distributions. To deal with such problems, one needs the notion of *convergence of random variables.* Random variables are *functions*, and there are several different types of convergence of RVs used in probability.

Let $X_0, X_1, X_2, \ldots$ be RVs on a common probability space. One says that X_n converge to X_0 **almost surely** (a.s.), or with probability 1, if $\lim_{n\to\infty} X_n(\omega) = X_0(\omega)$ holds on an event of probability 1, i.e., for all $\omega \in A$ for some A with $\mathbf{P}(A) = 1$. More generally, one says that a certain property holds a.s. if it holds on an event of probability 1. Note that the a.s. convergence is equivalent to the following assertion[29]:

$$\text{for any } \varepsilon > 0, \quad \mathbf{P}\left(\sup_{m>n} |X_m - X_0| > \varepsilon\right) \to 0 \quad \text{as } n \to \infty. \tag{2.38}$$

This is the *strongest* mode of convergence one deals with in probability theory.

A weaker form is **convergence in probability**:

$$\text{for any } \varepsilon > 0, \quad \mathbf{P}(|X_n - X_0| > \varepsilon) \to 0 \quad \text{as } n \to \infty. \tag{2.39}$$

It is obvious from (2.38) that a.s. convergence always implies convergence in probability.

[29]For a numeric sequence $\{a_n\}$, the **supremum** $\sup_n a_n$ is the *least upper bound* of the sequence defined as the number a such that all $a_n \le a$, but, for any $\varepsilon > 0$, there exists an $a_n > a - \varepsilon$. The difference between supremum and *maximum* is that the latter is to be *attained* on one of the a_n's (which is not necessarily the case for a general sequence $\{a_n\}$), while the former exists always. For example, $\sup_{n>1}(1 - 1/n) = 1$, but the maximum of those values does not exist. Of course, supremum is defined for any numeric set, not only sequences. A dual notion is that of the **infimum**, the *greatest lower bound.*

More often, the so-called **weak convergence**, or **convergence in distribution**, is used. Since this is *convergence of distributions*, we do not need all the X_n's to be defined on a common probability space; only their distributions matter. One says that the sequence of RVs X_n converges in distribution to X_0 (equivalently, distributions P_{X_n} converge weakly to P_{X_0}), if, for any bounded continuous function g on $\mathbf{R}$,

$$\lim_{n\to\infty} \mathbf{E}\, g(X_n) \equiv \lim_{n\to\infty} \int g(x)\, dP_{X_n} = \int g(x)\, dP_{X_0} \equiv \mathbf{E}\, g(X_0). \tag{2.40}$$

This is a general definition of weak convergence of distributions that can be used not only for real-valued RVs, but also for random elements with values in quite general spaces. For RVs, the above definition is equivalent to convergence of the DFs $F_{X_n}(t) \to F_{X_0}(t)$ as $n \to \infty$ at each point t where the limiting DF $F_{X_0}(t)$ is continuous.

To illustrate why we do not require convergence at discontinuity points, consider a simple example where the RVs $X_n \equiv x_n$ are constant and $x_n \searrow x_0$. Then the DFs $F_{X_n}(t) = \mathbf{1}_{[x_n,\infty)}(t) \to \mathbf{1}_{[x_0,\infty)}(t) = F_{X_0}(t)$ everywhere except for the point x_0 where F_{X_0} has a unit jump. Thus, with that kind of convergence, two discrete distributions will be close to each other when the jump points of the DF of one of them are close to those of the other, and the respective jumps sizes are close, too. We do not require that the points to which the distributions assign positive probabilities *should coincide* for the two distributions.

As the reader might have already correctly guessed, since this type of convergence is referred to as *weak*, there must be modes of convergence of distributions that are stronger than (2.40). This is true indeed. In particular, both "point-wise convergence" of probabilities P_{X_n} to P_{X_0}:

$$\text{for any } B \in \mathcal{B}, \quad P_{X_n}(B) \to P_{X_n}(B) \quad \text{as } n \to \infty,$$

and *convergence in total variation:*

$$\sup_{B\in\mathcal{B}} |P_{X_n}(B) - P_{X_n}(B)| \to 0 \quad \text{as } n \to \infty$$

imply (2.40), with the total variation convergence implying the point-wise one (but not the other way around).

It is easy to see that convergence in probability always implies convergence in distribution. The converse is, of course, not true. However, there is a remarkable *Skorokhod*[30] *theorem* stating that, for any sequence of RVs $\{X_n\}$ converging in distribution to an RV X_0, one can define, on a common probability space $(\Omega', \mathcal{F}', \mathbf{P}')$, a sequence of their "copies" X_n' (X_n' has the same distribution as X_n, $n = 0, 1, 2, \dots$) such that $X_n' \to X_0'$ a.s. as $n \to \infty$. It is very helpful to keep this result in mind when dealing with convergence in distribution.

---

[30] Anatolii Volodymyrovych Skorokhod (10.09.1930–03.01.2011), a Ukrainian mathematician, one of the most prominent probabilists of the 20th century.

2.5 Expectations

We again begin with appealing to the relative frequency interpretation (1.2) of probability. Assume there is an RV X related to our random experiment $\mathcal{E}$, which can take finitely many different values $x_1, \ldots, x_d$ with respective probabilities p_j, $j = 1, \ldots, d$. Such an RV is called **simple**. Of what value can one expect our X to be "on the average"? Denote by X_i the realisation of the RV in the ith independent replication of the random experiment $\mathcal{E}$. Then the average of the X-values in a series of n replications of $\mathcal{E}$ is

$$\frac{1}{n}\sum_{i=1}^{n} X_i = \sum_{j=1}^{d} x_j \frac{n_{\{X=x_j\}}}{n} \rightsquigarrow \sum_{j=1}^{d} x_j p_j \quad \text{since } \frac{n_{\{X=x_j\}}}{n} \rightsquigarrow p_j \text{ as } n \to \infty$$

from (1.2). That is, one can *expect* that the "time average" $n^{-1}\sum_{i=1}^{n} X_i$ will be close to the "space average" (given by the sum on the right-hand side of the above formula). A general result of that sort is referred to as the **law of large numbers**; for the time being, just note that the above relation provides a motivation for defining the *expectation* of an RV as the sum $\sum_{j=1}^{d} x_j p_j$ and, in the general case, as an integral of the RV.

As we have already mentioned in Section 2.4, since RVs are measurable, they can be integrated in the sense discussed briefly in Section 2.2 (Lebesgue integrals). Let $X = X(\omega)$ be an RV given on a probability space $(\Omega, \mathcal{F}, \mathbf{P})$. If $X \geq 0$, the value of the (Lebesgue) integral

$$\mathbf{E}\, X = \mathbf{E}\,(X) = \int_{\Omega} X(\omega)\, d\mathbf{P}\,(\omega) \tag{2.41}$$

is always defined (it can be infinite) and is called the **expectation** (or expected/mean value, or simply the mean) of the random variable X. In the general case, one says that the expectation of X exists if at least one of $X^+ = \max\{X, 0\}$ and $X^- = -\min\{X, 0\}$ has a finite mean (note that $X^\pm \geq 0$ and $X = X^+ - X^-$), and the value of the expectation is then

$$\mathbf{E}\, X = \mathbf{E}\, X^+ - \mathbf{E}\, X^-. \tag{2.42}$$

If the difference is finite, one says that the expectation is finite; in that case X is said to be **integrable**. If both terms on the right-hand side are infinite, the expectation does not exist (what is $\infty - \infty$?).

It is important to note that if our RV is the indicator of an event: $X = \mathbf{1}_A$ for some $A \in \mathcal{F}$, then its expectation is exactly the probability of the event A:

$$\mathbf{E}\, \mathbf{1}_A = 1 \times \mathbf{P}\,(A) + 0 \times \mathbf{P}\,(A^c) = \mathbf{P}\,(A). \tag{2.43}$$

Also, we will use notation $\mathbf{E}(X; A) = \mathbf{E}(X\mathbf{1}_A)$ for expectations over events (i.e., integrals not over the whole Ω, but over a part thereof).

Changing the variables and making use of the notion of the distribution of an RV, one can show that the expectation of an RV can be computed as an integral over the real line:

$$\mathbf{E}X = \int_{\mathbf{R}} x\, dP_X(x) = \int_{\mathbf{R}} x\, dF_X(x), \tag{2.44}$$

the last notation (which can often be encountered in texts on probability) standing for the so-called Lebesgue-Stieltjes[31] integral. The meaning is identical to the respective Lebesgue integral, the only difference is that formally this integral is with respect to a *function* (DF in our case), not measure. In the most typical cases when X is either discrete or absolutely continuous, the integral becomes, respectively, the sum

$$\sum_j x_j \mathbf{P}(X = x_j) \tag{2.45}$$

and the integral

$$\int_{-\infty}^{\infty} x\, f_X(x)\, dx, \tag{2.46}$$

which can usually be understood and computed as a conventional (Riemann) integral.

It can be convenient (although not always correct: beware of discontinuous integrands!) to think about the integral $\int g(x)\, dF_X(x)$ as a sort of conventional integral, which is defined as a result of the following limiting procedure. To define the Riemann integral of g, we partition the integration interval into a collection of small subintervals Δ_i, take the sum of the products {"typical value" of g on Δ_i} × {length of Δ_i} and pass to the limit over a sequence of such "refining" partitions (when the length of the largest subinterval goes to zero). Now, instead of the lengths of the Δ_i's, we substitute into the products the values of the *increments* of the function F on these subintervals! The limit will give us the value of that integral. This explains how we get the special cases (2.45) and (2.46).

If one has to compute the expectation of the RV $g(X)$ for a known (measurable) function g and an RV X, there is no need to first find the distribution P_Y of the RV $Y := g(X)$ and then apply (2.44) to compute $\mathbf{E}Y$. Changing variables, one gets

$$\mathbf{E}g(X) = \int g(x)\, dF_X(x); \tag{2.47}$$

---

[31] Thomas Johannes Stieltjes (29.12.1856–31.12.1894), a Dutch mathematician who worked in many areas of mathematics.

in the special cases when X is discrete or absolutely continuous, the value of the integral is

$$\sum_j g(x_j)\mathbf{P}(X = x_j) \quad \text{or} \quad \int_{-\infty}^{\infty} g(x)\, f_X(x)\, dx,$$

respectively. If X has a mixture distribution of the form (2.12), then

$$\mathbf{E}\, g(X) = \sum_i a_i \int g(x)\, dF_i(x),$$

where F_i is the DF of the distribution $\mathbf{P}_i$. Thus, if X has the distribution $(1-\rho)I_0 + \rho Exp(\mu - \lambda)$ of the waiting time in the simple queueing system from Section 7.2.1, then

$$\mathbf{E}\, g(X) = (1-\rho)g(0) + \rho(\mu - \lambda) \int_0^{\infty} g(x)\, e^{-(\mu-\lambda)x} dx.$$

Making use of the general properties of integrals, we see that expectation has the following properties:

(i) If a and b are constants, X and Y are integrable RVs, then

$$\mathbf{E}\,(aX + bY) = a\mathbf{E}\, X + b\mathbf{E}\, Y.$$

That is, the operation $\mathbf{E}$ is *linear.*

(ii) If $X(\omega) \le Y(\omega)$ a.s., then $\mathbf{E}\, X \le \mathbf{E}\, Y$, so that $\mathbf{E}$ is *monotone.*

(iii) If $X(\omega) \equiv c =$ const, then $\mathbf{E}\, X = c$.

When approximating RVs (say, random characteristics of a complex stochastic system) with RVs having known (and simpler) distributions, it is often important to know whether the expected values of the RVs will also be close. To this end, one can use several key results on convergence of expectations.

Monotone convergence theorem. Let $0 \le X_n \nearrow X_0$ a.s. as $n \to \infty$. Then $\mathbf{E}\, X_n \nearrow \mathbf{E}\, X_0$.

Dominated convergence theorem. Let $X_n \to X_0$ a.s. as $n \to \infty$, and $|X_n| \le Y$, $n \ge 1$, where $\mathbf{E}\, Y < \infty$. Then $\mathbf{E}\, X_n \to \mathbf{E}\, X_0$, and the last integral is finite.

When a sequence of integrable RVs $\{X_n\}$ converges in distribution to an (integrable) limit X_0, to ensure that $\mathbf{E}\, X_n \to \mathbf{E}\, X_0$, one has to require that the X_n's are **uniformly integrable**:

$$\lim_{N\to\infty} \sup_{n>0} \mathbf{E}\,(|X_n|;\, |X_n| > N) = 0. \tag{2.48}$$

Note that the domination condition $|X_n| \le Y$, $\mathbf{E}\, Y < \infty$, implies uniform integrability: by virtue of (2.49) below, for any $n > 0$,

$$\mathbf{E}\,(|X_n|;\, |X_n| > N) \le \mathbf{E}\,(Y;\, Y > N) \le \frac{\mathbf{E}\, Y}{N} \to \infty \quad \text{as } N \to \infty,$$

so that (2.48) holds.

Without any additional conditions, the convergence of X_n to X_0 (a.s. or in distribution) as $n \to \infty$ does not guarantee that $\mathbf{E}\, X_n \to \mathbf{E}\, X_0$. Indeed, let $X_0 \equiv 0$ and $X_n = n^3$ w.p. n^{-2} and $= 0$ w.p. $1 - n^{-2}$. Then $X_n \to X_0 (\equiv 0)$ a.s. (moreover, one can easily show that, with probability 1, $X_n \ne 0$ for only finitely many n's), but clearly $\mathbf{E}\, X_n = n \not\to \mathbf{E}\, X_0 = 0$. (Verify that the sequence $\{X_n\}$ is **not** uniformly integrable!)

The uniform integrability condition (2.48) ensures that such "sharp spikes" as in the above example, which comply with convergence in distribution but can destroy convergence of moments, are impossible.

Along with expectations, one often deals with the **moments** of higher orders of RVs (expectation itself is called the *first moment*; the term was borrowed from mechanics due to the analogy with the moment of a mass distributed along a rod). For an RV X, the expectation $\mathbf{E}\, X^k$ of the RV X^k is called the kth *moment* of X (or its distribution), $\mathbf{E}\,(X - \mathbf{E}\, X)^k$ the kth *central moment*, and $\mathbf{E}\,|X|^k$ the kth *absolute moment*. It is not hard to show that if the kth absolute moment is finite, then all the moments of orders $r \le k$ are finite, too.

The most important of higher order moments is the **variance**, or the second central moment of an RV:

$$\mathrm{Var}\,(X) := \mathbf{E}\,(X - \mathbf{E}\, X)^2 = \mathbf{E}\, X^2 - (\mathbf{E}\, X)^2.$$

The variance (or, rather, the square root thereof called the **standard deviation** of the RV) roughly outlines the "spread" of the distribution of X (and also appears in the central limit theorem, characterising the limiting law for sums of independent copies of X). Recall the following basic properties of variance: $\mathrm{Var}\,(X) = 0$ iff $X = c$ a.s. for a constant c, $\mathrm{Var}\,(cX) = c^2 \mathrm{Var}\,(X)$, and $\mathrm{Var}\,(X + c) = \mathrm{Var}\,(X)$.

The moments of a distribution, being simple numerical characteristics, enable one to describe, to some extent, the basic features of a far more complex object—the distribution itself. The very existence of finite moments indicates the rate at which the **tails** $F(-x)$ and $1 - F(x)$ of the distribution vanish as $x \to \infty$: since for $x > 0$ one always has $1 < X/x$ on the event $\{X > x\}$ so that $\mathbf{1}_{\{X>x\}} < (X/x)\mathbf{1}_{\{X>x\}}$, the following *Chebyshev*[32]

[32]Pafnuty Lvovich Chebyshev (16.05.1821–8.12.1892), an outstanding Russian mathematician, the founder of the St. Petersburg mathematical school. He is famous for his work on the theory of prime numbers, theoretical mechanics, theory of approximation

inequality holds due to (2.43):

$$\mathbf{P}(X > x) = \mathbf{E}\,\mathbf{1}_{\{X>x\}} < \frac{1}{x}\mathbf{E}(X; X > x) \le \frac{1}{x}\mathbf{E}(X; X \ge 0). \qquad (2.49)$$

Now if $g(x) \ge 0$ is an increasing function, we get from (2.49) that

$$\mathbf{P}(X > x) = \mathbf{P}(g(X) > g(x)) < \frac{\mathbf{E}\,g(X)}{g(x)}. \qquad (2.50)$$

Therefore when, say, $\mathbf{E}(X^+)^k < \infty$ for some $k > 0$, then, taking $g(x) := (x^+)^k \equiv x^k \mathbf{1}_{\{x>0\}}$, we get

$$\mathbf{P}(X > x) \le x^{-k}\mathbf{E}(X^+)^k.$$

Applying relation (2.49) to the RV $|X|$, and (2.50) to $|X - \mathbf{E}\,X|$ and X with $g(x) = x^2$, $x > 0$, and $g(x) = e^{aX}$, $a > 0$, respectively, we get the following popular versions[33] of the Chebyshev inequality: for $x > 0$,

$$\mathbf{P}(|X| > x) < \frac{\mathbf{E}\,|X|}{x}, \quad \mathbf{P}(|X - \mathbf{E}\,X| > x) < \frac{\mathrm{Var}(X)}{x^2},$$
$$\mathbf{P}(X > x) < \frac{\mathbf{E}\,e^{aX}}{e^{ax}}. \qquad (2.51)$$

Note that in all the versions of the inequality, we can simultaneously replace both strict inequality signs with the non-strict ones.

Recall also the important Cauchy–Bunyakovskii[34] inequality: for any RVs X_j with $\mathbf{E}\,X_j^2 < \infty$,

$$\mathbf{E}(X_1 X_2) \le \sqrt{\mathbf{E}\,X_1^2\,\mathbf{E}\,X_2^2}, \qquad (2.52)$$

where equality holds iff $X_2 = cX_1$ a.s. for some constant c.

The above inequality implies, in particular, that

$$\mathbf{E}\,|X - \mathbf{E}\,X| \le \sqrt{\mathrm{Var}(X)} \qquad (2.53)$$

and theory of probability. By the way, he appears to be the first to effectively use the notions of a random variable and its expectation and, in particular, to introduce the indicators of events.

[33]It is the middle inequality in (2.51) that was originally established and used to prove the law of large numbers by P.L. Chebyshev in 1867. Sometimes it is also called Bienaymé–Chebyshev inequality (for it can be found in I.J. Bienymaé's mémoire published in 1853).

[34]First established for finite sums by Augustin Louis Cauchy (21.08.1789–23.05.1857), a famous French mathematician who, in particular, introduced the clear notion of limit and continuity, and for integrals by the Russian mathematician Viktor Yakovlevich Bunyakovsky (16.12.1804–12.12.1889) in his paper *Sur quelques inegalités concernant les intégrales aux différences finies* published in 1859. For reasons unknown, the inequality is often called Cauchy–Schwarz or even simply Schwarz inequality, although it appeared in H.A. Schwarz' (25.01.1843-30.11.1921) work not earlier than in 1884.

(just set $X_1 := X - \mathbf{E}\,X$, $X_2 := \operatorname{sign} X_1$).

To prove (2.52), put $\widetilde{X}_j := X_j/\sqrt{\mathbf{E}\,X_j^2}$, note that $\mathbf{E}\,\widetilde{X}_j^2 = 1$ and use

$$0 \le \mathbf{E}\,(\widetilde{X}_1 - \widetilde{X}_2)^2 = \mathbf{E}\,\widetilde{X}_1^2 - 2\mathbf{E}\,\widetilde{X}_1\widetilde{X}_2 + \mathbf{E}\,\widetilde{X}_2^2$$
$$= 2(1 - \mathbf{E}\,\widetilde{X}_1\widetilde{X}_2) = 2\left[1 - \frac{\mathbf{E}\,X_1X_2}{\sqrt{\mathbf{E}\,X_1^2\,\mathbf{E}\,X_2^2}}\right].$$

In the case of equality in (2.52) one must have $\mathbf{E}\,(\widetilde{X}_1 - \widetilde{X}_2)^2 = 0$ from the last displayed formula, which is only possible when $\widetilde{X}_1 = \widetilde{X}_2$ a.s., i.e., $X_2 = cX_1$ for some constant c.

One more helpful result reduces computing Lebesgues–Stieltjes integrals for expectations to finding conventional ones. For an RV $X \ge 0$, since clearly $X = \int_0^\infty \mathbf{1}_{\{X>t\}}dt$, we have, swapping the order of integrals (which can be done due to (2.62)), that

$$\mathbf{E}\,X = \mathbf{E}\left[\int_0^\infty \mathbf{1}_{\{X>t\}}dt\right] = \int_0^\infty \left[\mathbf{E}\,\mathbf{1}_{\{X>t\}}\right] dt = \int_0^\infty (1 - F_X(t))\,dt \tag{2.54}$$

from (2.43). In the general case, since $X = X^+ - X^-$ is a difference of two non-negative RVs, we get from linearity of integrals in the same way that

$$\mathbf{E}\,X = \int_0^\infty (1 - F_X(t))\,dt - \int_{-\infty}^0 F_X(t)\,dt = \int_0^\infty \big(1 - F_X(t) - F_X(-t)\big)\,dt.$$

Higher order moments can also be computed in a similar way. Observing that, for an RV $X \ge 0$ and $k \ge 1$, the DF of the RV X^k is given by $F_{X^k}(t) = \mathbf{P}\,(X \le t^{1/k}) = F_X(t^{1/k})$, $t > 0$, we have from (2.54) that

$$\mathbf{E}\,X^k = \int_0^\infty (1 - F_X(t^{1/k}))\,dt = k\int_0^\infty s^{k-1}(1 - F_X(s))\,ds, \tag{2.55}$$

by changing the variables ($s = t^{1/k}$). So once we know how fast the tail of the distribution of X vanishes at the infinity, we can always say if this or that moment of X is finite or not.

We will now compute the first two moments for several standard distributions.

For a RV $X \sim Po(\lambda)$,

$$\mathbf{E}\,X = \sum_{j=0}^\infty je^{-\lambda}\frac{\lambda^j}{j!} = e^{-\lambda}\lambda\sum_{j=1}^\infty \frac{\lambda^{j-1}}{(j-1)!} = e^{-\lambda}\lambda e^{\lambda} = \lambda,$$

$$\mathbf{E}\,X^2 = \sum_{j=0}^\infty j(j-1)e^{-\lambda}\frac{\lambda^j}{j!} + \mathbf{E}\,X = e^{-\lambda}\lambda^2\sum_{j=2}^\infty \frac{\lambda^{j-2}}{(j-2)!} + \lambda = \lambda^2 + \lambda,$$

so that $\text{Var}(X) = \mathbf{E}\,X^2 - (\mathbf{E}\,X)^2 = \lambda$.

To compute the expectation of $X \sim U(a,b)$, we may first note that $X = a + (b-a)U$ in distribution, $U \sim U(0,1)$, and hence $\mathbf{E}\,X = a + (b-a)\mathbf{E}\,U$, $\text{Var}(X) = (b-a)^2\text{Var}(U)$. It remains to find

$$\mathbf{E}\,U = \int_0^1 x\,dx = \left[\frac{1}{2}x^2\right]_0^1 = \frac{1}{2},$$

$$\mathbf{E}\,U^2 = \int_0^1 x^2\,dx = \left[\frac{1}{3}x^3\right]_0^1 = \frac{1}{3},$$

so that

$$\text{Var}(U) = 1/3 - 1/4 = 1/12. \tag{2.56}$$

For a Cauchy distributed RV X with the density $1/(\pi(1+x^2))$, $x \in \mathbf{R}$, we clearly have both $\mathbf{E}\,X^{\pm} = \infty$, so that the expectation of X does not exist, while the variance is obviously infinite.

Quite often one has to compute *mixed moments*, i.e., the expectations of the form $\mathbf{E}\,XY$, X and Y being RVs (given on a common probability space). If X and Y are independent and integrable, then

$$\mathbf{E}\,XY = \mathbf{E}\,X\,\mathbf{E}\,Y \tag{2.57}$$

(which is a direct consequence of the definition of independence of events: $\mathbf{P}(AB) = \mathbf{P}(A)\,\mathbf{P}(B)$ is equivalent to $\mathbf{E}\,\mathbf{1}_A\mathbf{1}_B = \mathbf{E}\,\mathbf{1}_A\,\mathbf{E}\,\mathbf{1}_B$ due to (2.43), and from that it is easy to establish the above formula for *simple RVs*, and then extend it to the general case passing to the limit), and

$$\text{Var}(X \pm Y) = \text{Var}(X) + \text{Var}(Y). \tag{2.58}$$

Observe that the last expression does not imply that the RVs X and Y are independent; they will be just *uncorrelated* in that case, which means that the **covariance**

$$\text{Cov}(X,Y) := \mathbf{E}\,(X - \mathbf{E}\,X)(Y - \mathbf{E}\,Y) \equiv \mathbf{E}\,XY - \mathbf{E}\,X\,\mathbf{E}\,Y \tag{2.59}$$

is equal to zero. In the general case, setting $X_0 := X - \mathbf{E}\,X$, $Y_0 := Y - \mathbf{E}\,Y$, we see that the variance

$$\begin{aligned}\text{Var}(X \pm Y) &= \text{Var}(X_0 \pm Y_0) = \mathbf{E}\,(X_0 \pm Y_0)^2 \\ &= \mathbf{E}\,X_0^2 + \mathbf{E}\,Y_0^2 \pm 2\mathbf{E}\,X_0Y_0 = \text{Var}(X) + \text{Var}(Y) \pm 2\text{Cov}(X,Y).\end{aligned}$$

The covariance $\text{Cov}(X,Y)$ of X and Y—or, rather, the **correlation**

$$\rho(X,Y) := \frac{\text{Cov}(X,Y)}{\sqrt{\text{Var}(X)\text{Var}(Y)}}$$

between X and Y—is the standard measure of *linear dependence* between the RVs. It immediately follows from Cauchy–Bunyakovskii's inequality (2.52) that always $|\rho(X,Y)| \le 1$, and that $\rho(X,Y) = 0$ iff $Y = aX + b$ a.s. for some constants a and b.

In fact, $\mathbf{E}\,XY$ is nothing else but the *scalar* (or *inner*) product of the elements X and Y of the linear space L^2 of square integrable RVs on our $(\Omega, \mathcal{F}, \mathbf{P})$, generating the norm $\|X\| := \sqrt{\mathbf{E}\,X^2}$ in this space. Relation (2.58) is actually the Pythagoras theorem in that space (that X and Y are uncorrelated means that $X_0 = X - \mathbf{E}\,X$ and $Y_0 = Y - \mathbf{E}\,Y$ are orthogonal to each other in L^2: $\mathbf{E}\,X_0Y_0 = 0$). The correlation has the interpretation of the cosine of the angle between the "vectors" X and Y (in particular, the vectors are *orthogonal* to each other when it is equal to zero).

Note that it follows from Chebyshev inequality (2.51) that convergence in L^2 norm implies convergence in probability (2.39):

$$\mathbf{P}\left(|X_n - X_0| > \varepsilon\right) \le \varepsilon^{-2}\mathbf{E}\,(X_n - X_0)^2 \equiv \varepsilon^{-2}\|X_n - X_0\|^2,$$

which tends to zero for any $\varepsilon > 0$ provided that $\|X_n - X_0\| \to 0$ as $n \to \infty$, and hence convergence in distribution as well.

To illustrate (2.58), note that if $X_i \sim U(0,1)$ are independent and identically distributed (i.i.d.) RVs, then, for $Y = X_1 + \cdots + X_{12} - 6$, we clearly have $\mathbf{E}\,Y = 0$ and $\mathrm{Var}\,(Y) = 1$ from (2.56).

For a random vector $\boldsymbol{X} = (X_1, \ldots, X_d)$, its mean (or *mean vector*) is defined as the vector of the means of its components:

$$\mathbf{E}\,\boldsymbol{X} = (\mathbf{E}\,X_1, \ldots, \mathbf{E}\,X_d).$$

To characterise the "spread" of the components' distributions and (linear) dependence between them, one uses the **covariance matrix** of $\boldsymbol{X}$ defined as

$$C_{\boldsymbol{X}} := (\mathrm{Cov}\,(X_j, X_k))_{j,k=1,\ldots,d} = \mathbf{E}\,(\boldsymbol{X} - \mathbf{E}\,\boldsymbol{X})^T(\boldsymbol{X} - \mathbf{E}\,\boldsymbol{X}) \qquad (2.60)$$

(recall that ^{T} denotes transposition, so that $(\boldsymbol{X} - \mathbf{E}\,\boldsymbol{X})^T$ is a column vector, and the matrix product on the right-hand side of (2.60) is a $d \times d$-matrix). Observe that any covariance matrix is symmetric and *non-negative definite*: $C^T = C$ and, for any vector $\boldsymbol{a} = (a_1, \ldots, a_d) \in \mathbf{R}^d$,

$$\sum_{j,k=1}^{d} (C_{\boldsymbol{X}})_{ij}\, a_j a_k \equiv \boldsymbol{a} C_{\boldsymbol{X}} \boldsymbol{a}^T = \mathbf{E}\,\boldsymbol{a}(\boldsymbol{X} - \mathbf{E}\,\boldsymbol{X})^T(\boldsymbol{X} - \mathbf{E}\,\boldsymbol{X})\boldsymbol{a}^T$$

$$= \mathbf{E}\,\boldsymbol{a}(\boldsymbol{X} - \mathbf{E}\,\boldsymbol{X})^T(\boldsymbol{a}(\boldsymbol{X} - \mathbf{E}\,\boldsymbol{X})^T)^T = \mathbf{E}\,\left|\boldsymbol{a}(\boldsymbol{X} - \mathbf{E}\,\boldsymbol{X})^T\right|^2 \ge 0,$$

where we used the transposition rule for matrix products: $(AB)^T = B^T A^T$, and the observation that $\boldsymbol{a}(\boldsymbol{X} - \mathbf{E}\,\boldsymbol{X})^T = (\boldsymbol{a}(\boldsymbol{X} - \mathbf{E}\,\boldsymbol{X})^T)^T$ is a scalar. Note also that the diagonal entries of $C_{\boldsymbol{X}}$ are equal to the variances of the respective components of $\boldsymbol{X}$.

When a random vector $\boldsymbol{X} \in \mathbf{R}^d$ is subject to a linear transformation: $\boldsymbol{Y} := \boldsymbol{X}A$, where A is a $d \times m$-matrix for some $m \geq 1$ (so that $\boldsymbol{Y} \in \mathbf{R}^m$), the mean and covariance matrix of the new vector are easily seen to be given, respectively, by $\mathbf{E}\,\boldsymbol{Y} = (\mathbf{E}\,\boldsymbol{X})A$ and

$$\begin{aligned} C_{\boldsymbol{Y}} &= \mathbf{E}\,(\boldsymbol{Y} - \mathbf{E}\,\boldsymbol{Y})^T(\boldsymbol{Y} - \mathbf{E}\,\boldsymbol{Y}) = \mathbf{E}\,[(\boldsymbol{X} - \mathbf{E}\,\boldsymbol{X})A]^T[(\boldsymbol{X} - \mathbf{E}\,\boldsymbol{X})A] \\ &= A^T\big[\mathbf{E}\,(\boldsymbol{X} - \mathbf{E}\,\boldsymbol{X})^T(\boldsymbol{X} - \mathbf{E}\,\boldsymbol{X})\big]A = A^T C_{\boldsymbol{X}} A. \end{aligned}$$

Thus, if $\boldsymbol{X}$ has the unit covariance matrix: $C_{\boldsymbol{X}} = I$ (like the standard normal distribution with density (2.37)), then the covariance matrix of $\boldsymbol{Y}$ is merely $C_{\boldsymbol{Y}} = A^T A$. Combining this observation with (2.37) and (2.33) when, say $m = d$ and the matrix A is non-degenerate, one derives the following general form of the normal density in $\mathbf{R}^d$:

$$\frac{1}{\sqrt{(2\pi)^d \det C_{\boldsymbol{Y}}}} \exp\left\{-\frac{1}{2}(\boldsymbol{y} - \boldsymbol{\mu})C_{\boldsymbol{Y}}^{-1}(\boldsymbol{y} - \boldsymbol{\mu})^T\right\}. \qquad (2.61)$$

So the normal distribution in $\mathbf{R}^m$ is **uniquely determined** by its mean vector and covariance matrix. Moreover, the above formula implies the following important fact: for any vector $\boldsymbol{\mu} \in \mathbf{R}^m$ and any symmetric non-negative definite $m \times m$ matrix C there exist a random vector $\boldsymbol{Y} \in \mathbf{R}^m$ with mean $\boldsymbol{\mu}$ and covariance matrix $C_{\boldsymbol{Y}} = C$ (one can take the normal vector having density (2.61) with $C_{\boldsymbol{Y}}$ replaced by C).

One more related result to be mentioned here shows how to compute the expectations of functions of several independent RVs.

Fubini's[35] theorem. *Let $g(x, y)$ be a (Borel) measurable function, X and Y two independent RVs. If $\mathbf{E}\,g(X, Y)$ exists, then*

$$\begin{aligned} \mathbf{E}\,g(X,Y) &= \int g(x,y)\,dP_{(X,Y)}(x,y) \qquad (2.62) \\ &= \int\left[\int g(x,y)\,dP_X(x)\right]dP_Y(y) = \int\left[\int g(x,y)\,dP_Y(y)\right]dP_X(x). \end{aligned}$$

In the conclusion of this section, we will review a general rule for computing the distribution of the sums of independent RVs. Suppose X_1 and X_2 are independent RVs, $S = X_1 + X_2$. First let X_i, $i = 1, 2$, be integer-valued with distributions $\boldsymbol{p}_i = (p_i(j))_{j \in \boldsymbol{Z}}$, $p_i(j) := \mathbf{P}\,(X_i = j)$. Since

$$\mathbf{P}\,(S = k | X_2 = j) = \mathbf{P}\,(X_1 + j = k | X_2 = j) = p_1(k - j)$$

[35] Guido Fubini (19.01.1879–06.06.1943), an Italian mathematician; he established the theorem reducing multiple integration to the iterative one in 1907.

by independence, we have by the TPF that

$$\mathbf{P}(S=k) = \sum_j \mathbf{P}(S=k|X_2=j)\,\mathbf{P}(X_2=j) = \sum_j p_1(k-j)\,p_2(j). \tag{2.63}$$

The sequence given by the expression on the right-hand side is called the **convolution** of the sequences $\boldsymbol{p}_1$ and $\boldsymbol{p}_2$, and is denoted by $\boldsymbol{p}_1 * \boldsymbol{p}_2$; clearly, $\boldsymbol{p}_1 * \boldsymbol{p}_2 = \boldsymbol{p}_2 * \boldsymbol{p}_1$ as both sides give, due to (2.63), the distribution of one and the same RV S. Similarly, when the X_i's are absolutely continuous with densities f_i, the sum S is also absolutely continuous and has the density

$$f_S(x) = (f_1 * f_2)(x) := \int_{-\infty}^{\infty} f_1(x-y)\,f_2(y)\,dy; \tag{2.64}$$

the function is also called *convolution* (of densities), and $f_1 * f_2 = f_2 * f_1$.

Both (2.63) and (2.64) are special cases of the general convolution formula: for the DF F_S of the sum, one has

$$F_S(x) = (F_1 * F_2)(x) := \int_{-\infty}^{\infty} F_1(x-y)\,dF_2(y) = \int_{-\infty}^{\infty} F_2(x-y)\,dF_1(y). \tag{2.65}$$

Note that, although we are using the same symbol $*$ for all the convolutions, there is a difference between the convolutions of densities and those of DFs.

Example 2.1. *Poisson RVs.* If $X_i \sim Po(\lambda_i)$, $i = 1, 2$, are independent Poisson RVs, then (2.63) yields that, for $j = 0, 1, 2, \ldots,$

$$\mathbf{P}(S=j) = \sum_{k=0}^{j} e^{-\lambda_1}\frac{\lambda_1^{j-k}}{(j-k)!} \times e^{-\lambda_2}\frac{\lambda_2^k}{k!}$$

$$= e^{-(\lambda_1+\lambda_2)}\frac{1}{j!}\sum_{k=0}^{j}\frac{j!}{(j-k)!k!}\lambda_1^{j-k}\lambda_2^k = e^{-(\lambda_1+\lambda_2)}\frac{(\lambda_1+\lambda_2)^j}{j!},$$

by the binomial formula, so that S has the Poisson distribution with the parameter $\lambda_1 + \lambda_2$.

Example 2.2. Let $X_i \sim B_p$, $i = 1, \ldots, n$, be i.i.d. RVs. Then the sum $S_n = X_1 + \cdots + X_n$ will have the **binomial distribution** $B_{n,p}$: for $k = 0, \ldots, n$,

$$\mathbf{P}(S_n = k) = \binom{n}{k} p^k (1-p)^{n-k}, \quad \binom{n}{k} = \frac{n!}{k!\,(n-k)!}. \tag{2.66}$$

To see that, we will use **mathematical induction** in n. It is obvious that (2.66) is true for $n = 1$. For any $n \geq 1$, assuming that we have already proved (2.66) for that value of n, show that it holds for the value $n+1$. Indeed, (2.63) implies that, for $k>0$

$$\begin{aligned}\mathbf{P}(S_{n+1}=k) &= \mathbf{P}(S_n=k)\,\mathbf{P}(X_{n+1}=0)+\mathbf{P}(S_n=k-1)\,\mathbf{P}(X_{n+1}=1)\\ &= \left(\binom{n}{k}+\binom{n}{k-1}\right)p^k(1-p)^{n-k+1}\\ &= \frac{n!}{(k-1)!(n-k)!}\left(\frac{1}{k}-\frac{1}{n-k+1}\right)p^k(1-p)^{n-k+1}\\ &= \binom{n+1}{k}p^k(1-p)^{n+1-k},\end{aligned}$$

which completes the proof.

Example 2.3. *Uniform RVs.* Suppose $X_i \sim U(0,1)$, $i \geq 1$, are i.i.d. RVs. Applying (2.64) and using the observation that

$$f_{X_1}(x-y)\,f_{X_2}(y) = \mathbf{1}_{\{x-y\in[0,1]\}}\mathbf{1}_{\{y\in[0,1]\}},$$

we see that the sum X_1+X_2 has the "triangular density"

$$f_{X_1+X_2}(x) = \int_0^1 f_{X_1}(x-y)\,dy = \int_0^1 \mathbf{1}_{\{x-y\in[0,1]\}}\,dy = \begin{cases}0, & x\notin[0,2],\\ x, & x\in[0,1],\\ 2-x, & x\in[1,2].\end{cases}$$

The integral here is clearly the length of the intersection of the segments $[0,1]$ and $[x-1,x]$. The graph of the density of the sum $X_1+X_2+X_3$ will consist of three pieces of parabolas:

$$f_{X_1+X_2+X_3}(x) = \int_0^1 f_{X_1+X_2}(x-y)\,dy = \begin{cases}0, & x\notin[0,3].\\ \frac{x^2}{2}, & x\in[0,1].\\ 1-\frac{(2-x)^2}{2}-\frac{(x-1)^2}{2}, & x\in[1,2],\\ \frac{(3-x)^2}{2}, & x\in[2,3].\end{cases}$$

The shapes of the densities of X_1, X_1+X_2, and $X_1+X_2+X_3$ are shown in Fig. 2.3, the last one being in fact very close to that of the normal density with the same mean and variance (see Fig. 2.4). The graph of the density of the sum $X_1+X_2+X_3+X_4$ will consist of four pieces of cubic parabolas and so on. If we shift the origin to the point $n/2$, then, as n increases, the shape (up to a scaling transformation) of the density of the sum $X_1+\cdots+X_n$ will be approaching that of the function e^{-x^2}. Such a behaviour is a display of the *central limit theorem* to be discussed in some detail in Section 2.9.

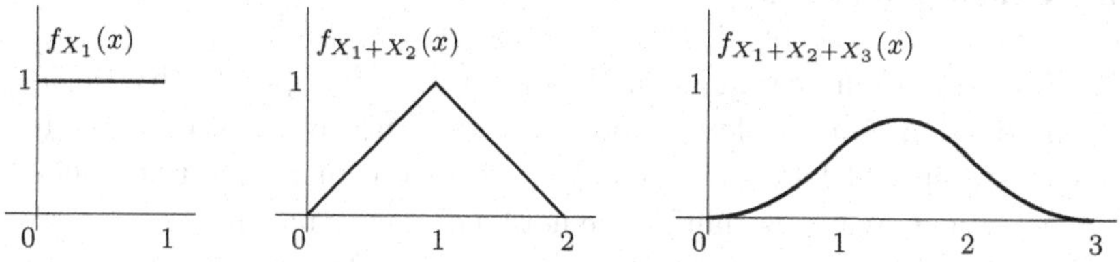

Fig. 2.3 Illustration to convolutions of uniform densities.

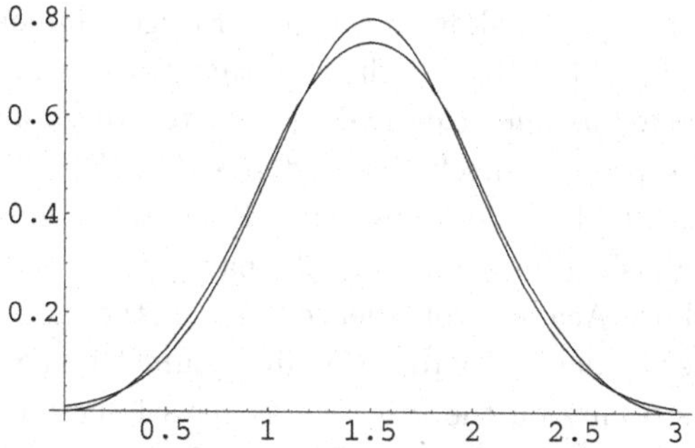

Fig. 2.4 The plots of $f_{X_1+X_2+X_3}$ and of the normal density with the same mean and variance.

The two examples above illustrate the general rule: the convolution of two distributions is at least as "smooth" as any of them (if at least one of the X_i's is absolutely continuous, then so is the sum S), and gets smoother and smoother when the number of terms increases. It is not hard to give an example when X_i are not absolutely continuous, while their sum is. Note only that convolutions of discrete distributions can only be discrete again, but the general tendency will persist: we will get more "spread out" distributions.

Computing convolutions is, however, impractical in non-trivial cases, and even when the number of random summands is moderate, finding the distribution of their sums is a very hard task. In that case, one relies more on various approximations provided by the *limit theorems* of probability theory, a few of which we will review in Section 2.9 below.

2.6 Utility Functions

From the very beginning, gambling has been providing probability theory with motivation and problems (note that insurance practice can also be viewed as a special kind of gambling), one of the main questions being when a particular game is "fair". The notion of the "expectation" of an RV was introduced to make the last term meaningful. Imagine two players, I and II, playing a game of chance. The outcome of each play is described by an RV X representing the amount of money player II pays to I (if $X < 0$, it means that in fact I pays II the amount $|X|$). One says that the game is *fair* if $\mathbf{E}\,X = 0$. Then, according to the law of large numbers, the average pay-off per play will tend to zero in the long run. However, as the following famous *St. Petersburg paradox* clearly demonstrates, this approach does not always work. The problem itself was first posed by N. Bernoulli[36] in his letter of 09.09.1713 to P.R. de Monmort[37]; it became widely known after D. Bernoulli[38] published his solution of the paradox in 1738 in the journal of the St. Petersburg Academy of Sciences—hence the name.

Suppose player II repeatedly tosses a fair (symmetric) coin until it lands heads up. If this occurs on the kth toss, II pays I the amount of $\$2^{k-1}$, $k = 1, 2, \ldots$ What is the fair price for I to pay for the game?

The probability that heads shows up for the first time on the kth toss is

$$\mathbf{P}(\underbrace{T \cdots T}_{k-1 \text{ times}} H) = \mathbf{P}(T)^{k-1}\mathbf{P}(H) = 2^{-k}$$

by independence. Therefore, if I pays the second player $\$b$ before each play, the payoff X has the distribution

$$\mathbf{P}(X = 2^{k-1} - b) = 2^{-k}, \quad k = 1, 2, \ldots, \tag{2.67}$$

so that

$$\mathbf{E}\,X = \sum_{k=1}^{\infty} 2^{k-1} \times 2^{-k} - b = \frac{1}{2}\sum_{k=1}^{\infty} 1 - b = \infty$$

---

[36]Nikolas Bernoulli (21.10.1687–29.11.1759), a Swiss mathematician, one of the many prominent mathematicians in the Bernoulli family, a nephew of J. Bernoulli.

[37]Pierre Rémond de Montmort (27.10.1678–07.10.1719), a French mathematician.

[38]Daniel Bernoulli (08.02.1700–17.03.1782), another nephew of J. Bernoulli; he investigated not only mathematics, but also medicine, biology, physiology, mechanics, astronomy and oceanography. He derived the famous Bernoulli law in fluid dynamics, which is the basis for many applications such as aircraft wing design.

for any b! It means, that, basing on the expectation criterion, the game is favourable for I whatever b is. The paradox is that, in reality, almost nobody would agree to pay more that \$7 to play such a game...

A possible resolution of the paradox is based on the observation that one and the same amount of money can have different value for different people. Hence the value of a thing should be determined not by its price (which is common for all people), but by its *utility* for a particular person. Thus, winning \$1,000 is much more important for a poor man than for a rich one. Therefore, we just cannot use the expectation criterion!

What one could do is to introduce a **utility function** $u(x)$ (which is often assumed to be concave[39], e.g., $x^{1/2}$ or $\log x$ for $x > 0$) describing the "utility" of the amount of money x for the respective individual (or company), and then declare that the game of chance is fair for the individual when the balance of expected utilities takes place. That is, the price our individual will consider to be fair to pay for playing the game will have utility (for that individual!) equal to the expected utility of the (random) amount of money he/she can get as a result of playing. Similarly, in optimisation problems, one maximises not the expected gain $\mathbf{E}\,X$, but the expected utility $\mathbf{E}\,u(X)$. Utility functions typically increase much slower than just x, so that (i) the expectation $\mathbf{E}\,u(X)$ can be finite, while $\mathbf{E}\,X = \infty$, and (ii) we will get answers quite different from what the expectation criterion would lead us to. This is due to the fact that, when computing expected utilities, the contribution of *very large* but *rather unlikely* values of X will be relatively smaller.

The last observation is true when one uses the so-called *risk-averse* utility functions characterised by the property that $u'' < 0$ (concavity). If $u'' = 0$ ($u'' > 0$) on an interval, the function u is called *risk-neutral* (*risk-seeking*) on the interval. The meaning is similar to what he have just explained: if, for example, an individual prefers an unlikely large gain to a steady income, such a behaviour is to be described using a risk-seeking utility function. Studies show that many people are risk-averse in the gains region and risk-seeking in the loss region.

One of the standard choices is the so-called *Bernoulli utility*

$$u(x) = k \log x + c. \tag{2.68}$$

[39]A real-valued function $u(x)$, $x \in \mathbf{R}^d$, is said to be **concave** if, for any $x, y \in \mathbf{R}^d$ and $\lambda \in (0,1)$, $u(\lambda x + (1-\lambda)y) \geq \lambda u(x) + (1-\lambda)u(y)$. When the inequality is inverse, the function $u(x)$ is said to be **convex**. If $u(x)$, $x \in \mathbf{R}$, is twice differentiable, concavity is equivalent to the condition $u''(x) < 0$. In the multidimensional space, the matrix of the second order partial derivatives must be negative-definite.

This choice can be justified as follows. Suppose that, for our individual, the utility of a small amount of money is *inversely proportional* to the amount he/she already has: for an increment dx of the amount of money, the utility of that increment to our individual is

$$du(x) = k\frac{1}{x}dx, \qquad \text{or} \qquad \frac{du}{dx} = \frac{k}{x}.$$

This relation immediately leads to (2.68)[40].

Return to the St. Petersburg paradox and derive the "fair price" for the game using the logarithmic utility function (2.68). Denote the initial capital of player I by a. Then, equating the utility of a to the expected utility (computed using (2.67)), we get the equation

$$\sum_{n=1}^{\infty} 2^{-n} \log(a - b + 2^{n-1}) = \log a.$$

Solving it for b, we will obtain the fair price (which will clearly depend on the initial capital a). For instance, if $a = 10$, then one has $b \approx 3$, while for $a = 1,000$, the one has $b \approx 6$.

2.7 Integral Transforms

Integral transforms are powerful tools for analysing probability distributions. They can be extremely helpful for computing important characteristics of the distributions and for proving various representations and limit theorems.

The **moment generating function** (MGF) of an RV X (or its distribution) is a function of the real variable t given by

$$\varphi_X(t) := \mathbf{E}\, e^{tX} = \int e^{tx}\, dF_X(x) = \begin{cases} \sum_j e^{tx_j}\, \mathbf{P}\,(X = x_j) \\ \int e^{tx}\, f_X(x)\, dx \end{cases}$$

when the RV is discrete or absolutely continuous with a density f_X, respectively. Of course, in the general case, the MGF does not need to be finite for any value of t except 0 (which is the case, for example, for Cauchy distributed X's). But when it is finite for some $t_0 > 0$, one can easily show (using integral inequalities; in fact, it suffices to apply the last relation from (2.51)) that it will also be finite for all $t \in [0, t_0]$ (similarly for $t_0 < 0$).

[40]The argument used by D. Bernoulli in his 1738 paper on utility. It was apparently the first instance when differential equations (then almost a new-born infant) were used in probability theory.

The term is due to the fact that differentiating the MGF at zero "generates" the moments of X: if, for some $\varepsilon > 0$, $\varphi_X(t) < \infty$ for $t \in (-\varepsilon, \varepsilon)$, then φ_X is infinitely many times differentiable in this interval[41], and

$$\frac{d}{dt}\varphi_X(t) = \frac{d}{dt}\mathbf{E}\,e^{tX} = \mathbf{E}\left[\frac{d}{dt}e^{tX}\right] = \mathbf{E}\left[Xe^{tX}\right], \quad \varphi_X'(0) = \mathbf{E}\,X,$$

$$\frac{d^2}{dt^2}\varphi_X(t) = \mathbf{E}\left[\frac{d^2}{dt^2}e^{tX}\right] = \mathbf{E}\left[X^2e^{tX}\right], \quad \varphi_X''(0) = \mathbf{E}\,X^2,$$

and so on, so that

$$\frac{d^k}{dt^k}\varphi_X(0) = \mathbf{E}\,X^k, \quad k = 1, 2, \ldots \tag{2.69}$$

There is a one-to-one correspondence between MGFs and distributions F having finite exponential moments $\int e^{tx}dF(x)$ for some $t \neq 0$ (i.e., to any such DF there corresponds one and only one MGF), and there are some (not very convenient) **inversion formulae** enabling one to find the distribution from its known MGF. Also, there is a very nice property of the MGFs that weak convergence of distributions is equivalent to (point-wise) convergence of their MGFs (it suffices to require that the convergence takes place for all t from an arbitrary small interval).

Another key property of the MGFs is that to convolutions of distributions (which is usually quite hard to compute) there correspond just the products of the respective MGFs : for independent RVs X and Y,

$$\varphi_{X+Y}(t) = \varphi_X(t)\,\varphi_Y(t). \tag{2.70}$$

Indeed, $\varphi_{X+Y}(t) = \mathbf{E}\left(e^{tX}e^{tY}\right)$ is the expectation of a product of independent RVs, and (2.70) follows from (2.57).

Example 2.4. The MGF of a standard normal RV X is equal to

$$\mathbf{E}\,e^{tX} = \frac{1}{\sqrt{2\pi}}\int e^{tx-x^2/2}dx = \frac{e^{t^2/2}}{\sqrt{2\pi}}\int e^{-(x-t)^2/2}dx = e^{t^2/2}.$$

So, for an RV $Y \sim N(\mu, \sigma^2)$, since $\mu + \sigma X$ has the same distribution as Y,

$$\mathbf{E}\,e^{tY} = e^{t\mu}\mathbf{E}\,e^{(\sigma t)X} = e^{\mu t + \sigma^2 t^2/2}.$$

For two independent RVs $Y_j \sim N(\mu_j, \sigma_j^2)$, $j = 1, 2$, due to (2.70), the MGF of the sum $Y_1 + Y_2$ is equal to

$$\mathbf{E}\,e^{t(Y_1+Y_2)} = e^{\mu_1 t + \sigma_1^2 t^2/2} \times e^{\mu_2 t + \sigma_2^2 t^2/2} = e^{(\mu_1+\mu_2)t + (\sigma_1^2+\sigma_2^2)t^2/2},$$

so that $Y_1 + Y_2 \sim N(\mu_1 + \mu_2, \sigma_1^2 + \sigma_2^2)$.

[41]Moreover, it will be analytic (as a function of the complex variable t) in the whole band $\mathrm{Re}\,t \in (-\varepsilon, \varepsilon)$ on the complex plane.

For RVs $X \geq 0$, the standard analysis tool is the **Laplace transform**

$$\varphi_X(-t) = \mathbf{E}\, e^{-tX} = \int e^{-tx} dF_X(x); \tag{2.71}$$

this expectation is always finite for $t \geq 0$. All the properties one can establish for MGFs will stay true, with obvious changes, for Laplace transforms.

The right-hand side of (2.71) is also called the **Laplace-Stieltjes transform** of the DF F_X (the second name is due to the fact that this is a Stieltjes integral). One often uses this transform for more general functions as well; say, the Laplace-Stieltjes transform of the function $\max\{0, x\}$ is $\int_0^\infty e^{-tx} dx = 1/t$. The most important for us property is still that to convolutions (of the form (2.65)) of functions there correspond products of their Laplace-Stieltjes transforms.

For *integer-valued* RVs, it is often more convenient to deal with their **generating functions** (GFs) which are obtained by substituting $z = e^t$ into the MGF:

$$g(z) \equiv g_X(z) := \varphi_X(\log z) = \mathbf{E}\, z^X = \sum_k z^k\, \mathbf{P}\,(X = k). \tag{2.72}$$

When $X \geq 0$, $g(z)$ is an analytic function (of the complex variable z) inside the unit disk $\{z : |z| < 1\}$; in the general case, we can only assert that $g(z)$ always exists on the unit circle $\{z : |z| = 1\}$.

Similarly to (2.70), to the convolution of sequences there corresponds the product of the respective GFs. Differentiating GFs at $z = 0$ produces the probabilities of the particular values of the distribution:

$$\frac{d^k}{dz^k} g(0) = k!\, \mathbf{P}\,(X = k), \quad k = 1, 2, \ldots, \tag{2.73}$$

while differentiating GFs at $z = 1$ yields the so-called *factorial moments* of X:

$$\frac{d^k}{dz^k} g(1) = \mathbf{E}\,[X(X-1)\cdots(X-k)], \quad k = 1, 2, \ldots \tag{2.74}$$

To illustrate the notion by a simple example, note that the GF of the Bernoulli distribution B_p is $g(z) = 1 - p + pz$. Now the sum S_n of n independent B_p-distributed RVs will have the GF

$$g_n(z) = (g(z))^n = ((1-p) + pz)^n = \sum_{k=1}^{n} \binom{n}{k} p^k (1-p)^{n-k} z^k$$

by the binomial formula. Comparing the expression on the right with the right-hand side of (2.72), we immediately get (2.66).

For Poisson RVs $X_j \sim Po(\lambda_j)$, $j = 1, 2$, the GFs are

$$g_{X_j}(z) = \sum_{k\geq 0} z^k e^{-\lambda_j} \frac{\lambda_j^k}{k!} = e^{-\lambda_j} \sum_{k\geq 0} \frac{(\lambda_j z)^k}{k!} = e^{\lambda_j(z-1)}. \tag{2.75}$$

Therefore, for independent Poisson X_1 and X_2, the GF of their sum is just

$$g_{X_1}(z)\, g_{X_2}(z) = e^{(\lambda_1+\lambda_2)(z-1)},$$

which corresponds to $Po(\lambda_1 + \lambda_2)$, a much simpler derivation of the result of Example 2.1.

The most general integral transform is the (complex-valued) **characteristic function** (ChF)

$$\varphi_X(it) = \mathbf{E}\, e^{itX} = \int e^{itx} dF_X(x), \quad -\infty < t < \infty,$$

also called the *Fourier*[42]*–Stieltjes transform* of the DF F_X. Since $|e^{itx}| = 1$ for real t and x, the integral is always finite (and $|\varphi_X(it)| \leq 1$). There is also a one-to-one correspondence between ChFs and DFs, and a more convenient inversion formula for computing the DF from its known ChF

We leave it to the reader to prove that any ChF is (i) uniformly continuous, i.e., $\varphi_X(i(t+h)) - \varphi_X(it)$ is uniformly (in t) small when h is small, and (ii) non-negative definite, i.e., for any integer $n > 0$, real $t_1, \ldots, t_n$, and complex $a_1, \ldots, a_n$,

$$\sum_{j,k=1}^{n} a_j \bar{a}_k \varphi_X(i(t_j - t_k)) \geq 0 \tag{2.76}$$

(see Problem 15 for hints). When a function $\varphi(it)$, $t \in \mathbf{R}$, with $\varphi(0) = 1$ has properties (i) and (ii), it is the ChF of some probability distribution (the assertion of the famous Bochner[43]–Khinchin[44] theorem).

Note also that when the distribution of X is symmetric (i.e., $F_{-X} \equiv F_X$), its ChF is always real and even. We will refer to this observation later, when discussing stationary processes in Chapter 9.

When F_X has density f_X, its ChF is nothing else but the *Fourier transform* of the density, and the powerful Fourier theory can be employed.

[42](Jean-Baptiste-)Joseph Fourier (21.03.1768–16.05.1830), a French mathematician (a.k.a. as an Egyptologist), whose work had a great influence on mathematical physics and the theory of functions of a real variable. For details about his life see, e.g., Grattan-Guinnes, I. *Joseph Fourier, 1768-1830*, MIT Press, Cambridge, 1972.

[43]Salomon Bochner (20.08.1899–02.05.1982), an Austrian-Hungarian-American mathematician, known for his work in analysis, probability theory and differential geometry.

[44]Aleksandr Yakovlevich Khinchin (19.07.1894-18.11.1959), a prominent Russian mathematician who, in particular, made significant contributions to analysis, number theory and probability theory.

We will only note here that there is the following general rule relating the behaviour of distributions to that of their ChFs:

The "smoother" the distribution, the faster the ChF $\varphi_X(it)$ decays as $t \to \pm\infty$. And vice versa, the smoother the ChF, the "lighter" the tails of the distribution.

Thus, if X is integer-valued (no "smoothness" in the distribution at all), then clearly $\varphi_X(2\pi ki) = 1$ for all integers k (and hence there is no decay as $|t| \to \infty$ at all). If it is absolutely continuous, then $\varphi_X(it) \to 0$ as $|t| \to \infty$ (the fact is know as the *Lebesgue theorem*). Further, if $|\varphi_X(it)| = 1$ for some $t \neq 0$, then X is *lattice-valued* (all possible values of X are of the form $a + bk$, $k = \ldots, -1, 0, 1, \ldots$, for some fixed a and b), while if $\varphi_X(it)$ is integrable, then X is absolutely continuous with the density

$$f_X(x) = \frac{1}{2\pi} \int_{-\infty}^{\infty} e^{-itx} \varphi_X(it)\, dt. \tag{2.77}$$

On the other hand, if X has a finite kth order absolute moment, then $\varphi_X(it)$ is k times differentiable. And the other way around: a bit loosely speaking, if $\varphi_X(it)$ is *smooth*, then X has finite higher order moments.

For Laplace transforms, one can get more detailed results of that sort (the so-called Abelian/Tauberian theorems).

In conclusion note that all the above integral transforms can be defined not only for probability distributions, but also for arbitrary measures on $\mathbf{R}$ (or even on $\mathbf{R}^d$; in that case, instead of e^{tX}, say, one takes $e^{\boldsymbol{t}\boldsymbol{X}^T}$, where $\boldsymbol{t}\boldsymbol{X}^T = \sum t_j X_j$ is the inner product of the vectors in that space).

2.8 Conditional Probabilities and Expectations

In Section 2.3 we introduced the notion of conditional probability given an event (of a positive probability). Now we will extend this notion.

For an event A, its **conditional probability** given an RV X (both A and X should refer to a common probability space, of course) is an *RV which is a function of* X given by

$$\mathbf{P}(A|X) := U(X), \quad \text{where } U(x) := \mathbf{P}(A|X = x) = \frac{\mathbf{P}(A \cap \{X = x\})}{\mathbf{P}(X = x)} \tag{2.78}$$

if $\mathbf{P}(X = x) > 0$; when $\mathbf{P}(X = x) = 0$, then, a bit loosely speaking, we can take

$$U(x) = \lim_{\varepsilon \searrow 0} \mathbf{P}\big(A|X \in (x - \varepsilon, x + \varepsilon)\big). \tag{2.79}$$

Note that, for any one-to-one function φ, $\mathbf{P}(A|X) = \mathbf{P}(A|\varphi(X))$.

For a discrete RV Y, its conditional distribution given an RV X is defined as the collection of the values $\mathbf{P}(Y = y|X)$. Note that these values are random!

Example 2.5. Let $X_i \sim Po(\lambda_i)$, $i = 1, 2$, be independent RVs. What is the conditional distribution of X_1 given the sum $S = X_1 + X_2$?

Since $S \sim Po(\lambda_1 + \lambda_2)$ (see Example 2.1), by independence of the X_i's, for $0 \le j \le k$:

$$\mathbf{P}(X_1 = j|S = k) = \frac{\mathbf{P}(X_1 = j,\, S = k)}{\mathbf{P}(S = k)} = \frac{\mathbf{P}(X_1 = j,\, X_2 = k - j)}{\mathbf{P}(S = k)}$$

$$= \frac{\mathbf{P}(X_1 = j)\,\mathbf{P}(X_2 = k - j)}{\mathbf{P}(S = k)} = e^{-\lambda_1}\frac{\lambda_1^j}{j!}e^{-\lambda_2}\frac{\lambda_2^{k-j}}{(k-j)!}e^{\lambda_1+\lambda_2}\frac{k!}{(\lambda_1+\lambda_2)^k}$$

$$= \frac{k!}{j!(k-j)!}p^j(1-p)^{k-j} \quad \text{with } p = \frac{\lambda_1}{\lambda_1+\lambda_2},$$

which is the binomial distribution $B_{n,p}$ given by (2.66).

For absolutely continuous RVs it is convenient to deal with **conditional densities**. Let $(X, Y) \in \mathbf{R}^2$ be an absolutely continuous random vector with density $f_{(X,Y)}(x, y)$. The conditional density of Y given $X = x$ will then be given by the ratio

$$f_{Y|X}(y|x) := \frac{f_{(X,Y)}(x, y)}{f_X(x)}. \tag{2.80}$$

To illustrate this definition, note that, for small intervals $\Delta_x \ni x$ and $\Delta_y \ni y$, one has, according to (2.26) and (2.32), that

$$\mathbf{P}(Y \in \Delta_y|X \in \Delta_x) = \frac{\mathbf{P}(Y \in \Delta_y,\, X \in \Delta_x)}{\mathbf{P}(X \in \Delta_x)}$$

$$\approx \frac{f_{(X,Y)}(x, y)\,|\Delta_x| \times |\Delta_y|}{f_X(x)\,|\Delta_x|} = f_{Y|X}(y|x)\,|\Delta_y|,$$

which makes definition (2.80) look very natural indeed (here, as in (2.26), $|\Delta_z|$ denotes the length of the interval Δ_z, $z = x, y$).

The **conditional expectation** $\mathbf{E}(Y|X)$ of an integrable RV Y given X can be computed as the expectation under the respective conditional distribution. Thus, $\mathbf{E}(Y|X) = V(X)$ is a function of X defined as

$$V(x) := \mathbf{E}(Y|X = x) = \begin{cases} \sum_j y_j\,\mathbf{P}(Y = y_j|X = x), \\ \int y\, f_{Y|X}(y|x)\, dy. \end{cases} \tag{2.81}$$

in the discrete and absolutely continuous cases, respectively. Note that, for any one-to-one function φ, $\mathbf{E}(Y|X) = \mathbf{E}(Y|\varphi(X))$.

Conditional probabilities and expectations given a random vector are defined in exactly the same way.

List the main properties of conditional expectations:

(i) *Linearity.* If a_j are constants, Y_j are integrable RVs, $j = 1, 2$, then

$$\mathbf{E}(a_1Y_1 + a_2Y_2|X) = a_1\mathbf{E}(Y_1|X) + a_2\mathbf{E}(Y_2|X) \quad \text{a.s.}$$

(ii) *Monotonicity.* If $Y_1(\omega) \le Y_2(\omega)$ a.s., then $\mathbf{E}(Y_1|X) \le \mathbf{E}(Y_2|X)$ a.s.

(iii) If $Y = c =$ const a.s., then $\mathbf{E}(Y|X) = c$ a.s.

(iv) *The total probability formula.* If Y is an integrable RV, then

$$\mathbf{E}\left[\mathbf{E}(Y|X)\right] = \mathbf{E}\,Y.$$

In words, the formula says that the average of averages equals the overall average. In the integral form, using the function V defined in (2.81), the formula takes the form

$$\mathbf{E}\,Y = \int V(x)\,dF_X(x) = \begin{cases} \sum_i \mathbf{E}(Y|X = x_i)\,\mathbf{P}(X = x_i), \\ \int \mathbf{E}(Y|X = x) f_X(x)\,dx \end{cases}$$

in the cases where X is discrete and absolutely continuous, respectively.

(v) For any (measurable) function g, if the RV $g(X)Y$ is integrable, then $\mathbf{E}(g(X)Y|X) = g(X)\mathbf{E}(Y|X)$.

Conditional probability and expectation given an RV as described above are simplified notions being special cases of the general conditional probability and expectation given a σ-field $\mathcal{F}_0$ (generated, in the above special case, by the RV X: such an $\mathcal{F}_0$ is the class of events of the form $\{X \in B\}$, $B \in \mathcal{B}$). Formally, for an integrable RV Y on a probability space $(\Omega, \mathcal{F}, \mathbf{P})$, its conditional expectation given a σ-field $\mathcal{F}_0 \subset \mathcal{F}$ is defined as an $\mathcal{F}_0$-measurable RV Z (meaning that $\{Z \le x\} \in \mathcal{F}_0$ for any $x \in \mathbf{R}$) such that, for any event $A \in \mathcal{F}_0$, one has $\mathbf{E}(Y; A) = \mathbf{E}(Z; A)$. By the *Radon-Nikodym theorem*, such a Z always exists and is unique (up to its values on an event of probability zero). If $\mathcal{F}_0$ is generated by an RV X, all $\mathcal{F}_0$-measurable RVs are functions of X. Conditional probability of an event A given a σ-field $\mathcal{F}_0$ is defined as the conditional expectation $\mathbf{P}(A|\mathcal{F}_0) = \mathbf{E}(\mathbf{1}_A|\mathcal{F}_0)$ of its indicator (so that with this strict formal approach, the primary notion is that of conditional expectation, not probability). These more general objects have properties analogous to the above-listed ones.

Note the following important fact: viewed as an operator on the space $L_1(\Omega, \mathcal{F}, \mathbf{P})$ of all integrable RVs Y on $(\Omega, \mathcal{F}, \mathbf{P})$, the conditional expectation $\mathbf{E}(Y|\mathcal{F}_0)$ is a *projection*[45] onto the subspace of all $\mathcal{F}_0$-measurable RVs, since from the definition it follows that $\mathbf{E}(\mathbf{E}(Y|\mathcal{F}_0)|\mathcal{F}_0) = \mathbf{E}(Y|\mathcal{F}_0)$ (cf. (v)). This implies, in particular, that when $\mathbf{E}\,Y^2 < \infty$, the conditional expectation $\mathbf{E}(Y|X)$ is the *best* (in the mean-quadratic sense) *predictor*

[45] An operator A is called a *projection* if, for any x from the domain of A, the element Ax also belongs to the domain of A and $A^2x = A(Ax) = Ax$.

for Y from X, i.e., of all functions $g(X)$ of X, the conditional expectation $g(X) := \mathbf{E}(Y|X)$ has the smallest mean quadratic error $\mathbf{E}(g(X) - Y)^2$ as a predictor for Y.

In conclusion, we note the following simple fact to be often used in the sequel: Let X and Y be RVs (or random vectors), $g(x, y)$ a (measurable) function, and A an event such that $\mathbf{P}(A) > 0$ and $X \equiv x_0$ on A. Then, for any measurable set B,

$$\mathbf{P}(g(X,Y) \in B|\, A) = \frac{\mathbf{P}(g(X,Y) \in B;\, A)}{\mathbf{P}(A)} = \frac{\mathbf{P}(g(x_0,Y) \in B;\, A)}{\mathbf{P}(A)} = \mathbf{P}(g(x_0,Y) \in B|\, A). \quad (2.82)$$

2.9 Limit Theorems

Even for very simple stochastic systems it is rather hard to obtain closed form expressions for the distributions of interest. Thus, sums of independent RVs are often components of stochastic models. One could compute the distribution of such a sum using the convolution formulae ((2.63) and (2.64)), or inverting the Laplace transforms or ChFs of the sums which can easily be found as products (2.70). However, even for rather small numbers of summands, it becomes a very tedious task. Fortunately, there are mathematical results called *limit theorems* giving (often very good) approximations to the distributions which cannot be computed directly, and also the overall asymptotic behaviour of the systems of interest (a.s., in probability, or in distribution). The term refers to situations where a certain parameter(s) (e.g., the number of random summands) tends to a limiting value (infinity in that example). However, quite often limit theorems (especially when they are accompanied by refinements such as the so-called asymptotic expansions) work very well even for the values of the parameters that are still far away from the limiting ones (see, e.g., Fig. 2.3).

The first key limit theorem of probability theory is the **law of large numbers** (LLN) which is basically a mathematical fact showing that our model for a series of independent (or even "weakly dependent") random experiments does have the crucial property (1.2).

Let $X_1, X_2, \ldots$ be i.i.d. RVs and $a = \mathbf{E}\, X_j$ exist (it does not need to be finite). Call the sequence of partial sums

$$S_0 = 0, \quad S_n = X_1 + \cdots + X_n, \quad n = 1, 2, \ldots, \quad (2.83)$$

a **random walk** (RW) (with jumps X_j) starting at 0. Then the (strong)

LLN says that

$$\overline{X}_n := \frac{S_n}{n} = \frac{1}{n}\sum_{j=1}^{n} X_j \to a \quad \text{a.s. as } n \to \infty. \tag{2.84}$$

The converse is true as well: if relation (2.84) with a finite a holds for an i.i.d. sequence $\{X_j\}$, then $\mathbf{E}\,X_1$ exists and is equal to a.

Note that the statements hold not only for i.i.d. sequences, but also under much weaker assumptions. Roughly speaking, what is needed is that the distributions of the X_j's are "approximately the same", and that the dependence between X_j and X_{j+k} decays fast enough as k increases.

While proving the strong LLN would require a few pages of exposition, we can easily prove the *weak LLN*: $\overline{X}_n \to a$ *in probability*, under the additional assumption that $\sigma^2 = \text{Var}(X_j) < \infty$. Indeed, from the Chebyshev inequality (the second version in (2.51)) and from the properties of variance, one gets, for any $\varepsilon > 0$,

$$\mathbf{P}(|\overline{X}_n - a| > \varepsilon) < \frac{\text{Var}(\overline{X}_n)}{\varepsilon^2} = \frac{\sum_{j=1}^n \text{Var}(X_j)}{n^2\varepsilon^2} = \frac{\sigma^2}{n\varepsilon^2} \to 0$$

as $n \to \infty$.

A illustration to the strong LLN is given in Fig. 2.5: whatever $\varepsilon > 0$ is, with probability 1 there exists a finite random number $n(\omega)$ such that the trajectory of our RW, after time $n(\omega)$, will stay forever inside the cone between the boundary lines $x = (a \pm \varepsilon)n$.

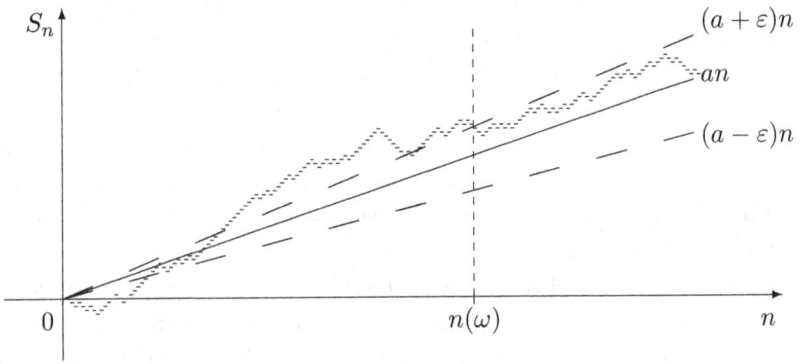

Fig. 2.5 The law of large numbers.

Example 2.6. Let $Y_1, Y_2, \ldots$ be a sequence of i.i.d. random observations with a common DF F. For the *sample* of the first n RVs Y_j, $j = 1, \ldots, n$, the average

$$F_n^*(t) = \frac{1}{n}\sum_{j=1}^{n} \mathbf{1}_{[Y_j,\infty)}(t), \quad -\infty < t < \infty, \tag{2.85}$$

is called the **empirical** (or **sample**) **distribution function** (EDF) for the sample. Note that $F_n^*(t)$ is a DF indeed: the respective (random) distribution assigns the probabilities of $1/n$ to each of the n points of the sample (so that they are equally likely; see Fig. 2.6). So, for any t, the EDF estimates the probability $F(t) = \mathbf{P}(Y \le t)$ by the relative frequency of the event $\{Y_j \le t\}$ in the first n trials. From the statistical point of view, the EDF contains all the information contained in the sample (given, of course, that the order of the observations does not matter), i.e., is a *sufficient statistic.*

Now, for any fixed t, $X_j = \mathbf{1}_{[Y_j,\infty)}(t)$ are clearly i.i.d. RVs following the Bernoulli distribution B_p with

$$p = \mathbf{E}\, X_j = \mathbf{E}\, \mathbf{1}_{[Y_j,\infty)}(t) = \mathbf{P}(Y_j \le t) = F(t), \tag{2.86}$$

so that by the strong LLN, $F_n^*(t) \to F(t)$ a.s. as $n \to \infty$. Moreover, using the fact that both $F_n^*(t)$ and $F(t)$ are DFs (and hence non-decreasing), it is not hard to show that the following *Glivenko*[46]*–Cantelli*[47] *theorem* holds true:

$$\sup_{-\infty<t<\infty} |F_n^*(t) - F(t)| \to 0 \quad \text{a.s. as } n \to \infty. \tag{2.87}$$

This is a fundamental result of mathematical statistics: it shows that, observing a sequence of i.i.d. RVs, one can precisely reconstruct the unknown distribution of the elements of the sequences. Moreover, most of *statistics* (functions of the observations) used in statistics are actually (relatively simple) *functionals* of the EDF. Thus, the sample mean $\overline{Y}_n = n^{-1}\sum_{j=1}^n Y_j$ is clearly the mean of the EDF F_n^* (likewise, sample moments are the respective moments of the EDF), the sample median is the median of F_n^*, and so on. From this point of view, it is often very easy to prove *consistency* of various estimators (i.e., their convergence to the true values of the parameters as the sample size increases) *etc.*, which would follow immediately from (2.87). Using this approach, we can also give bounds for the convergence rates and compute the distribution of the error terms, which is very important for statistical inference (it is needed to find confidence intervals and tests with given characteristics for moderate/large samples). See Example 12.9 (p. 378) for a somewhat more detailed discussion.

If, for our RW $\{S_n\}$, one assumes more than just that the mean $a = \mathbf{E}\, X$ is finite, it is possible to get refinements of the LLN. The most profound

[46]Valery Ivanovich Glivenko (02.01.1897–15.02.1940), a Russian mathematician who worked in foundations of mathematics, real analysis and probability theory.

[47]Francesco Paolo Cantelli (20.12.1875–21.07.1966), an Italian mathematician who mostly worked in probability theory (including its foundations) and its applications.

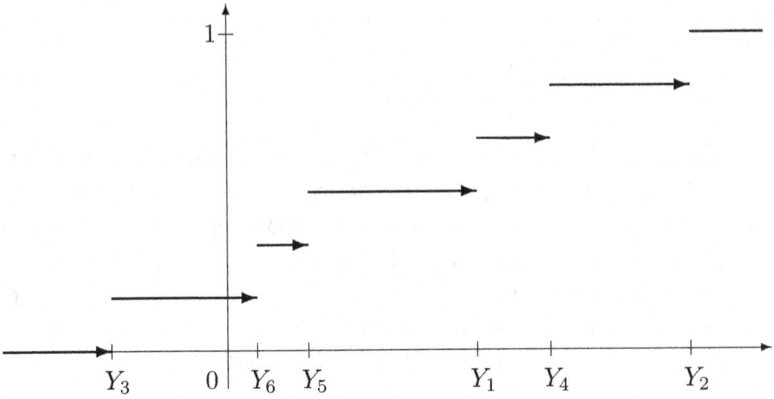

Fig. 2.6 The EDF for the sample $Y_1, \ldots, Y_6$.

and often used result is the celebrated **central limit theorem** (CLT): if $\text{Var}\,(X_j) = \sigma^2 < \infty$, then the distribution of

$$\frac{n^{1/2}}{\sigma}(\overline{X}_n - a) \equiv \frac{S_n - an}{\sigma n^{1/2}}$$

converges to the standard normal one as n increases. In other words, for any x,

$$\left|\mathbf{P}\left(\frac{S_n - an}{\sigma n^{1/2}} \le x\right) - \Phi(x)\right| \to 0 \quad \text{as } n \to \infty. \tag{2.88}$$

This convergence is actually uniform in x: $\sup_x |\cdots| \to 0$ (which is always the case when the limiting DF is continuous).

Thus, the CLT is a sort of "magnifying glass" showing in detail what happens when, due to the LLN, $\overline{X}_n$ approaches the mean a: not only the RW stays inside the cone outlined by $(a \pm \varepsilon)n$ (Fig. 2.5), but its "typical deviation" from the drift an is actually much smaller—of the order $n^{1/2}$ only. There exists an a.s. result saying precisely how far from the drift line our RW can be; the exact asymptotic bounds are $an \pm \sigma\sqrt{2n \log\log n}$ (the so-called *law of the iterated logarithm*).

If the variance is infinite, but a moment of a lower order is finite: $\mathbf{E}\,|X_j|^r < \infty$ for some $r \in (1, 2)$, there can still exist a limiting distribution for appropriately scaled (not by $n^{1/2}$ this time!) sums S_n. However, unlike the case of the CLT, the above condition is not sufficient for the result to hold. Roughly speaking, the tails $F(x)$ and $1 - F(-x)$ need to vanish, as $x \to \infty$, at the same rate and in a "regular fashion". (Moreover, the CLT itself holds under slightly more general conditions than just $\text{Var}\,(X_j) < \infty$.)

In the general case of independent, but not necessarily identically distributed X_j's, conditions for convergence to the normal law are given by the so-called *Lindeberg*[48]–*Feller*[49] *theorem*; roughly speaking, the variance of the sum should increase unboundedly with the number of summands, while the summands' individual contributions should be negligible compared to their sum, and the tails of the summands are to be, in a sense, uniformly small.

If we go further and assume, say, that the third absolute moment is finite: $\beta := \mathbf{E}\,|X_1 - a|^3 < \infty$, then we can estimate the rate of convergence in the CLT as follows: the left-hand side of (2.88) does not exceed $C\beta\sigma^{-3}n^{-1/2}$ for an absolute constant[50] C. In the general case, this is the true order of the convergence rate when $\beta < \infty$.

Example 2.6 (continued). Using the observation (2.86) [which also implies that $\mathrm{Var}\,(\mathbf{1}_{[Y_j,\infty)}(t)) = p(1-p) = F(t)(1-F(t))$], we see that, for any t, the distribution of that value of the so-called *empirical process*

$$E_t^{(n)} := n^{1/2}(F_n^*(t) - F(t)) \tag{2.89}$$

at point t converges weakly to $N(0, F(t)(1-F(t)))$ as $n \to \infty$.

One more important result we need to mention here is the **Poisson limit theorem** (a sort of the "law of small numbers"). Suppose we have a collection $X_1, \ldots, X_n$ of independent Bernoulli random variables with

$$\mathbf{P}\,(X_i = 1) = 1 - \mathbf{P}\,(X_i = 0) = p_i.$$

Then, if all the p_i's are small, the distribution of the sum $S_n = X_1 + \cdots + X_n$ will be close to that of the Poisson RV Y with parameter $\lambda := \sum_{j=1}^n p_j$. Moreover, we can quantitatively estimate how good the approximation is. Namely, the following bound is true:

$$\sup_B |\mathbf{P}\,(S_n \in B) - \mathbf{P}\,(Y \in B)| \le \frac{1 - e^{-\lambda}}{\lambda} \sum_{i=1}^n p_j^2, \tag{2.90}$$

which becomes simply $(1 - e^{-\lambda})\lambda/n$ in the special case when all $p_j = \lambda/n$, $j = 1, \ldots, n$. Thus, when we have "rare events" (a large number of trials, each of them being unlikely successful), the distribution of the total number of "successes" is to be close to the Poisson one.

[48]Jarl Waldemar Lindeberg (04.08.1876–12.12.1932), a Finnish mathematician known, in particular, for his work on the central limit theorem.

[49]William Feller (born Vilibald Srećko Feller, 07.07.1906–4.01.1970), an (eventually) American mathematician, a prominent probabilist, the author of the famous classical two-volume textbook on probability theory and its applications.

[50]This is the assertion of the *Berry–Esseen theorem.* As of 2012, the best upper estimate for that constant is $C < 0.4748$, while it has been known since 1956 that C cannot be smaller than 0.4097.

Note also that there is actually a *continuous transition* from the Poisson limit theorem to the CLT: when λ is large (and hence the variance $\text{Var}\,(S_n) = \sum_{j=1}^{n} p_j(1-p_j)$ of the sum S_n is also large), the approximating Poisson law $Po(\lambda)$ is, in turn, well approximated by the respective normal distribution (of which the mean and the variance are both equal to λ).

As it is the case with the LLN and CLT, the assumptions of the Poisson theorem can be substantially relaxed.

2.10 Stochastic Processes

We already used the term "stochastic (or random) process" (SP) several times in the previous sections. The formal definition thereof is as simple as that: a collection $\{X_t,\, t \in T\}$ of RVs given on a common probability space $(\Omega, \mathcal{F}, \mathbf{P})$ is called a **stochastic/random process**. The *parameter* t is usually called the "time" (although its meaning may be position in space *etc.*). The parameter set T is in most cases either

• discrete: $T = \{0, 1, 2, \dots\}$ or $T = \{\dots, -2, -1, 0, 1, 2, \dots\}$, and then one speaks of *random sequences* (or *time series*), or

• continuous: $T = [0, \infty)$ or $T = \mathbf{R}$, and then one speaks of *continuous time processes*.

Sometimes one has to deal with $T \subset \mathbf{R}^d$ (e.g., the temperature at point $t \in T = \{\textit{map of Victoria}\}$), and then one speaks of *spatial processes* (or *random fields*).

Note that a random process $X = X_t(\omega)$ is actually a function of two variables, t and ω. Thus, we have two sources of variability: the chance ω and the "time" t. Correspondingly, one can analyse the process as a function of each of the two variables separately.

(i) When t is *fixed*, we just get a single RV $X_t(\cdot)$ completely characterised by its distribution F_t (of course, F_t can vary with time t). The latter shows how likely are, at the time t, particular values of the variable modelled by this random process. Note that knowing the F_t's only is insufficient for describing the process, for they say nothing about the character of dependence between the values X_t for different t's.

(ii) When ω is fixed, we get a **realisation** (**trajectory**, **path**) of the process, which is just a function of t. *Path-wise* properties of an SP are the properties of these functions (different for different $\omega \in \Omega$). It is often convenient to identify a functional space[51] containing (with probability 1) the trajectories of the SP of interest with the sample space Ω itself. The probability distribution induced on the functional space is called the **distribution of the process.** It is (usually) defined on the σ-field of Borel sets, i.e., the smallest σ-field containing all open sets in the space, and can be uniquely characterised by the family of the so-called **finite-dimensional distributions** (FDD) of the process, which is the collection all joint distributions $F_{t_1,\dots,t_n}(x_1,\dots,x_n)$ of the RVs $X_{t_1},\dots,X_{t_n}$, $n \ge 1$, $t_1,\dots,t_n \in T$. Note that, for any SP, the family of the FDDs is *consistent* in the sense that, for any values of the parameters,

$$F_{t_1,\dots,t_n,t_{n+1}}(x_1,\dots,x_n,\infty) = F_{t_1,\dots,t_n}(x_1,\dots,x_n).$$

It is a standard approach to define an SP by specifying its FDDs, all the more so since for *any consistent family of FDDs* there always exists an SP (more precisely: there exists a distribution on the space $\mathbf{R}^T$ of *all real-valued functions* on T) whose FDDs coincide with that family. The last statement is again a consequence of the *Carathéodory theorem.*[52]

One says that an SP $\{X_t\}$ is **stationary** if, for any values $n \ge 1$ and $t_1,\dots,t_n \in T$, its FDDs $F_{s+t_1,\dots,s+t_n}$ are independent of s (we assume, of course, that all $s + t_k \in T$). This means that the distributional (or statistical) properties of the process remain unchanged as time elapses. There exists a lot of theory and strong results for such processes. In particular, the strong LLN (possibly with a random limiting value in the general case!) and, under additional "mixing" conditions, the CLT will hold for the sums (or integrals) of the values of the process.

Example 2.7. Let $X_t \equiv X \sim N(0,1)$, $t \in T$. Then the trajectories of the process $\{X_t\}$ are just horizontal lines (positioned at different—random—heights), $F_t = \Phi$ for all t, and

$$F_{t_1,\dots,t_n}(x_1,\dots,x_n) = \mathbf{P}(X_{t_1} \le x_1,\dots,X_{t_n} \le x_n) = \Phi(\min\{x_1,\dots,x_n\}).$$

[51]In the case of continuous time, when the process is "regular", it is usually either the space of continuous functions $C[0,T]$ on the interval $[0,T]$ (with the uniform metric on it) or the space $D[0,T]$ of left-continuous functions on $[0,1]$ having right limits at each point of the interval (this space is usually endowed with the so-called *Skorokhod metric*; according to this metric, functions g_1 and g_2 are close to each other if, for some change of time $\tau = \tau(t)$ which is close to the identity function t, the uniform distance between the functions $g_1(\tau(t))$ and $g_2(t)$ is small. For more detail see, e.g., Billingsley (1968)).

[52]Also, there exist simple conditions on two- and three-dimensional distributions of SPs in continuous time ensuring that the process has a version whose trajectories a.s. belong to the spaces $C[0,T]$ or $D[0,T]$, respectively. For more detail see, e.g., Section 18.2 in Borovkov (2013), or Billingsley (1968), or Gikhman and Skorokhod (1969).

Now assume that all the X_t's are i.i.d. RVs following the same distribution $N(0,1)$, so that again $F_t = \Phi$ for all t. However, for dimensions higher than one, we will have different FDDs:

$$F_{t_1,\dots,t_n}(x_1,\dots,x_n) = \Phi(x_1) \times \dots \times \Phi(x_n).$$

Clearly, both processes are stationary.

The trajectories of the second process are, however, extremely irregular (they are discontinuous everywhere!). The process itself is referred to as the **Gaussian white noise**[53] and is a very popular element of various stochastic systems where it plays the role of a "stochastic driver", X_t representing "new randomness" arising (fed into the system) at time t.

Stationarity is actually a very stringent condition and can be very hard to test in practice. Quite often one deals with the far less restrictive assumption of **weak stationarity**, which means that the moments of the first two orders of the process values are time invariant:

(i) the mean function $m_t := \mathbf{E}\, X_t = \text{const}$, and
(ii) the (auto)covariance function

$$r(s,t) := \operatorname{Cov}(X_s, X_t) = \mathbf{E}\,(X_s - m_s)(X_t - m_t) = \mathbf{E}\, X_s X_t - m_s m_t = \gamma(t-s)$$

depends on the time difference $t-s$ only. The theory of such processes is also very well developed (and has a nice geometric interpretation) and is often used in applications. In particular, there exists a variant of the LLN, a spectral representation for the process *etc.* Elements of this theory are presented below in Chapter 9.

In the general case, one says that $\{X_t\}$ is a **Gaussian SP** if all the FDDs of the process are Gaussian. Note that if this is the case, then the distribution of the process is *completely determined* by the means $m_t = \mathbf{E}\, X_t$ and the *(auto)covariance function* $r(s,t)$ (since any multivariate Gaussian distribution is determined by its mean vector and covariance matrix). For non-Gaussian processes, the last statement is not true in the general case.

We have already dealt with a couple of random processes: an RW $\{S_n\}$ is an example of an SP in discrete time, while the EDF $F_n^*(t)$ is an SP in

[53] *White light* is sort of a "mixture of all colours" (electromagnetic waves with frequencies from certain ranges). Likewise, a stochastic *white noise* is a (random) "mixture of harmonics" of all frequencies (a more precise statement can be made using Fourier series/integrals), cf. Section 9.1 below. In the continuous time case, to make the Gaussian white noise a useful mathematical modelling tool, one needs a more complicated formal definition for the object: it is actually defined as the so-called *generalised derivative* of the *Brownian motion process* (the latter will be introduced in the present section and discussed in more detail in Chapter 11).

continuous time. We also saw that, for the one-dimensional distributions of these processes, one has certain limit theorems. It turns out that all three major limit theorems discussed in the previous section hold in the *functional setup* as well. First consider an RW $\{S_n\}$ with i.i.d. jumps.

LLN. This theorem provides the "first order approximation" to the trajectory of the RW. If we compress both time and space with the same factor n by setting $A_t^{(n)} := n^{-1}S_{\lfloor nt \rfloor}$, $t \in [0,1]$, then $\sup_{t\in[0,1]} |A_t^{(n)} - at| \to 0$ a.s. as $n \to \infty$. In words, the sequence of random functions $A_t^{(n)}$ converges to the linear limit at in space the $C[0,1]$ of continuous functions on $[0,1]$ with the uniform distance.

CLT. Like the "ordinary" CLT, its functional version provides the "second order approximation" to the RW. Centering the RW by subtracting its trend (i.e., the first order approximation) and then compressing the time axis n times and the space $\sigma n^{1/2}$ times, we obtain the process $B_t^{(n)} := (S_{\lfloor nt \rfloor} - at)/\sigma n^{1/2}$, $t \in [0,1]$. The so-called *functional CLT* (a.k.a. the (Donsker[54]–Prokhorov[55] *invariance principle*[56]) claims that the process converges in distribution to the so-called standard **Brownian**[57] **motion process** (a.k.a. the **Wiener**[58] **process**), which is defined as a Gaussian process $\{W_t, t \ge 0\}$ with independent increments such that $W_t \sim N(0,t)$. A more precise definition is as follows:

W1. For any $0 \le s_1 < t_1 \le s_2 < t_2 \le \cdots \le s_k < t_k$, the increments $W_{t_1} - W_{s_1}, W_{t_2} - W_{s_2}, \ldots, W_{t_k} - W_{s_k}$ are mutually independent RVs.

W2. For any $t > 0$, the RV $W_t \sim N(0,t)$. In particular, $W_0 = 0$ a.s.

W3. For any $\omega \in \Omega$, the path $W_t(\omega)$, $t \ge 0$, is a continuous function.

Note that, using MGFs or ChFs, one can immediately deduce from W1

---

[54]Monroe David Donsker (17.10.1924-08.06.1991), an American mathematician known for his work in probability theory.

[55]Yuri Vasilyevich Prokhorov (15.12.1929–16.07.2013), a Russian mathematician who, in particular, did the pioneering work on the theory of convergence of distributions in function spaces.

[56]Due to the fact that the convergence takes place regardless of the particular form of the distribution of the jumps X_j: we just need a finite second moment.

[57]After Robert Brown (21.12.1773–10.06.1858), a Scottish botanist, the discoverer of the cell's nucleus, who was the first to report his observations of highly irregular movement of pollen particles suspended in liquid. Also, he was among the first naturalists to study New Holland (nowadays called Australia). In 1801–1805, he circumnavigated the continent aboard the Matthew Flinders' ship *Investigator* and gathered and classified approx. 3,900 plant species.

[58]After Norbert Wiener (26.11.1884–18.03.1964), an American mathematician who, in particular, established the science of *cybernetics* (common factors of control and communication in living organisms, automatic machines and organisations).

and W2 that, for any $0 < s < t$, $W_t - W_s \sim N(0, t-s)$. Indeed, the two terms on the right-hand side of $W_t = W_s + (W_t - W_s)$ are independent RVs, so, for any $u \in \mathbf{R}$, the MGF of W_t is equal to

$$e^{u^2 t/2} = \mathbf{E}\, e^{uW_t} = \mathbf{E}\, e^{uW_s} \times \mathbf{E}\, e^{u(W_t - W_s)} = e^{u^2 s/2} \mathbf{E}\, e^{u(W_t - W_s)}$$

by (2.70), which implies that $\mathbf{E}\, e^{u(W_t - W_s)} = e^{u^2(t-s)/2}$, and the assertion follows.

That $\{W_t\}$ is Gaussian follows now from the obvious fact that, for any $0 = t_0 < t_1 < \cdots < t_n$, the vector $(W_{t_1}, \ldots, W_{t_n})$ is a linear transformation of $(W_{t_1} - W_{t_0}, \ldots, W_{t_n} - W_{t_{n-1}})$, which is clearly a normal random vector from W1 and the above observation.

Note that W1 and W2 specify a consistent family of FDDs and hence ensure the existence of an SP with trajectories in the space $\mathbf{R}^{[0,\infty)}$ of all functions $[0, \infty) \mapsto \mathbf{R}$. That there exists an SP with such FDDs *and* continuous trajectories is a very non-trivial fact requiring a proof (there are simple sufficient conditions for existence of continuous processes with given FDDs that involve only bivariate distributions, i.e., those of (X_s, X_t) only; see, e.g., Section 18.2 in Borovkov (2013)).

This is a very important and interesting process, that has been the subject of numerous studies over several decades. Among its curious properties is that although its trajectories are continuous, they are a.s. nowhere differentiable![59]

The above-mentioned functional CLT enables one to find distributional approximations to various characteristics of RWs by taking the distributions of the respective variables for the Brownian motion. In particular, if the (i.i.d.) jumps in our RW have zero mean and unit variance, then, as $n \to \infty$,

$$\mathbf{P}\left(n^{-1/2} \max_{k \le n} S_k \le x\right) \to \mathbf{P}\left(\max_{0 \le t \le 1} W_t \le x\right) = 2\Phi(x) - 1, \quad x \ge 0.$$

Example 2.6 (continued). The Glivenko-Cantelli theorem is already a sort of a functional LLN: as the sample size increases, EDFs converge uniformly to the theoretical DF. A functional CLT is stated in terms of the empirical process (2.89): it converges in distribution to the *Gaussian process* $W^0_{F(t)}$,

[59] For those with a broader mathematical background: as we have already said, the white noise process is in fact the so-called *generalised derivative* of W_t. Also, the trajectories of process have unbounded variation on any interval, and hence one cannot define the (path-wise) Stieltjes integral $\int g(t)\, dW_t$. Yet integrals of that form are widely used in probability theory, but they are the so-called *stochastic integrals* and are defined in a different way, which is the subject of *stochastic calculus*. We will discuss how that can be done later, in Chapter 11.

where W_t^0 is the so-called *Brownian bridge*[60]:

$$W_t^0 = W_t - tW_1, \quad t \in [0,1], \tag{2.91}$$

W_t being the standard Brownian motion. Equivalently, W_t^0 can be defined as a Gaussian process such that $\mathbf{E}\, W_t^0 = 0$ and $\mathbf{E}\, W_s^0 W_t^0 = s \wedge t - st$, $s, t \in [0,1]$ (note that the variance of W_t^0 is $\mathrm{Var}\,(W_t^0) = t(1-t)$, and hence that of $W_{F(t)}^0$ is $F(t)(1 - F(t))$, which coincides with the variance of the empirical process (2.89)).

The meaning of the fact is that one has the following distributional approximation to the EDF:

$$F_n^*(t) \approx F(t) + n^{-1/2} W_{F(t)}^0. \tag{2.92}$$

In words, the EDF differs from the (limiting) theoretical DF by a random term which is basically $n^{-1/2} \times$ *Gaussian process* with known parameters. Since, as we have already noted, many statistical estimators are (smooth) functionals of $F_n^*(\cdot)$, one can derive from the last relation their asymptotic behaviour for large sample sizes n (in particular, their errors are typically of the order $n^{-1/2}$ and are asymptotically normally distributed). For more detail, see Example 12.9.

Another important implication is that, for *any continuous* DF F, the distribution of the statistic

$$n^{1/2} \sup_t |F_n^*(t) - F(t)| \tag{2.93}$$

tends to that of $\sup_t |W_{F(t)}^0| \equiv \sup_t |W_t^0|$, and the latter (a.k.a. the *Kolmogorov distribution*) does not depend on F! Therefore, if one wishes to test the hypothesis that the theoretical DF is F, one can form statistic (2.93) and compute the type I error probability using the fact that, for large enough n, (2.93) will have a known distribution (the procedure is called the *Kolmogorov test* and is one of the so-called *distribution-free tests*).

Example 2.8. *The "functional Poisson theorem".* Now let $X_1, \ldots, X_n$ be i.i.d. Bernoulli RVs with $\mathbf{P}\,(X_j = 1) = \lambda/n$ for some $\lambda \in (0, \infty)$ (note that the distribution of the X_j's depends on n, so that it would be more appropriate to use notation $X_{n,1}, \ldots, X_{n,n}$ here). Assume that the X_j's are the indicators of events which may occur or not occur during the times slots $((j-1)/n, j/n]$, respectively. We know from (2.90) that, for any $t \in [0,1]$,

[60] Called so because $W_0^0 = W_1^0 = 0$: the trajectory of the processes forms a (very bumpy) bridge between the points with coordinates (0,0) and (0,1), resp. Sometimes the process is also called the *tied-down Brownian motion.* We will further discuss it in Section 12.5.3.

the distribution of the number $S_{\lfloor nt \rfloor}$ of events occurred by the time t is close to $Po(\lambda t)$. It turns out that the *whole process* $\{S_{\lfloor nt \rfloor},\, t \in [0,1]\}$ will then be close in distribution to the so-called *Poisson process* $\{N_t\}$ to be studied in more detail in Chapter 5 below. For the time being, we just say that, as you might have already correctly guessed, the Poisson process also has *independent increments* (cf. W1 above; the increments of $S_{\lfloor nt \rfloor}$ are just sums of independent RVs over disjoint subsets of indices and hence are independent themselves), while $N_t - N_s \sim Po(\lambda(t-s))$, $s < t$; cf. also Example 2.1 which confirms that this definition is consistent.

2.11 Recommended Literature

Texts of "technically lower level" are asterisked. Other books are more advanced and more suitable for "mathematically minded" students.

BILLINGSLEY, P. *Convergence of probability measures.* Wiley, New York, 1968. [*A good advanced level text on convergence in distribution. The 2nd edition appeared in 1999, but the first one seems to be better.*]

BILLINGSLEY, P. *Probability and measure.* Wiley, New York, any edition (1979–2012). [*A good rather modern reference for both probability and measure theory.*]

BOROVKOV, A.A. *Probability theory.* Springer, New York, 2013. [*A good high level probability textbook.*]

*FELDMAN, R.M. AND VALDEZ-FLORES, C. *Applied probability and stochastic processes.* PWS, Boston, 1996. [*A rather elementary text, with numerous examples and exercises.*]

*FELLER, W. *An introduction to probability theory and its applications.* Wiley, New York. 2 V. 2nd-3rd edns, 1966-70. [*A classical reading on probability theory.*]

GIKHMAN, I.I. AND SKOROKHOD, A.V. *Introduction to the theory of random processes.* Saunders, Philadelphia, 1969. [*A classical rather high level text on stochastic processes.*]

GRIMMETT, G.R. AND STIRZAKER, D.R. *Probability and Random Processes.* Clarendon Press, Oxford, 1981 (or any later edition). [*A good probability textbook.*]

GRIMMETT, G.R. AND STIRZAKER, D.R. *One Thousand Exercises in Probability.* 2nd edn. Oxford University Press, Oxford, 2001. [*A companion volume to the previous book.]*

HEYMAN, D.P. AND SOBEL, M.J. *Stochastic models: A handbook in operations research and management science.* North-Holland, New York, 1990. [*A handbook covering many topics from the present course.*]

*KARR, A.F. *Probability.* Springer-Verlag, New York, 1993. [*An intermediate level probability textbook.*]

MAISTROV, L.E. *Probability theory; a historical sketch.* Academic Press, New York, 1974. [*An interesting study of the history of probability theory.*]

*PARZEN, E. *Stochastic Processes.* Holden Day, San Francisco, 1962. [*A well-written introductory level text on random processes.*]

*ROSS, S.M. *Introduction to Probability Models.* 4th edn. Academic Press, New York, 1989. [*A text close to the present course. The 10th edition appeared in 2009.*]

SHIRYAEV, A.N. *Probability.* Springer-Verlag, New York, 1984. [*A good probability textbook of a rather high level. The 2nd edition appeared in 1995.*]

2.12 Problems

The first ten problems are simple exercises aimed at reviewing some elements of probability calculus. Further problems are less elementary.

Begin solving problems 1–8 by specifying an appropriate sample space in each particular situation.

1. A box contains 25 parts, of which 10 are defective. Two parts are drawn at random from the box. What is the probability that

   (i) both are good?

   (ii) both are defective?

   (iii) one is good and one is defective?

   Hints. (i) Consider events G_i := "the ith part is good", $i = 1, 2$. What is the event G_1G_2? What is the probability $\mathbf{P}(G_1)$? [We have $25 - 10 = 15$ good parts of the 25 parts in the box!] What is the conditional probability $\mathbf{P}(G_2|G_1)$? [We have $24 - 10 = 14$ good parts of the 24 parts remaining in the box, if the first part drawn is good!]

   Recall the definition of conditional probability. How can one use this notion to calculate the probability $\mathbf{P}(G_1G_2)$?

   (ii) Similar reasoning.

(iii) Let E_1 := "both are good", E_2 := "both are defective", E_3 := "one is good and one is defective".

Are these events mutually exclusive? What is the probability of the event $E_1 \cup E_2 \cup E_3$? How to use these facts and the results of (i) and (ii) above to find $\mathbf{P}(E_3)$?

An alternative way is to consider the events H_1 := "the 1st is good and 2nd is defective", H_2 := "the 1st is defective and the 2nd is good". Are these events mutually exclusive? What is $H_1 \cup H_2$?

2. The ten digits $0, 1, 2, \ldots, 9$ are arranged in a row in a random order. [Any such arrangement is called a *permutation* of these digits.]

   (i) What is the probability of the outcome 0123456789?

   (ii) What is the probability of the outcome 9876543210?

   (iii) What is the probability that the digit 3 is at the first place?

   (iv) What is the probability that the digit 3 is at the first place and the digit 4 is at the second place?

   (v) What is the probability that the digits 3 and 4 are neighbours?

   Hints. (i, ii) What is the sample space for this random experiment? How many possible outcomes does it contain? Will all possible outcomes be equally likely? What is the probability of a particular outcome?

   (iii) How many outcomes does this event contain? We could also consider a "reduced" random experiment consisting in observing only the first digit in the random permutation. What sample space could you propose for this case? What are possible outcomes?

   (iv) Similar to (iii).

   (v) List all (mutually exclusive) possibilities where 3 and 4 appear as neighbours. How many such possibilities do we have?

3. Given the information that a family consists of two children and that the second child is a boy, find the probability that both are boys.

4. Given the information that a family consists of two children and that at least one of these two children is a boy, find the probability that both are boys.

   Hint. This is a problem on conditional probability. The sample space consists of 4 points: BB, BG, GB, GG.

5. Three marksmen A, B, and C each fire one shot at a target. The probabilities of each hitting the target are:

$$A: \ 0.30; \qquad B: \ 0.25; \qquad C: \ 0.10.$$

   If one bullet is found in the target, find the probability that it came from A's gun; from B's gun; from C's gun.

Hints. There are 8 possible outcomes of a trial consisting in three shots. List them. What are their probabilities? (Independence!) List the outcomes in the event "one hits, two miss". What is the probability of the event? Recall the definition of conditional probability and calculate the conditional probability of "A hits" given "one hits, two miss".

6. Suppose we choose at random one of the marksmen from Problem 5. He fires one shot and hits the target. Find the probability that we have chosen A; we have chosen B; we have chosen C.

   Hint. Use the Bayes formula.

7. Paul has 3 attempts to pass a test. He is not well prepared so that the probability of his passing the test from the first attempt is 0.6. We know that if Paul fails his first attempt, he starts working hard and increases his success probability up to 0.75 before the second attempt. Moreover, if he fails again, he improves further to achieve the probability of 0.9 to pass the test before he makes his third attempt.

   (i) What is the probability that Paul will pass the test?

   (ii) What is the probability that he will pass the test and do this from the third attempt?

8. The number of accidents on XYZ highway each day is a Poisson random variable with mean 3. We know that these numbers are independent for different days.

   (i) What is the probability that no accidents occur today?

   (ii) What is the probability that there will be exactly 2 accidents on XYZ highway during this weekend? There will be at least 2 accidents?

   (iii) We have not got complete data for yesterday yet, but it is already known that at least 2 accidents occurred. What is the probability that there were exactly 3 accidents yesterday given this information?

9. The length (in kilobytes) of a message to be transmitted to a computer is an RV having the *gamma distribution with parameter* 3/2. This distribution has the density

$$f(x) = \frac{2}{\sqrt{\pi}}\, x^{1/2}\, e^{-x}, \qquad x > 0.$$

   Any message is compressed before transmission so that instead of the original message of length x, a compressed version will be sent. Compression rate is random (it depends on the contents of the message). We assume that the length of the compressed message, given the original one was of length x, can be represented by an RV of the form $\frac{1}{10}(1 + 2U)\, x^{1/2}$, U being a uniformly distributed over $(0, 1)$ RV. Find the expected value and the variance of the length of a compressed message.

   Hint. You may wish to make use of the gamma function (see its definition in Problem 13 below) and its values at some special points.

10. Let (the input) X be an RV with a DF $F(x) = \mathbf{P}(X \le x)$. We know that (the output) $Y = g(X)$, where g is an increasing function with an inverse $h = g^{-1}$.

    (i) Find the DF of Y.

    (ii) Answer the same question when g is decreasing. [There is a small trap here!]

    (iii) Answers parts (i) and (ii) in the particular case when F is the DF of $U(0,2)$ (the uniform on $(0,2)$ distribution), (i) $g(x) = x^2$ and (ii) $g = 1/x$, respectively.

11. Find the analytic form of the DFs depicted in Fig. 2.2.

12. A water reservoir R having the shape of an upside-down cone with height $h = 5$ m and base radius $r = 20$ m is used to store rainfall water collected from an area of 1 km^2. At the beginning of a certain week, the reservoir was empty. With probability 0.2 there will be no rain during that week, otherwise the amount of rainfall is uniformly distributed between 0 and 40 mm. Assuming that only 10% of the rainfall comes from the area to R and that any excessive water (for which there is no space in R) is lost, find the distribution and expected value of the water level at the reservoir by the end of the week (zero level corresponds to empty reservoir).

13. Suppose that the probability distribution $\{p_k\}$ of an integer-valued RV $X \ge 0$ satisfies the so-called *Panjer's recurrence relation*:

$$p_k = \left(a + \frac{b}{k}\right) p_{k-1}, \quad k = 1, 2, \ldots,$$

    where $a < 1$ and $b \in \mathbf{R}$ are some constants. Show that

    (i) if $a = 0$, then $b > 0$ and $X \sim Po(b)$;

    (ii) if $a < 0$, then $b = -a(n+1)$ for some natural n, and $X \sim B_{n,p}$ with $n = -b/a - 1$ and $p = a/(a-1)$;

    (iii) if $a \in (0,1)$, then $a+b > 0$ and X follows the *negative binomial distribution*:

$$p_k = \frac{\Gamma(n+k)}{\Gamma(n)k!} a^k (1-a)^n \equiv \frac{(n+k-1)\cdots(n+1)n}{k!} a^k (1-a)^n, \quad k = 0, 1, \ldots,$$

    where $n = 1 + b/a$ and Γ is the gamma function $\Gamma(x) = \int_0^\infty t^{x-1} e^{-t} dt$. Note that the last law is a favoured claim number distribution in insurance applications.

14. John repeatedly rolls a symmetric die till all the six faces show up. Denote by T the (random) number of trials he needs for that.

    (i) Find the GF of T.

    (ii) Find $\mathbf{E}\,T$ and $\mathrm{Var}\,(T)$.

Hint. What is the distribution of the number of trials between the times when the kth and the $(k+1)$st faces appeared for the first time? (We number the faces in the order in which they appear in the experiment.)

15. Prove that any ChF is
(i) *uniformly continuous*, i.e., $\sup_t |\varphi_X(i(t+h)) - \varphi_X(it)| \to 0$ as $h \to 0$, and
(ii) *non-negative definite* (see (2.76)).
(iii) If the distribution of X is symmetric (i.e., $F_{-X} \equiv F_X$), its ChF is always real and even: $\varphi_X(it) = \varphi_X(-it)$.

Hints. (i) Write down the difference $\varphi_X(i(t+h)) - \varphi_X(it)$ as a single integral and then split it into two parts: $\int_{|x|\le N}$ and $\int_{|x|>N}$. Bound the two parts separately, choosing an arbitrary large N. (ii) $\mathbf{E}\,|\sum_j a_j e^{it_jX}|^2 \ge 0$.

16. Show that the class of Borel subsets of $\mathbf{R}$ contains all closed intervals $[a,b]$.

17. Let $\{\mathcal{F}_i\}_{i\in I}$ be an arbitrary collection of σ-fields on a sample space Ω. Show that $\bigcap_{i\in I}\mathcal{F}_i$ is always a σ-field, while $\bigcup_{i\in I}\mathcal{F}_i$ does not need to be one.

18. Show that the σ-field generated by open balls in $\mathbf{R}^n$ coincides with the product σ-field $\mathcal{B}_n$.

19. Show that the field generated by the class of all semi-open intervals $(a,b] \subset \mathbf{R}\cup\{-\infty,+\infty\}$ (the "extended real line") consists of all finite unions of such intervals.

20. Show that both definitions of random vectors given in Section 2.4 are equivalent.

21. Confer the definitions of independence of events (2.23) and RVs (2.34) and explain why in the former, the relation was required for all subcollections of indices, while in the latter it was not the case.

22. Show that property P3 of probability (p. 17) is equivalent to the combination of finite additivity and continuity of probability (2.9).

23. Prove (2.57) for independent simple RVs.

24. Show that if, for two integrable RVs X and Z on a common probability space $(\Omega,\mathcal{F},\mathbf{P})$, one has $\mathbf{E}(X;A) = \mathbf{E}(Z;A)$ for any $A\in\mathcal{F}$, then $X = Z$ a.s.

25. Prove that $\mathrm{Var}(X) = \min_{a\in\mathbf{R}} \mathbf{E}(X-a)^2$, and the minimum is attained at the point $a = \mathbf{E}X$.

Hint. $\mathbf{E}(X-a)^2$ is a quadratic function of a.

26. A slightly harder question: what is the value of a at which the minimum of $\mathbf{E}\,|X-a|$ is attained?

Hint. Use (2.54) to show that $\mathbf{E}\,|X-a| = \int_{-\infty}^a F(x)\,dx + \int_a^\infty (1-F(x))\,dx$. Analyse this expression as a function of a (is it differentiable?).

27. Let X and Y be RVs given on a common probability space, $\mathbf{E}Y^2 < \infty$. Show that the conditional expectation $\mathbf{E}(Y|X)$ is the *best* (in the mean-quadratic sense) predictor for Y from X. This means that, of all the functions $g(X)$

of X, the conditional expectation $g(X) := \mathbf{E}(Y|X)$ has the smallest mean quadratic error $\mathbf{E}(g(X) - Y)^2$.

Hint. Use properties (iv) and (v) of conditional expectations.

28. Prove that (2.38) is equivalent to a.s. convergence of X_n to X.

    Hint. Note that $\{\sup_{m>n} |X_n - X_0| > \varepsilon\} = \bigcup_{m>n}\{|X_m - X_0| > \varepsilon\}$.

29. Show that (2.40) is equivalent to convergence of the DFs $F_{X_n}(t) \to F_{X_0}(t)$ as $n \to \infty$ at each point t where F_{X_0} is continuous.

30. Let $\{X_n\}$ be a sequence of RVs converging in distribution to a limit X_0, G a continuous function. Show that $G(X_n)$ converges in distribution to $G(X_0)$.

31. Verify (2.73) and (2.74).

32. Using W1 and W2 (p. 65), compute the mixed moment $\mathbf{E}\, W_s W_t$, $t > s > 0$. Compute also the covariance function

$$\mathrm{Cov}\,(N_s, N_t) = \mathbf{E}\, N_s N_t - \mathbf{E}\, N_s\, \mathbf{E}\, N_t$$

    of the Poisson process $\{N_t\}$ with parameter $\lambda > 0$ and compare the answer with that function for the Brownian motion process.

33. Prove that if $\{W_t\}$ is the standard Brownian motion process, then so is the process $X_t = tW_{1/t}$, $t > 0$.

34. Using the Berry-Esseen theorem, give an upper bound for $\max_x |\mathbf{P}(Y \le x) - \Phi(\lambda + x\sqrt{\lambda})|$, where $Y \sim Po(\lambda)$, $\lambda > 0$.

35. Prove the Glivenko-Cantelli theorem (2.87).

36. Let $0 < t_1 < t_2$. Using W1 and W2, find the joint density of the random vector (W_{t_1}, W_{t_2}) for the standard Brownian motion process $\{W_t\}$. Extend the result to the n-dimensional vector $(W_{t_1}, \dots, W_{t_n})$ with $0 < t_1 < \dots < t_n$.

    Hint. You may wish to use conditional densities.

37. Let $0 \le t_1 < s < t_2$. For the standard Brownian motion process $\{W_t\}$, find the conditional distribution of W_s given W_{t_1} and W_{t_2}.

    Hint. Use Problem 36. Or you may wish to use the Brownian bridge process to derive it.

38. Let $U_1, U_2, \dots$ be independent $U(0,1)$ RVs, and $U_{k:n}$ be the kth *order statistic* of the sample $(U_1, \dots, U_n)$ (i.e., $U_{1:n}$ is the smallest of the first n RVs U_j, $U_{2:n}$ the second smallest *etc*).

    (i) Find the limiting distribution for $nU_{1:n}$. That is, find

$$\lim_{n\to\infty} \mathbf{P}(nU_{1:n} \le t), \quad t \ge 0.$$

    (ii) Find the limiting distribution for $nU_{2:n}$. How is it called?

(iii) Extend the result of (ii) to $nU_{k:n}$ for any *fixed* value of $k = 1, 2, \dots$ What can you say about the process limiting (as $n \to \infty$) to the sequence of counting processes $\{N_t^{(n)}\}_{t\geq 0}$ having jumps at the points $nU_{k:n}$, $k = 1, 2, \dots$? [Note that by using the scaling $\times n$ for the order statistics, we are simply "viewing through a magnifying glass" the left-most piece of the EDF for our sample.]

Hint. (i) It is easier to deal with distribution tails rather than with the DFs when considering the minima of independent RVs (why?). You may wish to use the fundamental limit $(1 + a/m)^m \to e^a$ as $m \to \infty$.

Chapter 3

Markov Chains

3.1 Definitions

Recall the first simple SPs from our Example 2.7: $X_t \equiv X$ for all $t \in T$, and the white noise process. In both cases, dependence between the values of the process at different times is quite primitive: all the values either coincide with each other or are independent. The former is not a process at all (as nothing changes in it as time passes), while the latter may be a reasonable model for noise in electronic systems. However, there are very few situations where it can be used as it is, in its "raw" form.

Example 3.1. Let $\{X_j\}$ be a sequence of symbols from an alphabet $\mathcal{A}$ (or their codes such as ASCII codes, if we want to have a numeric sequence). It could be a text from a manuscript one wants to attribute, or a DNA sequence (then the alphabet $\mathcal{A}$ will only have four letters: A, C, G and T standing for the respective nucleobases). Such sequences can be very long; for instance, the human genome is about 3 billion nucleobase pairs long. Is it a good idea to assume that neighbouring letters in our sequence are independent? We again appeal to the frequency interpretation (1.2). To verify the independence assumption, take a (long enough) empirical text $\mathcal{T} = \{a_1\, a_2 \cdots a_n\}$, $a_j \in \mathcal{A}$, and denote by

$$n_{l_i} = \#\{j \le n : a_j = l_i\}, \quad n_{l_1 l_2} = \#\{j \le n-1 : a_j a_{j+1} = l_1 l_2\}$$

the frequencies of given letters $l_i \in \mathcal{A}$, $i = 1, 2$, and their collocations in the text. Assuming that $\mathcal{T}$ can be thought of as a *random text* (which is often reasonable—as we noted in the Introduction—when the algorithm forming the text is complex enough), we expect that

$$\frac{n_{l_i}}{n} \approx \mathbf{P}\,(\text{letter} = l_i), \quad \frac{n_{l_1 l_2}}{n} \approx \mathbf{P}\,(\text{two consecutive letters} = l_1 l_2).$$

If the letters were independent, we would have from (2.22) that

$$\frac{n_{l_1 l_2}}{n} \approx \frac{n_{l_1}}{n} \times \frac{n_{l_2}}{n}.$$

But this almost never holds! For instance, for a plain English text, setting $l_1 := l_2 :=$ "a" would yield $\mathbf{P}(\text{letter} = l_i) \approx 0.08$, while the pair "aa" is extremely rare (effectively, $\mathbf{P}(\text{two consecutive letters} = \text{"aa"}) = 0$[1]).

So using independent sequences is unjustified in most situations: random processes typically have more complicated dependence structure. However, one can notice that in many situations, the outcome of a particular trial depends basically on what occurred on the previous trial *only*. When the outcome of the latter is known, there is (almost) no dependence on the outcomes of the preceding trials. This is the so-called **Markov[2] property** which can be formally stated as follows.

A random sequence $\{X_n\}$ taking values in a measurable space $(S, \mathcal{S})$ (referred to as the *state space* of the chain) is called a **Markov chain** (MC) if, for any sets $B, A_j \in \mathcal{S}$, any $x \in S$ and any time n, one has

$$\mathbf{P}(X_{n+1} \in B | X_n = x, X_{n-1} \in A_{n-1}, X_{n-2} \in A_{n-2}, \ldots) \\ = \mathbf{P}(X_{n+1} \in B | X_n = x). \quad (3.1)$$

Less formally, one has the following equality for conditional probabilities:

$$\mathbf{P}(\text{Future}|\{\text{Exact Present}\} \ \& \ \text{Past}) = \mathbf{P}(\text{Future}|\text{Exact Present}) \quad (3.2)$$

or, equivalently,

$$\mathbf{P}(\text{Future} \ \& \ \text{Past}|\text{Exact Present}) \\ = \mathbf{P}(\text{Future}|\text{Exact Present})\,\mathbf{P}(\text{Past}|\text{Exact Present}). \quad (3.3)$$

That is, when the *present* is fixed, the *future* and the *past* are (conditionally) independent.

The importance of the class of MCs (and also continuous time Markov processes with a subclass of which we will deal later) is due to their wide applicability and the existing powerful well-developed theory of such processes. Before discussing any results of the theory, consider a couple of simple examples, bearing in mind the main question of how to see if a particular random sequence is an MC.

---

[1]Unless it is a book about the mammals of southern Africa (including aardvarks and aardwolfs).

[2]Called after Andrei Andreevich Markov (06.06.1856–20.07.1922), a prominent Russian mathematician who, in particular, introduced and studied SPs having that property.

Example 3.2. A sequence of independent RVs $\{X_n\}$ is clearly an MC, since by independence

$$\mathbf{P}(X_{n+1} \in B | X_n = x, X_{n-1} \in A_{n-1}, \dots) \\ = \mathbf{P}(X_{n+1} \in B) = \mathbf{P}(X_{n+1} \in B | X_n = x).$$

Example 3.3. *Random walks.* Let $Y_1, Y_2, \dots$ be i.i.d. RVs. Set

$$X_0 := 0, \quad X_{n+1} := X_n + Y_{n+1} \equiv Y_1 + Y_2 + \cdots + Y_{n+1}, \quad n \geq 0. \tag{3.4}$$

Noting that the condition $C = \{X_n = x, X_{n-1} \in A_{n-1}, X_{n-2} \in A_{n-2}, \dots\}$ can also be written as $(Y_1, \dots, Y_n) \in A$ for a set $A \in \mathbf{R}^n$, we have

$$\mathbf{P}(X_{n+1} \in B | C) = \mathbf{P}(x + Y_{n+1} \in B | (Y_1, \dots, Y_n) \in A) = \mathbf{P}(x + Y_{n+1} \in B)$$

by (2.82) and independence of the Y_j's. The same argument shows that $\mathbf{P}(X_{n+1} \in B | X_n = x) = \mathbf{P}(x + Y_{n+1} \in B)$ as well, so that the RW is an MC indeed.

Example 3.4. *A queueing system.* Imagine a situation where we have a stream of "customers" (they could be shoppers, phone calls *etc.*) arriving at a "server" (a checkout point, telephone exchange *etc.*). The times between arrivals of consecutive customers are i.i.d. RVs τ_j, so that the jth customer arrives at time $\tau_1 + \cdots + \tau_j$. If the server is idle when a customer arrives, it starts providing service to that customer immediately. If the server is busy, the arriving customer joins the queue, and his/her/its service will start only after the service of all the customers arrived prior to him/her/it has already been completed. Once the server starts providing a customer with service, it works with that customer only till the demand of that customer has been met. The service times required to meet the demands of the customers are i.i.d. RVs ξ_j. What is the *waiting time* X_n of the nth customer (i.e., the time from his/her/its arrival till the service starts)?

Note that $X_n + \xi_n$ (= waiting time + service time) is the time from the arrival of the nth customer till his/her/its service is completed. Now if $X_n + \xi_n \leq \tau_{n+1}$, then the service will be finished *before* the arrival of the next customer. Therefore, at the time of that arrival the server will already be idle, and hence $X_{n+1} = 0$. If $X_n + \xi_n > \tau_{n+1}$, then the time the $(n+1)$st customer will have to wait will be less than $X_n + \xi_n$ by the interarrival time, so that $X_{n+1} = X_n + \xi_n - \tau_{n+1}$.

Thus, assuming the server was idle when the first customer arrived, we have $X_1 = 0$ and

$$X_{n+1} = \max\{0, X_n + Y_{n+1}\}, \quad Y_{n+1} := \xi_n - \tau_{n+1}, \quad n \geq 1,$$

where the Y_j's are i.i.d. The sequence $\{X_n\}$ is an MC for the same reason as the RW above:

$$\mathbf{P}(X_{n+1} \in B | X_n = x, X_{n-1} \in A_{n-1}, \dots) \\ = \mathbf{P}(\max\{0, x + Y_{n+1}\} \in B) = \mathbf{P}(X_{n+1} \in B | X_n = x).$$

Summarising what was common in the above examples, it is not hard to show that one actually has the following

General rule: *Any MC $\{X_n\}$ can be represented in the form*

$$X_{n+1} = f_n(X_n, Y_{n+1}) \tag{3.5}$$

for a sequence of non-random functions $\{f_n(\cdot\,,\,\cdot)\}$ and i.i.d. RVs $\{Y_n\}$.

Thus, if one can find such a representation for $\{X_n\}$, it means that the sequence is an MC.

In what follows we will mostly restrict ourselves to the case where the state space $S \ni X_n$ is either finite or countable, i.e., without loss of generality (w.l.o.g.), we may assume that either

$$S = \{s_1, \dots, s_m\} = \{1, \dots, m\}$$

(and then the MC is said to be **finite**) or

$$S = \{s_1, s_2, \dots\} = \{1, 2, \dots\} \quad \text{or} \quad = \{\dots, -2, -1, 0, 1, 2, \dots\}.$$

In that case, we can take only one-point sets A_j in the definition of MCs which will then take the following equivalent form.

A random sequence $\{X_n\}_{n\geq 0}$ forms a **Markov chain** if, for any $k, j, j_0, \dots, j_{n-1}$,

$$\mathbf{P}(X_{n+1} = k | X_n = j, X_{n-1} = j_{n-1}, \dots, X_0 = j_0) \\ = \mathbf{P}(X_{n+1} = k | X_n = j) =: p_{jk}(n). \tag{3.6}$$

If the **transition probabilities** $p_{jk}(n) = p_{jk}$ do not depend on time n, the MC $\{X_n\}$ is called **homogeneous**.

The **transition matrix** (or the *matrix of transition probabilities*) for the MC will be denoted by

$$P = (p_{jk}) = \begin{pmatrix} p_{11} & p_{12} & \cdots & p_{1m} \\ p_{21} & p_{22} & \cdots & p_{2m} \\ \vdots & \vdots & \ddots & \vdots \\ p_{m1} & p_{m2} & \cdots & p_{mm} \end{pmatrix};$$

this is an $m \times m$-matrix when the state space S consists of m elements. When S is countable, you may think of P as a sort of "square infinite matrix".

The transition matrix P is often "sparse": most of its entries are zeros, i.e., the respective direct (one-step) transitions are impossible. When this is the case or the total number of possible states of the chain is small, it may be helpful—for better understanding the character of the process—to "visualise" the matrix by drawing the transition diagram of the MC; a fragment of such a diagram is shown in Fig. 3.1. Positive transition probabilities are indicated by arrows; there are no arrows for impossible one-step transitions.

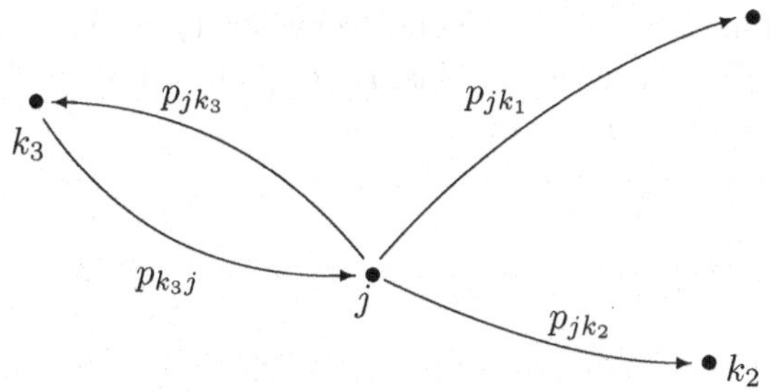

Fig. 3.1 Part of a transition diagram.

Note that always

(i) $p_{jk} \geq 0,\ j,k \in S$;

(ii) all the row sums are equal to one: $\sum_k p_{jk} = 1,\ j \in S$.

Indeed, the sum is just $\sum_k \mathbf{P}(X_1 = k|X_0 = j) = \mathbf{P}(X_1 \in S|X_0 = j) = 1$.

Any square matrix P having properties (i) and (ii) is called a **stochastic matrix**, and for any such matrix, there always exists an MC whose transition matrix coincides with the given stochastic matrix. This is because, as we will see below, the Markov property implies that the stochastic matrix, together with the distribution of the initial value X_0, defines a consistent family of FDDs, cf. Section 2.10.

Example 3.2 (continued). Let $\{X_n\}$ be a sequence of i.i.d. RVs with a common distribution $\mathbf{P}(X_n = k) = \pi_k,\ k = 1, \ldots, m$. Since by independence

$$p_{jk} = \mathbf{P}(X_1 = k|X_0 = j) = \mathbf{P}(X_1 = k) = \pi_k,$$

its transition matrix is

$$P = (p_{jk}) = \begin{pmatrix} \pi_1 & \pi_2 & \dots & \pi_m \\ \pi_1 & \pi_2 & \dots & \pi_m \\ \vdots & \vdots & \ddots & \vdots \\ \pi_1 & \pi_2 & \dots & \pi_m \end{pmatrix}. \tag{3.7}$$

Example 3.5. In a simple linear communication system, the digits 0 and 1 are transmitted from node to node. On each transmission, there is a probability p that the digit will pass unchanged from node n to node $n+1$; with the complementary probability $1-p$, it will be changed. Setting $X_n = 1$ if the value of the signal is 0 at node n and $X_n = 2$ if it is 1 (we could, of course, retain the symbols 0 and 1 to denote the states of the chain; sometimes, however, it is not convenient), we get a sequence of random digits $\{X_n\}$ showing the "history of transformations" of a single digit as it has been transmitted from node to node (see Fig. 3.2).

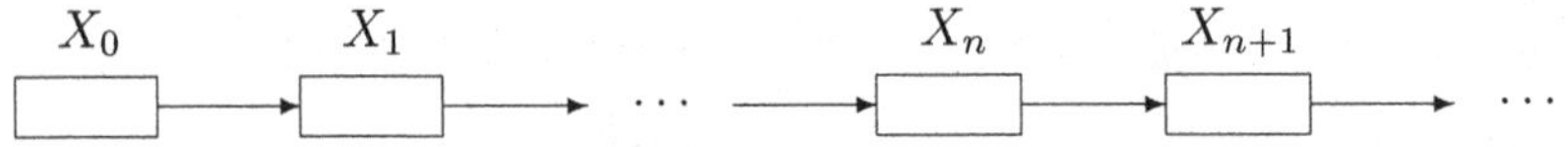

Fig. 3.2 Linear communication system: Example 3.5.

It is easy to see that we have got a two-state MC with

$$\mathbf{P}(X_{n+1} = k | X_n = j) = \begin{cases} p & \text{if } j = k, \\ 1-p & \text{if } j \neq k, \end{cases} \quad P = \begin{pmatrix} p & 1-p \\ 1-p & p \end{pmatrix}. \tag{3.8}$$

Example 3.6. *A weather model.* Set

$$X_n := \begin{cases} 1 & \text{if it rains on day } n, \\ 2 & \text{if it does not rain on day } n. \end{cases}$$

Suppose that the chance of rain tomorrow depends on the previous weather conditions only "through today" (whether it rains or not today):

1. Rain today $\longrightarrow$ rain tomorrow w.p. p.
2. No rain today $\longrightarrow$ rain tomorrow w.p. q.

Then $\{X_n\}$ is an MC with the transition matrix

$$P = \begin{pmatrix} p & 1-p \\ q & 1-q \end{pmatrix}.$$

Example 3.7. *A simple random walk.* Assume that, in the RW (3.4), the jumps have the following common distribution:

$$Y_n = \begin{cases} +1 & \text{w.p. } p, \\ -1 & \text{w.p. } 1 - p =: q. \end{cases}$$

Direct transitions are possible to the neighbouring states of the state space $\mathbf{Z} = \{\text{all integers}\}$ only: the transition probabilities are

$$p_{jk} = \begin{cases} p & \text{if } k = j + 1, \\ q & \text{if } k = j - 1, \\ 0 & \text{otherwise,} \end{cases}$$

so that the (doubly-infinite) transition matrix has the form

$$P = \begin{pmatrix} \ddots & \cdot & \cdot & \cdot & \cdot & \cdot & \cdots \\ \cdots & 0 & p & 0 & 0 & 0 & \cdots \\ \cdots & q & 0 & p & 0 & 0 & \cdots \\ \cdots & 0 & q & 0 & p & 0 & \cdots \\ \cdots & 0 & 0 & q & 0 & p & \cdots \\ \cdots & 0 & 0 & 0 & q & 0 & \cdots \\ \cdots & \cdot & \cdot & \cdot & \cdot & \cdot & \ddots \end{pmatrix}. \tag{3.9}$$

As we have already said, for any (homogeneous) MC $\{X_n\}$, its distribution is *uniquely* determined by its

(i) *initial distribution* $\mathbf{p} = \{p_j\}$:

$$p_j = \mathbf{P}(X_0 = j), \quad j \in S,$$

and

(ii) transition matrix $P = (p_{jk})$.

Indeed, first note that, for any events A, B and C,

$$\mathbf{P}(AB|\,C) = \frac{\mathbf{P}(ABC)}{\mathbf{P}(C)} = \frac{\mathbf{P}(A|\,BC)\,\mathbf{P}(BC)}{\mathbf{P}(C)}$$
$$= \mathbf{P}(A|\,BC)\,\mathbf{P}(B|\,C). \tag{3.10}$$

Now by the TPF, for any $l = 1, 2, \ldots, n-1$, the n-**step transition probabilities** will be given by

$$p_{jk}^{(n)} := \mathbf{P}(X_{t+n} = k|X_t = j) = \mathbf{P}(X_n = k|X_0 = j)$$
$$= \sum_i \mathbf{P}(X_n = k, X_l = i|X_0 = j)$$
$$= \sum_i \mathbf{P}(X_n = k, |X_l = i, X_0 = j)\,\mathbf{P}(X_l = i|X_0 = j) \qquad \text{by (3.10)}$$
$$= \sum_i \mathbf{P}(X_n = k|X_l = i)\,\mathbf{P}(X_l = i|X_0 = j), \qquad \text{by Markov property}$$

so that we arrive at the so-called **Chapman[3]-Kolmogorov equation:**

$$p_{jk}^{(n)} = \sum_i p_{ji}^{(l)} p_{ik}^{(n-l)}. \tag{3.11}$$

In words, the relation means that, in an n-step transition of our MC from state j to state k, it will pass through some state i at time l, and, for any given i, the probability of such transition from j to i in l steps and then further to state k in $n-l$ steps is given by $p_{ji}^{(l)} p_{ik}^{(n-l)}$. In matrix form, setting $P^{(n)} = (p_{jk}^{(n)})$, we see that (3.11) is equivalent to

$$P^{(n)} = P^{(l)} P^{(n-l)}. \tag{3.12}$$

This implies that the matrix of the n-step transition probabilities is just the nth power of the transition matrix P:

$$P^{(n)} = P^n. \tag{3.13}$$

Now we will show how to find the FDDs of our MC. To get one-dimensional distributions, note that by the TPF, for any $n > 0$ and $k \in S$,

$$\begin{aligned}\mathbf{P}(X_n = k) &= \sum_j \mathbf{P}(X_n = k | X_0 = j)\, \mathbf{P}(X_0 = j) \\ &= \sum_j p_j p_{jk}^{(n)} = (\boldsymbol{p} P^{(n)})_k = (\boldsymbol{p} P^n)_k,\end{aligned}$$

the kth element of the vector $\boldsymbol{p} P^n$.

In the general case, for $r > 1$, $0 < n_1 < \cdots < n_r$ and $k_1, \ldots, k_r \in S$, we have by Markov property that

$$\begin{aligned}&\mathbf{P}(X_{n_1} = k_1, \ldots, X_{n_r} = k_r) \\ &= \mathbf{P}(X_{n_r} = k_r | X_{n_{r-1}} = k_{r-1}, \ldots, X_{n_1} = k_1) \\ &\qquad \times \mathbf{P}(X_{n_{r-1}} = k_{r-1}, \ldots, X_{n_1} = k_1) \\ &= \mathbf{P}(X_{n_r} = k_r | X_{n_{r-1}} = k_{r-1})\, \mathbf{P}(X_{n_{r-1}} = k_{r-1}, \ldots, X_{n_1} = k_1) \\ &= p_{k_{r-1}k_r}^{(n_r - n_{r-1})}\, \mathbf{P}(X_{n_{r-1}} = k_{r-1}, \ldots, X_{n_1} = k_1) \\ &= p_{k_{r-1}k_r}^{(n_r - n_{r-1})} p_{k_{r-2}k_{r-1}}^{(n_{r-1} - n_{r-2})}\, \mathbf{P}(X_{n_{r-2}} = k_{r-2}, \ldots, X_{n_1} = k_1) \\ &\ldots \\ &= \sum_j p_j p_{jk_1}^{(n_1)} p_{k_1k_2}^{(n_2 - n_1)} \cdots p_{k_{r-1}k_r}^{(n_r - n_{r-1})}.\end{aligned} \tag{3.14}$$

[3]Sydney Chapman (29.01.1888–16.06.1970), an English mathematician and physicist. Among his contributions are a modification of Maxwell's kinetic theory which led to predicting thermal diffusion (confirmed later experimentally) and discovery that the geomagnetic field is at least partially generated in the atmosphere.

In particular, the (conditional) probability of a given trajectory $k_1k_2\cdots k_n$ given the initial state was $X_0 = j$ is

$$\mathbf{P}(X_1 = k_1, X_2 = k_2, \ldots, X_n = k_n | X_0 = j) = p_{jk_1}p_{k_1k_2}\cdots p_{k_{n-1}k_n}. \tag{3.15}$$

Example 3.2 (continued). For an i.i.d. sequence $\{X_n\}$,

$$P^{(2)} = P^2 = \begin{pmatrix} \pi_1 & \pi_2 & \ldots & \pi_m \\ \pi_1 & \pi_2 & \ldots & \pi_m \\ \cdot & \cdot & \cdots & \cdot \\ \pi_1 & \pi_2 & \ldots & \pi_m \end{pmatrix} \times \begin{pmatrix} \pi_1 & \pi_2 & \ldots & \pi_m \\ \pi_1 & \pi_2 & \ldots & \pi_m \\ \cdot & \cdot & \cdots & \cdot \\ \pi_1 & \pi_2 & \ldots & \pi_m \end{pmatrix} = P.$$

Example 3.5 (continued). For our simple communication system,

$$P^{(2)} = P^2 = \begin{pmatrix} p & 1-p \\ 1-p & p \end{pmatrix} \begin{pmatrix} p & 1-p \\ 1-p & p \end{pmatrix}$$
$$= \begin{pmatrix} p^2 + (1-p)^2 & 2p(1-p) \\ 2p(1-p) & p^2 + (1-p)^2 \end{pmatrix}.$$

Example 3.7 (continued). Here we have

$$P^2 = \begin{pmatrix} \ddots & \cdot & \cdot & \cdot & \cdot & \cdots \\ \cdots & 0 & p & 0 & 0 & \cdots \\ \cdots & q & 0 & p & 0 & \cdots \\ \cdots & 0 & q & 0 & p & \cdots \\ \cdots & 0 & 0 & q & 0 & \cdots \\ \cdots & \cdot & \cdot & \cdot & \cdot & \ddots \end{pmatrix} \times \begin{pmatrix} \ddots & \cdot & \cdot & \cdot & \cdot & \cdots \\ \cdots & 0 & p & 0 & 0 & \cdots \\ \cdots & q & 0 & p & 0 & \cdots \\ \cdots & 0 & q & 0 & p & \cdots \\ \cdots & 0 & 0 & q & 0 & \cdots \\ \cdots & \cdot & \cdot & \cdot & \cdot & \ddots \end{pmatrix}$$
$$= \begin{pmatrix} \ddots & \cdot & \cdot & \cdot & \cdot & \cdots \\ \cdots & 2pq & 0 & p^2 & 0 & \cdots \\ \cdots & 0 & 2pq & 0 & p^2 & \cdots \\ \cdots & q^2 & 0 & 2pq & 0 & \cdots \\ \cdots & 0 & q^2 & 0 & 2pq & \cdots \\ \cdots & \cdot & \cdot & \cdot & \cdot & \ddots \end{pmatrix}.$$

Indeed, the last matrix corresponds to the two-step transitions in the simple RW, and the jumps in the MC $\{X_{2n}\}$ have the distribution of the sum

$$Y_1 + Y_2 = \begin{cases} -2 & \text{w.p. } q^2, \\ 0 & \text{w.p. } 2pq, \\ +2 & \text{w.p. } p^2, \end{cases}$$

as one can easily verify using the convolution formula (2.63) or computing the square of the GF of Y_1 (cf. (2.66)).

3.2 Classification of States

To understand what sort of long-term behaviour an MC can display, one first has to *classify* the states of the MC.

A state $k \in S$ is said to be **accessible** from $j \in S$ (we write $j \mapsto k$) if, for some $n > 0$, one has $p_{jk}^{(n)} > 0$. So the relation $j \mapsto k$ means that there exists a "path" $j = i_0, i_1, \ldots, i_n = k$ such that $p_{i_0 i_1} p_{i_1 i_2} \cdots p_{i_{n-1} i_n} > 0$.

A state j is called **non-essential** if there exists a state k such that $j \mapsto k$, but $k \not\mapsto j$. That is, for some $n > 0$, $p_{jk}^{(n)} > 0$, but $p_{kj}^{(m)} = 0$ for all $m > 0$: with a positive probability, we can reach k from j, but there is no way to j from k. In terms of the transition matrix P, it means that, for some $n > 0$, the power P^n has a non-zero (j,k)-entry, while the (k,j)-entry is zero for all P^m, $m > 0$. The collection of all non-essential states in our MC we denote by S_{NEss}.

Otherwise, the state j is said to be **essential**; in that case, if $j \mapsto k$, then always $k \mapsto j$. If two states j and k satisfy the above relations, we say that they **communicate** and write $j \leftrightarrow k$. In that case, for some $n_1, n_2 > 0$, $p_{jk}^{(n_1)} > 0$ and $p_{kj}^{(n_2)} > 0$. Notation S_{Ess} will be used for the set of all essential states.

Example 3.8. Consider an MC with four states and transition matrix P shown below:

$$P = \begin{pmatrix} 0 & 1/2 & 1/2 & 0 \\ 1/2 & 0 & 0 & 1/2 \\ 0 & 0 & 1/2 & 1/2 \\ 0 & 0 & 1/2 & 1/2 \end{pmatrix}$$

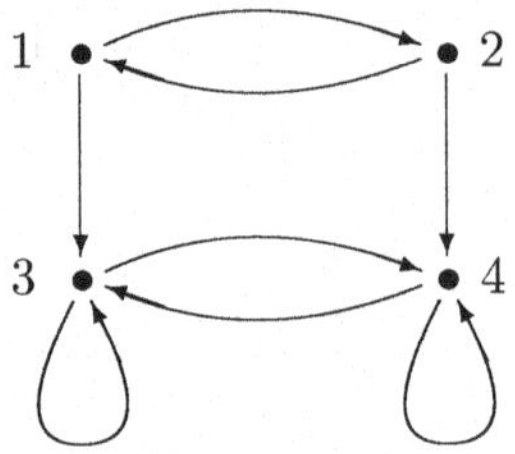

Here states 1 and 2 are non-essential, 3 and 4 are essential. This is obvious from the transition diagram.

For any *non-essential* state j, the MC *never* visits j **after** some *random time*. An opposite, in a sense, of a non-essential state is a so-called absorbing state. A state j is called **absorbing** if, having arrived at the state j, the MC never leaves it. There can be more than one absorbing state in a chain.

Example 3.9. Suppose the transition matrix of our MC is of the form

$$P = \begin{pmatrix} p_{11} & p_{12} & \cdots & p_{1j} & \cdots & p_{1m} \\ \cdots & \cdots & \cdots & \cdots & \cdots & \cdots \\ p_{j-1,1} & p_{j-1,2} & \cdots & p_{j-1,j} & \cdots & p_{j-1,m} \\ 0 & 0 & \cdots & 1 & \cdots & 0 \\ p_{j+1,1} & p_{j+1,2} & \cdots & p_{j+1,j} & \cdots & p_{j+1,m} \\ \cdots & \cdots & \cdots & \cdots & \cdots & \cdots \\ p_{m1} & p_{m2} & \cdots & p_{mj} & \cdots & p_{mm} \end{pmatrix},$$

where all $p_{ik} > 0$ for $i \neq j$. Then j is absorbing and is the only essential state of the MC, while all $k \neq j$ are non-essential ones.

For a fixed essential state $j \in S_{Ess}$, denote by $S(j)$ the class of all states *communicating* with j. Note that the relation $\leftrightarrow$ ("communicate") on S_{Ess} has the following properties:

(i) *Reflexivity*: $j \leftrightarrow j$. Indeed, either j is absorbing (and then $p_{jj} = 1$, so that $j \leftrightarrow j$), or there exists a k such that $j \leftrightarrow k$. If the latter is the case, for some n_i one has $p_{jk}^{(n_1)} > 0$ and $p_{kj}^{(n_2)} > 0$, so that $p_{jj}^{(n_1+n_2)} \geq p_{jk}^{(n_1)} p_{kj}^{(n_2)} > 0$.

(ii) *Symmetry*: $j \leftrightarrow k$ iff $k \leftrightarrow j$. Obvious.

(iii) *Transitivity*: If $j \leftrightarrow k$ and $k \leftrightarrow i$, then $j \leftrightarrow i$. Use an argument similar to the one in (i).

In mathematics, a *binary relation* satisfying the conditions (i)–(iii) above is called an **equivalence**, and the "points" (states in our case) j and k related to each other ($j \leftrightarrow k$) are said to be *equivalent*. The general fact is that whenever one has an equivalence given on a set, the set can be partitioned into disjoint classes of equivalent points.

Indeed, one can easily see that the classes $S(j)$ for different $j \in S_{Ess}$ are either *identical* or *disjoint*. If $S(j) \cap S(k) \neq \emptyset$, it means that there is a common to both classes state $i \in S_{Ess}$: $j \leftrightarrow i$ and $k \leftrightarrow i$. Now if $v \in S(j)$, i.e., $v \leftrightarrow j$, then by (iii) above, $v \leftrightarrow i$, so that $v \in S(i)$ and hence $S(j) \subset S(i)$. Similarly, if $u \in S(i)$, then $u \leftrightarrow j$, so that $S(i) \subset S(j)$, and hence $S(i) = S(j)$. We can argue in the same way for k to show that $S(i) = S(k)$, and therefore we conclude that $S(j) = S(k)$.

Accordingly, we can divide the state space into a number of disjoint classes. Firstly, there is the class S_{NEss} of all non-essential states. If the initial state $X_0 \in S_{NEss}$, the MC will eventually leave the class and never be back again. The complement of S_{NEss} is the class S_{Ess} of essential states, which can further be partitioned into **closed classes** $S_1, S_2, \ldots$ of communicating states. They are closed in the sense that once our chain enters any of them, it will never leave it: if $X_n \in S_i$ for some n, then $X_m \in S_i$ for all $m > n$. The part of the transition matrix of the MC corresponding to states from S_{Ess} will clearly be a transition matrix as well. It can be decomposed into blocks corresponding to closed classes of its states as illustrated in Fig. 3.3 (note that the block sizes can all be different in the general case); all the blocks will be stochastic (sub)matrices.

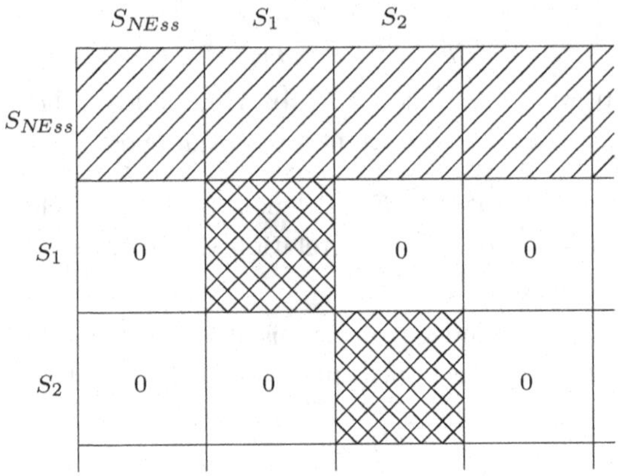

Fig. 3.3 Transition matrix: blocks correspond to classes.

If an MC has only one class, i.e., *all its states communicate with each other*, then it is said to be **irreducible**.

Example 3.10. In a five-state MC with the transition matrix

$$P = \begin{pmatrix} 1/2 & 1/2 & 0 & 0 & 0 \\ 1/2 & 1/2 & 0 & 0 & 0 \\ 0 & 0 & 0 & 0 & 1 \\ 1/5 & 1/5 & 1/5 & 1/5 & 1/5 \\ 0 & 0 & 1 & 0 & 0 \end{pmatrix},$$

we have $S_{NEss} = \{4\}$, and there are two closed classes of essential states: $S_1 = S(1) = S(2) = \{1, 2\}$ and $S_2 = S(3) = S(5) = \{3, 5\}$. But note the difference between the classes S_i: although each of them consists of two states only, the behaviour of the MC upon entering one of them is very different for S_1 and S_2! If the chain first enters S_1, from that time on the only values it can take are 1 and 2, and, at each step, one of them is chosen independently of the past with probability 1/2 (in layman's language, "at random"). But if the MC enters the class S_2 first, from that time on there will be no randomness in its behaviour: the states 3 and 5 will be alternating all the time. This shows that we need further classification.

A state j is said to be **periodic** with a **period** $d > 1$, if $p_{jj}^{(n)} = 0$ for all $n \neq kd$, $k = 1, 2, \ldots$, and d is the *largest* integer with that property. In other words, d is the *greatest common divisor* (GCD) of all n such that $p_{jj}^{(n)} > 0$. If the GCD of the values n such that $p_{jj}^{(n)} > 0$ is equal to one, the state j is called **aperiodic**.

Example 3.11. (i) In this example, all the states $j = 1, 2, 3$ clearly have one and the same period $d = 2$:

$$P = \begin{pmatrix} 0 & 1/2 & 1/2 \\ 1 & 0 & 0 \\ 1 & 0 & 0 \end{pmatrix}$$

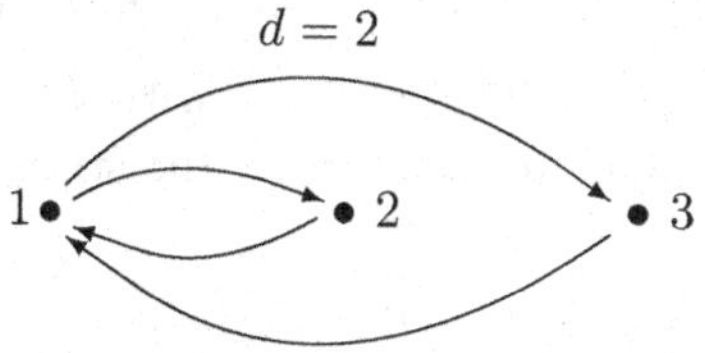

(ii) In this case, we can see that $d = 3$ for all four states:

$$P = \begin{pmatrix} 0 & 0 & 1/2 & 1/2 \\ 1 & 0 & 0 & 0 \\ 0 & 1 & 0 & 0 \\ 0 & 1 & 0 & 0 \end{pmatrix}$$

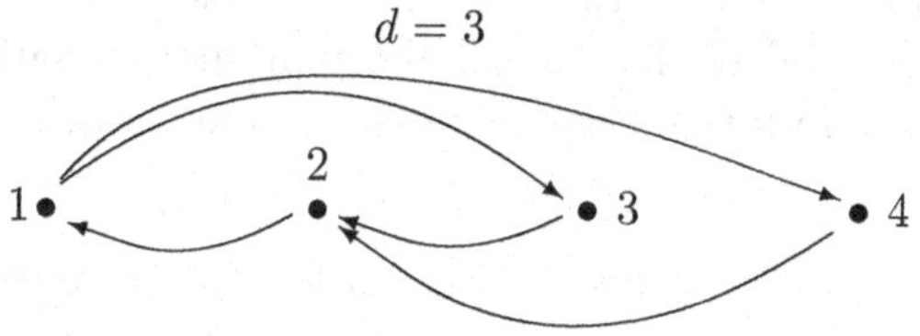

This is actually a general fact: as we will see later, communicating states always have a common period.

If we now slightly change the transition probability values and put, for an arbitrary small $\varepsilon > 0$, $p_{21} := 1 - \varepsilon$, $p_{22} := \varepsilon$, then the MC will become *aperiodic*:

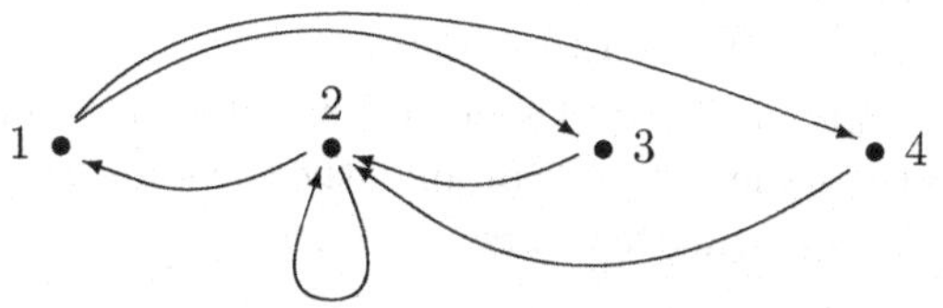

Note that the presence of such a "loop" (meaning that the respective diagonal element of P is positive) *always implies that this state is aperiodic.* Moreover, all the other states that state communicates with will also be aperiodic (why?).

Remark: The above classification *does not depend on the particular values* of the transition probabilities p_{jk}; all what matters is whether $p_{jk} = 0$ or $p_{jk} > 0$. As the next example shows, when the state space of the MC is infinite, a further, "more delicate" classification is needed.

Example 3.7 (continued). In a simple RW with transition matrix (3.9), all the states are clearly essential and the chain is irreducible. Indeed, assume that $j < k$. Then

$$p_{jk}^{(k-j)} = p^{k-j} > 0 \quad \text{and} \quad p_{kj}^{(k-j)} = q^{k-j} > 0,$$

so that $j \leftrightarrow k$. Also, it is evident that our walk can return to the starting point in an *even number of steps* only:

$$p_{jj}^{(2n+1)} = 0 \qquad \text{for any } \ j \in S, \ n \geq 0, \tag{3.16}$$

and hence the MC is periodic with $d = 2$.

Let $p > q$. Recall that, by the strong LLN, as $n \to \infty$,

$$\frac{X_n}{n} \to \mathbf{E}\, Y_1 = p - q > 0 \quad \text{w.p. } 1.$$

Therefore the walk escapes to $+\infty$. Hence, for any fixed $j \in S$, our MC will visit j only finitely many times, as if j were non-essential! Yet all the states are essential, and the chain is irreducible. So one needs further classification to distinguish between such situations and the cases when MCs keep visiting certain states as time elapses.

Let

$$f_j := \mathbf{P}\,(X_n = j \text{ for some } n > 0 |\, X_0 = j) \tag{3.17}$$

denote the probability of *ever returning* to the initial state given it was j. A state j is said to be **recurrent** if $f_j = 1$, and **transient** if $f_j < 1$.

In words, state j is recurrent if, after each time the MC visits j, it will almost surely return to it again. If j is transient, there is a positive probability that, upon leaving j, the MC will never be back to j (note that, for a homogeneous MC, this probability of not returning to j will be one and the same value $1 - f_j$ each time we visit j).

There is a simple criterion of recurrence stated in terms of transition probabilities; at the same time, the proof of the theorem is a nice illustration of the method of generating functions.

Theorem 3.1. *Put* $q_j := \sum_{n=1}^{\infty} p_{jj}^{(n)} \leq \infty$. *Then*

$$f_j = \frac{q_j}{1 + q_j} \qquad (\textit{with } f_j = 1 \textit{ when } \ q_j = \infty).$$

Therefore, state j *is recurrent iff* $q_j = \infty$.

Proof Introducing the probabilities

$$f_j^{(n)} := \mathbf{P}\left(X_n = j, X_{n-1} \neq j, \dots, X_1 \neq j | X_0 = j\right)$$

of returning to j at time n for the first time, we have by the TPF that

$$p_{jj}^{(n)} = f_j^{(1)} p_{jj}^{(n-1)} + f_j^{(2)} p_{jj}^{(n-2)} + \dots + f_j^{(n-1)} p_{jj}^{(1)} + f_j^{(n)}$$

(note that the sum on the right-hand side corresponds to the partition of the event $\{X_n = j\}$ according to the value of the *first time* $k = 1, \dots, n$ when $X_k = j$). Now multiply both sides by z^n and take the sum $\sum_{n=1}^{\infty}$. Then, for the GF $\widetilde{q}_j(z)$ of the sequence $\{p_{jj}^{(n)}\}_{n\geq 1}$, we will get the relation

$$\begin{aligned}
\widetilde{q}_j(z) &:= \sum_{n=1}^{\infty} z^n p_{jj}^{(n)} \\
&= \sum_{n=1}^{\infty} (z f_j^{(1)} \cdot z^{n-1} p_{jj}^{(n-1)} + z^2 f_j^{(2)} \cdot z^{n-2} p_{jj}^{(n-2)} + \dots + z^n f_j^{(n)}) \\
&= \sum_{n=1}^{\infty} \sum_{k=1}^{n} z^k f_j^{(k)} \cdot z^{n-k} p_{jj}^{(n-k)} = \sum_{k=1}^{\infty} \sum_{n=k}^{\infty} z^k f_j^{(k)} \cdot z^{n-k} p_{jj}^{(n-k)} \\
&= \sum_{k=1}^{\infty} z^k f_j^{(k)} \sum_{n-k=0}^{\infty} z^{n-k} p_{jj}^{(n-k)} = \widetilde{f}_j(z)(1 + \widetilde{q}_j(z)),
\end{aligned}$$

where $\widetilde{f}_j(z)$ is the GF of $\{f_j^{(n)}\}_{n\geq 1}$, so that

$$\widetilde{f}_j(z) = \frac{\widetilde{q}_j(z)}{1 + \widetilde{q}_j(z)}.$$

The assertion of the theorem follows if we take the limits on both sides as $z \nearrow 1$ and make use of the observation that, for the GF $\widetilde{a}(z) := \sum_{n=1}^{\infty} a_n z^n$ of a (non-negative) sequence $\{a_n\}$, one has $\lim_{z\nearrow 1} \widetilde{a}(z) = \sum_{n=1}^{\infty} a_n$. □

The theorem has the following important

Corollary 3.1. *If state j is recurrent and $j \leftrightarrow k$, then k is also recurrent.*

Proof By definition, $j \leftrightarrow k$ means that, for some fixed s and t, both $p_{jk}^{(s)} > 0$ and $p_{kj}^{(t)} > 0$, so that $\alpha := p_{jk}^{(s)} p_{kj}^{(t)} > 0$ and

$$\begin{aligned}
p_{jj}^{(s+n+t)} &= \mathbf{P}\left(X_{s+n+t} = j | \, X_0 = j\right) \\
&\geq \mathbf{P}\left(X_{s+n+t} = j, X_{s+n} = k, X_s = k | \, X_0 = j\right) \\
&= p_{jk}^{(s)} p_{kk}^{(n)} p_{kj}^{(t)} = \alpha p_{kk}^{(n)}.
\end{aligned}$$

Similarly,

$$p_{kk}^{(s+n+t)} \geq p_{kj}^{(t)} p_{jj}^{(n)} p_{jk}^{(s)} = \alpha p_{jj}^{(n)}.$$

This means that, for $n > u = s + t$,

$$\alpha p_{jj}^{(n-u)} \leq p_{kk}^{(n)} \leq \frac{1}{\alpha} p_{jj}^{(n+u)}, \tag{3.18}$$

so that the series $\sum_n p_{jj}^{(n)}$ and $\sum_n p_{kk}^{(n)}$ converge/diverge simultaneously! Hence if state j is recurrent, then by Theorem 3.1 the first series diverges, and therefore the second one diverges as well. Again applying Theorem 3.1, we see that k is also recurrent. □

In our Example 3.7, *all the states* are transient when $p \neq q$. This is absolutely impossible for finite MCs. Moreover, the recurrent/transient classification is only meaningful for infinite MCs: in the case of a finite S, once a state if essential, it is always recurrent. This fact is an immediate consequence of Corollary 3.1 and the following assertion to be proved below: in a finite MC, it is impossible that *all* states are transient. Indeed, that implies that at least one state is recurrent, and therefore all states communicating with it are also recurrent. This argument holds for any closed class of essential states, and therefore there can be no transient essential states in a finite MC.

The above-made assertion can be proved using the following general fact: the conditional distribution of the total number $V_j := \#\{m \geq 0 : X_m = j\}$ of visits to a *transient* state j, given it was visited, is **geometric**:

$$\mathbf{P}(V_j = n | V_j > 0) = (1 - f_j) f_j^{n-1}, \quad n = 1, 2, \ldots \tag{3.19}$$

To see that, note that after the MC has visited j for the first time (which does occur on the conditioning event $\{V_j > 0\}$), it will never be back to j w.p.

$$\mathbf{P}(V_j = 1 | V_j > 0) = \mathbf{P}(X_n \neq j \text{ for all } n > 0 | X_0 = j) = 1 - f_j$$

(cf. (3.17)). This is due to the Markov property: when the MC enters j for the first time, it evolves further as it has just started from the state j at that time.[4]

Now if the MC does return to j for the second time (which occurs w.p. f_j given we start at it), then it again "starts anew" from j and will never

[4]More precisely, this is due to the so-called *strong Markov property*, which is basically the Markov property that holds at the so-called *stopping times*. For more detail, please refer to the more advanced probability texts listed in Section 2.11.

be back w.p. $1-f_j$, so that the conditional probability of exactly two visits to j is $(1-f_j)f_j$. More formally,

$$\begin{aligned}
&\mathbf{P}\left(V_j=2 \mid V_j>0\right)=\mathbf{P}\left(V_j=2 \mid X_0=j\right) \\
&=\sum_{n=1}^{\infty} \mathbf{P}\left(X_m \neq j, m>n \mid X_n=j\right) \mathbf{P}\left(X_n=j, X_k \neq j, 0<k<n \mid X_0=j\right) \\
&=\underbrace{\mathbf{P}\left(X_m \neq j, m>0 \mid X_0=j\right)}_{1-f_j} \underbrace{\sum_{n=1}^{\infty} \mathbf{P}\left(X_n=j, X_k \neq j, 0<k<n \mid X_0=j\right)}_{f_j} \\
&=(1-f_j)f_j.
\end{aligned}$$

Similarly, after the chain's nth visit to j (which occurs w.p. f_j^{n-1} given that $V_j>0$) it will never be back to j w.p. $1-f_j$, hence (3.19).

Thus, the total number of visits to any given transient state is a finite RV. Now if all the states in a finite MC are transient, it follows that each of them will be visited *finitely many times.* This is a contradiction, for there are only *finitely many* states in the chain, while the time interval is infinite.

Now we will illustrate the recurrence criterion from Theorem 3.1 by one of our examples.

Example 3.7 (continued.) Let us apply Corollary 3.1 to our simple RW example. Compute the m-step transition probabilities $p_{jj}^{(m)}$. For odd values $m=2n+1$ of steps, all of them are zeros, see (3.16). For even times $m=2n$, they will just be the binomial probabilities of having exactly n successes in $2n$ trials:

$$p_{jj}^{(2n)}=p_{00}^{(2n)}=\binom{2n}{n}p^n q^n=\frac{(2n)!}{n!n!}(pq)^n. \tag{3.20}$$

To evaluate the last expression, make use of the famous *Stirling*[5] *formula*[6]: as $n\to\infty$,

$$n!=\sqrt{2\pi n}n^n e^{-n}(1+o(1)),$$

[5]After James Stirling (1692–05.12.1770), a Scottish mathematician who contributed to the theory of infinite series and complex numbers. He also uncovered the trade secrets of the Venetian glassmakers (he went to Venice to study after having been expelled from Oxford). In fact, the formula was obtained earlier by Abraham De Moivre (26.05.1667–27.11.1754), a French Huguenot who had to leave France for Britain (after having been jailed for being a Protestant) where he became one of the leading personalities in science. He applied the formula to get the first version of the CLT (for sums of i.i.d. indicators with success probabilities 1/2, i.e., for the total number of successes in a series of n trials, when success and failure are equally likely).

[6]The formula is very sharp: in fact, $n!=\sqrt{2\pi n}n^n e^{-n+\theta/12n}$, $0<\theta<1$ (and one can get further refinements). For "mathematically minded": note that the formula gives an approximation to the gamma function $\Gamma(z)$ for large $\mathrm{Re}\, z$ as well.

where used a popular in mathematics notation $o(1)$ representing *any quantity* that tends to zero. More generally, for sequences $\{a_n\}$, $\{b_n\}$ one writes $a_n = o(b_n)$ ($a_n = O(b_n)$) as $n \to \infty$ iff $a_n/b_n \to 0$ ($|a_n/b_n| < C < \infty$, respectively). The notation is also used for functions (say, $f(x) = o(g(x))$ as $x \to x_0$ means that $f(x)/g(x) \to 0$ as $x \to x_0$). Note the convenience of the notation: for instance, $(1 + o(1))(1 + o(1)) = 1 + o(1)$ *etc.*

Now for asymmetric walks ($p \neq q$), (3.20) yields

$$p_{jj}^{(2n)} = \frac{\sqrt{2\pi \cdot 2n}(2n)^{2n}e^{-2n}}{(\sqrt{2\pi n}n^n e^{-n})^2}(pq)^n(1 + o(1)) = \frac{1 + o(1)}{\sqrt{\pi n}}(4pq)^n.$$

Since $4pq = 4p(1-p) < 1$ for $p \neq q$, the last expression vanishes geometrically fast, so that $\sum_m p_{jj}^{(m)} < \infty$ and therefore the MC is transient. In fact, we have already derived that earlier from the LLN.

In the symmetric case $p = q = 1/2$ we get

$$p_{jj}^{(2n)} = \frac{1 + o(1)}{\sqrt{\pi n}}, \tag{3.21}$$

so that $\sum_m p_{jj}^{(m)} = \infty$, and hence the MC is recurrent.

It is interesting to note that, for the two-dimensional symmetric simple RW (on the integer grid, with equally likely jumps of the form $(0, \pm 1)$ and $(\pm 1, 0)$), one has $p_{jj}^{(2n)} = (1 + o(1))/\pi n$. Indeed, if we rotate the coordinate system on the plane by $\pi/4$ (45°), the components of the RW in the new coordinate system will be independent copies of the symmetric simple RW, and returning to the starting point after $2n$ steps would mean that both components' increments are simultaneously equal to zero at that time. So by independence the probability of this event will be $((1 + o(1))/\sqrt{\pi n})^2$. Hence the MC is again recurrent since $\sum_{n=1}^{\infty} n^{-1} = \infty$ (the harmonic series). For dimensions ≥ 3, the symmetric RW will be transient. For more detail, see, e.g., Feller (1970).

In the last example, all the states were periodic and transient/recurrent simultaneously, which was quite natural because transition probabilities were "space-homogeneous". However, such a solidarity of states is a general property.

Theorem 3.2. (The solidarity theorem) *In any closed class $S_r \subset S$ of an MC $\{X_n\}$ with the state space S, all the states $j \in S_r$ are*

(i) *either recurrent or transient, and*

(ii) *either periodic with a common period $d > 1$ or aperiodic.*

The first part follows from Corollary 3.1, while (ii) is a simple exercise left to the reader.

If the states from class S_r are periodic with a period $d > 1$, then one has a partition (we use "+" instead of "∪" to indicate that the sets are disjoint)

$$S_r = S_r^{(1)} + S_r^{(2)} + \cdots + S_r^{(d)}$$

such that from the subclass $S_r^{(i)}$ the MC goes w.p. 1 to subclass $S_r^{(i+1)}$, $i = 1, 2, \ldots, d-1$, and from $S_r^{(d)}$ it proceeds to $S_r^{(1)}$. Thus, the submatrix of P corresponding to S_r (as we have already noted, that part of P is a proper stochastic matrix since S_r is closed) will have a block structure shown in Fig. 3.4; note again that, in the general case, the block sizes can be different! The sets $S_r^{(i)}$ are referred to as **cyclic subclasses**.

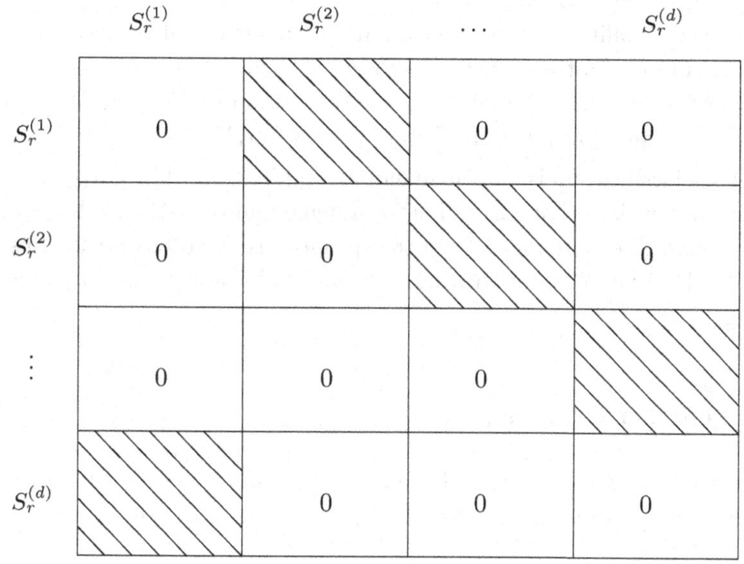

Fig. 3.4 A periodic transition matrix.

Note that, for a chain with the transition matrix P^d (which corresponds to sampling the original chain at times n from a lattice with the span d: $n = m + kd$, $k = 1, 2, 3, \ldots$), each of the subclasses $S_r^{(i)}$ will become a closed aperiodic class. So we can reduce the study of periodic MCs to that of aperiodic ones.

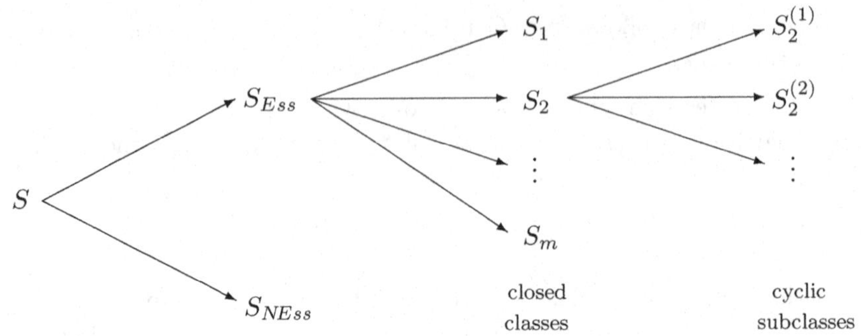

Fig. 3.5 The MC classification scheme.

One can briefly summarise the classification of the states of an MC as follows (see also Fig. 3.5):

(i) accessibility of states $\longrightarrow$ essential/non-essential states;

(ii) communicating states $\longrightarrow$ closed classes;

(iii) the "intraclass" analysis: periodicity and recurrence/transience (the latter is meaningful for infinite state classes only).

The classification is, of course, not an end in itself. The purpose of doing it is to determine what sort of long-run behaviour our MC will have. Before proceeding to discussing the limiting properties of MCs, we will consider a few further examples illustrating how one could model certain phenomena using MCs.

3.3 Further Examples

Example 3.12. Suppose we have an infinite supply of light bulbs, and

$$Z_i := \text{lifetime of the } i\text{th bulb (in days)}, \quad i = 1, 2, \dots,$$

are i.i.d. RVs with distribution

$$\mathbf{P}(Z_i = k) = a_k > 0, \quad k = 1, 2, \dots, \quad \sum_{k \geq 1} a_k = 1 \tag{3.22}$$

(so the time is discrete). At time $n = 0$, the first bulb is turned on; when it fails (at time Z_1), it is immediately replaced by the second one which is also switched on, and so on.

Let X_n be the age of the bulb which is on at time n; we set $X_n := 0$ if a failure occurs at time n. Is $\{X_n\}$ an MC? If yes, find its transition matrix and classify the states.

The state space in our example is clearly $S = \{0, 1, 2, \dots\}$ (it slightly differs from our usual choice $S = \{1, 2, \dots\}$, but in this case it is more convenient), and possible transitions in the process are $k \to 0$ (failure) and $k \to k+1$ (no failure) only. We do have the Markov property for our sequence $\{X_n\}$, since if we know the age of the currently working bulb, all the "past" is irrelevant from the point of view of the future evolution. Also, we can find from (3.22) the conditional probabilities

$$p_{k0} = \mathbf{P}(Z_1 = k+1 | Z_1 > k) = \frac{a_{k+1}}{\sum_{j \geq k+1} a_j},$$

$$p_{k,k+1} = \mathbf{P}(Z_1 > k+1 | Z_1 > k) = 1 - p_{k0}.$$

The transition matrix of the MC has the form

$$P = \begin{pmatrix} p_{00} & p_{01} & 0 & 0 & 0 & \cdots \\ p_{10} & 0 & p_{12} & 0 & 0 & \cdots \\ p_{20} & 0 & 0 & p_{23} & 0 & \cdots \\ p_{30} & 0 & 0 & 0 & p_{34} & \cdots \\ p_{40} & 0 & 0 & 0 & 0 & \cdots \\ \cdots & \cdots & \cdots & \cdots & \cdots & \ddots \end{pmatrix}. \tag{3.23}$$

It is not hard to see that the MC is irreducible, aperiodic (note that we assumed that all $a_k > 0$) and recurrent (since Z_j are finite RVs, sooner or later we will visit 0 due to a failed bulb).

The situation is often referred to as the *regeneration scheme*: the process starts anew at the regeneration times, and evolves afterwards as an independent copy of the original process. We will discuss certain aspects of this scheme later, in Chapter 8 devoted to renewal processes.

Observe the following interesting thing: if the Z_i's have the geometric distribution:

$$a_k = (1-q)q^{k-1}, \quad k = 1, 2, \dots,$$

for some $q \in (0, 1)$, then

$$\sum_{j \geq k+1} a_j = (1-q)(q^k + q^{k+1} + \cdots) = (1-q)q^k \times \frac{1}{1-q} = q^k,$$

so that the transition probabilities

$$p_{k0} = \frac{a_{k+1}}{\sum_{j \geq k+1} a_j} = \frac{(1-q)q^k}{q^k} = 1 - q, \quad p_{k,k+1} = q$$

do not depend on k. Therefore, in that case, the failures are "generated" by an i.i.d. Bernoulli sequence with "success" probability $1 - q$.

Example 3.13. Under the assumptions of Example 3.12, let

$$Y_n := \text{the time from } n \text{ till the first failure after time } n.$$

We ask ourselves the same questions as in the previous example—but now about the process $\{Y_n\}$.

Now the state space is $S = \{1, 2, \dots\}$, while possible transitions are $1 \to k$ w.p. a_k and $k(> 1) \to k-1$ w.p. 1. The probabilities do not depend on the past, so this is indeed an MC. The transition matrix has the form

$$P = \begin{pmatrix} a_1 & a_2 & a_3 & a_4 & \cdots \\ 1 & 0 & 0 & 0 & \cdots \\ 0 & 1 & 0 & 0 & \cdots \\ 0 & 0 & 1 & 0 & \cdots \\ \dots & \dots & \dots & \dots & \ddots \end{pmatrix}. \tag{3.24}$$

It is also easy to see that this MC is irreducible, recurrent and aperiodic. Note that if we didn't assume that all $a_k > 0$, the chain could have a period $d = \mathrm{GCD}\,\{k : a_k > 0\} > 1$.

Example 3.14. Under the assumptions of Example 3.12, set

$$N_n := \text{the number of bulbs failed by the time } n.$$

Is $\{N_n\}$ an MC?

Now the state space is $S = \{0, 1, 2, \dots\}$. Possible transitions are $k \to k$ and $k \to k+1$ only. But, in the general case, $\{N_n\}$ is *not* an MC, for the past does matter for the future evolution of the process when the present is known.

Indeed, we obviously have

$$\begin{aligned}\mathbf{P}\,(N_{n-1} = n-1, N_n = n-1, N_{n+1} = n-1)& \\ = \mathbf{P}\,(Z_1 = \cdots = Z_{n-1} = 1, Z_n > 2) &= a_1^{n-1}(1 - a_1 - a_2)\end{aligned}$$

and

$$\begin{aligned}\mathbf{P}\,(N_{n-1} = n-1, N_n = n-1)& \\ = \mathbf{P}\,(Z_1 = \cdots = Z_{n-1} = 1, Z_n > 1) &= a_1^{n-1}(1 - a_1).\end{aligned}$$

Dividing the former equality by the latter, we get the conditional probability

$$\mathbf{P}\,(N_{n+1} = n-1 | N_n = n-1, N_{n-1} = n-1) = \frac{1 - a_1 - a_2}{1 - a_1}.$$

On the other hand,

$$\mathbf{P}\,(N_{n+1} = n-1 | N_n = n-1, N_{n-1} = n-2) = \mathbf{P}\,(Z_{n-1} > 1) = 1 - a_1$$

(note that the condition means that a new bulb was switched on at the time n). Now if $\{N_n\}$ were an MC, each of the two conditional probabilities above would be equal to $\mathbf{P}(N_{n+1} = n-1 | N_n = n-1)$, so that the two would have to coincide with each other. Therefore, if $(1 - a_1 - a_2) \neq (1 - a_1)^2$, our process cannot be an MC.

If, however, the Z_i's are *geometric* RVs: $a_k = (1-q)q^{k-1}$, $k = 1, 2, \ldots$, for some $q \in (0, 1)$, then $\{N_n\}$ is an MC! Recall that, in this case, as we saw in Example 3.12, $N_n = \xi_1 + \cdots + \xi_n$ is an RW with i.i.d. jumps $\xi_i \sim B_{1-q}$. Note that here we do have $(1 - a_1 - a_2) = (1 - a_1)^2$ (as both sides are equal to q^2).

Example 3.15. (an extension of Example 3.6). The weather condition on day n is either

$$\text{sunny: } X_n = 1, \quad \text{or rainy: } X_n = 2.$$

Suppose that the weather on day $n+1$ depends upon the weather conditions on days $n-1$ and n only. More precisely, one has the following conditional probabilities for the weather conditions on day $n+1$ given those on days $n-1$ and n (and regardless of what happened prior to day $n-1$):

Day: $n-1$	n		$n+1$
rain	rain	$\longrightarrow$	rain w.p. 0.6
sunny	sunny	$\longrightarrow$	sunny w.p. 0.8
sunny	rain	$\longrightarrow$	rain w.p. 0.5
rain	sunny	$\longrightarrow$	sunny w.p. 0.75

Could one model the situation using an MC?

Clearly $\{X_n\}$ is not an MC. What one can do in the case of such a "finite-range" dependence is to extend the state space by forming finite sequences of states from the original space ($\{1, 2\}$ in our case). New states can be so "extensive" in time that dependence on the "past" would become "incorporated" into the "present".

So let $Y_n := (X_{n-1}, X_n)$, $n \geq 1$. There are four possible states: 11, 12, 21 and 22. Note the "overlapping" in time: the first symbol in Y_{n+1} must coincide with the last one in Y_n, and this is an MC now! Possible transitions are:

$$\begin{array}{lll} 11 \to 11 \text{ w.p. } 0.8 & & 21 \to 11 \text{ w.p. } 0.75 \\ \quad \searrow & & \quad \searrow \\ \quad\quad 12 \text{ w.p. } 0.2 & & \quad\quad 12 \text{ w.p. } 0.25 \\ & & \\ 12 \to 21 \text{ w.p. } 0.5 & & 22 \to 21 \text{ w.p. } 0.4 \\ \quad \searrow & & \quad \searrow \\ \quad\quad 22 \text{ w.p. } 0.5 & & \quad\quad 22 \text{ w.p. } 0.6 \end{array}$$

so that the transition matrix is

$$P = \begin{pmatrix} 0.8 & 0.2 & 0 & 0 \\ 0 & 0 & 0.5 & 0.5 \\ 0.75 & 0.25 & 0 & 0 \\ 0 & 0 & 0.4 & 0.6 \end{pmatrix}.$$

The MC is seen to be irreducible and aperiodic (note the "loops": the first and the last diagonal elements are positive!).

Example 3.16. In this example we give a simplified description of a telecommunication scheme known as the *slotted ALOHA*[7] *protocol.*

The idea is roughly as follows. Suppose there is a group of people in a room, and from time to time, each of them has something to say to another person. It may happen that two or more people will start talking simultaneously, and then nobody will understand anything. If frustrated people tried again and again to say what they wanted to, the whole conversation would be blocked forever. To avoid such a "clash", one suggests that each of the people, once he or she has something to say, flips a coin. If it lands heads up, the person starts speaking, otherwise s/he tosses the coin again *etc.* This can make the conversation possible (all the "clashes" will be resolved sooner or later).

More formally, suppose there are N users in a region, each having a transmitter and a receiver and communicating via a satellite. When, say, user u has a message for user v, he sends it to the satellite which retransmits it to the whole region so that each of the users receives it (the "header" of the message will tell them that this is a message for user v). When u receives the message, he knows that the message was sent successfully.

Again, it may happen that more than one user were transmitting their messages simultaneously, so that the users will get a garbled message (and

[7]The protocol was originally implemented at the University of Hawaii in 1970 to network computers that were located on different islands; the word *aloha* means "love" in Hawaiian, used as greeting or farewell.

then u will know that he failed). To formally describe how ALOHA suggests to overcome this problem, we will make a few assumptions:

- All users have synchronised clocks.
- The users send packets of information in time slots of equal length (say, at times $n = 0, 1, 2, \dots$).
- If a user started transmitting a packet at time n, the reception would be completed at time $n+1$ and everybody would know if it was successful or not (a "clash" occurred).
- If there was a clash, the protocol prescribes for each user involved to keep re-transmitting his message as follows: in each subsequent time slot, each user independently decides, with probability $r \in (0,1)$, to attempt re-transmission of his packet until he succeeds.

To completely specify the operation of the system, we need to say how the packets are generated. Let us assume that if in time slot n a user was idle or successfully transmitted a packet, then in time slot $n+1$ he will have another packet w.p. $p \in (0,1)$. Otherwise this user will have no new packet for transmission.

Now denote by X_n the number of "backlogged" users (i.e., the users who still have a packet to re-transmit) at the beginning of the nth time slot.

When $X_n = 0$, all what can happen during this time slot is that users generate and try to transmit new packets. When there is no or one new packet, no clash occurs and $X_{n+1} = 0$, otherwise all the users who generated a packet "backlog". So

$$\mathbf{P}(X_{n+1} = j \,|\, X_n = 0) = \begin{cases} (1-p)^N + Np(1-p)^{N-1}, & j = 0, \\ \binom{N}{j} p^j (1-p)^{N-j}, & j = 2, \dots, N. \end{cases}$$

On the other hand, the value $X_n > 0$ can *decrease by one only*, and this occurs when *exactly* one message is re-transmitted and none of the idle $N - X_n$ users generates a new packet. Therefore, by independence, for $m > 0$,

$$\mathbf{P}(X_{n+1} = m-1 \,|\, X_n = m) = mr(1-r)^{m-1} \times (1-p)^{N-m}.$$

The value of X_n *increases by one* iff at least one backlogged user re-transmits and exactly one new packet is generated:

$$\mathbf{P}(X_{n+1} = m+1 \,|\, X_n = m) = [1 - (1-r)^m] \times (N-m)p(1-p)^{N-m-1}.$$

It will increase by $j \geq 2$ whenever there are j newly generated packets (so there will certainly be a clash):

$$\mathbf{P}(X_{n+1} = m+j \mid X_n = m) = \binom{N-m}{j} p^j (1-p)^{N-m-j}, \quad j = 2, \ldots, N-m.$$

If nothing of the above occurs (i.e., either only one new packet is generated and no backlogged users tried to re-transmit, or no new packets and at least two attempts to re-transmit), then $X_{n+1} = X_n$, and hence

$$\mathbf{P}(X_{n+1} = m \mid X_n = m) = 1 - \text{sum of the above probabilities}, \quad m > 0.$$

This defines a homogeneous MC which can be used to analyse the performance of ALOHA systems.

3.4 The Limiting Behaviour of Markov Chains

Let $\{X_n\}$ be an MC starting at a fixed state j. Denote by $\tau_i := \tau_i(j)$, $i = 1, 2, \ldots$, the times between successive visits of our MC to the state j. As we have already noted, due to the Markov property, these visits "cut" the trajectory of the MC into **independent cycles** which are i.i.d. random elements, while the τ_i's themselves (the lengths of the cycles) are i.i.d. RVs. By definition, if j is *transient* then $\tau_i = \infty$ with a positive probability, while if j is *recurrent* then $\tau_i < \infty$ a.s. Thus if the last relation holds, we have an infinite sequence of "regenerations" (or "renewals") of our MC, i.e., the times when $\{X_n\}$ returns to the initial state j, so that the MC starts "anew" at those times. Therefore one may expect some kind of "statistical regularity" effects like (1.2): the relative frequency of visits to j will tend to a certain number. For this limiting frequency to be positive, we need to require in addition that the *mean time* between successive visits to j is also finite: $\mu_j = \mathbf{E}\,\tau_i < \infty$.

Example 3.7 (continued.) As we have already seen, a symmetric simple RW is recurrent. However, the expected number of visits to j till time $2k$ is, by virtue of (3.21),

$$\mathbf{E} \sum_{n=1}^{k} \mathbf{1}_{\{X_{2n}=j\}} = \sum_{n=1}^{k} p_{jj}^{(2n)} = \sum_{n=1}^{k} \frac{1+o(1)}{\sqrt{\pi n}} = (2+o(1))\sqrt{\frac{k}{\pi}}.$$

We conclude that the *relative frequency* of visits to j during the first k time units is $\approx 2\sqrt{k/\pi}/(2k) = 1/\sqrt{\pi k}$ which goes to zero as time passes, i.e., visits to any given state become more and more rare events. As we will see below, this means that $\mathbf{E}\,\tau_i(j) = \infty$.

Example 3.17. On the contrary, for any finite irreducible MC, the RV $\tau = \tau_i(j)$ not only has a finite mean for any state j, but also has an exponentially fast decaying tail:

$$\mathbf{P}(\tau > n) < Ca^n \quad \text{for some } a < 1,\ C < \infty. \tag{3.25}$$

Indeed, assume that $S = \{1, 2, \ldots, m\}$ and set

$$q_k(n) := \mathbf{P}(X_n \neq j, \ldots, X_1 \neq j | X_0 = k).$$

Note that

$$\mathbf{P}(\tau > n) \equiv \mathbf{P}(X_n \neq j, \ldots, X_1 \neq j | X_0 = j) \leq q(n) := \max_k q_k(n).$$

By (3.10) and the Markov property, for $t \geq 1$,

$$\begin{aligned} q_k(n+t) &= \sum_{r \neq j} \mathbf{P}(X_{n+t} \neq j, \ldots, X_{n+1} \neq j, X_n = r, \\ &\qquad\qquad X_{n-1} \neq j, \ldots, X_1 \neq j |\, X_0 = k) \\ &= \sum_{r \neq j} q_r(t)\, \mathbf{P}(X_n = r, X_{n-1} \neq j, \ldots, X_1 \neq j |\, X_0 = k) \\ &\leq q(t) \sum_{r \neq j} \mathbf{P}(X_n = r, X_{n-1} \neq j, \ldots, X_1 \neq j |\, X_0 = k) \\ &= q(t) q_k(n) \leq q(t) q(n), \end{aligned}$$

and therefore $q(n+t) \leq q(t)q(n)$. Hence, for $n = rm$,

$$q(n) \leq (q(m))^r = (q(m)^{1/m})^n.$$

Now it only remains to demonstrate that $q(m) < 1$, so that one can put $a := q(m)^{1/m}$ in (3.25). That can be done as follows.

Since the MC is irreducible, for any $k \in S$ there exists an $n_k < \infty$ and an n_k-step path $k = i_0, i_1, \ldots, i_{n_k} = j$ from k to j of a positive probability: $p_{ki_1} p_{i_1 i_2} \cdots p_{i_{n_k - 1} j} > 0$, which is equivalent to $q_k(n_k) < 1$. If $n_k > m$ then, among the n_k states $i_1, \ldots, i_{n_k}$ forming the path, there will be *repetitions* (as there are only m different states!), i.e., there will be a *cycle* inside the path. Removing it will only increase the probability of the (shorter) path still leading from k to j. Therefore the minimum n_k having the above-stated property is less than m, and hence $q_k(m) < 1$, $k = 1, \ldots, m$, so that $q(m) < 1$ as well.

When μ_j is finite, by the LLN the time of the vth visit of the MC to j is $\tau_1 + \cdots + \tau_v \approx \mu_j v$. Therefore, after n steps in our MC the number of visits v_n to j will apparently satisfy the relation $n \approx \mu_j v_n$, so that $v_n \approx n/\mu_j$, and the relative frequency of being at j is approx. $1/\mu_j > 0$. This is why recurrent states j with $\mu_j < \infty$ are called **positive**, and those with $\mu_j = \infty$ **null** states: the limiting relative frequencies of visiting the states are positive and null, respectively.

If this is the case, one could also expect that the *probability* of being at state j after a large number of steps must be close to that value as well. The standard approach to proving these facts is based on *renewal theory* to be discussed later in Chapter 8. A formal proof of the following key result can be found, e.g., in Feller (1970) or Kulkarni (1995).

Theorem 3.3. *If state j is aperiodic and $\mu_j = \mathbf{E}\,\tau_k(j) < \infty$ (and hence the state j is recurrent!), then, for any i such that $i \leftrightarrow j$,*

$$p_{ij}^{(n)} \to \frac{1}{\mu_j} \quad as \quad n \to \infty.$$

If $\mu_j = \infty$ then $p_{ij}^{(n)} \to 0$.

What happens is that once the MC first visits the state j (reaching this state when the MC starts at i can take a long time, but we only need this time to be a.s. *finite*), we get that cyclic behaviour: the MC keeps returning to state j at the average rate $1/\mu_j$, and in the long run it becomes insignificant how long it took the chain to get to j for the first time.

If, in an MC $\{X_n\}$, the limits

$$\pi_j := \lim_{n\to\infty} p_{ij}^{(n)}, \qquad \sum_j \pi_j = 1, \tag{3.26}$$

exist and *do not depend* on i, then the MC $\{X_n\}$ is said to be **ergodic**[8]. That is, for ergodic MSc, the n-step transition matrices P^n converge, as $n \to \infty$, to a transition matrix of the form (3.7), which corresponds to independent trials, so that the RVs X_t and X_{t+n} are "almost independent" for large n. The distribution $\boldsymbol{\pi} = (\pi_j)$ is called the **stationary** (or **steady-state**, or **equilibrium**) **distribution** of $\{X_n\}$ (or its transition matrix P).

For any initial distribution $\boldsymbol{p}$, it follows from (3.26) that the MC with $X_0 \sim \boldsymbol{p}$ will also have the same limiting distribution as $n \to \infty$:

$$\mathbf{P}(X_n = j) = (\boldsymbol{p}P^n)_j = \sum_i p_i p_{ij}^{(n)} \to \sum_i p_i \pi_j = \pi_j, \tag{3.27}$$

[8]From Greek *ergon* "work" and *hodos* "way, path" (cf. electr*ode*).

i.e., $\boldsymbol{p}P^n \to \boldsymbol{\pi}$. That is, an ergodic MC "forgets" about its initial state.

When the distribution of X_0 coincides with $\boldsymbol{\pi}$, one says that the MC (system) is in the *stationary regime*, or *equilibrium*, or in the *steady state*. What happens is that the process X_t, $t \geq 0$, is then *strictly stationary*: its FDDs are invariant with respect to time shifts. More precisely, for any k and $s_1 < s_2 < \cdots < s_k$, the distribution of $(X_{t+s_1}, X_{t+s_2}, \ldots, X_{t+s_k})$ does not depend on t. As the time passes, the distributional characteristics of the stationary process do not vary. We stress that the process itself, however, remains *random*: being in equilibrium does not mean settling down at a particular state!

Note also that relation (3.27) implies that

$$\boldsymbol{\pi} = \lim_{n\to\infty} \boldsymbol{\pi} P^n = \lim_{n\to\infty} (\boldsymbol{\pi} P^{n-1} \cdot P) = \Big(\lim_{n\to\infty} \boldsymbol{\pi} P^{n-1}\Big) P = \boldsymbol{\pi} P. \tag{3.28}$$

Any $\boldsymbol{\pi}$ satisfying the relation $\boldsymbol{\pi} P = \boldsymbol{\pi}$ is called **invariant** for the MC $\{X_n\}$ (or transition matrix P).

Theorem 3.4. *An MC* $\{X_n\}$ *is* **ergodic** *iff*

(i) *there is a single non-empty closed class of essential states, and the class is aperiodic;*

(ii) *there exists a state* j_0 *such that, for the recurrence time* $\tau_1(j_0)$ *to the state* j_0*, one has*

$$\mathbf{E}\,(\tau_1(j_0)|\, X_0 = j_0) < \infty.$$

For an ergodic MC, the limits $\pi_j = \lim_{n\to\infty} p_{ij}^{(n)}$ *form a unique solution to the system of linear equations*

$$\begin{cases} \boldsymbol{\pi} = \boldsymbol{\pi} P \quad \Longleftrightarrow \quad \pi_j = \sum_i \pi_i p_{ij}, \quad j \in S; \\ \sum_j \pi_j = 1. \end{cases} \tag{3.29}$$

Moreover,

$$\pi_j = \frac{1}{\mu_j}, \quad \text{where } \mu_j = \mathbf{E}\,(\tau_1(j)|\, X_0 = j). \tag{3.30}$$

On the other hand, if an MC satisfies (i) *and* (3.29) *has a unique solution* $\pi_j \geq 0$*, then the MC is ergodic and the unique solution to* (3.29) *is the chain's stationary distribution.*

The standard approach to proving this theorem uses renewal theory, see Theorem 8.4(iii) below and our comment afterwards. For a formal proof of the theorem, see, e.g., Chapter 15 in Feller (1970), Borovkov (2013) or Kulkarni (1995).

Corollary 3.2. *For a* **finite** *MC* $\{X_n\}$*, ergodicity is equivalent to condition* (i) *from Theorem 3.4.*

Indeed, as we saw in Example 3.17, in a finite irreducible MC, the recurrence times always have exponentially fast decaying tails and hence have finite means (and even exponential moments). Moreover, in a finite ergodic MC, convergence $\boldsymbol{p}P^n \to \boldsymbol{\pi}$ is exponentially fast: for some $C < \infty$ and $\rho \in (0,1)$,

$$|p_{ij}^{(n)} - \pi_j| < C\rho^n, \qquad n = 1, 2, \ldots,$$

for any i and j.

Note also the following: for a finite MC with m states, there are $m+1$ equations in the system (3.29) for the stationary distribution. However, the rank of the homogeneous system of linear algebraic equations $\boldsymbol{\pi}P = \boldsymbol{\pi}$ is at most $m-1$ (since P is a stochastic matrix, all row sums are equal to one, and hence the vector sum of the columns of the matrix $P - I$, I being the unity matrix, is zero; it is the condition of *irreducibility* that ensures that the rank is actually exactly $m-1$). So we need an additional equation (the sum of the π_j's is one, which makes $\boldsymbol{\pi}$ a probability distribution) to ensure that there is a *unique solution* to the system, and we can always discard one of the equations from the main system $\boldsymbol{\pi} = \boldsymbol{\pi}P$.

In the case of finite MCs, one can visualise the ergodic theorem as follows. The set of all probability distributions on $\{1, \ldots, m\}$ is an $(m-1)$-dimensional *simplex* C_m (the geometric meaning of the conditions $p_i \in [0,1]$, $\sum_i p_i = 1$), its vertices being the degenerate distributions (one $p_i = 1$, all the others are zeros). The *linear operator* P maps C_m into itself: the vector $\boldsymbol{p}P$ is again a distribution and hence belongs to C_m (due to the fact that P is a stochastic matrix). The remarkable fact is that P is a *contraction* on C_m (but not on the whole $\mathbf{R}^m$!): the image $C_m P = \{\boldsymbol{p}P : \boldsymbol{p} \in C_m\}$ is strictly "smaller" than C_m itself, and applying the operator P to it again and again (which corresponds to finding all possible distributions of the values of our MC after two steps, three steps and so on) keeps "contracting" the set C_m to a single point which is just the stationary distribution of the chain—and the stationary point of the operator P: $\boldsymbol{\pi} = \boldsymbol{\pi}P$.

Example 3.5 (continued.) The transition matrix (3.8) from our linear communication system example is clearly irreducible and aperiodic. To find the stationary distribution, we take the first equation of the main system in (3.29): $p\pi_1 + (1-p)\pi_2 = \pi_1$, which yields $\pi_1 = \pi_2$, so that the stationary distribution is $(0.5, 0.5)$.

Note that transition matrix (3.8) is a special case of what is called a **doubly stochastic** matrix: not only all the row sums, but also all the *column sums are equal to one*, i.e., for any j, $\sum_i p_{ij} = 1$. The uniform distribution

$$\boldsymbol{\pi} = \left(\frac{1}{m}, \ldots, \frac{1}{m}\right) \tag{3.31}$$

is always invariant for an MC with a doubly stochastic transition matrix (verify that!).

Example 3.15 (continued.) What is the long-run proportion of rainy days in our second weather model? Drawing the transition diagram of the MC, we easily see that it is irreducible and aperiodic (loops!), and hence ergodic. The first three equations of the system (3.29) are (here 1 corresponds to state "11", 2 to "12" *etc.*):

$$\pi_1 = 0.8\pi_1 + 0.75\pi_3,$$
$$\pi_2 = 0.2\pi_1 + 0.25\pi_3,$$
$$\pi_3 = 0.5\pi_2 + 0.4\pi_4,$$

and together with $1 = \pi_1 + \pi_2 + \pi_3 + \pi_4$ they yield

$$\pi_1 = \frac{15}{28}, \quad \pi_2 = \pi_3 = \frac{1}{7}, \quad \pi_4 = \frac{5}{28}.$$

The proportion of rainy days is equal to the proportion of the time spent by our MC in the set $\{2, 4\}$ ($= \{$"12", "22"$\}$; equivalently, we could take $\{3, 4\}$—why?) given by $\pi_2 + \pi_4 = 9/28 \approx 0.3214$.

Example 3.13 (continued.) Consider the general case of our "recurrent events" example with transition matrix (3.24) (here, we do not assume that all $a_k > 0$). One can assume w.l.o.g. that $d = \text{GCD}\,\{k : a_k > 0\} = 1$ (otherwise we can just *enlarge* the time steps making them equal to d). Then, as we observed, the MC will be aperiodic and irreducible. The recurrence time $\tau(1)$ for state 1 has the distribution of Z_i, so that the MC is ergodic iff

$$\mu = \mathbf{E}\,\tau(1) = \sum_{k=1}^{\infty} k a_k < \infty.$$

System (3.29) for the stationary distribution takes in this example the form

$$\begin{aligned} \pi_1 &= a_1\pi_1 + \pi_2, \\ \pi_2 &= a_2\pi_1 + \pi_3, \\ \pi_3 &= a_3\pi_1 + \pi_4, \\ &\ldots \end{aligned}$$

From here we infer that

$$\begin{aligned} \pi_2 &= (1 - a_1)\pi_1, \\ \pi_3 &= \pi_2 - a_2\pi_1 = \pi_1(1 - a_1 - a_2), \\ \pi_4 &= \pi_3 - a_3\pi_1 = \pi_1(1 - a_1 - a_2 - a_3), \\ &\ldots \end{aligned}$$

so that $\pi_k = r_k \pi_1$, where

$$r_k = a_k + a_{k+1} + \cdots = \mathbf{P}(Z_1 > k-1) = 1 - F_Z(k-1)$$

is the tail of the distribution of Z_1. Since

$$1 = \sum_{k\geq 1} \pi_k = \pi_1 \sum_{k\geq 1} r_k = \pi_1 \underbrace{\int_0^\infty (1 - F_Z(x))\,dx}_{\mathbf{E}\,Z_1} = \pi_1 \mu$$

from (2.54), we get $\pi_1 = 1/\mu$ (which actually is an immediate consequence of (3.30)), and therefore the stationary distribution has the form

$$\pi_k = \frac{1}{\mu}(1 - F_Z(k-1)), \quad k = 1, 2, \ldots \tag{3.32}$$

Example 3.18. A general random walk with jumps $0, \pm 1$. Consider an MC on non-negative integers whose transitions can occur to the neighbouring states only. In other words, the transition matrix of the chain has the form

$$P = \begin{pmatrix} p_{00} & p_{01} & 0 & 0 & 0 & \cdots \\ p_{10} & p_{11} & p_{12} & 0 & 0 & \cdots \\ 0 & p_{21} & p_{22} & p_{23} & 0 & \cdots \\ 0 & 0 & p_{32} & p_{33} & p_{34} & \cdots \\ 0 & 0 & 0 & p_{43} & p_{44} & \cdots \\ \cdot & \cdot & \cdot & \cdot & \cdot & \ddots \end{pmatrix}. \tag{3.33}$$

Assume that $p_{j,j\pm 1} > 0$ for all $j > 0$, $p_{01} > 0$, and at least one of the diagonal elements is positive: $p_{jj} > 0$ for some $j \geq 0$. Then the MC is irreducible and aperiodic. The system (3.29) again has a simple form:

$$\begin{aligned} \pi_0 &= p_{00}\pi_0 + p_{10}\pi_1, \\ \pi_1 &= p_{01}\pi_0 + p_{11}\pi_1 + p_{21}\pi_2, \\ \pi_2 &= p_{12}\pi_1 + p_{22}\pi_2 + p_{32}\pi_3, \\ &\cdots \end{aligned}$$

The first equation yields

$$\pi_1 = \frac{1 - p_{00}}{p_{10}} \pi_0 = \frac{p_{01}}{p_{10}} \pi_0.$$

Substituting this expression for π_1 in the second equation, we get

$$\pi_2 = \frac{1}{p_{21}} \Big(\underbrace{(1 - p_{11})}_{p_{10}+p_{12}} \pi_1 - p_{01}\pi_0 \Big) = \frac{p_{01}((p_{10} + p_{12}) - p_{10})}{p_{21}p_{10}} \pi_0 = \frac{p_{01}p_{12}}{p_{21}p_{10}} \pi_0,$$

and so on. The general formula we derive in this way is

$$\pi_j = K_j \pi_0, \quad K_j = \frac{p_{01}p_{12}\cdots p_{j-1,j}}{p_{j,j-1}\cdots p_{21}p_{10}}, \quad j \geq 1. \tag{3.34}$$

Since $1 = \sum_{j=0}^{\infty} \pi_j = \pi_0 \sum_{j=0}^{\infty} K_j$ (setting $K_0 := 1$), we obtain that

$$\pi_j = \frac{K_j}{\sum_{r=0}^{\infty} K_r} \quad \text{if} \quad \sum_{r=0}^{\infty} K_r < \infty. \tag{3.35}$$

Then the MC is *ergodic* with the stationary distribution given by relations (3.34)–(3.35). If the last series in (3.35) diverges, the MC is either *transient* and will escape to infinity (this is the case, for instance, when all $p_{jj} = 0$, $p_{j,j+1} = p > 1/2$, $j \geq 1$, which corresponds to the reflected simple RW with a positive drift) or *null-recurrent* (which will be the case for a reflected RW with $p = 1/2$: in that case the drift is equal to zero, and the chain keeps visiting all the states, but does it *so rarely* that the proportion of the time spent at each of them vanishes in the long run). If the drift in such an RW is negative ($p < 1/2$), the MC is always ergodic.

Direct calculation of mean recurrence times is typically a very hard problem. Verifying if the system (3.29) has a unique solution is not a simple task either. Therefore it is important to have (relatively simply verified) conditions sufficient (along with irreducibility) for ergodicity of MCs. In particular, that can be the presence of a "mean drift" towards the origin once the value of the MC is large enough. It is interesting that such a condition can be stated even when adding/subtracting the values assumed by the MC is meaningless (the encoding numbers do not represent any numeric characteristics to which one can apply arithmetic operations); the condition is similar to the classical Ljapunov stability condition for differential equations.

Theorem 3.5. (Foster criterion) *Let $\{X_n\}$ be an irreducible MC and there exist a non-negative function V on S and a finite subset $C \subset S$ such that, for some $\varepsilon > 0$,*

$$\mathbf{E}\left(V(X_1) - V(j) | X_0 = j\right) < \begin{cases} \infty & \text{for } j \in C, \\ -\varepsilon & \text{for } j \notin C. \end{cases}$$

Then $\{X_n\}$ is ergodic.

Checking such conditions can be a much easier task, and then the criterion is easy to apply.

Why does the criterion work? Simply put, it implies that C is a "positive recurrent" set for the MC which, in turn, implies ergodicity. Indeed, when X_n is *outside* C, there is a "uniformly strong" negative trend in the sequence of the values $V(X_n)$. So, if X_n stayed outside C all the time, that would eventually make $V(X_n)$ negative! Since that is impossible (as $V \geq 0$!), the chain must "regularly" visit C to break the "negative trend tradition".

Another powerful approach to establishing ergodicity is based on the so-called *coupling*. Suppose there exists a stationary MC with a given transition matrix P. Suppose we can "run" an MC with the same transition matrix P and an arbitrary initial distribution on a common sample space with the stationary MC. If we can show that the trajectories of the two MCs will meet w.p. 1, that would imply that our MC is ergodic, for we can "couple" the trajectories of the two MCs after the (random) time when they meet. A detailed exposition of the approach can be found, e.g., in Lindvall (1992).

There exist other popular conditions as well, including the so-called Doeblin condition and Harris recurrence. But discussing them is beyond the scope of this text.

Computing the stationary distributions is often (especially when the state space is not "small") a tedious problem. A discussion of some approaches to doing that can be found, e.g., in Sections 3.6 and 3.7 of Kullkarni (1995) and in Chapter 5 of Heyman and Sobel (1990). For an algebraic treatment of the problem see, e.g., Chapter 16 in Feller (1970).

So far we have been dealing with the limiting behaviour of *irreducible chains*. What could one say about that of **reducible** ones?

Firstly, as we know, once an MC enters the class S_{Ess} of essential states, it stays in it forever. Secondly, if there are more than one closed class of essential states, the first class S_r the MC enters becomes the only set of states it can visit from that time on.

Inside the class S_r, if the corresponding *stochastic submatrix* P_r of P (consisting of all entries p_{jk}, $j, k \in S_r$) is aperiodic and recurrent, with a finite mean recurrence time for at least one $j \in S_r$, then this "subchain" is ergodic, and

$$\mathbf{P}(X_n = j | X_0 \in S_r) \to \pi_j^{(r)} \quad \text{as} \quad n \to \infty \quad \text{for} \quad j \in S_r,$$

for some $\pi_j^{(r)} \geq 0$ with $\sum_{j \in S_r} \pi_j^{(r)} = 1$, and

$$\mathbf{P}(X_n \notin S_r | X_0 \in S_r) = 0.$$

The limiting distribution of the MC will then be a mixture of the distributions $\boldsymbol{\pi}^{(r)}$ on the disjoint sets S_r, with the weights $a_r = \mathbf{P}(X_\tau \in S_r)$, where $\tau := \min\{n \geq 0 : X_n \in S_{Ess}\}$.

If, however, P_r corresponds to a *periodic* MC with a period $d > 1$, which is yet recurrent with a finite mean recurrence time for at least one state $j \in S_r$, then (given $X_0 \in S_r$), for any k, the MC X_{nd+k}, $n = 0, 1, 2, \ldots$ is ergodic. Let $S_r^{(l)}$, $l = 1, \dots, d$, be the cyclic subclasses of S_r. Then, using the operation of addition modulo d defined by

$$x + y \pmod d = x + y - \lfloor (x+y)/d \rfloor d, \quad x, y \in \{0, 1, \dots, d-1\},$$

where $\lfloor z \rfloor$ denotes the integer part of z, we can state that, for any $m, k = 0, 1, \dots, d-1$ and $j \in S_r^{((m+k \pmod d))+1)}$,

$$\mathbf{P}\left(X_{nd+k} = j \,\middle|\, X_0 \in S_r^{(m+1)}\right) \to \widetilde{\pi}_j^{(r)} \quad \text{as } n \to \infty,$$

for some $\widetilde{\pi}_j^{(r)} \geq 0$ with $\sum_{j \in S_r^{(l)}} \widetilde{\pi}_j^{(r)} = 1$ for any $l = 1, 2, \dots, d$. On the other hand, clearly

$$\mathbf{P}\left(X_{nd+k} \notin S_r^{((m+k \pmod d))+1)} \,\middle|\, X_0 \in S_r^{(m+1)}\right) = 0.$$

Yet one more possibility is that S_r is *null recurrent*, i.e., there is no state $j \in S_r$ with a finite mean recurrence time (as in the symmetric RW of Example 3.7). Then there is no limiting probability distribution for $\{X_n\}$: as $n \to \infty$, all the values $p_{ij}^{(n)} \to 0$, $i, j \in S_r$.

In conclusion of this section we will give two more related results for MCs, of which the first, in fact, has already been informally stated before Theorem 3.3. To understand why the second one takes place, it may help to compare it with Theorem 8.2 from Chapter 8 on renewal theory.

Theorem 3.6. (i) *Suppose that $\{X_n\}$ is an ergodic MC with stationary distribution $\boldsymbol{\pi}$. Then, for the number of visits $V_j(n) = \#\{t \leq n : X_t = j\}$ to any fixed state j of the MC, one has*

$$\frac{V_j(n)}{n} \to \pi_j \quad a.s. \text{ as } n \to \infty. \tag{3.36}$$

(ii) *Moreover, if $\pi_j > 0$ and $\sigma^2 := \operatorname{Var}(\tau_1(j) | X_0 = j) < \infty$, then we also have the CLT for the number of visits to j: for any $k \in S$, uniformly in $x \in \mathbf{R}$,*

$$\mathbf{P}\left(\frac{V_j(n) - \pi_j n}{\sigma \pi_j^{3/2} \sqrt{n}} \leq x \,\middle|\, X_0 = k\right) \to \Phi(x) \quad \text{as } n \to \infty.$$

Thus the stationary probabilities give us the proportions of the time the MC spends in particular states (or sets of states). Using this information, one can, in particular, optimise the long-run behaviour of systems modelled by MCs in the sense of minimising the relevant average costs.

Suppose that, for the modelled system, its staying in state j for one time unit incurs a cost of $f(j)$, f being a function defined on the state space S.

What is the average/expected cost per time unit? When such a question is asked, it is usually tacitly assumed that it refers to the stationary regime of the MC (or, which is basically the same for ergodic chains, to the long-run behaviour of the MC). The cost related to the nth time unit is $f(X_n)$, and provided that $X \sim \boldsymbol{\pi}$, the expected cost is

$$\mathbf{E}\, f(X_n) = \sum_{j\in S} f(j)\,\pi_j. \tag{3.37}$$

Note also that if the MC is ergodic, the long-run time average cost will, by Theorem 3.6, coincide with the "space" average (3.37).

Example 3.19. Suppose that maintaining a certain component A of a complex system during the kth time unit of the component's lifetime costs c_k, $k = 1, 2, \ldots$ The replacement of A by a new component costs R. We know that a component of that type whose age is $j-1$ (i.e., a component that has kept functioning during the first $j-1$ time units of its life) survives the jth time unit w.p. $r_j \in (0,1)$ (and fails w.p. $1-r_j$). The adopted replacement policy is as follows:

(i) if A fails, we immediately replace it by a new component;

(ii) we never use a component which is more than N time units old.

The question is, what value of N minimises the average costs?

First we have to describe the modelling process. To evaluate the average costs, we need to find the modelling process' stationary distribution, and to this end we must first show that the process is actually an ergodic MC.

Denote by X_n the age of the component A functioning at time n, so that $X_n = j$ if the nth time unit is the jth one in the life of the currently working A. It is easy to see from the problem conditions that $\{X_n\}$ is an MC with the state space $S = \{1, 2, \ldots, N\}$ and transition matrix

$$P = \begin{pmatrix} 1-r_1 & r_1 & 0 & 0 & \cdots & 0 \\ 1-r_2 & 0 & r_2 & 0 & \cdots & 0 \\ 1-r_3 & 0 & 0 & r_3 & \cdots & 0 \\ \vdots & \vdots & \vdots & \vdots & \ddots & \vdots \\ 1-r_{N-1} & 0 & 0 & 0 & \cdots & r_{N-1} \\ 1 & 0 & 0 & 0 & \cdots & 0 \end{pmatrix}.$$

Clearly, this finite MC is irreducible and aperiodic, and hence ergodic by Corollary 3.2. From the system $\boldsymbol{\pi} = \boldsymbol{\pi} P$ we immediately obtain that, for $j \geq 1$,

$$\pi_{j+1} = r_j \pi_j = r_j r_{j-1} \pi_{j-1} = \cdots = \underbrace{r_j r_{j-1} \cdots r_1}_{v_{j+1}} \pi_1, \quad v_1 = 1.$$

The last equation in (3.29) yields

$$1 = \sum_{j=1}^{N} \pi_j = \pi_1 \sum_{j=1}^{N} v_j, \quad \pi_1 = \left(\sum_{j=1}^{N} v_j\right)^{-1}, \quad \pi_j = v_j \pi_1, \quad j = 2, \ldots, N.$$

Now the long run average costs per time unit will be equal to the expected costs:

$$\mathbf{E}\,(\text{costs per time unit}) = \sum_{j=1}^{N} c_j \pi_j + R\pi_1 = \frac{\sum_{j=1}^{N} c_j v_j + R}{\sum_{j=1}^{N} v_j},$$

and we have to *minimise* this expression in N.

In the general case, one can only solve the problem numerically. We will now consider the special case when, for some $a > 1$ and $r \in (0, 1)$,

$$c_j = a^{j-1}, \quad j \geq 1, \qquad r_j = r, \quad j \geq 1.$$

In words, we are assuming that the maintenance costs increase exponentially fast with ageing, while the reliability of the maintained components remains at the same level.

Then $v_j = r^{j-1}$, $j \geq 1$, and

$$\sum_{j=1}^{N} v_j = \sum_{j=0}^{N-1} r^j = \frac{r^N - 1}{r - 1}, \quad \sum_{j=1}^{N} c_j v_j = \sum_{j=0}^{N-1} (ar)^j = \frac{(ar)^N - 1}{ar - 1},$$

so that

$$\mathbf{E}\,(\text{cost per time unit}) = \frac{r - 1}{r^N - 1}\left(\frac{(ar)^N - 1}{ar - 1} + R\right) =: C(N),$$

which is to be minimised in N.

It is not hard to see that the right-hand side here is a strictly convex function $C(N)$ of the (continuous) variable N, and hence (i) if the minimum exists, it is unique, and (ii) to find it, one can just differentiate $C(N)$ and find the solution N^* to $C'(N) = 0$. As we are only interested in integer values N, the only possible candidates for the solution are the two neighbouring integers $\lfloor N^* \rfloor$ and $\lfloor N^* \rfloor + 1$.

Letting, for example, $a = 1.1$, $r = 0.9$ and $R = 5$, and solving $C'(N) = 0$ for N yields $N^* \approx 8.9$. Checking the values 8 and 9, we find that the minimum is attained at $N = 9$.

Example 3.20. Hans-Jürgen writes feature stories for a daily newspaper. On average, it takes him a day to come up with a new topic. He shows it to his boss, and if the latter does not like the proposal (and he does not like 70% of Hans-Jürgen's new ideas), Hans-Jürgen abandons the topic and starts looking for a new one.

If the boss approves the proposal, Hans-Jürgen starts writing the story up (which requires him, on average, three days). The boss approves 60% of Hans-Jürgen's finished stories (of which the topics have been previously approved). Each approved story is published, earning Hans-Jürgen \$200.

How much does Hans-Jürgen earn p.a by writing for the newspaper? (Assume that each year Hans-Jürgen works 300 days.)

Here we simply have to find the average duration T of a "successful writing cycle": how long it takes Hans-Jürgen to earn his \$200 payment. Once this is done, the answer is $\$200 \times 300/T$.

One can model the situation using an MC with the following states: 1 = "looking for a new topic", 2 = "writing up", and 3 = "successful finish". From the conditions we infer the following transition matrix:

$$P = \begin{pmatrix} 0.7 & 0.3 & 0 \\ 0.4 & 0 & 0.6 \\ 1 & 0 & 0 \end{pmatrix},$$

with the associated "costs" (which are now times Hans-Jürgen "pays" at the respective stages) $f(1) = 1$, $f(2) = 3$, and $f(3) = 0$.

The MC is clearly irreducible and aperiodic, and hence ergodic. The stationary distribution $\boldsymbol{\pi}$ satisfies (3.29) which now takes the form (discarding the first equation in $\boldsymbol{\pi} = \boldsymbol{\pi} P$):

$$\pi_2 = 0.3\pi_1, \quad \pi_3 = 0.6\pi_2, \quad \pi_1 + \pi_2 + \pi_3 = 1.$$

This yields

$$\pi_1 = \frac{25}{37}, \quad \pi_2 = \frac{15}{74}, \quad \pi_3 = \frac{9}{74}.$$

Therefore the average "cost" (i.e., time!) per step is

$$\sum_{j=1}^{3} f(j)\,\pi_j = 1 \times \frac{25}{37} + 3 \times \frac{15}{74} + 0 \times \frac{9}{74} = \frac{95}{74}.$$

The average length of a "successful cycle" is equal to the mean number of steps required for state 3 to recur, which is given by the reciprocal of the stationary probability (see Theorem 3.3):

$$\mu_3 = \frac{1}{\pi_3} = \frac{74}{9} \quad \text{(steps)},$$

so the mean "cost" per such cycle is $T = \frac{95}{74} \times \frac{74}{9} = \frac{95}{9}$ (days). Therefore the average annual earnings are

$$200 \times \frac{300}{T} \approx 5,684.21.$$

Alternatively, we could model the situation using an MC with the transition matrix

$$P = \begin{pmatrix} 0.7 & 0.3 & 0 \\ 0.4 & 0 & 0.6 \\ 0 & 0 & 1 \end{pmatrix},$$

so that finishing a story means absorption at state 3. Using the same cost function f, we see that the total average "cost" till absorption (which is nothing else but the average duration of the "story cycle") is

$$T = \mathbf{E}\left(\sum_{n=0}^{\infty} f(X_n) \middle| X_0 = 1 \right) = \sum_{n=0}^{\infty} \mathbf{E}\left(f(X_n)| X_0 = 1\right)$$
$$= \sum_{n=0}^{\infty} \sum_{j=1}^{3} f(j)\, p_{1j}^{(n)} = \sum_{j=1}^{2} f(j) \sum_{n=0}^{\infty} p_{1j}^{(n)} \quad \text{since } f(3) = 0.$$

To find $p_{1j}^{(n)}$, $j = 1, 2$, we note that they are the first two entries in the first row of the matrix $\sum_{n=0}^{\infty} P^n$ or, which is the same, the entries in the first row of the matrix

$$\sum_{n=0}^{\infty} \widetilde{P}^n = (I - \widetilde{P})^{-1}, \qquad \widetilde{P} = \begin{pmatrix} 0.7 & 0.3 \\ 0.4 & 0 \end{pmatrix}.$$

Here $\widetilde{P}$ is the submatrix of P consisting of the elements on the intersection of the first two rows and two columns of P. We used here the fact that the corresponding submatrix of P^n coincides with $\widetilde{P}^n$, and all this was done because $(I - P)^{-1}$ does not exist, while $(I - \widetilde{P})^{-1}$ does, the reason being that P is a *stochastic matrix* and hence has the eigenvalue 1, whereas $\widetilde{P}$ is *defective* (the second row sum is equal to $0.4 < 1$, and hence the maximum in absolute value eigenvalue of $\widetilde{P}$ is less than one—why?).

One easily finds that $\det(I - \widetilde{P}) = 0.18$, and hence

$$(I - \widetilde{P})^{-1} = \frac{1}{0.18} \begin{pmatrix} 1 & 0.3 \\ 0.4 & 0.3 \end{pmatrix},$$

so that $T = 1 \times 1/0.18 + 3 \times 0.3/0.18 = 95/9$, the same value as obtained using the first model.

3.5 Random Walks

Recall that an RW is an MC of the form $X_{n+1} = X_n + Y_{n+1}$ with Y_n being (not necessary integer-valued) i.i.d. RVs (so in the general case the state space is the entire $\mathbf{R}$); we also assume (w.l.o.g.) that $X_0 = 0$. Random walks form a very important class of SPs widely used in various applications. In this section, we will discuss a few general results for RWs (for more detail the reader is referred to the books by Feller (Vol.2, 1970) and Borovkov (2013)) and solve the classical "gambler's ruin problem".

Denote by

$$W_+ := \sup_{n\geq 0} X_n, \quad W_- := \inf_{n\geq 0} X_n$$

the global maximum and minimum of the RW, respectively, and by

$$T_+ := \inf\{n > 0 : X_n > 0\}, \quad T_- := \inf\{n > 0 : X_n < 0\}$$

the times of the first positive and negative sums, respectively. Note that $W_+ = 0$ ($W_- = 0$) iff $T_+ = \infty$ ($T_- = \infty$), i.e., when the RW never takes a positive (negative) value.

The following trichotomy holds true: either

(i) $W_+ < \infty$ and $W_- = -\infty$ a.s., $\mathbf{E}\, T_- < \infty$, or
(ii) $W_+ = \infty$ and $W_- = -\infty$ a.s., $\mathbf{E}\, T_+ = \mathbf{E}\, T_- = \infty$, or
(iii) $W_+ = \infty$ and $W_- > -\infty$ a.s., $\mathbf{E}\, T_+ < \infty$.

Thus, in case (ii) we have an *oscillating* ("between $\pm\infty$") RW, while in cases (i) and (iii) the walk merely drifts away to $-\infty$ and ∞, respectively.

Moreover, (i) is equivalent to each if the two relations $\mathbf{P}\,(W_+ = 0) > 0$ and $\sum_{n=1}^{\infty} n^{-1}\mathbf{P}\,(X_n > 0) < \infty$. Furthermore, one can prove that

$$\mathbf{P}\,(W_+ = 0) = \mathbf{P}\,(T_+ = \infty) = \exp\left\{-\sum_{n=1}^{\infty} n^{-1}\mathbf{P}\,(X_n > 0)\right\} \qquad (3.38)$$

and

$$\mathbf{E}\, T_- = \exp\left\{\sum_{n=1}^{\infty} n^{-1}\mathbf{P}\,(X_n \geq 0)\right\}.$$

Note that, due to the (strong) Markov property, when the first positive sum appears in our RW, it sort of "starts anew" at the point with coordinates (time $= T_+$, location $= S_{T_+}$):

$$\{X_n^*\}_{n\geq 0} := \{X_{T_+ + n} - X_{T_+}\}_{n\geq 0}$$

is an independent copy of $\{X_n\}_{n\geq 0}$ (given $T_+ < \infty$). In this new RW $\{X_n^*\}$, there again may appear (or not appear) the first positive sum *etc.* When $\mathbf{P}(W_+ = 0) > 0$, there will only be finitely many such "first positive sums". Using an argument similar to the one employed to establish (3.19), one can easily see that the total number of such "ladder epochs" is geometrically distributed. Accordingly, the distribution of W_+ is actually a *geometric mixture* of the form:

$$\mathbf{P}(W_+ \leq x) = (1-b)\sum_{n=0}^{\infty} b^n \mathbf{P}(Z_n \leq x), \quad b := \mathbf{P}(W_+ > 0),$$

where $Z_0 = 0$, $Z_n = \zeta_1 + \cdots + \zeta_n$, $n \geq 1$, and ζ_j are i.i.d. RVs having the same distribution as the first positive sum given $W_+ > 0$ (or, which is the same, given $T_+ < \infty$):

$$\mathbf{P}(\zeta_j \leq x) = \mathbf{P}(S_{T_+} \leq x |\, T_+ < \infty).$$

In particular, when the RW $\{X_n\}$ is integer-valued with $Y_n \leq 1$ a.s. (the case of the so-called "skip-free" RW), so that the first positive sum is always equal to 1: $\mathbf{P}(\zeta_j = 1) = 1$, the global maximum W_+ has the geometric distribution:

$$\mathbf{P}(W_+ \geq m) = b^m, \quad m = 0, 1, 2, \ldots \tag{3.39}$$

A sufficient condition for (i) to hold is that $a = \mathbf{E}\,Y_1 < 0$ (a negative mean drift in the RW).

Similar statements hold for (iii).

A sufficient condition for (ii) is that $a = 0$ (no drift). Since when the jumps Y_n have a finite expectation a, one can easily "remove" the drift by forming a new RW $X_n' = X_n - an$, the numerous limit theorems for the "zero drift" case apply, in an appropriately modified form, to the general case as well. So, in many aspects, one can restrict oneself to considering the zero drift case only. We have already discussed the LLN and CLT and the functional variants thereof in Sections 2.9 and 2.10.

As we said, in case (ii) both T_+ and T_- have infinite means. Note, however, that their minimum has not only a finite mean, but even a finite exponential moment! Indeed,

$$\begin{aligned}\mathbf{P}(\min\{T_+, T_-\} > n) &= \mathbf{P}(X_1 = X_2 = \cdots = X_n = 0)\\ &= \mathbf{P}(Y_1 = Y_2 = \cdots = Y_n = 0) = \mathbf{P}(Y_1 = 0)^n\end{aligned}$$

by independence of the Y_j's.[9] That $\mathbf{E}\,T_+ = \infty$ means that, on average, one has to wait for a long time till the RW will take a positive value for the first time (but this will

[9]Note that such a situation would be impossible should the RVs $T_\pm$ be independent. As an exercise, show that if T_1 and T_2 are independent RVs with $\mathbf{E}\,T_j = \infty$, $j = 1, 2$, then $\mathbf{E}\,\min\{T_1, T_2\} = \infty$ as well.

certainly happen). Interpreting X_n as the capital of one of the two players in a gambling example (in a "fair game", when the outcomes of successive plays are represented by the Y_n's with the mean $a = \mathbf{E}\, Y_n = 0$), we can say that the time intervals during which one of the players will be *ahead all the time* tend to increase (but note that, alongside with that general tendency, there will be infinitely many "very short" intervals of that kind!). For the "skip-free" walks, when Y_n are integer-valued and their maximum possible value is one, we have the following remarkable relation for the distribution of T_+:

$$\mathbf{P}\,(T_+ = n) = \frac{1}{n}\mathbf{P}\,(X_n = 1), \quad n = 1, 2 \ldots$$

This is a special case of the more general formula

$$\mathbf{P}\,(\min\{m:\ X_m \ge k\} = n) = \frac{k}{n}\mathbf{P}\,(X_n = k), \quad 1 \le k \le n = 1, 2, \ldots, \tag{3.40}$$

for the skip-free RWs .

Another important special case where explicit formulae for the distribution of T_+ are available is when the "right tail" of Y_n is geometric or exponential. This is closely related to the fact that in that case the time of the first crossing a fixed level and the value of the RW at that time turn out to be independent RVs. The right tail of the global maximum W_+ is than also geometric or exponential, respectively.

One more interesting (and counter-intuitive) related result describes the character of the *fluctuations* of oscillating RWs . Denote by

$$\theta_n := \min\{k \ge 0:\ X_k = \max_{m \le n} X_m\},$$

$$\nu_n := \#\{k \le n:\ X_k > 0\} \equiv \sum_{k=1}^{n} \mathbf{1}_{\{X_k > 0\}}$$

the time of the (first) maximum of the RW on $[0, n]$ and the time spent by the RW above zero on that time interval, respectively. Than one can derive from the functional CLT that the following **arcsine law** holds: if $\mathbf{E}\, Y_j = 0$ and $\mathrm{Var}\,(Y_j) < \infty$, then, for any $t \in [0, 1]$, as $n \to \infty$,

$$\mathbf{P}\left(n^{-1}\theta_n \le t\right) \to \frac{2}{\pi}\arcsin\sqrt{t}, \quad \mathbf{P}\left(n^{-1}\nu_n \le t\right) \to \frac{2}{\pi}\arcsin\sqrt{t}. \tag{3.41}$$

Another sufficient condition for (3.41) is that the distribution of the jumps Y_j is symmetric. The common limit is called the *arcsine distribution.* This is the *exact distribution* of the respective times for the (limiting for the RW) Brownian motion process on $[0, 1]$.

Note that the density of the arcsine law[10]

$$f(t) = \frac{1}{\pi\sqrt{t(1-t)}}, \qquad t \in (0, 1),$$

[10]The arcsine law is a special case (corresponding to the parameter values $a = b = 1/2$) of the so-called *beta distribution* that has density of the form $C_{a,b}t^{a-1}(1-t)^{b-1}$, $t \in (0, 1)$, with parameters $a, b > 0$. The name of the distribution is due to the fact that $C_{a,b}^{-1} = \mathrm{B}(a, b)$ is the popular *beta function.*

is symmetric with respect to the point $t = 1/2$ and "U-shaped" (with $f(t) \to \infty$ as $t \to 0$ or $t \to 1$). This means that, say, in our gambling example, it is more likely that one of the players is ahead most of the time than that the proportion of the time when the first player is ahead (or behind) is close to 1/2 (in other words, troubles seldom come singly). Also, the maximum "lead" in a finite series of plays of a fixed length is achieved either quite soon after beginning or close to the end.

In conclusion of this section, we will analyse the behaviour of an RW with *absorbing barriers.*

Example 3.21. *The gambler's ruin problem.* Suppose our gambler has the initial capital of N dollars and aims to win (extra) M dollars and then stop playing. Of course, if he goes bankrupt before achieving the goal, he also stops playing.

He finds a slot machine which gives "odds" of p of winning a dollar for every dollar one puts in (and w.p. $q = 1 - p$ the dollar is lost). We assume that successive games are independent.

What is the probability that the gambler will win the desired extra M dollars before he goes bankrupt? What is the expected duration of the game?

We already know that the process modelling such a game is an RW; the only new thing is that now there are constraints on the possible values of the walk.

Let X_n be the gambler's capital at time n. The natural state space is $S = \{0, 1, \ldots, M + N\}$, the states 0 and $M + N$ being *absorbing.* It is easy to see that all other states are non-essential.

To answer the first question, denote by

$$u_i := \mathbf{P}\,(\text{absorption at } N + M | X_0 = i)$$

the probability of achieving the gambler's goal given that his initial capital is i dollars. We have to find u_N.

Conditioning on the outcome of the first play, we use the TPF to obtain that, for $i = 1, 2, \ldots, M + N - 1$,

$$\begin{aligned} u_i &= \mathbf{P}\,(\text{absorption at } N + M | X_1 = i + 1)\,\mathbf{P}\,(X_1 = i + 1 | X_0 = i) \\ &\quad + \mathbf{P}\,(\text{absorption at } N + M | X_1 = i - 1)\,\mathbf{P}\,(X_1 = i - 1 | X_0 = i) \\ &= pu_{i+1} + qu_{i-1}. \end{aligned}$$

Thus we have the following *boundary problem* for the derived *linear difference equation*:

$$\begin{cases} u_i = pu_{i+1} + qu_{i-1}, & 0 < i < N + M, \\ u_0 = 0, \; u_{N+M} = 1. \end{cases} \tag{3.42}$$

We could solve it using the standard techniques for difference equations.[11] However, in our case, it is perhaps easier to notice that $u_i = pu_i + qu_i$ and hence the first equation in (3.42) is equivalent to the equation

$$qv_i = pv_{i+1}, \quad 0 < i < N + M,$$

for the differences $v_i = u_i - u_{i-1}$. Therefore, putting $a := q/p$, we obtain that $v_{i+1} = av_i = a^2 v_{i-1} = \cdots = a^i v_1$, and hence from the first boundary condition ($u_0 = 0$) one gets

$$u_i = \sum_{j=1}^{i} v_j = v_1 \sum_{j=1}^{i} a^{j-1} = \begin{cases} \dfrac{a^i - 1}{a - 1} v_1 & \text{if } a \neq 1, \\ i v_1 & \text{if } a = 1. \end{cases} \tag{3.43}$$

Using the second boundary condition ($u_{N+M} = 1$), we can now find v_1:

$$1 = u_{N+M} = \begin{cases} \dfrac{a^{N+M} - 1}{a - 1} v_1 & \Longrightarrow u_1 = v_1 = \dfrac{a - 1}{a^{N+M} - 1}, \quad a \neq 1, \\ (N + M) v_1 & \Longrightarrow u_1 = v_1 = \dfrac{1}{N + M}, \quad a = 1, \end{cases}$$

so that

$$u_i = \begin{cases} \dfrac{a^i - 1}{a^{N+M} - 1}, & p \neq q, \\ \dfrac{i}{N + M}, & p = q = 1/2. \end{cases} \tag{3.44}$$

Note that our gambler's *gain* is an RV G such that

$$G = \begin{cases} +M & \text{w.p. } u_N, \\ -N & \text{w.p. } 1 - u_N, \end{cases}$$

and hence the expected gain is given by

$$\mathbf{E}\, G = M u_N - N(1 - u_N) = (M + N) u_N - N.$$

Therefore, one has $\mathbf{E}\, G = 0$ iff $a = 1$, i.e., $p = q = \frac{1}{2}$. A "more realistic" figure is $p = 0.45$ (playing with slot machines is always unfavourable). When the gambler starts with $N = \$50$ and aims to win extra $M = \$10$, we get

$$a = \frac{q}{p} \approx 1.2222, \quad u_{50} \approx 0.1344.$$

[11] See, e.g., Mickens, R.E. *Difference Equations,* Van Nostrand, New York, 1987, or Kelly W.G., and Peterson, A.C. *Difference Equations: An Introduction with Applications,* Academic Press, New York, 1991.

Note that if the initial capital were $N = \$100$ and the gambler wanted to win the same proportion (one fifth) of his fortune before quitting (so that $M = \$20$), the success probability would be much lower:

$$u_{100} \approx 0.0181.$$

What happens is that we have an RW with a negative drift and two absorbing barriers. In the first case, the higher probability of winning is due to the fact that there is a higher chance of hitting the upper boundary first because of the *random fluctuations* in our walk. In the second case, although the ratio of the distances to the boundaries from the starting point is the same, the upper barrier is farther away, while the "scale" of the random fluctuations (the "volatility" of the process) remains the same! And so there is a fat chance that the "volatility" will now be strong enough to cause such a deviation from the "destiny"—the bankruptcy of the gambler which, because of the linear negative drift, will be inevitable should the gambler keep playing—that we will reach the upper boundary before hitting the lower one.

Note also that the results are rather "sensitive" (for obvious reasons, cf. the form of (3.44)) to the changes in the value of p. Thus, for a "fairer" machine with $p = 0.49$, one has $a \approx 1.041$ and $u_{50} \approx 0.6374$ in the first case.

To compute the expected duration

$$\mathbf{E}\,(T|X_0 = N), \quad T := \min\{n : X_n = 0 \ \text{ or } \ N + M\},$$

of the game, put

$$w_i := \mathbf{E}\,(T|X_0 = i).$$

A similar argument using the relation

$$\mathbf{E}\,(T|X_1 = i + 1) = 1 + \mathbf{E}\,(T|X_0 = i + 1), \quad 0 < i < M + N,$$

shows that, in the above range of i-values,

$$\begin{aligned} w_i &= \mathbf{E}\,(T|X_1 = i + 1)\,\mathbf{P}\,(X_1 = i + 1|X_0 = i) \\ &\quad + \mathbf{E}\,(T|X_1 = i - 1)\,\mathbf{P}\,(X_1 = i - 1|X_0 = i) \\ &= (1 + w_{i+1})p + (1 + w_{i-1})q = 1 + pw_{i+1} + qw_{i-1} \end{aligned}$$

with the boundary conditions $w_0 = w_{N+M} = 0$. It is not hard to verify that that boundary problem has a unique solution

$$w_i = \begin{cases} \dfrac{i}{q-p} - \dfrac{M+N}{q-p}\dfrac{a^i - 1}{a^{N+M} - 1}, & p \neq q, \\ i(M + N - i), & p = q = 1/2. \end{cases} \tag{3.45}$$

In our numerical illustration, for $p = 0.45$ we get

$$w_{50} \approx 419.34 \quad \text{when} \quad N = 50, \; M = 10;$$
$$w_{100} \approx 978.31 \quad \text{when} \quad N = 100, \; M = 20.$$

Note that in this case the drift $a = \mathbf{E}\,Y_1 = p - q = -0.1$, and the times required for the deterministic trend of that rate to reach the lower boundaries are 500 and 1,000, respectively. The latter value is much closer to the figure $w_{100} \approx 978.31$ we found for the RW than the value 500 to w_{50} in the first case. This is due to the above-mentioned fact that the presence of random fluctuations is less significant in the second case.

For a slot machine with $p = 0.49$, in the second case ($N = 100$) one obtains $w_{100} \approx 2331.72$, while for the deterministic trend with the same rate $a = \mathbf{E}\,Y_1 = -0.02$, the respective hitting time is 5,000. Thus the difference between stochastic and deterministic models in the case of the "fairer" machine is much bigger—which is due to the greater role of randomness in "almost fair" games.

We will encounter a differential equation analogue of the above approach in Chapter 12, when discussing boundary hitting problems for *diffusion processes*.

In conclusion note that the problems from this example admit a much shorter solution exploiting the notion of *martingale* to be discussed in Chapter 11 as well. The interested reader is also referred to any advanced probability textbook listed in Section 2.11.

3.6 Recommended Literature

The first two books below have already been listed in the previous chapter.

BOROVKOV, A.A. *Probability theory.* Springer, New York, 2013. [*A good high level probability textbook.*]

FELLER, W. *An Introduction to Probability Theory and Its Applications.* Wiley, Vol. 1. 1970 [There are several editions; *a classical text on probability.*]

HEYMAN, D.P. AND SOBEL, M.J. *Stochastic models: A handbook in operations research and management science.* North-Holland, New York, 1990. [*Chapter 5 discusses numerical methods for finding the stationary distributions of MCs.*]

KEMENY, J.G. AND SNELL, J.L. *Finite Markov Chains.* Springer, New York, 1976. [*A technically relatively low level textbook on MCs.*]

KULKARNI, V.G. *Modeling and Analysis of Stochastic Systems.* Chapman & Hall, New York, 1995. [*A large number of examples and exercises, of which a few were adapted for the present chapter. Also, the book contains a discussion of numerical methods for computing stationary distributions.*]

LINDVALL, T. *Lectures on the coupling method.* Wiley, New York, 1992.

TAYLOR, H.M. AND KARLIN, S. *An Introduction to Stochastic Modeling.* 3rd edn. Academic Press, New York, 1998. [*Other editions are fine as well.*]

3.7 Problems

1. Show that, in the general case, the Markov property (3.1) *does not* imply that, for any sets A, B and C,

$$\mathbf{P}\left(X_{n+1} \in B | X_n \in A, X_{n-1} \in C\right) = \mathbf{P}\left(X_{n+1} \in B | X_n \in A\right)$$

   (it is not generally true when A consists of more than one point!).

2. Assuming a discrete state space S, restate relations (3.2) and (3.3) in mathematical terms and prove that they are equivalent to (3.1).

3. Clearly, (3.6) is a special case of (3.1). Show that, when the state space is finite or countable, (3.1) follows from (3.6).

4. Prove part (ii) of Theorem 3.2.

5. Explain why the MC from Example 3.12 is recurrent and aperiodic.

6. Show that the discrete uniform distribution (3.31) is invariant for any doubly stochastic matrix.

7. *A gambling example.* Two players A and B own between them the amount of a dollars and play for unit stakes with each other with the agreement that every time a player loses his last dollar, his adversary immediately returns it (the capitals of the players would remain 1 and $a - 1$ after such a play), so that the game can continue forever. In each play, A wins with probability p, $0 < p < 1$, and B wins with probability $q = 1 - p$. Outcomes of all the plays are independent of each other.

   (i) Model the situation as an MC. Draw a transition diagram. Find the transition matrix P. Classify the states of the MC. Say if the MC is ergodic, and if yes, find the stationary distribution.

   (ii) What would change in the model if the players were not so generous and the game stopped when one of them lost his last dollar? Draw a transition diagram and find the transition matrix for the modified MC. Classify the states of the latter.

Hints. (i) Take the state space of the form $S = \{1, 2, \ldots, a-1\}$ and say that the MC is in state k if the capital of A is k (then the fortune of B will, of course, be $a - k$).

8. A little spider lives in a rectangular box of which the sides are 3 cm and 4 cm long. It can only sit in one of the four corners marked with the numbers 1, 2, 3 and 4, as shown on the diagram:

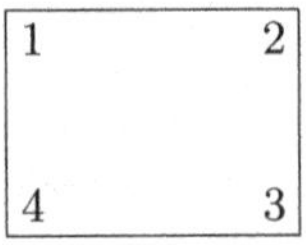

From time to time the spider runs from the corner it occupies to another one, chosen at random with probabilities inversely proportional to the distances to the corner from the current position of the spider. Denote by X_n the number of the corner the spider is at after the nth run. Comment on why the sequence $\{X_n\}$ is a Markov chain.

(i) Find the transition probabilities matrix for the Markov chain $\{X_n\}$.

(ii) Is this chain reducible? Periodic? Ergodic? Explain. If the chain is ergodic, find the stationary distribution.

(iii) Let the initial distribution of X_0 be uniform: $\boldsymbol{p} = (0.25, 0.25, 0.25, 0.25)$. Find the probability $\mathbf{P}(X_1 = 1, X_2 = 4, X_4 = 2)$.

(iv) Assume now that the spider has changed its tactics and, on any given transition, never returns directly back to the corner where it came from on the previous step. Show that the sequence $\{X_n\}$ is not a Markov chain any more. Suggest a Markov chain model for the modified system.

9. A system can be in one of the states 1, 2, 3 and 4. If the system is at state j, $j < 4$, then, on the next step, it passes to state $j + 1$. From state 4, the system passes either to 2 or to 3 with equal probabilities $p_{42} = p_{43} = 1/2$.

(i) Draw a transition diagram for the MC modelling the system. Classify the states of the MC.

(ii) Find the transition matrix P.

(iii) Calculate the n-step transition probabilities for $n = 2$ and $n = 16$.

10. Let

$$P := \begin{pmatrix} 0 & 0 & 0 & ? & ? & 0 & 0 & 0 & 0 & 0 \\ 0 & 0 & 0 & 0 & 0 & 0 & 0 & ? & 0 & 0 \\ 0 & 0 & ? & 0 & 0 & 0 & 0 & 0 & 0 & 0 \\ ? & 0 & 0 & 0 & 0 & 0 & 0 & 0 & 0 & 0 \\ ? & 0 & 0 & 0 & 0 & 0 & 0 & 0 & 0 & 0 \\ 0 & 0 & 0 & 0 & 0 & 0 & ? & 0 & 0 & 0 \\ 0 & 0 & 0 & 0 & 0 & 0 & 0 & 0 & ? & 0 \\ ? & 0 & ? & 0 & 0 & ? & 0 & 0 & 0 & ? \\ 0 & 0 & 0 & 0 & 0 & ? & ? & 0 & 0 & 0 \\ 0 & ? & 0 & 0 & 0 & 0 & 0 & ? & 0 & 0 \end{pmatrix},$$

where "?" denotes positive entries.

(i) Draw a transition diagram for an MC with the transition matrix P.

(ii) Classify the states of the MC.

(iii) What will change in your answers to (i) and (ii) if we additionally assume that (a) $p_{25} > 0$ or (b) $p_{52} > 0$?

11. *Moving averages.* Let $\{Y_n\}$ be a sequence of i.i.d. RVs each assuming the values ± 1 with probability $p = 1/2$. Put $X_n := (Y_n + Y_{n+1})/2$.

(i) Find the transition probabilities

$$p_{jk}(m,n) = \mathbf{P}(X_n = k | X_m = j), \quad m < n; \quad j,k = -1,0,1.$$

(ii) Show that $\{X_n\}$ is **not** an MC.

Hints. (ii) Calculate $\mathbf{P}(X_n = k | X_{n-1} = j, X_{n-2} = m)$ for reasonably chosen k, j, and m.

12. Let $\{X_n\}$ be an MC with the state space $\{1,2,3\}$, transition matrix

$$P = \begin{pmatrix} 0 & 1/3 & 2/3 \\ 1/4 & 3/4 & 0 \\ 2/5 & 0 & 3/5 \end{pmatrix},$$

and initial distribution $\boldsymbol{p} = (2/5, 1/5, 2/5)$. Compute the following probabilities:

(i) $\mathbf{P}(X_1 = 2,\ X_2 = 2,\ X_3 = 1 | X_0 = 1)$;
(ii) $\mathbf{P}(X_1 = 2,\ X_2 = 2,\ X_3 = 1)$;
(iii) $\mathbf{P}(X_1 = 2,\ X_4 = 2,\ X_6 = 2)$.

13. Let $\{X_n\}$ be a three-state MC with state space $\{1, 2, 3\}$, transition matrix

$$P = \begin{pmatrix} 0 & 1/2 & 1/2 \\ 1 & 0 & 0 \\ 0 & 1 & 0 \end{pmatrix},$$

and initial distribution $\boldsymbol{p} = (1/3, 1/3, 1/3)$. Find

(i) $\mathbf{P}(X_1 = 2)$;

(ii) $\mathbf{P}(X_2 = 2)$;

(iii) $\mathbf{P}(X_3 = 2 | X_0 = 1)$, and

(iv) classify the states of the MC;

(v) compute $\lim_{n\to\infty} \mathbf{P}(X_n = 2)$;

(vi) find the stationary distribution of $\{X_n\}$.

14. Let the state space be $\{1, 2\}$. Classify the states and compute the stationary distribution of $\{X_n\}$ and the limits $\lim_{n\to\infty} \mathbf{P}(X_n = 1)$ when the initial distribution is $\boldsymbol{p} = (1/4, 3/4)$ for the cases when the transition matrix is:

(i)

$$P = \begin{pmatrix} 1/2 & 1/2 \\ 1/4 & 3/4 \end{pmatrix};$$

(ii)

$$P = \begin{pmatrix} 1/2 & 1/2 \\ 0 & 1 \end{pmatrix};$$

(iii)

$$P = \begin{pmatrix} 0 & 1 \\ 1 & 0 \end{pmatrix}.$$

15. A machine produces two items per day. The probability that an item is non-defective is p (all the items are independent), and defective items are thrown away instantly. The demand is one item per day, and any demand that cannot be satisfied by the end of the day is lost, while any extra item is stored. Let X_n be the number of items in storage just before the beginning of the nth day. Model $\{X_n\}$ as an MC.

(i) Draw the transition diagram for $\{X_n\}$.

(ii) Find the transition matrix for this MC.

(iii) When is this MC ergodic? Compute the steady-state distribution of $\{X_n\}$ when it exists.

(iv) Suppose it costs \$$c$ to store an item for one night and \$$d$ for every demand that cannot be fulfilled. Compute the long-run cost rate of the production facility when it is stable.

16. Suppose A and B decided to flip pennies, the one coming closest to the wall wins. Player B has a probability of 0.6 of winning on each flip (the outcomes of the flips are assumed to be mutually independent). If B starts with 5 pennies, and A with 10, then what is the probability that B will wipe A out? What if B starts with 10 and A with 20 pennies?

17. On any given day John is either cheerful (**C**), so-so (**S**), or glum (**G**). If he is cheerful today, then he will be **C** tomorrow w.p. 0.6 and **S** w.p. 0.2. If John is

feeling so-so today, then he will be **S** tomorrow w.p. 0.3 and **G** w.p. 0.5. If he is glum today, then w.p. 0.5 he will be **G** and w.p. 0.5 he will be **C** tomorrow.

Let X_n denote John's mood on the nth day. Then $\{X_n\}_{n\geq 0}$ is a three-state MC (state 1 = **C**, state 2 = **S**, state 3 = **G**).

(i) Draw the transition diagram and find the transition matrix for the chain.

(ii) Find the probability that this week John will be cheerful on Tuesday and Thursday and glum on Sunday given he was so-so on Monday.

(iii) In the long run, what proportion of time is John in each of his three moods?

18. In Example 3.21, if the slot machine were fair ($p = q = 1/2$), what would be the probability of winning extra (i) \$10 when starting with \$50; (ii) \$20 when starting with \$100?

19. Derive the expressions (3.45) for the mean game duration.

20. The so-called *bonus-malus* premium calculation principle is widely spread in car insurance practice. There is a finite number of classes (tariff groups), and the premium a policy holder pays depends on what class s/he belongs to. For each year, the policy holder's class is determined basing on what class s/he belonged to last year and on the number of claims made by this person last year. The policy holder gets a bonus (transfer to a lower class) for no claims and a malus (shifting to a higher class) if there were claims.

    The table[12] on the next page shows the premium scale (in % to the basic premium paid by class 13 drivers) and bonus rules for the German bonus-malus system. There are 18 classes labelled from 1 to 18.

    In practice, one usually assumes that, for any policy holder, the number of claims per year is a random variable following the Poisson distribution whose parameter λ can depend on the policy holder. The random variables are supposed to be independent for different years and different policy holders. In this problem, assume that, for a policy holder Herr Z, the value $\lambda = 1.6$.

[12] Borrowed from T.Rolski et al., *Stochastic Processes for Insurance and Finance*. Wiley, 1999 (p.280).

Class	Premium scale	Class after one year (per no. of claims)				
		0	1	2	3	≥ 4
18	200	13	18	18	18	18
17	200	13	18	18	18	18
16	175	13	17	18	18	18
15	175	13	16	17	18	18
14	125	13	16	17	18	18
13	100	12	14	16	17	18
12	85	11	13	14	16	18
11	70	10	13	14	16	18
10	65	9	12	13	14	18
9	60	8	11	13	14	18
8	55	7	11	13	14	18
7	50	6	11	13	14	18
6	45	5	11	13	14	18
5	40	4	10	12	13	18
4	40	3	9	11	13	18
3	40	2	8	11	13	18
2	40	1	7	11	13	18
1	40	1	7	11	13	18

(i) Find the transition matrix governing the Markov chain describing the tariff group which Herr Z belongs to in consequent years. Classify the states of the chain.

(ii) In 2010 Herr Z belonged to class 11. Find the probability distribution of the class number which he belongs to in 2020. Plot this distribution.

(iii) Redo part (ii) assuming now that Herr Z drove twice as safely as in the first scenario (that is, assume a new $\lambda' := \lambda/2$) and compare the two distributions.

(iv) Suppose that the basic premium (paid by class 13 drivers) is €800 p.a. For the premium Herr Z would have to pay in year 2020 in the scenarios in (ii) and (iii) above, find its expected value and standard deviation. What do you think is likely to happen to these quantities during the years from 2021 to 2030?

(v) In 2010, in the small town of Xburg, there were 400 policy holders from class 11 characterised by the value $\lambda = 1.6$. Find approximate values of the upper and lower quartiles of the distribution of the total amount of the premium those 400 drivers have to pay in 2020.

(vi) In addition to the 400 policy holders from part (v), in 2010 there were also 200 drivers from class 11 in Xburg who are characterised by the value λ' from part (iii). Answer the same question as in (v) but for the total group of the 600 policy holders.

Chapter 4

Markov Decision Processes

In this chapter we will discuss mathematical models enabling one to construct and follow optimal (in a certain sense) strategies in environments evolving in time in the presence of uncertainty. We assume that we have a process describing a system evolving in discrete time, and at each step, one is required to take an action. The consequences of the action are not certain, but the distribution of possible outcomes depends only on the state at which the action was taken. A possible action could simply be the termination of the process at some stage. Graphically, this scheme can be illustrated by an example shown on the diagram in Fig. 4.1.

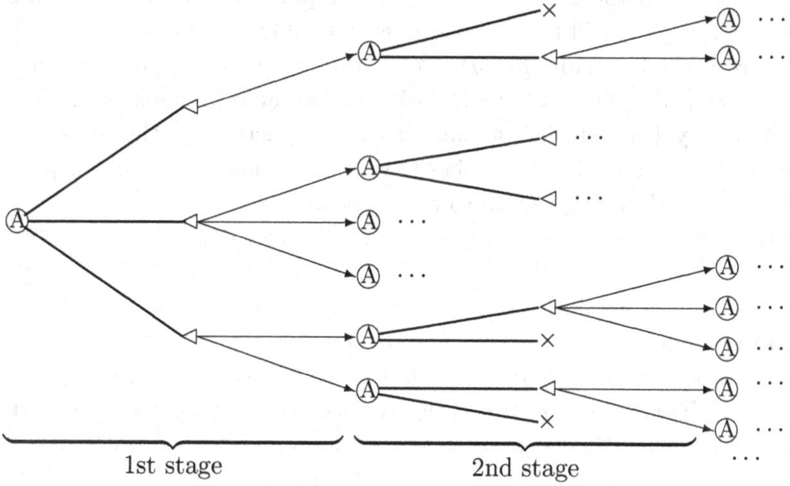

Fig. 4.1 A decision tree.

The circles are "action nodes"; at any such node, we need to take one of a number of possible actions shown by thick edges. After an action is taken, we proceed (along the respective edge) to a "chance node" (they are represented by the symbols $\lhd$). At these nodes, there exists uncertainty in the further evolution of the system, the arrows showing alternative ways. The system proceeds along an arrow chosen at random (according to a distribution specific to the particular chance node). When the evolution is terminated, we have a cross on our diagram. If the decision process is multi-stage, the cycle is repeated: we again have to take an action, and so on. At each action node, we have to base our decision on the information about the evolution of the system up to that node only. The task is to choose a sequence of actions optimising a given objective function.

4.1 Finite-Stage Models

The simplest stochastic decision model involves:

• A **process** $\{X_t\}$ observed over a discrete time period $t = 1, \ldots, T$ (sometimes it will be more convenient for us to consider the time period $t = 0, 1, \ldots, T$), whose states are enumerated by integer numbers $i \geq 0$ (later we will also be dealing with cases where the state space is continuous).

• A (finite) set A of **actions** $a \in A$. If $X_t = i$ and an action $a \in A$ is chosen (basing on the observed values of the process at times $\leq t$), then the next state X_{t+1} will be j with a given probability $p_{ij}(a)$.

• A **reward function** $R(i, a)$, i is a state, a an action. If action a is chosen when $X_t = i$, the reward $R(i, a)$ is earned at the current stage.

• A **policy** $\{a_t\}$ which is a *rule* for choosing actions at the respective times: at time t, the policy prescribes to take the action a_t (that can depend on the history of the process up to that time t).

If the policy is *stationary* (one's action at time t depends on X_t only: $a_t = f(i)$ given $X_t = i$), then $\{X_t\}$ is a time-homogeneous Markov chain with transition probabilities $p_{ij}(f(i))$, and the process is called a **Markov decision process**. One's action may depend on the time t as well; the resulting process will be a MC again, but possibly a non-homogeneous one.

• The **objective** is to maximise the expected value of the sum of rewards earned over a given time span of length T (a "finite horizon" problem):

$$\mathbf{E}\left[\sum_{t=1}^{T} R(X_t, a_t)\right] \longrightarrow \max_{\{a_1,\ldots,a_T\}}.$$

We will achieve this goal if, for any initial value i, we solve the following maximisation problem:

$$\mathbf{E}\left[\sum_{t=1}^{T} R(X_t, a_t)\middle| X_1 = i\right] \longrightarrow \max_{\{a_1,\dots,a_T\}} =: V_T(i), \tag{4.1}$$

and then compute $\mathbf{E}\,V_T(X_1)$.

The maximum values $V_n(i)$ for different n and i form what is called the **optimum value function**.

Sometimes the objective is to *minimise* the expected total "reward" (e.g., costs), but we can always reduce a minimisation problem to a maximisation one by switching the new reward function $\widetilde{R}(\cdot\,,\,\cdot) := -R(\cdot\,,\,\cdot)$.

The **problem** is to find the value $V_n(i)$ for $n = T$ and the optimal policy for which this value is attained. This problem can be solved *recursively* in n using the so-called *dynamic programming* technique. The idea is to split the "global maximisation problem" (when we are looking for the best sequence of actions) into simpler "one-step" maximisation sub-problems (in each of which we just have to find the best action for a single stage). Starting at the end of the planning period ($n = 0$), we will work backwards in time using recursive relations to derive $V_1, V_2, \dots$ and finally the desired V_T.

The main tool here is the **optimality equation** which is a consequence of the TPF: for $n = 1, 2, \dots, T$,

$$\begin{aligned} V_n(i) &= \max_a \Big[R(i,a) + \mathbf{E}_a(V_{n-1}(X_2)|X_1 = i) \Big] \\ &\equiv \max_a \Big[R(i,a) + \sum_j p_{ij}(a) V_{n-1}(j) \Big], \end{aligned} \tag{4.2}$$

where $\mathbf{E}_a$ stands for the expectation under the distribution corresponding to the immediate action a. In words, the total expected reward when there are n stages to go, is given by the maximum (over all actions we can take at the current stage) of the sum of (i) the *immediate reward* (earned *now*) and (ii) the total expected reward over the remaining—after the present step—time interval given that we follow, during that time period, the optimal policy (the second term can be referred to as the *maximum expected future reward*).

Note that, in (4.2), n denotes the *number of steps to go*, so that i represents the value of the process $\{X_t\}$ at time $t = T - n + 1$. Formally, the conditional expectation on the right-hand side of the first line in (4.2) is $\mathbf{E}_a(V_{n-1}(X_{T-n+2})|X_{T-n+1} = i)$, but since the process is homogeneous, we can use X_2 and X_1 instead of X_{T-n+2} and X_{T-n+1}, respectively.

Now to find the optimal sequence of actions, we start at the very end of the time horizon and set $V_0(i) := 0$. Using the optimality equation (4.2), we move *backwards* and find

$$V_1(i) = \max_{a \in A} R(i, a),$$

and then, recursively, all the remaining values of the optimality function:

$$V_T \longleftarrow V_{T-1} \longleftarrow \cdots \longleftarrow V_2 \longleftarrow V_1 \longleftarrow V_0.$$

The *optimal policy* chooses the actions which maximise the expression in the square brackets in (4.2) when the process state is i and there are n time periods to go (that is, $t = T - n + 1$). Note that the expression in the square brackets depends on the current state i of the process only. Therefore the action maximising the expression in $[\cdots]$ in (4.2) will also depend on i only, and hence $\{X_t\}$ will indeed be a Markov process (as observed at the beginning of the present section). For the optimal policy, the expected value of the total reward, as a function of the initial state, will be given by the optimal value function $V_T(\cdot)$.

We will begin illustrating the outlined approach by the following simple three-stage problem.

Example 4.1. *Selling a house.* A person moving overseas has to sell her house urgently. Three different buyers are going to offer her, one after another, their prices, which are believed to be independent and identically distributed random variables Z_j, $j = 1, 2, 3$, with

$$\mathbf{P}(Z_j = 100) = 0.3, \quad \mathbf{P}(Z_j = 110) = 0.5, \quad \mathbf{P}(Z_j = 120) = 0.2$$

(Z_j are given in thousand dollars). If the seller rejects an offer, the offer is lost.

The seller aims at maximising the expected price. The problem is to derive the optimal policy for selling the house and find the maximum expected value of the selling price.

First we need to construct an appropriate process $\{X_t\}$ containing all the information used to make decisions. In particular, it is not enough to know the price offered at the moment; we need to somehow indicate if the house has already been sold or not. We can do that by setting

$$X_t := \begin{cases} Z_t & \text{if not sold yet,} \\ 0 & \text{otherwise,} \end{cases} \qquad t = 1, 2, 3 \ (= T).$$

At each step, there are only two possible actions which will be denoted as follows: $a = 1$ means selling, $a = 0$ means doing nothing. The respective transition probabilities are now

$$p_{x0}(1) = 1, \quad p_{00}(a) = 1 \quad \text{for any } a, \quad p_{xy}(0) = \mathbf{P}(Z_j = y) \quad \text{for any } x \neq 0.$$

Selecting the reward function to be

$$R(x, 1) := x, \quad R(x, 0) := 0,$$

we see that the total additive reward $\sum_{t=1}^{3} R(X_t, a_t)$ will simply be the selling price (only one term in the sum—the one corresponding to the time when the owner sells the property—will be non-zero).

We start by letting $V_0(x) := 0$ which means that if all three buyers' offers have already been refused, one can gain nothing. Next observe that when $x > 0$, from (4.2) we have

$$V_1(x) = \max_a R(x, a) = x,$$

with the maximum corresponding to $a = 1$, and therefore this is the optimal action in that case. So if one has not sold the house to the first two buyers, the property should be sold to the last one: in that case, $a_3 = 1$ whatever the price Z_3. Also, clearly $V_1(0) = 0$ (if the house is already sold, nothing can be gained during the remaining time interval).

To find $V_2(x)$, we again use (4.2). Clearly, $V_2(0) = 0$ ($x = 0$ means that there is nothing to sell when there remain two time periods to go). So one only has to deal with the case $x > 0$, and in this case, as $\mathbf{E}\, V_1(Z_3) = 100 \times 0.3 + 110 \times 0.5 + 120 \times 0.2 = 109$, one has

$$\mathbf{E}_a(V_1(X_3)|X_2 = x) = \begin{cases} \mathbf{E}\, V_1(Z_3) = 109, & a = 0; \\ \mathbf{E}\, V_1(0) = 0, & a = 1, \end{cases}$$

and therefore

$$V_2(x) = \max_a \Big[R(x, a) + \mathbf{E}_a(V_1(X_3)|X_2 = x) \Big] = \max\{x,\, 109\}.$$

Hence the optimal action when the second offer has been made is to sell when $x > 109$ and wait otherwise. That is, $a_2 = 1$ if $X_2 = 110$ or 120 and $a_2 = 0$ if $X_2 = 100$. Note that

$$\mathbf{E}\, V_2(Z_2) = \mathbf{E}\, \max\{Z_2,\, 109\} = 109 \times 0.3 + 110 \times 0.5 + 120 \times 0.2 = 111.7.$$

It remains to find the optimal action at time $t = 1$ and the function $V_3(x)$ (and hence the maximum expected selling price $\mathbf{E}\, V_3(X_1)$). One always has $X_1 = Z_1 > 0$. Therefore, similarly to the previous step,

$$V_3(x) = \max_a \Big[R(x, a) + \mathbf{E}_a(V_2(X_2)|X_1 = x) \Big] = \max\{x,\, 111.7\},$$

so that the optimal action is $a_1 = 1$ (sell) if $X_1 > 111.7$ (i.e., the price is 120) and wait otherwise. The maximum expected selling price is hence

$$\mathbf{E}\, V_3(X_1) = \mathbf{E}\, \max\{Z_1,\, 111.7\} = 111.7 \times (0.3 + 0.5) + 120 \times 0.2 = 113.36.$$

Example 4.2. *An American call option model.* Let X_t denote the price of a given stock on the tth day. Assume that the dynamics of the price are given by the simple (*absolute*) *random walk model*[1]:

$$X_{t+1} = X_t + Y_{t+1} = X_0 + \sum_{j=1}^{t+1} Y_j, \tag{4.3}$$

Y_j being i.i.d. RVs with a common DF F having a finite mean $\mu = \mathbf{E}\, Y_1$.

An **American call** is an option entitling the holder to buy a block of shares ("exercise the option") of a given company at a stated price at any time during a stated time interval[2]. One of the main uses of such options is in *hedging risks*, and pricing them is one of the basic tasks of mathematical finance. To illustrate how a call option could be used, consider the following two simple situations:

• An investor hopes that the price of the stock he/she wants to buy *may drop* in the near future, but is not sure. Hence the investor wants to wait and purchases a call to protect against a near future price rise:

$$\text{Price} \searrow \quad \Longrightarrow \quad \text{buy and } \textit{ignore} \text{ the option.}$$

$$\text{Price} \nearrow \quad \Longrightarrow \quad \textit{exercise} \text{ the option.}$$

• A speculator expects a sharp price *rise* to occur soon, but is not sure. Rather than making a purchase of the stock itself, for much less money he/she buys a call on that stock:

$$\text{Price} \searrow \quad \Longrightarrow \quad \textit{do not exercise} \text{ the option.}$$

$$\text{Price} \nearrow \quad \Longrightarrow \quad \textit{exercise} \text{ the option—and resell the stock.}$$

Now let us state our problem. Suppose you have an option to buy one share of a given stock at a fixed price c and you have T days in which to exercise the option. What policy maximises the speculator's *expected profit*?

[1] A problem with this model is that the values of the price given by (4.2) can, in the general case, be negative. An alternative is the geometric RW model $X_{t+1} = X_t Y_{t+1}$, to be discussed in more detail in Chapter 13. However, over a short time interval, when the jumps in the walk are small (which is typically the case), its behaviour is very close to that of (4.3), and the analysis is similar to what we do here.

[2] There exist various types of call options. The simplest one is the so-called *European* (or vanilla) call that can only be exercised at the terminal point of the time interval (at the option's "maturity time"). See Chapter 13 and bibliography therein for more detail.

Theorem 4.1. *The optimal*[3] *policy has the following form: there are increasing numbers* $c =: s_1 \le s_2 \le \cdots \le s_T$ *such that if there are* n *days to go, then one should exercise the option if the* present price $\ge s_n$ (Fig 4.2).

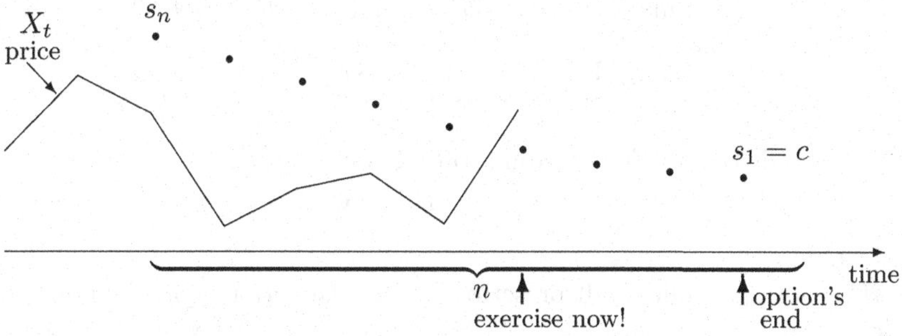

Fig. 4.2 The optimal policy for exercising a call option.

Proof We begin by stating the model. The process describing the state of the system is given by (4.3) (note that the state space is now, generally speaking, $\mathbf{R}$)—at least, until the time we exercise the option. Once we have done that, it is convenient to set $X_t = -\infty$ from that time on (thus—formally speaking—extending the state space to $\mathbf{R} \cup \{-\infty\}$). So the transition probabilities will be those of the RW (4.3) before exercising the option, and then the process will get absorbed at $-\infty$.

On each day, there are two possible actions:

$$\begin{array}{l} a = 1 \text{ -- exercise the option} \\ a = 0 \text{ -- do not exercise the option} \end{array}, \quad A = \{0, 1\}.$$

If we define the reward function by

$$R(s, a) := \begin{cases} 0 & \text{if } a = 0, \\ s - c & \text{if } a = 1, \end{cases}$$

the only non-zero term in the sum in (4.1) will be that corresponding to the day when one exercises the option. That is, there will be no "expected

[3]In fact, there is a theoretical result in mathematical finance stating that it is *never optimal* to exercise an American call option prematurely (see Theorem 13.5 on p. 415). But that result is based on the assumption that one can always sell/buy any options at the so-called "no-arbitrage prices". Here we deal with a different situation: we do not suppose we can sell our call option, we do not discuss what the price of the option could be *etc.* We can only *exercise* the option, and the question is *when* it is best to do that.

future gain" given the option is exercised *now*: $\mathbf{E}_1(V_{n-1}(X_2)|X_1 = s) = 0$, and hence the optimality equation (4.2) becomes

$$\begin{aligned} V_n(s) &= \max_{a\in\{0,1\}}\Big[R(s,a) + \mathbf{E}_a(V_{n-1}(X_2)|X_1 = s)\Big] \\ &= \max\Big\{R(s,0) + \mathbf{E}\,V_{n-1}(s+Y_1),\ R(s,1)+0\Big\} \qquad (4.4) \\ &= \max\Big\{\underbrace{\mathbf{E}\,V_{n-1}(s+Y_1)}_{\text{when } a=0},\ \underbrace{s-c}_{a=1}\Big\}. \end{aligned}$$

Starting with $V_0(s) \equiv 0$, according to (4.4) we get

$$V_1(s) = \max\{s-c, 0\},$$

and so on.

The optimal policy will prescribe us to take actions corresponding to the maximum terms in (4.4). Therefore, it is as follows: if there are n days to go and the current price is s, then we do not exercise the option if $V_n(s) > s - c$ (the term corresponding to $a = 1$ *is not the maximum* one) or, equivalently, if

$$V_n(s) - s > -c. \qquad (4.5)$$

Next we show that the left-hand side in (4.5) is a non-increasing function of s and hence the set of all s satisfying this relation is clearly a half-line of the form $s < s_n$. We will prove that using mathematical induction.

Case $n = 1$: we have $V_1(s) - s = \max\{-c, -s\}$, which is a non-increasing function.

Induction step: given we have already proved the property for $n-1$, show that it holds for n as well. To do that, note that

$$\begin{aligned} V_n(s) - s &= \max\Big\{\mathbf{E}\,V_{n-1}(s+Y_1) - s,\ -c\Big\} \\ &= \max\Big\{\mathbf{E}\,\underbrace{[V_{n-1}(s+Y_1) - (s+Y_1)]}_{\text{non-increasing in } s} + \mu,\ -c\Big\}, \end{aligned}$$

and hence the expectation (a "weighed sum" of the expressions in the square brackets for different values of Y_1 with positive weights) is also a non-increasing function of s. The maximum of two non-increasing functions is again non-increasing, so that the induction step is completed.

It remains to observe that the function on the left-hand side of (4.5) is increasing in n (additional time cannot harm!), and hence the sequence $\{s_n\}$ is non-increasing, which completes the proof of the theorem. □

To calculate the explicit values of the s_n's, we will clearly need to specify the distribution of the jumps Y_j.

Example 4.3. *The secretary problem.* Suppose we are presented with T offers, one after another (with the traditional interpretation being job applications for a secretary's position). After looking at an offer, we *must* either accept it (and thus *terminate* the process) or reject it—and then the offer is lost forever! The only information we have at any time is the relative rank of the present offer compared to the previous ones (so we assume that offers can be ranked). The order at which the offers are presented is random, i.e., all $T!$ different sequences (permutations) are equally likely. The objective is to **maximise the probability** of selecting the *best offer.*

It is important that we know in advance *how many offers* will be presented: it is intuitively clear that the information about how many are still ahead is crucial and should be included into the process $\{X_t\}$ (indeed, if we are at the very beginning, we could wait for a better opportunity, while at the end...). So set

$$X_t := (t, r_t),$$

r_t being the rank of the tth offer among the first t offers. We will also agree to set $X_t := (t, 0)$ if, by the time t, the choice has already been made.

Observe that the relative ranks r_t are in fact independent RVs with

$$\mathbf{P}(r_t = k) = \frac{1}{t}, \quad k = 1, \ldots, t. \tag{4.6}$$

Indeed, denote by S_t the (random) set of the *absolute* ranks of the first t offers among all T offers. For any $1 \le k_1 < k_2 < \cdots < k_{t-1} \le T$, given the event $S_{t-1} = \{k_1, \ldots, k_{t-1}\}$, all possible orders of the first $t-1$ offers are *equally likely by symmetry*, and hence that event and the RVs $r_1, \ldots, r_{t-1}$ are independent:

$$\mathbf{P}(S_{t-1} = \{k_1, \ldots, k_{t-1}\} | r_1, \ldots, r_{t-1}) = \mathbf{P}(S_{t-1} = \{k_1, \ldots, k_{t-1}\}). \tag{4.7}$$

Now we get by the TPF that, for any $k \in \{1, \ldots, t\}$,

$$\begin{aligned}
&\mathbf{P}(r_t = k | r_1, \ldots, r_{t-1}) \\
&\overset{\text{TPF}}{=} \sum_{1 \le k_1 < \cdots < k_{t-1} \le T} \mathbf{P}(r_t = k | \underbrace{r_1, \ldots, r_{t-1}}_{\text{now redundant}}, S_{t-1} = \{k_1, \ldots, k_{t-1}\}) \\
&\qquad \times \mathbf{P}(S_{t-1} = \{k_1, \ldots, k_{t-1}\} | r_1, \ldots, r_{t-1}) \\
&= \sum_{\cdots} \mathbf{P}(r_t = k | S_{t-1} = \{\cdots\}) \mathbf{P}(S_{t-1} = \{\cdots\}) \qquad \text{from (4.7)} \\
&\overset{\text{TPF}}{=} \mathbf{P}(r_t = k),
\end{aligned}$$

so that we have the desired independence. To find the value of the probability, we again use the TPF (this time with conditions of the form $S_t = \{\cdots\}$ instead of $S_{t-1} = \{\cdots\}$) and symmetry considerations to get

$$\mathbf{P}(r_t = k) = \sum_{\cdots} \underbrace{\mathbf{P}(r_t = k | S_t = \{\cdots\})}_{= 1/t \text{ by symmetry}} \mathbf{P}(S_t = \{\cdots\}) = \frac{1}{t}.$$

In the secretary problem, we have only two possible actions at each step:

$$a = \begin{cases} 1 \\ 0 \end{cases} \text{meaning that we } \begin{matrix} \text{accept} \\ \text{reject} \end{matrix} \text{ the offer,}$$

and if $a = 1$ at time t, then $X_{t+j} = (t + j, 0)$ for all $j \geq 1$.

Since we have to maximise the probability of choosing the best offer, we define the reward function as

$$R((t, r), a) := \begin{cases} \mathbf{P}(t\text{th offer is the best}) & \text{if } a = 1, \\ 0 & \text{if } a = 0. \end{cases}$$

Clearly, $R((t, r), 1) = 0$ if $r \geq 2$, and

$$R((t, 1), 1) = \mathbf{P}(t\text{th offer best of } T | \, t\text{th offer best of first } t) = \frac{1/T}{1/t} = \frac{t}{T}.$$

Furthermore, since we put $R(\cdot, 0) := 0$, it is obvious that the sum of the rewards $\sum_{t=1}^{T} R(X_t, a_t)$ equals the only term for which $a_t = 1$.

The optimality equation for the *optimal value function* has now the form

$$\begin{aligned} V_{T-t}(t, r) &= \text{maximum probability of choosing the absolutely best offer} \\ &\quad\ \text{among the last } T - t \text{ ones given the rank of the } t\text{th offer} \\ &\quad\ \text{among the first } t \text{ is } r \\ &= \max_a \Big[R((t, r), a) + \mathbf{E}_a V_{T-t-1}(t + 1, r_{t+1}) \Big] \end{aligned}$$

since r_{t+1} is independent of r. Now if $r \neq 1$, the only reasonable action is $a = 0$ (the current offer is definitely not the best one!), while for $r = 1$ the maximum becomes

$$\max \Big\{ \overbrace{\frac{t}{T} + 0}^{a=1}, \overbrace{0 + \underbrace{\mathbf{E}\, V_{T-t-1}(t + 1, r_{t+1})}_{=: H(t)}}^{a=0} \Big\} = \max \Big\{ \frac{t}{T}, H(t) \Big\}, \tag{4.8}$$

the function $H(t)$ being clearly decreasing in t (this is the "best we can do" when presented the last $T - t$ offers). On the other hand, t/T increases in t, and hence, setting

$$j_0 := \max \Big\{ t : \frac{t}{T} < H(t) \Big\},$$

we see that the first term in the maxima in (4.8) (the one corresponding to the action $a = 1$) is smaller than the second one iff $t \le j_0$.

Hence the optimal policy has the following form: **reject** the first j_0 offers and then **accept** *the first offer to appear after that which is better than any of its predecessors.* To find the probability of choosing the best offer under the optimal policy, note first that, for $t > j_0$,

$$\begin{aligned}
&\mathbf{P}_{\text{opt}}(t\text{th accepted}) \\
&\quad = \mathbf{P}\left([\text{best of first } j_0] = [\text{best of first } t-1],\ t\text{th} = [\text{best of first } t]\right) \\
&\qquad = \mathbf{P}\left(r_{j_0+1} > 1, r_{j_0+2} > 1, \ldots, r_{t-1} > 1, r_t = 1\right) \\
&\qquad = \mathbf{P}\left(r_{j_0+1} > 1, r_{j_0+2} > 1, \ldots, r_{t-1} > 1\right)\mathbf{P}\left(r_t = 1\right) = \frac{j_0}{t-1} \times \frac{1}{t}
\end{aligned}$$

due to independence and the fact that the position of the best of the first $t-1$ offers is uniformly distributed on $\{1, \ldots, t-1\}$, and therefore, for the event $B := \{\text{the best offer was chosen}\}$, we have by the TPF that

$$\begin{aligned}
\mathbf{P}_{\text{opt}}(B)) &= \sum_{t=j_0+1}^{T} \underbrace{\mathbf{P}_{\text{opt}}(B \mid t\text{th accepted})}_{\mathbf{P}(\text{best of } T \mid \text{best of } t) = t/T} \mathbf{P}_{\text{opt}}(t\text{th accepted}) \\
&= \sum_{t=j_0+1}^{T} \frac{t}{T} \times \frac{j_0}{t-1} \times \frac{1}{t} = \frac{j_0}{T} \sum_{t=j_0+1}^{T} \frac{1}{t-1} \\
&\approx \frac{j_0}{T} \int_{j_0}^{T} \frac{1}{x} dx = \frac{j_0}{T} \log x \Big|_{j_0}^{T} = \frac{j_0}{T} \log \frac{T}{j_0}.
\end{aligned}$$

To find the optimal policy explicitly, it remains to determine the value of j_0, i.e., $\arg\max_x g(x)$ for $g(x) = x \log(T/x)$, cf. the end of Example 3.19. Equating the derivative $g'(x) = \log(T/x) - 1$ to zero, we get $x = T/e$. Hence, for large T,

$$j_0 \approx \frac{T}{e}, \quad \mathbf{P}_{\text{opt}}(B) \approx \frac{1}{e} \approx 0.3679.$$

4.2 Discounted Dynamic Programming

Quite often (especially in situations related to economics) one needs to take into account the *time factor* by introducing *discounting*. Thus, for some reasons (inflation *etc.*), for an individual, the value of \$1 **now** and, say, in five years' time is not one and the same thing. Such effects can be incorporated into models by introducing the so-called *discounted return*

$$\sum_{t=0}^{T} \alpha^t R(X_t, a_t),$$

where $\alpha \in (0,1)$ is called the *discount factor*, and T can now be either finite (a finite horizon) or infinite (an infinite horizon, which makes sense due to discounting as the series can now be convergent).

The *optimality criterion* is

$$\mathbf{E}_\pi \left[\sum_{t=0}^{T} \alpha^t R(X_t, a_t) \middle| X_0 = i \right] \longrightarrow \max_\pi,$$

where $\mathbf{E}_\pi$ stands for the expectation under the policy π (we know that the evolution of the process $\{X_t\}$ depends on actions taken, so we will have different distributions of $\{X_t\}$ for different policies). For notational convenience, in this section we will include the time instance $t = 0$ into the time horizon.

It is not hard to see that the *optimality equation* in the finite horizon case is now

$$\begin{aligned} V_n(i) &= \max_a \left[R(i,a) + \alpha \mathbf{E}_a (V_{n-1}(X_1) | X_0 = i) \right] \\ &= \max_a \left[R(i,a) + \alpha \sum_j p_{ij}(a) V_{n-1}(j) \right]. \end{aligned}$$

In the case of the *infinite horizon*, there is no such thing as the "number of time periods to go"—it is always one and the same infinity—so one writes $V(\cdot)$ instead of both $V_n(\cdot)$ and $V_{n-1}(\cdot)$:

$$\begin{aligned} V(i) &= \max_a \left[R(i,a) + \alpha \mathbf{E}_a (V(X_1) | X_0 = i) \right] \\ &= \max_a \left[R(i,a) + \alpha \sum_j p_{ij}(a) V(j) \right]. \end{aligned} \tag{4.9}$$

Example 4.4. *Lifetime portfolio selection* (optimal consumption-saving). To understand better what happens in this model, we will begin with considering a simpler deterministic situation.

1. *"Non-random environment"*. Denote by X_t the wealth of a particular individual at the beginning of the tth time period, $t = 0, 1, 2, \ldots, T$. Of these X_t units, the individual consumes an amount of C_t units during that time period and invests $X_t - C_t$ units into a *non-risky asset* for a certain fixed rate $r > 1$ of yield, so that at the end of the tth period, the individual's wealth is $X_{t+1} = r(X_t - C_t)$.

For the individual, the *utility* is in consumption: consuming c units of wealth leads to utility $u(c)$, $u(\cdot)$ being a utility function (see Section 2.6). In the so-called *Ramsey model* the objective is to maximise the *total discounted utility*:

$$\sum_{t=0}^{T-1} \alpha^t u(C_t) \longrightarrow \max_{\{C_t\}}$$

for prescribed initial and terminal values X_0 and X_T. The role of the discount factor α can be interpreted as reflecting the "subjective time preference" specific to our individual. Thus, if α is close to 0, one does not care much about the distant future (the respective contribution to the sum to be maximised is very small), while if α is close to 1, one does care about the whole time horizon.

Since $C_t = X_t - r^{-1}X_{t+1}$, the problem is equivalent to finding the maximum

$$\max_{\{X_t\}} \sum_{t=0}^{T-1} \alpha^t u\Big(X_t - \frac{1}{r}X_{t+1}\Big).$$

Denoting the last sum by $\phi = \phi(X_1, \ldots, X_{T-1})$ (the values X_0 and X_T are fixed, so that the sum only depends on the listed arguments), we get the following equations for the stationary points of the function:

$$0 = \frac{\partial \phi}{\partial X_t} = -\frac{\alpha^{t-1}}{r} u'\Big(X_{t-1} - \frac{X_t}{r}\Big) + \alpha^t u'\Big(X_t - \frac{X_{t+1}}{r}\Big), \quad t = 1, \ldots, T-1,$$

or

$$u'\Big(X_{t-1} - \frac{X_t}{r}\Big) = \alpha r u'\Big(X_t - \frac{X_{t+1}}{r}\Big), \quad t = 1, \ldots, T-1. \tag{4.10}$$

In the special case of the *Bernoulli utility* $u(x) = \log x$ (similar results hold for $u(x) = x^\gamma$, $0 < \gamma < 1$, as well), we can solve the equations explicitly. Now $u'(x) = 1/x$, and (4.10) becomes

$$\frac{1}{X_{t-1} - r^{-1}X_t} = \alpha r \frac{1}{X_t - r^{-1}X_{t+1}}.$$

The case when the denominators are zeros can be excluded: no consumption at all, clearly not optimal according to the chosen criterion. Therefore the above equation is equivalent to

$$X_{t+1} - r(1+\alpha)X_t + \alpha r^2 X_{t-1} = 0. \tag{4.11}$$

To solve this linear difference equation[4], let us look for its *partial solutions* of the form $X_t = \lambda^t$ for some fixed λ. Substituting this into (4.11), we get

$$\lambda^{t+1} - r(1+\alpha)\lambda^t + \alpha r^2 \lambda^{t-1} = 0.$$

Excluding the case $\lambda = 0$ (corresponding to $X_t = 0$), we obtain a quadratic equation (the so-called *characteristic equation* of the difference equation (4.11)) of the form

$$\lambda^2 - r(1+\alpha)\lambda + \alpha r^2 = 0.$$

Solving for λ, we get the roots $\lambda_1 = r$ and $\lambda_2 = \alpha r$. The general solution to (4.11) will now be given by the linear combination

$$X_t = b_1\lambda_1^t + b_2\lambda_2^t = \underbrace{b_1 r^t}_{\text{"deposit"}} + \underbrace{b_2 \alpha^t r^t}_{\text{"consume \& deposit"}}.$$

The constants b_i are to be determined from the boundary conditions:

$$\begin{cases} b_1 + b_2 = X_0, \\ b_1 r^T + b_2 \alpha^T r^T = X_T. \end{cases}$$

The system is easily seen to have the solution

$$b_2 = \frac{X_0 - r^{-T} X_T}{1 - \alpha^T}, \quad b_1 = X_0 - b_2.$$

Therefore, the optimal policy has the form

$$C_t = X_t - \frac{1}{r} X_{t+1} = b_2(1-\alpha)(\alpha r)^t.$$

If, for instance, $T = \infty$ (the individual is going to live a *very long* life), in the "no-bequest" situation ($X_T = 0$), we get $b_2 = X_0$, $b_1 = 0$, and hence $C_t = (1-\alpha)X_t$. That is, in that case the optimal behaviour for the individual would be to spend each year a certain fixed proportion of his/her wealth. Note that the proportion does not depend on the interest rate r!

2. *A stochastically-risky alternative asset.* Assume now that, along with the *safe asset*:

$$\text{invest \$1 at time } t \rightarrow\rightarrow \text{ get \$}r \text{ at time } t+1,$$

there exists a *risky asset*:

$$\text{invest \$1 at time } t \rightsquigarrow\rightsquigarrow \text{ get \$}Z_t \text{ at time } t+1,$$

[4]For general theory of such equations see references on p. 120.

where $Z_t > 0$, $t \geq 0$, are i.i.d. RVs.

For the tth time period, the individual decides to consume C_t units of wealth, invest a fraction w_t of the remaining part $X_t - C_t$ into the risky asset, and the fraction $1 - w_t$ of $X_t - C_t$ into the safe asset, so that

$$X_{t+1} = (X_t - C_t)((1 - w_t)r + w_t Z_t), \quad t = 0, 1, \ldots, T-1; \quad X_0, X_T \text{ are given.}$$

The optimisation problem is now

$$\mathbf{E} \sum_{t=0}^{T-1} \alpha^t u(C_t) \longrightarrow \max_{\{C_t, w_t\}} \quad \text{subject to} \quad C_t = X_t - \frac{X_{t+1}}{(1 - w_t)r + w_t Z_t}.$$

We assume that the initial wealth X_0 is given, and, for simplicity's sake, that $X_T = 0$ (no bequest). We can see that this is a discounted Markov decision model, with action at time t being specified by the values of C_t and w_t. The optimality equation takes the form: for $n = 1, 2, \ldots, T$,

$$V_n(x) = \max_{C_{T-n}, w_{T-n}} \left[u(C_{T-n}) + \alpha \mathbf{E} \left(V_{n-1}(X_{T-n+1}) | X_{T-n} = x \right) \right].$$

Starting, as usual, at the end of the planning period with $V_0(x) = 0$ (as there is no bequest, nothing is left for the future at time T), we get

$$V_1(x) = u(C_{T-1}) = u(X_{T-1}) \quad \text{since} \quad C_{T-1} = X_{T-1} \text{ (no bequest!).}$$

Further,

$$V_2(x) = \max_{C_{T-2}, w_{T-2}} \left[u(C_{T-2}) + \alpha \mathbf{E} \left(u(X_{T-1}) | X_{T-2} = x \right) \right]$$

$$= \max_{C_{T-2}, w_{T-2}} \left[u(C_{T-2}) + \alpha \mathbf{E}\, u\big(((1 - w_{T-2})r + w_{T-2} Z_{T-2})(x - C_{T-2})\big) \right].$$

To maximise the expression in the square brackets, we solve the equations

$$\frac{\partial[\cdots]}{\partial C_{T-2}} = \frac{\partial[\cdots]}{\partial w_{T-2}} = 0$$

(for the stationary points) for C_{T-2} and w_{T-2} (given $X_{T-2} = x$). On the next step, we find $V_3(x)$, and so on.

Once again, in the important special case of the *Bernoulli utility* $u(x) = \log x$, one can find the solution explicitly. The optimality equation for $n = 2$ becomes (we set $C := C_{T-2}$, $w := w_{T-2}$ and $Z := Z_{T-2}$):

$$\begin{aligned} V_2(x) &= \max_{C,w} \left[\log C + \alpha \mathbf{E} \log\{((1 - w)r + wZ)(x - C)\} \right] \\ &= \max_{C,w} \left[\log C + \alpha \log(x - C) + \alpha \mathbf{E} \log((1 - w)r + wZ) \right] \\ &= \max_{C} \left[\log C + \alpha \log(x - C) \right] + \alpha \max_{w} \mathbf{E} \log((1 - w)r + wZ), \end{aligned}$$

so that we can find the maxima separately. For the first term, we get

$$0 = \frac{\partial}{\partial C}\left[\log C + \alpha \log(x - C)\right] = \frac{1}{C} - \frac{\alpha}{x - C} \implies C = \frac{x}{1+\alpha}.$$

As for the second one, note first that the maximum exists and is unique (for the expectation is a concave function of w, being a linear combination of concave functions $\log((1-w)r + wz)$ for different z's). To find the point at which it is attained, we have to solve for w the equation

$$0 = \frac{\partial}{\partial w}\mathbf{E}\,\log((1-w)r + wZ) = \mathbf{E}\,\frac{Z - r}{(1-w)r + wZ}.$$

If the equation has no solution, the derivative is either positive or negative for all $w \in (0,1)$. In the former case, the maximum is clearly attained at $w^* = 1$ (i.e., the safe asset is not attractive at all), while in the latter at $w^* = 0$ (forget about the risky asset).

In any case, the unique solution w^* of the maximization problem is clearly seen to be independent of x and can be called the *optimal portfolio decision* (for it gives the value of w saying how to best split the invested capital between the risky and safe assets).

So the optimal action is

$$C_{T-2} = \frac{X_{T-2}}{1+\alpha}, \qquad w_{T-2} = w^*,$$

and the optimal value is

$$\begin{aligned}
V_2(x) &= \log\frac{x}{1+\alpha} + \alpha\log\left(x - \frac{x}{1+\alpha}\right) + \alpha\log r^* \\
&= (1+\alpha)\log x + \alpha\log\alpha - (1+\alpha)\log(1+\alpha) + \alpha\log r^* \\
&\equiv (1+\alpha)\log x + K_1, \qquad (4.12)
\end{aligned}$$

where r^* is defined by the relation $\log r^* = \mathbf{E}\,\log((1-w^*)r + w^*Z)$ and can be called (for obvious reasons) the *risk-corrected yield*: the expected performance of our stochastic model is the same as the performance of the deterministic one with that choice of r.

Now note that the expression in the last line in (4.12) is basically the same as the Bernoulli utility function (the additive constant K_1 does not matter for maximisation purposes)! Hence one can expect that the optimal decisions on C_{T-3} and w_{T-3} will again be independent and have a similar form.

Indeed,

$$V_3(x) = \max_{C_{T-3},w_{T-3}}\left[u(C_{T-3}) + \alpha\mathbf{E}\left(V_1(X_{T-2})|X_{T-3} = x\right)\right].$$

Setting $C := C_{T-3}$, $w := w_{T-3}$, $Z := Z_{T-3}$, and recalling that $X_{T-2} = (X_{T-3} - C)((1-w)r + wZ)$, we get

$$V_3(x) = \max_{C,w}\left[\log C + \alpha\mathbf{E}\left[(1+\alpha)\log\left\{((1-w)r + wZ)(x-C)\right\} + K_1\right]\right]$$
$$= \max_{C}\left[\log C + \alpha(1+\alpha)\log(x-C)\right]$$
$$+ \alpha(1+\alpha)\underbrace{\max_{w}\mathbf{E}\,\log((1-w)r + wZ)}_{\log r^*} + \alpha K_1,$$

with the optimal value $w = w_{T-3} = w^*$. To find the first maximum, we solve the equation

$$0 = \frac{\partial}{\partial C}\left[\log C + \alpha(1+\alpha)\log(x-C)\right] = \frac{1}{C} - \frac{\alpha(1+\alpha)}{x-C}$$

to get $C = C_{T-3} = x/(1+\alpha+\alpha^2)$. Now substituting the optimal decisions w_{T-3}, C_{T-3} into the formula for $V_3(x)$, we get the maximum value

$$V_3(x) = \log\frac{x}{1+\alpha+\alpha^2} + \alpha(1+\alpha)\log\left(x - \frac{x}{1+\alpha+\alpha^2}\right)$$
$$+ \alpha(1+\alpha)\log r^* + \alpha K_1$$
$$= (1+\alpha+\alpha^2)\log x + \alpha\log\alpha - (1+\alpha)\log(1+\alpha) + \alpha\log r^*$$
$$\equiv (1+\alpha+\alpha^2)\log x + K_2,$$

which is again similar to $u(x)$, and so on.

Thus the **optimal consumption decision** is

$$C_t = \frac{X_t}{1+\alpha+\cdots+\alpha^{T-t-1}} = \frac{1-\alpha}{1-\alpha^{T-t}}X_t, \quad t = 0, 1, \ldots, T-1.$$

The **optimal portfolio decision** is always

$$w_t = w^*, \quad t = 0, 1, \ldots, T-1.$$

Note that, in a sense, we could "split" the problem into two parts: one about the optimum choice of C_t, and the other about the optimal w_t which proved to be one and the same value w^* for all t (producing the maximum risk-corrected yield r^*). Observe also that, as T increases, C_t tends to $(1-\alpha)X_t$, our solution in the previous (deterministic) case when we assumed that $T = \infty$.

4.3 Further Examples

Example 4.5. *Component replacement.* A certain component of a machine can be in any one of a continuum of states which are represented by points from $[0,1]$. The smaller the numerical value, the better the state of the component (state $x = 0$ corresponds to a new component, $x = 1$ to a "very old" one). At the beginning of each time period, the component is inspected, its current state is determined and a decision is made whether or not to replace the component at a fixed cost $R > 0$ by a new one (at state $x = 0$).

The expected cost of having the component at state x for a single time period is $C(x) \geq 0$; the function $C(\cdot)$ is increasing on $[0,1]$. Given the state of the component at the beginning of a time period is x, its state at the end of the period is represented by an RV $Y_x \in [0,1]$ (the RVs are independent for different time periods) of which the DF is denoted by $F(y|x)$. We assume that $F(0|\,x) = 0$ and, for any $y \in [0,1]$,

$$F(y|\,x_1) \geq F(y|\,x_2) \quad \text{for} \quad 0 \leq x_1 < x_2 \leq 1. \tag{4.13}$$

That is, for any given time period and for any fixed level y, of two working components the one which was worse at the beginning of the period will always be more likely to be worse at the end as well (i.e., to have the x-value exceeding y at the end). In such a case, one says that Y_{x_2} is *stochastically greater* than Y_{x_1}.

Assuming a discount factor $\alpha \in (0,1)$ and an *infinite time horizon, find the optimal replacement policy* (recall a similar problem from Example 3.19 on component replacement).

First of all, we select a process describing the state of our system. It suffices to set $X_t :=$ the state of the current component at the beginning of the tth time interval. No other information is relevant and should be incorporated.

At each step, only two actions are possible: $a = 1$ means that we replace the component, and $a = 0$ means that we do not.

The evolution of the process $\{X_t\}$ is as follows. If $a_t = 1$, then X_{t+1} will be distributed as Y_0. If $a_t = 0$, then X_{t+1} will have the DF $F(y|\,x)$ given the current value of the process is $X_t = x$.

Since we understand optimality in the sense that the expected total discounted costs are minimal, we have to define the reward function by

$$R(x,1) := -R - C(0), \quad R(x,0) := -C(x).$$

The optimality equation (4.9), for $V(x) = -$expected total discount costs, becomes

$$V(x) = \max_a [R(x,a) + \alpha \mathbf{E}_a(V(X_1)|X_0 = x)]$$
$$= \max\{\underbrace{-R - C(0) + \alpha \mathbf{E}\, V(Y_0)}_{a=1}, \underbrace{-C(x) + \alpha \mathbf{E}\, V(Y_x)}_{a=0}\}.$$

Setting for convenience $U(x) := -V(x)$, we get

$$U(x) = \min\{R + C(0) + \alpha \mathbf{E}\, U(Y_0),\ C(x) + \alpha \mathbf{E}\, U(Y_x)\}. \tag{4.14}$$

Hence the optimal policy is as follows: given the current state of the component is x, take action $a = 1$ (replace) if the first term in the curly brackets in (4.14) is smaller; otherwise $a = 0$.

For what values of x does one take $a = 1$? To answer the question, let us analyse the function $U(\cdot)$. To find it, we can (i) solve the problem for a finite horizon n and find the respective optimal value function $U_n(x)$ using the optimality equation (4.15) below, and then (ii) let $U(x) := \lim_{n\to\infty} U_n(x)$.

Clearly, $U_0(x) \equiv 0$. Further, $U_1(x) = \min\{R + C(0), C(x)\}$ is non-decreasing since $C(x)$ is increasing. Next we note that we can continue this argument and claim by induction that

$$U_n(x) = \min\{R + C(0) + \alpha \mathbf{E}\, U_{n-1}(Y_0), C(x) + \alpha \mathbf{E}\, U_{n-1}(Y_x)\} \tag{4.15}$$

is non-decreasing, since $C(x)$ and $U_{n-1}(x)$ are both non-decreasing and we assumed (4.13).

Thus all $U_n(x)$ are non-decreasing in x, and so is the limit $U(x)$. Hence, due to (4.13), $\mathbf{E}\, U(Y_x)$ is also a non-decreasing function of x.

Indeed, stochastic monotonicity (4.13) implies that, for any non-decreasing function $g(y)$,

$$\mathbf{E}\, g(Y_{x_1}) \le \mathbf{E}\, g(Y_{x_2}), \quad x_1 < x_2. \tag{4.16}$$

To see why this is true, assume for simplicity's sake that g is differentiable. Integration by parts yields

$$\mathbf{E}\, g(Y_{x_i}) = \int_0^1 g(y)\, dF(y|\, x_i) = g(y)F(y|\, x_i)\Big|_0^1 - \int_0^1 F(y|\, x_i) g'(y)\, dy.$$

Since $F(1|\, x_i) = 1$ and $F(0|\, x_i) = 0$, we get

$$\mathbf{E}\, g(Y_{x_2}) - \mathbf{E}\, g(Y_{x_1}) = \int_0^1 (F(y|\, x_1) - F(y|\, x_2)) g'(y)\, dy.$$

The inequality in (4.16) is due to (4.13) and the fact that $g'(y) \ge 0$ by assumption.

Setting $f(x) := C(x) + \alpha \mathbf{E}\, U(Y_x)$, we see from the optimality equation (4.14) that the optimal policy is as follows: replace if

$$x \ge x^* := \min\{x :\ f(x) \ge R + f(0)\},$$

and do not replace otherwise.

Example 4.6. *The optimal disposal of an asset.* Suppose that a person has an asset (e.g., a block of land) she must dispose of during a time interval consisting of a given number T periods. For this asset, she is being offered a certain amount of money from period to period. Assume that these offers are i.i.d. RVs $Z_1, Z_2, \dots, Z_T$ following a known DF F. If she accepts an offer, she can invest the money at a fixed rate of interest for the remaining time periods. Put $r := 1 +$ interest rate (so that \$1 becomes \$$r$ at the end of the period *etc.*). Further, we assume that no offer is renewed.

The task is to maximise the expected total return over T periods of time. Note that the person must sell at time $t = T$ the latest.

First we need to construct a process $\{X_t\}$ modelling the situation. The process should clearly incorporate the following information: (i) if the person still has the asset, and (ii) what the offered price for the current time period is. It is convenient to put

$$X_t := \begin{cases} (0, Z_t) & \text{if the asset has not been sold by time } t, \\ (1, y) & \text{otherwise; } y = \text{investment at the beginning of the period.} \end{cases}$$

We again have only two possible actions: $a = 1$ means accepting the offer, and $a = 0$ rejecting it.

The evolution of the process $\{X_t\}$ can be described as follows:

$$(0, Z_t) \begin{array}{l} \nearrow \ (0, Z_{t+1}) \ \text{ if } \ a_t = 0, \\ \searrow \ (1, rZ_t) \ \text{ if } \ a_t = 1; \end{array}$$

$$(1, y) \ \to (1, ry) \ \text{ always}$$

(in the latter case, there is actually only one action available: $a = 0$).

To remain within the framework we set at the beginning of the present chapter, we want the return at the *end* of the planning period (our objective function) to be the sum of one-step increments (which will hence be the capital increments), so we put

$$\text{for } \ X_t = (0, x): \quad \begin{cases} R(X_t, 0) := 0, \\ R(X_t, 1) := rx; \end{cases}$$

$$\text{for } \ X_t = (1, x): \quad R(X_t, a) := rx - x = (r - 1)x.$$

With thus defined return function, by the so-called "telescoping argument" one has

$$\underset{\substack{\uparrow\\ \text{sell here}}}{0+\cdots+0+rx}+(r-1)rx+(r-1)r^2x+\cdots+(r-1)r^{T-t-1}x=r^{T-t}x,$$

i.e., everything cancels out except for the term appearing on the right-hand side of the last equality.

Now the optimal value function $V_n((i,x))$ = the expected return the owner of the asset will get when the current state of the process is (i,x) and she uses the *optimal policy* over the remaining n time periods.

Clearly, $V_0((i,x)) = 0$ since the asset is to be sold at the beginning of the last period the latest, and the optimality equation is

$$V_n((i,x)) = \max_a \big[R((i,x),a) + \mathbf{E}_a(V_{n-1}(X_1)|\, X_0 = (i,x))\big], \quad n = 1,\ldots,T.$$

From this it immediately follows that $V_1((i,x)) = rx$ for any (i,x).

If $a = 1$, then $[\cdots] = r^n x$ (we just invest the money for n periods of time); you can verify that by substituting explicit expressions for R.

If $a = 0$ and the current state is (i,x) with $i = 0$, then $[\cdots] = \mathbf{E}\, V_{n-1}((0,Z))$, where Z is an RV having the same distribution as Z_t (note that we can consider the case $i = 0$ only as it is in that case that we have to decide which action to take, while when $i = 1$ we cannot change the future evolution of the process any more!).

Putting $v_n(x) := V_n((0,x))$ for simplicity, we get $v_0(x) = 0$ and

$$v_n(x) = \max\{\underbrace{r^n x}_{a=1},\ \underbrace{\mathbf{E}\, v_{n-1}(Z)}_{a=0}\}. \tag{4.17}$$

Here the optimal action is $a = 1$ when the first term is the maximum and $a = 0$ otherwise. Our aim is to maximise $\mathbf{E}\, v_T(Z)$.

Now note that if we set $u_n(x) := r^{-n} v_n(x)$, then the optimality equation (4.17) can be re-written as

$$u_n(x) = \max\{x, \alpha \mathbf{E}\, u_{n-1}(Z)\}, \quad u_0(x) = 0, \tag{4.18}$$

where $\alpha = 1/r$. This clearly is the optimality equation for *discounted dynamic programming*, which is quite natural as it is just another interpretation of the problem: from the optimisation viewpoint, the opportunity of putting money into a fixed interest savings account is equivalent to discounting—with the reciprocal rate.

Since to the maximum $u_n(\cdot)$ there always corresponds the maximum $v_n(\cdot)$, we infer from (4.18) that the **optimal policy** is as follows: if there are $n \geq 1$ time periods to go and you have not sold the asset yet, accept the offer iff it is $\geq \mu_{n-1} := \alpha \mathbf{E}\, u_{n-1}(Z)$.

It remains to observe that the quantities $\mu_n := \alpha \mathbf{E}\, u_n(Z)$ can be *computed recursively.* We have:

$$\begin{aligned}
\mu_0 &= 0; \\
\mu_1 &= \alpha \mathbf{E} \max\{Z, 0\} = \alpha \mathbf{E}\, Z; \\
\mu_n &= \alpha \mathbf{E} \max\{Z, \mu_{n-1}\} = \alpha \mathbf{E}\,(Z;\, Z > \mu_{n-1}) + \alpha\, \underbrace{\mathbf{E}\,(\mu_{n-1};\, Z \leq \mu_{n-1})}_{\mu_{n-1}\mathbf{P}\,(Z \leq \mu_{n-1})} \\
&= \alpha G(\mu_{n-1}) + \alpha \mu_{n-1} F(\mu_{n-1}),
\end{aligned}$$

where

$$G(y) = \mathbf{E}\,(Z;\, Z > y) = \int_y^\infty x\, dF(x) = \int_y^\infty x f(x) dx,$$

the last equality holding if the DF F has the density f.

In conclusion, note that, for $n \geq 1$,

$$|\mu_{n+1} - \mu_n| \leq \alpha |\mu_n - \mu_{n-1}|, \quad \alpha < 1, \tag{4.19}$$

so that there exists a limit $\mu := \lim_{n\to\infty} \mu_n$ as $n \to \infty$, and the convergence here is actually geometrically fast (showing that is a good exercise for "mathematically-minded"; the fact of convergence itself is usually referred to in mathematics as the contraction principle). The limiting value μ will satisfy the relation $\mu = \alpha G(\mu) + \alpha \mu F(\mu)$ or, equivalently,

$$\mu(r - F(\mu)) = G(\mu).$$

Solving this for μ and denoting the root of the equation by μ^*, we see that the optimal policy for long planning intervals will be close to the following simple rule: sell if the offer $\geq \mu^*$; reject any offer $< \mu^*$.

4.4 Recommended Literature

ROSS, S.M. *Introduction to stochastic dynamic programming.* Academic Press, New York, 1983.

WHITE, D.J. Real applications of Markov decision processes. *Interfaces,* 15 (1985), 73–83.

4.5 Problems

1. Derive the optimality equation (4.2) using the total probability formula (you may consider the case of the discrete state space only).
2. A person must buy a block of land during the next three weeks. The lowest prices he can be offered on particular weeks are independent random variables $\$100,000 \times Z_j$, $j = 1, 2, 3$, distributed according to the following table:

x	$\mathbf{P}(Z_1 = x)$	$\mathbf{P}(Z_2 = x)$	$\mathbf{P}(Z_3 = x)$
2.2	0.3	0.2	0.2
2.3	0.5	0.6	0.5
2.4	0.2	0.2	0.3

Each week the person has to make a decision: either to buy or not to buy. If he does not buy for the best price (this week), the opportunity is lost (he cannot return to the offer later).

(i) Set this as a stochastic dynamic decision problem: define the decision process, possible actions, reward function *etc.*

(ii) Write down the optimality equation for the optimal value function.

(iii) Draw a decision tree for this problem. Derive the optimum policy for the buyer. What is the expected price when one follows the optimum policy?

3. *A Gambling Model.* At each play of the game, a gambler can bet any non-negative amount up to his present fortune and will either win or lose that amount with probabilities p and $q = 1-p$ respectively. The gambler is allowed to make T bets, and his objective is to maximise the expected Bernoulli utility of his final fortune. What is the optimal strategy?

Hints. [First try to solve the problem without reading the hints!] The gambler's goal is to maximise the expectation of the logarithm of his final fortune. Take the process X_t = the gambler's fortune at time t. Take actions to be the fractions of the gambler's fortune that he bets (so now the set of possible actions is $A = [0, 1]$). Given $X_{t-1} = x$, we have $X_t = x + axZ_t$, where $Z_t = \pm 1$ w.p.'s p and q, respectively. The optimality equation for $V_n(x)$—the maximal expected return if the gambler has a present fortune of x and is allowed n more gambles—takes now this form:

$$V_n(x) = \max_a \mathbf{E}_a(V_{n-1}(X_{N-n+1}) | X_{N-n} = x) = \max_a \mathbf{E}_a(V_{n-1}(X_1) | X_0 = x)$$

(as the process $\{X_t\}$ is homogeneous in time). Note that $V_0(x) = \log x$ (the Bernoulli utility of the fortune x).

(i) When $p \leq 1/2$, show that $V_n(x) = \log x$ and the optimum strategy is always to bet 0. [So if a game unfavourable for you, never play it!]

(ii) When $p > 1/2$, derive a general formula for $V_n(x)$ and show that the optimal decision is to bet each time the fraction $p - q$ of one's fortune.

4. A person has to sell a block of shares during the next four days. He believes that the prices Z_j, $j = 1, \dots, 4$, of the block on particular days are independent $U(0,1)$-RVs. The objective is to maximize the expected selling price.

   (i) Set this as a stochastic dynamic problem: define the decision process $\{X_t\}$, possible actions, reward function *etc.*

   (ii) Write down the optimality equation for the optimal value function. Use it to find $V_n(x)$ for $n = 1, \dots, 4$.

   (iii) What is the optimal policy? What is the maximum expected price (i.e., the value $\mathbf{E}\, V_4(X_1)$)?

   Hint. (ii) You may well wish to use the formula $\mathbf{E}\max(Z_j, c) = \frac{1}{2}(1 + c^2)$, $c \in [0,1]$. Verify it.

5. For the option model from Example 4.2, show that when $\mu = \mathbf{E}\, Y_j > 0$, one has $s_n = \infty$ for $n > 1$. In other words, it is never optimal to exercise the option before maturity when $\mu > 0$.

   Hints. It suffices to show that $s_2 = \infty$. (Why?) Using the optimality equation for $n = 2$, write down the explicit expression for $V_2(s) - s$ and recall that s_2 is the minimum value for which $V_2(s) - s \le -c$ (the LHS decreases in s). Does such a value exist? Recall that $\mathbf{E}\, V_1(s + Y_1) = \mathbf{E}\max\{s + Y_1 - c, 0\}$.

6. *Diversification pays, or do not put all eggs in one basket.*

   (i) Show that putting a fixed total of wealth equally into independent identically distributed investments will yield the same mean gain as any other portfolio, but will minimise the variance. [Thus such an investment portfolio is, in a sense, the most "reliable" one: the uncertainty is then minimal!] In other words, if $X_1, \dots, X_n$ are i.i.d. RVs (representing profits from investments) with finite mean $\mu = \mathbf{E}\, X_1$ and variance $\sigma^2 = \mathrm{Var}\,(X_1) < \infty$, then the mean of the random variable

   $$Y := \lambda_1 X_1 + \cdots + \lambda_n X_n$$

   (the total gain), where the values

   $$\lambda_j \ge 0,\ j = 1, \dots n, \quad \lambda_1 + \cdots + \lambda_n = 1$$

   represent proportions of one's wealth invested into different assets, does not depend on the choice of λ_j, while the minimum of $\mathrm{Var}\,(Y)$ is attained on the portfolio $\lambda_j = 1/n$, $j = 1, \dots, n$.

   (ii) However, if you are using a strictly concave[5] utility function $u(x)$ (such as the Bernoulli utility $u(x) = \log x$), then the investment portfolio $\lambda_j = 1/n$,

[5] That u is "strictly concave" means that, for any $x_1 < x_2 \in \mathbf{R}$ and $\alpha \in (0,1)$, one has $u(\alpha x_1 + (1-\alpha)x_2) > \alpha u(x_1) + (1-\alpha)u(x_2)$. In other words, if you draw a straight line through the points $(x_1, u(x_1))$ and $(x_2, u(x_2))$, then the graph of the function $u(x)$ will be *strictly above* that straight line on the interval $x \in (x_1, x_2)$. For a smooth u, strict concavity is equivalent to the condition that $u''(x) < 0$ everywhere.

$j = 1, \ldots, n$, is the optimal choice—it maximises the expected utility $\mathbf{E}\, u(Y)$. Prove that assertion (you may prove it in the case of the Bernoulli utility only).
(iii) Moreover, in (ii) above one can relax the assumption of having i.i.d. X's and require only that the X's are *exchangeable,* which means that, for any *permutation* of the indices $i_1, \ldots, i_n$, the distribution of the random vector $(X_{i_1}, \ldots, X_{i_n})$ is the same as that of the original $(X_1, \ldots, X_n)$. [In particular, i.i.d. RVs are exchangeable, and $X_1 = X_2 = \cdots = X_n$ are also exchangeable.]

7. Prove (4.19).

Chapter 5

The Exponential Distribution and Poisson Process

5.1 Properties of the Exponential Distribution

Recall that an RV $\tau \geq 0$ is said to have the exponential distribution with parameter $\lambda > 0$ (we write symbolically $\tau \sim Exp(\lambda)$) if its density is $\lambda e^{-\lambda t}$ for $t > 0$ and 0 otherwise; in that case, the DF of τ is

$$F_\tau(t) = \mathbf{P}(\tau \leq t) = \begin{cases} 1 - e^{-\lambda t}, & t \geq 0, \\ 0, & t < 0. \end{cases} \tag{5.1}$$

A very important observation is that the tail of $Exp(\lambda)$ is just the exponential function:

$$\overline{F}_\tau(t) := 1 - F_\tau(t) = \mathbf{P}(\tau > t) = e^{-\lambda t}, \quad t \geq 0. \tag{5.2}$$

Note also that, for any $a > 0$, $\tau_a = a\tau$ is exponentially distributed as well: $\tau_a \sim Exp(\lambda/a)$. Indeed,

$$\mathbf{P}(\tau_a > t) = \mathbf{P}(\tau > t/a) = e^{-(\lambda/a)t}, \quad t \geq 0.$$

Therefore λ is just the *scale parameter*: if $\tau \sim Exp(1)$, then $\tau/\lambda \sim Exp(\lambda)$.

The expectation of an RV $\tau \sim Exp(\lambda)$ is

$$\mathbf{E}\,\tau = \int_0^\infty \overline{F}_\tau(t)\,dt = \int_0^\infty e^{-\lambda t} dt = \frac{1}{\lambda} \tag{5.3}$$

by (2.54) and (5.2).

Below we discuss three characteristic properties of the exponential distribution which are extensively used in stochastic modelling (it is actually due to these properties that the distribution is so popular). We denote them by E1 to E3.

E1. *Lack of memory* (or *memoryless property*). If $\tau \sim Exp(\lambda)$ then, for any $t, s > 0$,

$$\mathbf{P}(\tau > t + s \,|\, \tau > t) = \mathbf{P}(\tau > s). \tag{5.4}$$

Indeed, since one has $\{\tau > t+s,\, \tau > t\} = \{\tau > t+s\}$, the left-hand side of (5.4) is, by the definition of conditional probability,

$$\frac{\mathbf{P}(\tau > t+s,\, \tau > t)}{\mathbf{P}(\tau > t)} = \frac{\mathbf{P}(\tau > t+s)}{\mathbf{P}(\tau > t)} = \frac{e^{-\lambda(t+s)}}{e^{-\lambda t}} = e^{-\lambda s}$$

from (5.2).

In words, devices whose lifetimes are exponentially distributed *do not age*: given that such a device has already operated for a given time s, the conditional distribution of its residual lifetime coincides with the unconditional lifetime distribution of a brand new device. This property is *characteristic* for the exponential distribution: no other distribution law is memoryless.[1] We will illustrate E1 by a simple example.

Example 5.1. Suppose that the time a computer operates without malfunction (e.g., crashing its recently updated operating system) is exponentially distributed with a mean of 10 hours. A person starts doing a 5-hour job on the computer which has already been working for 3 hours without failures. What are the chances for the person to complete the job in due time?

Denoting the time of failure-free operation of the computer by τ, we first infer from (5.3) that the parameter of the (exponential) distribution of τ is $\lambda = 1/\text{mean} = 0.1$ (hour^{-1}).

It remains to note that, by property E1,

$$\mathbf{P}\,(\text{residual failure-free oper. time } \geq 5 \,|\, \text{already survived 3 hours})$$

$$= \frac{\mathbf{P}(\tau \geq 3+5)}{\mathbf{P}(\tau > 3)} = \mathbf{P}(\tau \geq 5) = \overline{F}_\tau(5) = e^{-5/10} = \frac{1}{\sqrt{e}} \approx 0.607.$$

An important notion in the context of lifetime modelling is that of the so-called **hazard** (or **failure**) **rate function** (a.k.a. the **mortality rate function** in life insurance mathematics). For an absolutely continuous RV $\tau \geq 0$ having a DF F with density f, the hazard rate function $r(t)$ is defined by

$$r(t) := \frac{f(t)}{1 - F(t)} = \frac{f(t)}{\overline{F}(t)}.$$

[1] In discrete time when t and s can only take integer values, the only memoryless distribution is the *geometric* one.

This can be interpreted as follows: if T is the random life-time of a certain device, and the device has "survived" for the first t time units of its operation, then $r(t)$ gives the density of the "immediate failure" probability in the sense that

$$\mathbf{P}(\tau \in (t, t+dt) | \tau > t) = \frac{\mathbf{P}(\tau \in (t, t+dt))}{\mathbf{P}(\tau > t)} = \frac{f(t)\,dt}{\overline{F}(t)} = r(t)\,dt.$$

For an exponential RV $\tau \sim Exp(\lambda)$, the failure rate

$$r(t) = \lambda e^{-\lambda t}/e^{-\lambda t} = \lambda$$

is constant (no ageing!) and is often called just the **rate** of the RV τ. Note again that $rate = 1/mean$.

Of course, the property that the rate is constant is also characteristic for the exponential distribution. Indeed, if $r(t) = \lambda = \text{const}$ for $t > 0$, then, by the definition of $r(t)$,

$$\lambda = \frac{F'(t)}{1 - F(t)} = -(\log \overline{F}(t))' \Longrightarrow \log \overline{F}(t) = -\lambda t + C,$$

so that $\overline{F}(t) = e^{-\lambda t + C}$, $t > 0$. Now since $\tau > 0$ a.s., $\overline{F}(0) = 1$, and hence we infer that $C = 0$, so that F is just the DF of $Exp(\lambda)$.

E2. The minimum of several independent exponentially distributed RVs also has the exponential distribution (of which the rate is equal to the sum of the summands' rates). That is, if $\tau_1, \ldots, \tau_n$ are independent with $\tau_j \sim Exp(\lambda_j)$, then

$$M = \min(\tau_1, \ldots, \tau_n) \sim Exp\Big(\sum_{j=1}^{n} \lambda_j\Big).$$

Indeed, by independence

$$\begin{aligned}\mathbf{P}(M > t) &= \mathbf{P}(\tau_1 > t, \ldots, \tau_n > t) = \mathbf{P}(\tau_1 > t) \times \cdots \times \mathbf{P}(\tau_n > t) \\ &= e^{-\lambda_1 t} \cdots e^{-\lambda_n t} = \exp\Big\{-\Big(\sum_{j=1}^{n} \lambda_j\Big) t\Big\},\end{aligned}$$

which is the tail of $Exp(\lambda_*)$ with $\lambda_* = \sum_{j=1}^{n} \lambda_j$.

Thus, if the T_j's are the lifetimes of independently functioning vital components of a certain device, and all of them are exponentially distributed, then the lifetime of the device itself is also exponentially distributed. Moreover, one can easily derive the probability that it was the jth component whose failure caused the device to malfunction. For notational convenience, we will do that now for $j = n$ (the same argument works for any other value $j < n$ as well).

First note that $\tau := \min(\tau_1, \ldots, \tau_{n-1})$ and τ_n are independent and, as we saw above, $\tau \sim Exp\Big(\sum_{j=1}^{n-1} \lambda_j\Big)$. Therefore the probability that τ_n is the smallest of all the τ_j's is, by the TPF,

$$\begin{aligned}\mathbf{P}(\tau > \tau_n) &= \int_0^\infty \mathbf{P}(\tau > \tau_n | \tau_n = t)\, \mathbf{P}(\tau_n \in dt) \\ &= \int_0^\infty \mathbf{P}(\tau > t)\, \lambda_n e^{-\lambda_n t} dt = \int_0^\infty \exp\Big\{-t \sum_{j=1}^{n-1} \lambda_j\Big\} \lambda_n e^{-\lambda_n t} dt \\ &= \lambda_n \int_0^\infty e^{-\lambda_* t} dt = \frac{\lambda_n}{\lambda_*} \equiv \frac{\lambda_n}{\sum_{k=1}^n \lambda_k}, \qquad (5.5)\end{aligned}$$

which is just proportional to the rate of τ_n.

E3. *Connection with the Poisson process* (cf. Example 2.8). The exponential distribution describes the times between jumps in the Poisson process. We will begin with an argument showing how the process (and the exponentially distributed RVs as well) can arise in applications.

Suppose the time half-axis $(0, \infty)$ is partitioned into disjoint intervals ("time slots") $I_j = ((j-1)/n, j/n]$, $j = 1, 2, \ldots$, of length $1/n$, and during each time slot I_j, an independent event of probability λ/n occurs or does not occur. We assume that λ is a constant while $n \to \infty$, which refers to the so-called "rare events" scheme. Denoting by τ_1 the time till the first event, we see that, for $t = k/n$,

$$\begin{aligned}\mathbf{P}(\tau_1 > t) &= \mathbf{P}\left(\bigcap_{j=1}^{nt} \{\text{no event in } I_j\}\right) = \Big(\mathbf{P}(\text{no event in } I_1)\Big)^{nt} \\ &= \left(1 - \frac{\lambda}{n}\right)^{tn} \to e^{-\lambda t} \quad \text{as } n \to \infty\end{aligned}$$

(i.e., when the "time quanta" tend to zero), so that the distribution of τ_1 converges to the exponential law with parameter λ.

What can one say about the distribution of $N_t :=$ *# of events occurred prior to time t*? It is clear that we are in the conditions of the Poisson limit theorem:

$$N_t = \sum_{j \le nt} \underbrace{\mathbf{1}_{\{\text{event in the } j\text{th interval}\}}}_{1 \text{ w.p. } \lambda/n} \overset{\text{distr}}{\longrightarrow} Po(\lambda t) \quad \text{as } n \to \infty.$$

Also, the increment $N_{t+s} - N_t = \#$ *of events in the time interval* $(t, t+s]$ converges in distribution to $Po(\lambda s)$, and for disjoint intervals such increments are clearly independent, and hence the joint limiting distribution also has independent components. We see that the limiting process (for simplicity's sake we will also denote it by N_t) is nothing else but the Poisson process introduced in Example 2.8. The formal definition of the process is as follows.

5.2 The Poisson Process

An integer-valued non-decreasing right-continuous[2] SP $\{N_t\}_{t\geq 0}$ in continuous time is said to be a **Poisson process** with *rate* (intensity, parameter) $\lambda > 0$ if the following two conditions are met:

N1. For any $0 \leq s_1 < t_1 \leq s_2 < t_2 \leq \cdots \leq s_k < t_k$, the increments $N_{t_1} - N_{s_1}, N_{t_2} - N_{s_2}, \ldots, N_{t_k} - N_{s_k}$ are mutually independent RVs.

N2. For any $t > 0$, the RV $N_t \sim Po(\lambda t)$:

$$\mathbf{P}\,(N_t = k) = \frac{(\lambda t)^k}{k!} e^{-\lambda t}, \quad k = 0, 1, \ldots$$

This is a very popular model in a variety of applications, e.g., it is often used for claim occurrence epochs in insurance models.

It is not hard to see that the above conditions specify a *consistent family* of FDDs (see Section 2.10), and hence such a process does exist. It is also clear that it is Markovian and has non-decreasing trajectories (it is actually a special case of the so-called *counting processes* that *count* the numbers of events related to the corresponding time (or space) intervals). From N1–N2 one can easily derive that $N_{t+s} - N_t \sim Po(\lambda s)$, $t, s \geq 0$ (use GFs or MGFs, cf. p. 66), and that, for any fixed $h > 0$,

$$N_t^* = N_{h+t} - N_h, \quad t \geq 0,$$

is again a *Poisson process* (with the same rate, of course). Moreover, the last statement remains true even when $h = h(\omega)$ is an RV "depending on the past only", i.e., on the values N_s, $s \leq h(\omega)$ (such RVs are called *Markov* (or *stopping*) *times*). Therefore, one can expect that, in a Poisson process, the exponential distribution will describe not only the time till the first event, but also all the times between consecutive events.

[2]Recall that a function f_t, $t \in D \subset \mathbf{R}$, is said to be right-continuous at point $s \in D$ if $f_s = f_{s+} := \lim_{t \searrow s} f_t$. A function is said to be right-continuous if it is right-continuous at each point of its domain D that has "right neighbourhood".

Let $T_j := \min\{t > 0 : N_t = j\}$ be the time of the jth jump in the process $\{N_t\}$, $j = 1, 2, \ldots$, $T_0 := 0$. The times between consecutive jumps are just the differences $\tau_j := T_j - T_{j-1}$, $j \geq 1$.

Remark 5.1. Note that, in general, the distribution of *any counting process* $\{N_t\}$ is uniquely determined by that of the sequence $\{\tau_j\}$.

Theorem 5.1. *The RVs τ_j, $j = 1, 2, \ldots$, are i.i.d. and follow $Exp(\lambda)$ iff $\{N_t\}$ is the Poisson process with rate λ.*

This statement can actually be derived from the motivation argument in Section 5.1, where we demonstrated that the Poisson process arises as a limiting one in the discrete "rare events" scheme. Now we will only show formally that if $\tau_j \sim Exp(\lambda)$, then $N_t \sim Po(\lambda t)$. Indeed,

$$\mathbf{P}(N_t = 0) = \mathbf{P}(\tau_1 > t) = e^{-\lambda t},$$

which is a Poisson (with the parameter value λt) probability of 0. To complete the proof by induction, assume that we have already shown that $\mathbf{P}(N_t = k-1) = e^{-\lambda t}(\lambda t)^{k-1}/(k-1)!$ for a $k \geq 1$. Then, conditioning on the value of the first jump time T_1, we use the TPF to obtain (recall that f_{s-} denotes the *left limit* of the function f_t at point s, so that if $f_{s-} \neq f_s$ for a right-continuous function f_t it means that f_t has a jump at s)

$$\begin{aligned}
\mathbf{P}(N_t = k) &= \int_0^t \mathbf{P}(N_t - N_s = k-1 \,|\, \tau_1 = s)\, \mathbf{P}(\tau_1 \in ds) \\
&= \int_0^t \mathbf{P}(N_t - N_s = k-1 \,|\, N_{s-0} = 0, N_s = 1)\, \lambda e^{-\lambda s} ds \\
&= \int_0^t \mathbf{P}(N_t - N_s = k-1)\, \lambda e^{-\lambda s} ds \quad \text{by independent increments} \\
&= \int_0^t \frac{(\lambda(t-s))^{k-1}}{(k-1)!} e^{-\lambda(t-s)} \lambda e^{-\lambda s} ds \\
&= \frac{\lambda^k e^{-\lambda t}}{(k-1)!} \int_0^t (t-s)^{k-1} ds = \frac{(\lambda t)^k}{k!} e^{-\lambda t},
\end{aligned}$$

which is the $Po(\lambda t)$-probability of k. (You may also wish to use relation (5.6) of Problem 2 below to give an alternative derivation of this implication.)

In the remaining part of this section we will discuss some important aspects of Poisson processes.

First of all, due to the memoryless property of the exponential distribution, for a given time t the time till the next jump in the Poisson process is *independent* of the *past*, i.e., of the values N_s, $s \le t$ (this also follows directly from the definition of the process). Moreover, this is true when, instead of constant t, one considers a random Markov time. On the other hand, if τ_j—the times between consecutive jumps—are still i.i.d., but not exponentially distributed, the time till the first jump after t will *depend on the past.*

It is not hard to find the distribution of the time T_k of the kth jump in the process. We know that $T_k = \tau_1 + \cdots + \tau_k$ is a sum of i.i.d. RVs, and hence can use the convolution formula to find the density of T_k. But one could also observe that, due to the independent increments property, for $t \ge 0$,

$$\mathbf{P}(T_k \in (t, t+dt)) = \mathbf{P}(N_t = k-1, N_{t+dt} - N_t = 1)$$
$$= \underbrace{\mathbf{P}(N_t = k-1)}_{\frac{(\lambda t)^{k-1}}{(k-1)!}e^{-\lambda t}} \underbrace{\mathbf{P}(N_{t+dt} - N_t = 1)}_{\mathbf{P}(N_{dt}=1)=\lambda dt} = \lambda e^{-\lambda t}\frac{(\lambda t)^{k-1}}{(k-1)!}dt, \quad (5.6)$$

i.e., T_k follows $\gamma(k, \lambda)$—the gamma distribution with parameters $\alpha = k$ and λ (a.k.a. the **Erlang distribution** when the shape parameter α is integer-valued).

One can also establish the conditional distribution of the jumps' times within the interval $[0, t]$ given that there were exactly k jumps during that time interval. But first we will give a simple definition.

For a *sample* $X_1, \ldots, X_n$ of n RVs, the **order statistics** $X_{(j)}$ are the ordered values of the observations:

$$X_{(1)} \le X_{(2)} \le \cdots \le X_{(n)}.$$

That is, $(X_{(1)}, \ldots, X_{(n)})$ is obtained by ordering the values from the sample $(X_1, \ldots, X_n)$.

Theorem 5.2. *The conditional distribution of the vector of the jumps' times $(T_1, \ldots, T_k)$ given $N_t = k$ coincides with the distribution of the vector of order statistics $(U_{(1)}, \ldots, U_{(k)})$ for a sample of i.i.d. uniform on $[0, t]$ RVs $U_1, \ldots, U_k$. The same representation holds for the conditional distribution of $(T_1, \ldots, T_k)$ given $T_{k+1} = t$.*

Proof By our Remark 5.1 above, it suffices to show that the FDDs of the respective counting processes coincide. For simplicity's sake, we will show this for one-dimensional distributions only (i.e., just for the distributions the of the RVs N_s); the argument in the general case is almost identical and simply requires a slight extension of the result of our Example 2.5. We saw there that, for two independent RVs $X_i \sim Po(\lambda_i)$, $i = 1, 2$, the conditional distribution of X_1 given the sum $X_1 + X_2 = k$ is binomial with parameters k and $\lambda_1/(\lambda_1 + \lambda_2)$. Now, for $s < t$, $X_1 := N_s$ and $X_2 := N_t - N_s$ are independent Poisson RVs with parameters λs and $\lambda(t - s)$, respectively, so that the conditional distribution of N_s given $N_t \equiv N_s + (N_t - N_s) = k$ is binomial $B_{k,s/t}$.

On the other hand,

$$M_s := \#\{j \le k : U_j \le s\} = \max\{j \le k : U_{(j)} \le s\} = \sum_{j=1}^{k} \mathbf{1}_{\{U_j \le s\}}$$

is a counting process with jumps at the points $U_1, \ldots, U_k$. Since the indicators $\mathbf{1}_{\{U_j \le s\}}$ are independent Bernoulli RVs with success probabilities equal to $\mathbf{P}(U_j \le s) = s/t$, we have $M_s \sim B_{k,s/t}$, which coincides with the conditional distribution we derived above.

In the general case, we will simply have to deal with multinomial distributions instead of the binomial ones. □

Quite often one has sums of several counting processes (imagine several production lines whose operation is controlled using a single device testing the quality of items produced on all of them) or, on the contrary, "thinned" versions of the original process (defective items are discarded *etc.*). It turns out that, in the case of Poisson processes, such operations again lead to (different) Poisson processes.

Theorem 5.3. *Let $\{N_t\}$ and $\{M_t\}$ be two independent Poisson processes with rates λ and μ, respectively. Then the sum $L_t := M_t + N_t$, $t \ge 0$, is a Poisson process with rate $\lambda + \mu$.*

A similar assertion holds true for the sum of any number of independent Poisson processes.

An important notion related to counting processes is that of the **point process**. Roughly speaking, a point process is a *random measure* assigning unit values to the points where the counting process has jumps. It is a more universal tool: it can be used when we have random points positioned not on the real line, but possibly on some abstract space, where there is no natural ordering, and hence the very idea of counting the points using a counting process does not work. The point process corresponding to $\{L_t\}$ is called then the *superposition* of the point processes generating $\{M_t\}$ and $\{N_t\}$.

Proof Firstly, by independence, for any $t > 0$, $N_t + M_t \sim Po((\lambda + \mu)t)$ from Example 2.1. And secondly, it is obvious that $\{L_t\}$ has independent increments, for both $\{M_t\}$ and $\{N_t\}$ have this property and are independent of each other. □

Remark 5.2. Observe that, in the new process $\{L_t\}$, the time till the first jump (t.t.f.j.) is equal to the minimum

$$\min\{\text{t.t.f.j. in } \{M_t\}, \text{ t.t.f.j. in } \{N_t\}\} = \min\{\tau_1^{(M)}, \tau_1^{(N)}\} \sim Exp(\lambda + \mu)$$

from E2. Of course, the last relation is also an immediate consequence of Theorem 5.3.

Example 5.2. A shop has two entrances, one from XY street and the other from YX street. Flows of customers arriving in the shop from these two entrances are represented by independent Poisson processes with rates $\lambda_1 = 0.5$ min^{-1} and $\lambda_2 = 1.5$ min^{-1}, respectively.

(i) What is the probability that no new customers will enter the shop during a fixed three-minute time interval?

(ii) What is the mean time between successive arrivals of new customers?

(iii) What is the probability that a given customer entered the shop from XY street?

To answer all these questions, we first note that the arrival of customers to the shop is described by the sum of two independent Poisson processes and hence is a Poisson process itself, with the rate $\lambda = \lambda_1 + \lambda_2 = 2$ (min^{-1}). Therefore the inter-arrival times $\tau_j \sim Exp(2)$, and the answer to (i) is given by

$$\mathbf{P}(\tau_1 > 3) = e^{-2\times 3} = e^{-6} \approx 0.0025.$$

(ii) The mean inter-arrival time is clearly $1/\lambda = 0.5$ min.

(iii) Recalling relation (5.5), we see that the desired probability is given by

$$\frac{\lambda_1}{\lambda_1 + \lambda_2} = \frac{0.5}{2} = 0.25.$$

Now consider an "inverse" situation: suppose that, in a Poisson process $\{N_t\}$, each jump ("arrival") is marked independently w.p. $p \in [0, 1]$, and denote by M_t the number of "marked arrivals" by the time t.

Theorem 5.4. *Under the above assumptions, $\{M_t\}$ and $\{N_t - M_t\}$ are independent Poisson processes with rates $p\lambda$ and $(1 - p)\lambda$, respectively.*

Proof We will again give the proof for the "one-dimensional" (i.e., corresponding to a single fixed time instance) distributions only. The argument in the general case is similar. One has

$$\begin{aligned}
&\mathbf{P}(M_t = j,\ (N-M)_t = k) \\
&\quad = \underbrace{\mathbf{P}(M_t = j,\ (N-M)_t = k \mid N_t = k+j)}_{B_{k+j,p}(\{j\})}\ \underbrace{\mathbf{P}(N_t = k+j)}_{\frac{(\lambda t)^{k+j}}{(k+j)!}e^{-\lambda t}} \\
&\quad = \frac{(k+j)!}{k!j!}p^j(1-p)^k \times \frac{(\lambda t)^{k+j}}{(k+j)!}e^{-\lambda t} \\
&\quad = \underbrace{\frac{(p\lambda t)^j}{j!}e^{-p\lambda t}}_{Po(p\lambda)\ \text{of}\ \{j\}} \times \underbrace{\frac{((1-p)\lambda t)^k}{k!}e^{-(1-p)\lambda t}}_{Po((1-p)\lambda)\ \text{of}\ \{k\}},
\end{aligned}$$

so that M_t and $(N-M)_t$ are independent Poisson RVs with parameters $p\lambda t$ and $(1-p)\lambda t$, respectively. □

Example 5.3. Suppose that the customers' flow to a shop is described by a Poisson process with rate $\lambda = 25$ hour^{-1}. We know that each of the customers is a female w.p. $p = 0.8$. What is the probability that no male customer will enter the shop between 11.00 am and 11.15 am?

It follows from the assumptions and Theorem 5.4 that the arrival flows of male and female customers are independent with rates $\lambda^{(m)} = (1-p)\lambda = 0.2 \times 25 = 5$ hour^{-1} and $\lambda^{(f)} = p\lambda = 0.8 \times 25 = 20$ hour^{-1}, respectively. Hence during a fixed 15 min = 0.25 hour interval, the probability that there will be no events (= arrivals) from the first of these processes is

$$\mathbf{P}(N_{0.25}^{(n)} = 0) = e^{-\lambda^{(m)}t}\Big|_{t=0.25} = e^{-1.25} \approx 0.287.$$

As we said, the Poisson process model is often used for claim occurrence epochs in insurance applications. Since the claims themselves should typically be modelled by RVs as well, the following more general model is quite popular.

Let $\{N_t\}$ be a Poisson process with rate $\lambda > 0$, and $X_1, X_2, \ldots$ be i.i.d. RVs independent of the process. Then the SP

$$Y_t := \sum_{j \le N_t} X_j, \quad t \ge 0, \tag{5.7}$$

is said to be a **compound Poisson process**.

Like the Poisson process, the compound Poisson process clearly has independent increments, with the distribution of $Y_{t+s} - Y_t$ being independent of t ("stationary increments"). Moreover, one can readily derive the ChF (or any other integral transform we discussed earlier) of Y_t. Indeed, setting $S_k := X_1 + \cdots + X_k$ and conditioning on the value of N_t, we get from independence that

$$\begin{aligned}
\varphi_{Y_t}(iv) \equiv \mathbf{E}\, e^{ivY_t} &= \sum_{k=0}^{\infty} \mathbf{E}\,(e^{ivS_k} | N_t = k)\, \mathbf{P}\,(N_t = k) && \text{by the TPF} \\
&= \sum_{k=0}^{\infty} \big(\varphi_{X_1}(iv)\big)^k e^{-\lambda t} \frac{(\lambda t)^k}{k!} && \text{from (2.70)} \\
&= g_{N_t}\big(\varphi_{X_1}(iv)\big) = e^{\lambda t(\varphi_{X_1}(iv)-1)} && \text{from (2.75).}
\end{aligned} \qquad (5.8)$$

So the distribution of Y_t can be (relatively easily) evaluated. It is particularly easy when $X_j \geq 0$ are integer-valued. Then, equivalently to (5.8), the GF of Y_t is equal to

$$g_{Y_t}(z) = e^{\lambda t(g_{X_1}(z)-1)}.$$

Differentiating w.r.t. z, we get

$$g'_{Y_t}(z) = \lambda t g'_{X_1}(z) e^{\lambda t(g_{X_1}(z)-1)} = \lambda t g'_{X_1}(z)\, g_{Y_t}(z).$$

On the right-hand side we have a product of two functions of z. Further differentiating and using the relation

$$\frac{d^n}{dz^n}(a(z)b(z)) \equiv (a(z)b(z))^{(n)} = \sum_{j=0}^{n} \binom{n}{j} a^{(n-j)}(z)\, b^{(j)}(z)$$

(easily verifiable via mathematical induction), setting $z := 0$ yields

$$g_{Y_t}^{(k)}(0) = \lambda t \sum_{j=0}^{k-1} \binom{k-1}{j} g_{X_1}^{(k-j)}(0)\, g_{Y_t}^{(j)}(0).$$

Now (2.73) immediately gives us the following recursion relation for the probabilities $q_k := \mathbf{P}\,(Y_t = k)$: if we set $p_k := \mathbf{P}\,(X_1 = k)$, $k = 0, 1, 2, \ldots$, then

$$q_0 = e^{\lambda t(p_0 - 1)}, \quad q_k = \frac{\lambda t}{k} \sum_{j=0}^{k-1} (k-j) p_{k-j} q_j, \quad k \geq 1. \qquad (5.9)$$

To illustrate this scheme, note that when, say, $X_j \sim B_p$ (i.e., $p_0 = 1 - p$, $p_1 = p$), Y_t will simply be a "thinned" version of N_t and hence coincide with the Poisson process M_t with the rate $p\lambda$ from Theorem 5.4. On the other hand, (5.9) yields $q_0 = e^{-\lambda pt}$ and $q_k = \frac{\lambda t}{k} \times pq_{k-1}$ (only the last terms are non-zero in all the sums), so that $q_k = (\lambda pt)^k e^{-\lambda pt}/k!$, which is exactly the same result.

Compound Poisson processes play an important role in both applications and theory. For example, the classical compound Poisson risk model (also called the Cramér-Lundberg model) assumes that the risk process is given by

$$R(t) = u + \beta t - Y_t,$$

where u is the initial risk reserve, β is a constant premium collection rate, $Y_t = S_{N_t}$ is the aggregate amount of all claims arrived in the interval $[0, t]$. There exists a well developed theory for such risk processes.[3]

As for the importance of compound Poisson processes for theoretical probability, they have the remarkable property that, roughly speaking, any continuous time process with stationary independent increments (a.k.a. a Lévy[4] process) can be approximated arbitrary closely by such processes.

In the general case, any Lévy process $\{X_t\}$ can be decomposed as

$$X_t = X_0 + at + \sigma^2 W_t + J_t, \quad t \geq 0, \tag{5.10}$$

where a and $\sigma^2 \geq 0$ are constant trend and diffusion coefficients, $\{W_t\}$ the standard Brownian motion process, and $\{J_t\}$ a pure jump process independent of $\{W_t\}$. The process $\{J_t\}$ can be described as follows: there exists a *jump* (or *spectral*) *measure* Π on $(\mathbf{R}, \mathcal{B})$ such that, for any Borel set $A \subset \mathbf{R}$ (more precisely, $A \subset \mathbf{R} \setminus (-\varepsilon, \varepsilon)$ for an arbitrary $\varepsilon > 0$, to ensure that there will be finitely many jumps of sizes $\in A$ in any time interval $[0, t]$), the process

$$N_t^{(A)} := \#\{\text{times } s \in (0, t] : J_s - J_{s-0} \in A\},$$

counting the number of jumps whose values are in A (recall that $J_{s-0} = \lim_{u \nearrow s} J_u$ denotes the *left limit* of the process J_t at the point s; note that this can differ from J_s—this is the case when the process has a jump at that time s) is a Poisson process with rate $\Pi(A)$. For disjoint sets A and B, the processes $N_t^{(A)}$ and $N_t^{(B)}$ are independent of each other.

To illustrate the above-stated remarkable property, note that each of the components in decomposition (5.10) can be approximated by a compound Poisson process. Indeed, by the LLN, the term at can be obtained as a limit (as $n \to \infty$) of the sequence of processes $\{X_t^{(n)}\}$ having jumps of constant size a/n with rate n. These are clearly compound Poisson processes, with the ChF of $X_t^{(n)}$ being $\varphi(iv) = \exp\{nt(e^{iav/n} - 1)\} \to e^{iatv}$ as $n \to \infty$. The diffusion term $\sigma^2 W_t$ can also be obtained as a limit of compound Poisson

[3]For more detail, see, e.g., Rolski, T. *et. al.*, *Stochastic Processes for Insurance and Finance*, Wiley, New York, 1999.

[4]After Paul Pierre Lévy (15.09.1886–15.12.1971), an outstanding French mathematician, famous for his contributions to Probability Theory who can be considered (along with A.N. Kolmogorov) as one of the founders of the modern theory of stochastic processes.

processes: this time we assume that the jumps have rate n, but are equal to $\pm\sigma n^{-1/2}$ w.p. 1/2, and use the CLT. As for J_t, if the total mass $\Pi(\mathbf{R}) < \infty$ (in the general case, one can have $\Pi((-\varepsilon, \varepsilon)) = \infty$ for any $\varepsilon > 0$, which means that the total "intensity of jumps" is infinite, but most of them are so small that the values of J_t are finite), it is a compound Poisson process itself, with the rate $\lambda = \Pi(\mathbf{R})$ and "compounding distribution" $\mathbf{P}(X_j \in A) = \Pi(A)/\Pi(\mathbf{R})$.

5.3 Problems

1. Prove that if a non-negative RV τ has property (5.4), then it has the exponential distribution $Exp(\lambda)$ for some $\lambda > 0$.

2. Calculate the ChF of $\tau_j \sim Exp(\lambda)$ and find that of T_k. Using the inversion formula (2.77), derive a recursive relation for the densities f_{T_k} of T_k, $k > 1$, (expressing f_{T_k} in terms of $f_{T_{k-1}}$), and from that relation get an explicit expression for f_{T_k}.

3. *The waiting times paradox.* For a Poisson process $\{N_t\}$ with rate λ, denote by $T_{j(t)}$ the time of the first jump occurred *after* the time t; that is, the index $j(t)$ is determined from the relation $t \in [T_{j(t)-1}, T_{j(t)})$. We know that, for any given t, the time $T_{j(t)} - t$ till the next jump is $Exp(\lambda)$-distributed. On the other hand, the lengths of all the time intervals $[T_{j-1}, T_j)$ between consecutive jumps are also $Exp(\lambda)$-distributed. In particular, the length $T_{j(t)} - T_{j(t)-1}$ of the time interval which covers our t is also $Exp(\lambda)$-distributed. But clearly $T_{j(t)} - t < T_{j(t)} - T_{j(t)-1}$! Therefore the two cannot have the same distribution! What is wrong in the above argument?

4. Complete the proof of Theorem 5.2. Namely, show that, for any $0 < t_1 < t_2 < \cdots < t_m < t$, the conditional distribution of $(N_{t_1}, \ldots, N_{t_m})$ given that $N_t = k$ coincides with the distribution of $k(F_k^*(t_1), \ldots, F_k^*(t_m))$, where F_k^* is the EDF for a sample of k i.i.d. $U(0,t)$-RVs (cf. Example 2.6; the latter vector's components are just the values of the process "counting" the points from the sample).

5. Let $\{N_t\}$ be a Poisson process with rate $\lambda = 1$, and $N_{(t,t+h]} := N_{t+h} - N_t$ the increment of the process on the time interval $(t, t+h]$, $t, h \geq 0$ (the number of events in this time interval). Find

   (i) $\mathbf{P}(N_1 = 1)$;

   (ii) $\mathbf{P}(N_4 = 3 \mid N_2 = 1)$;

   (iii) $\mathbf{E}\, N_4$ and $\mathbf{E}\, N_{(1,5]}$;

   (iv) $\mathbf{P}(N_{(4,7]} = 2)$;

   (v) $\mathbf{P}(N_{(4,7]} = 2, N_{(3,6]} = 1)$;

   (vi) $\mathbf{P}(N_{(4,7]} = 2 \mid N_{(1,5]} = 2)$.

Hints. (vi) $N_{(1,7]}$ is the sum of three independent Poisson RVs which are the numbers of the events in the time intervals $(1, 4]$, $(4, 5]$, and $(5, 7]$, resp.

6. Consider a system of four identical machines each having an expected lifetime of 10 days. At time 0 one of these machines starts to work, the others are "put on stand-by". When the "active" machine fails, it is immediately replaced by a stand-by machine which then becomes active. Broken machines are not repaired and stand-by machines cannot fail. We say that the system breaks down, when the 4th machine breaks down.

   Assuming that the lifetimes of the machines are exponentially distributed independent RVs,

   (i) calculate the probability that the system is still operating after 40 days given that the first breakdown of a machine occurred after 3 days, and the second after 13 days of the system's operation;

   (ii) find the probability that the system will break down during the first week (i.e., in 7 days).

7. Ships pass a bird sanctuary according to a Poisson process with rate one per hour. Twenty per cent of the ships are oil tankers.

   (i) What is the probability that at least one oil tanker will pass during a (24-hour) day?

   (ii) If 30 ships have passed by in one day, what is the probability that 6 of them were oil tankers?

8. Let N_t be a Poisson process with rate $\lambda = 10$ per hour describing the arrivals of customers to a bank. Each customer brings some money in or withdraws some. Let X_j be the amount brought in by the jth customer; we assume that the X_j's are i.i.d. RVs independent of $\{N_t\}$, with the distribution

$$\mathbf{P}(X_j = k) = \frac{1}{10}, \quad k = -4, -3, \ldots, 5 \quad \text{(thousand dollars)}$$

   (when $X_j < 0$, the jth customer actually withdraws the amount $|X_j|$).

   Then the "balance" (the amount of money brought in/withdrawn) in the bank after t hours is given by the compound Poisson process

$$Y_t = X_1 + X_2 + \cdots + X_{N_t}$$

   (if $N_t = 0$, then the sum is empty and hence equals zero).

   (i) Draw a "typical" realisation (path, trajectory) of the process $\{Y_t\}$ (you can simulate it, or just depict what you think may be typical).

   (ii) Calculate the expected amount of money in the bank at the end of a working day (i.e., after 7 hours) and the variance of that quantity. Justify your calculation.

(iii) Basing on (ii), describe the long-term behaviour of Y_t.

9. There are 6 phones in the Carlton office of XYZ Company Pty Ltd. We know that, during a working day, phones 1, 2, 3 and 4 are called with rates 5 per hour each, and phones 5 and 6 with rates 10 per hour each (the "call flows" are assumed to be independent Poisson processes). The working day starts at 9 am.

   (i) What is the probability that there will be no calls to phone 1 during the time interval from 2:00 pm to 2:30 pm?

   (ii) What is the density (probability density function) of the time of the first call to the office?

   (iii) What is the probability that this first call will be to phone 5?

   (iv) Calls to phone 5 can be answered by two clerks. When the phone rings, they flip a fair coin to decide who is to answer the call. What is the probability that the first clerk will answer only one call during the first hour of his work?

Chapter 6

Jump Markov Processes

6.1 Definitions and Basic Results

A stochastic process $\{X_t\}$ in continuous time is said to be a **Markov process** (MP) if, for that process,

$$\mathbf{P}(\{\text{future}\}|\{\text{exact present}\}\ \&\ \{\text{past}\}) = \mathbf{P}(\{\text{future}\}|\{\text{exact present}\}),$$

or, a bit more formally, if, for any time t, state x and sets A and B (from appropriate σ-fields of sets)

$$\mathbf{P}(\{X_s,\ s > t\} \in A|X_t = x,\ \{X_u,\ u < t\} \in B) \\ = \mathbf{P}(\{X_s,\ s > t\} \in A|X_t = x).$$

We will consider **pure jump MPs** whose state space is $S = \{0, 1, 2, \dots\}$ (such processes are sometimes called *continuous time MCs*). The law of the evolution of such a process is completely determined by its initial distribution and **transition probabilities** $\mathbf{P}(X_{s+t} = k|\, X_s = j)$. If these probabilities do not depend on s: for any $j, k \in S$,

$$p_{jk}(s, s+t) := \mathbf{P}(X_{s+t} = k|\, X_s = j) = \mathbf{P}(X_t = k|\, X_0 = j) =: p_{jk}^{(t)},$$

one says that the MP is **homogeneous**. We will mainly deal with such MPs, and first we assume that the property holds to make the exposition more similar to that in the chapter on MCs.

Transition probabilities, as it was the case for discrete time MCs (and for the same reasons, see p. 84), also satisfy the *Chapman-Kolmogorov equations*: in matrix notation, for $P^{(t)} := (p_{jk}^{(t)})$ and $s, t > 0$, one has

$$P^{(t+s)} = P^{(t)}P^{(s)}. \tag{6.1}$$

From this relation one obtains, in particular, the following representation for the FDDs of the MP: for any $n \geq 1$, $t_0 < t_1 < \cdots < t_n$, $k_m \in S$,

$m = 1, \ldots, n$, and initial distribution $p_j = \mathbf{P}(X_0 = j)$, $j \in S$,

$$\mathbf{P}(X_{t_1} = k_1, X_{t_2} = k_2, \ldots, X_{t_n} = k_n) = \sum_j p_j p_{jk_1}^{(t_1)} p_{k_1 k_2}^{(t_2 - t_1)} \cdots p_{k_{n-1}k_n}^{(t_n - t_{n-1})} \tag{6.2}$$

similarly to (3.14).

The fact that t and s are now arbitrary (not necessarily integer) in (6.1) provides one with a convenient and efficient way of finding and/or analysing the behaviour of the transition probabilities based on differential equations.

Recall that, for discrete time MCs, the Chapman-Kolmogorov equation (3.12) implies that $P^{(t)} = P^t$, where $P = P^{(1)}$ (cf. (3.13)). Now we have from (6.1) that

$$P^{(t+h)} - P^{(t)} = P^{(t)}(P^{(h)} - I) = (P^{(h)} - I)\,P^{(t)}, \quad h > 0, \tag{6.3}$$

where $I = \text{diag}\{1, 1, \ldots\}$ is the identity matrix (operator).

An SP $\{X_t\}$ is said to be *stochastically continuous* at time point t if $X_{t+h} \to X_t$ in probability as $h \searrow 0$ (cf. (2.39)). In the case of homogeneous pure jump MPs, this property is implied by each of the two following relations: as $h \searrow 0$,

$$\mathbf{P}(X_{t+h} \neq X_t) \to 0 \quad \Longleftarrow \quad P^{(h)} \to P^{(0)} \equiv I \quad \text{(component-wise)}, \tag{6.4}$$

where the arrow "$\Longleftarrow$" indicates that the former relation, in turn, follows from the latter one.

When the above holds, under an additional technical condition on the process, we get, dividing the expressions in (6.3) by h and letting $h \searrow 0$, that

$$\frac{d}{dt} P^{(t)} = P^{(t)} A = A P^{(t)}, \tag{6.5}$$

where the derivative of the matrix function $P^{(t)}$ is understood in the component-wise sense (so it is the matrix $(\frac{d}{dt} p_{ij}^{(t)})$), and

$$A \equiv (a_{jk}) := \frac{d}{dt} P^{(t)}\Big|_{t=0} \equiv \lim_{h \searrow 0} \frac{1}{h}(P^{(h)} - I) \tag{6.6}$$

is a matrix with finite entries referred to as the **generator** (or *infinitesimal operator*) of the MP $\{X_t\}$ (or of the family $\{P^{(t)}\}$). Note that, for any generator A, all the row sums are zeros: $\sum_k a_{jk} = 0$, $j \in S$ (for this is true for the differences $P^{(h)} - I$).

The system (6.5) of linear differential equations with constant coefficients can be solved explicitly. Recall that, for a scalar-valued differentiable function $f(t)$, an analogue of (6.5):

$$\frac{d}{dt} f(t) = a f(t)$$

implies that $f(t) = f(0)e^{at}$. It turns out that, similarly, system (6.5) has a unique solution

$$P^{(t)} = \exp(tA) := \sum_{n=0}^{\infty} \frac{t^n}{n!} A^n, \qquad t \geq 0 \tag{6.7}$$

(note that the initial condition is $P^{(0)} = I$; also, we set $A^0 := I$). This implies that the generator A *completely determines* (like P in the discrete time case) all the transition probabilities $p_{jk}^{(t)}$ and hence the law of the evolution of the MP.

Although the series (6.7) gives an explicit representation for the transition probabilities, from the computational point of view, it may prove to be more practical to solve the system of differential equations (6.5) numerically. We will return to this question at the end of this section.

Conditions for ergodicity (existence of the limiting distribution independent of the initial state) are similar to those for discrete time MCs (they are even a bit simpler for now there is no such thing as periodicity). Thus, if there is only one class of essential states in our MP and there exists a state j_0 such that, starting at an arbitrary state j, our MP will eventually visit the state j_0 w.p. 1, and the recurrence time to j_0 has a finite mean, then there exist the limits

$$\lim_{t \to \infty} p_{jk}^{(t)} = \pi_k, \quad k \in S, \tag{6.8}$$

independent of j. The assertion can be proved employing an argument similar to the standard one for MCs and based on the use of renewal theory.

That is,

$$\lim_{t \to \infty} P^{(t)} = \Pi, \tag{6.9}$$

where the limiting matrix's rows are all identical to $\boldsymbol{\pi} = \{\pi_k\}$ (cf. (3.7)). In the same way as in the case of Markov chains (cf. (3.28)), it follows from Chapman-Kolmogorov equations (6.1) that the distribution $\boldsymbol{\pi}$ is *stationary* for the process $\{X_t\}$:

$$\boldsymbol{\pi} = \boldsymbol{\pi} P^{(t)} \quad \text{for any} \quad t \geq 0.$$

Taking derivatives on both sides and setting $t := 0$, we get the linear system

$$0 = \boldsymbol{\pi} A. \tag{6.10}$$

If (6.8) holds, the stationary distribution $\boldsymbol{\pi}$ is the unique solution to the system (6.10) with the additional equation $\sum_j \pi_j = 1$.

Thus knowing the explicit form of the generator A is very important for studying the MP. It turns out that the entries of A have a simple interpretation which can be very helpful for finding the generator itself.

Indeed, assume that $X_t = j$. Due to the Markov property, the time our MP will stay at the state j after time t *does not depend* on *how long* it has already been there. Since the MP is time homogeneous, the *transition rate* from j:

$$\lambda_j := \frac{\mathbf{P}(X_{t+dt} \neq j | X_t = j)}{dt} = \lim_{h \searrow 0} h^{-1} \mathbf{P}(X_{t+h} \neq j | X_t = j)$$

is constant (does not depend on time t) and depends on j only. This means that the "sojourn time" (the duration of the period of time the MP spends *this time* at the state j) is exponentially distributed, the value of the distribution's parameter being λ_j (for the hazard rate is constant, see our discussion of property E1 in Section 5.1).

Now, recalling the definition of A, we have $P^{(h)} = P^{(0)} + hA + o(h)$ as $h \searrow 0$, which is just a one-term Taylor expansion; recall that notation $o(h)$ (as $h \to 0$) is used in mathematics to represent any function $g(h)$ of h having the property that $g(h)/h \to 0$ as $h \to 0$ (in words, $g(h)$ *vanishes faster* than h as $h \to 0$, cf. p. 94). Equivalently,

$$\begin{aligned} p_{jk}^{(h)} &= \mathbf{P}(X_{t+h} = k | X_t = j) \\ &= (I + hA)_{jk} + o(h) = \begin{cases} ha_{jk} + o(h), & j \neq k, \\ 1 + ha_{jj} + o(h), & j = k. \end{cases} \end{aligned} \tag{6.11}$$

As the left-hand side is a value between 0 and 1, we must have $a_{jk} \geq 0$ for $j \neq k$, and $a_{jj} \leq 0$; the entry a_{jk} is nothing else but the rate of the transition $j \mapsto k$.

To find the *total transition rate* from j (which clearly coincides with the rate λ_j of the sojourn time), observe that, omitting for simplicity all the terms $o(h)$, one has

$$\mathbf{P}(X_{t+h} \neq j | X_t = j) = \sum_{k \neq j} \mathbf{P}(X_{t+h} = k | X_t = j) = h \sum_{k \neq j} a_{jk} = -h a_{jj}$$

since the row sums are all zeros in A. Therefore

$$\lambda_j = -a_{jj} = |a_{jj}|,$$

i.e., the rates of the sojourn times are just the absolute values of the elements on the main diagonal of A.

Given a transition from j occurred, the probability that the new value of the process is k is given by $a_{jk}/|a_{jj}|$. Indeed, for $k \neq j$ and $h \searrow 0$,

$$\begin{aligned}
&\mathbf{P}(X_{t+h} = k|\, X_{t+h} \neq j, X_t = j) \\
&\quad = \frac{\mathbf{P}(X_{t+h} = k, X_{t+h} \neq j, X_t = j)}{\mathbf{P}(X_{t+h} \neq j, X_t = j)} = \frac{\mathbf{P}(X_{t+h} = k, X_t = j)}{\mathbf{P}(X_{t+h} \neq j, X_t = j)} \\
&\quad = \frac{\mathbf{P}(X_{t+h} = k, X_t = j)}{\mathbf{P}(X_t = j)} \times \frac{\mathbf{P}(X_t = j)}{\mathbf{P}(X_{t+h} \neq j, X_t = j)} \\
&\quad = \frac{\mathbf{P}(X_{t+h} = k|\, X_t = j)}{\mathbf{P}(X_{t+h} \neq j|\, X_t = j)} = (1 + o(1))\frac{h a_{jk}}{h|a_{jj}|} \to \frac{a_{jk}}{|a_{jj}|}.
\end{aligned}$$

Thus the **evolution of the MP** $\{X_t\}$ can be described as follows: starting at some initial state $X_0 = j$ (chosen at random according to the initial distribution $\boldsymbol{p} = (p_j)$ on the state space S), the MP "sits" at j for a random time $\sim Exp(-a_{jj})$. Then it jumps to another state—and this new state will be $k \neq j$ with probability $a_{jk}/|a_{jj}|$—and sits there for an independent random time $\sim Exp(-a_{kk})$, and so on.

Example 6.1. *The Poisson process* $\{N_t\}$ *with rate* λ. We clearly have, for $j = 0, 1, 2, \ldots$ and $t \geq 0$,

$$\begin{aligned}
p_{j,j+k}^{(t)} &= \frac{(\lambda t)^k}{k!} e^{-\lambda t}, \qquad k = 0, 1, 2, \ldots; \\
p_{j,j+k}^{(t)} &= 0, \qquad\qquad\qquad k < 0.
\end{aligned}$$

Therefore $\frac{d}{dt} p_{j,j+k}^{(0)} = 0$ for $k \neq 0, 1$, while

$$\frac{d}{dt} p_{jj}^{(0)} = -\lambda e^{-\lambda t}\Big|_{t=0} = -\lambda, \quad \frac{d}{dt} p_{j,j+1}^{(0)} = \lambda e^{-\lambda t} - \lambda^2 t e^{-\lambda t}\Big|_{t=0} = \lambda,$$

so that the generator of the Poisson process has the form

$$A = \begin{pmatrix}
-\lambda & \lambda & 0 & 0 & 0 & \cdots \\
0 & -\lambda & \lambda & 0 & 0 & \cdots \\
0 & 0 & -\lambda & \lambda & 0 & \cdots \\
0 & 0 & 0 & -\lambda & \lambda & \cdots \\
0 & 0 & 0 & 0 & -\lambda & \cdots \\
\cdot & \cdot & \cdot & \cdot & \cdot & \ddots
\end{pmatrix}. \tag{6.12}$$

Now let us solve in this special case the inverse problem: how to find $p_{jk}^{(t)}$ from A? One could try to compute $\exp(tA)$; an equivalent approach is to solve the system of linear differential equations (6.5) which takes in this special case a particularly simple form. It is not hard to see from that system, that, for A given by (6.12), for any j, equations for $p_{j,j+k}$,

$k = 0, 1, 2, \ldots$, will coincide with those for p_{0k}, $k = 0, 1, 2, \ldots$, respectively. Hence it suffices to find $p_{0k}^{(t)}$, that is, the entries from the first row of $P^{(t)}$. To this end, take the initial distribution $\boldsymbol{p} = (1, 0, 0, \ldots)$ and note that the above-mentioned row is just the product $\boldsymbol{p}_0^{(t)} = \boldsymbol{p}P^{(t)}$. Multiplying relation (6.5) from the left by the constant vector $\boldsymbol{p}$, we see that $\boldsymbol{p}_0^{(t)}$ satisfies the system

$$\frac{d}{dt}\boldsymbol{p}_0^{(t)} = \boldsymbol{p}_0^{(t)} A$$

or, equivalently,

$$\begin{aligned} \frac{d}{dt}p_{00}^{(t)} &= -\lambda p_{00}^{(t)}, \\ \frac{d}{dt}p_{01}^{(t)} &= \lambda(p_{00}^{(t)} - p_{01}^{(t)}), \\ \frac{d}{dt}p_{02}^{(t)} &= \lambda(p_{01}^{(t)} - p_{02}^{(t)}), \\ &\ldots \end{aligned}$$

The first equation, together with the obvious initial condition $p_{00}^{(0)} = 1$, immediately implies that $p_{00}^{(t)} = e^{-\lambda t}$, $t \geq 0$. Substituting this into the second equation yields

$$\frac{d}{dt}p_{01}^{(t)} = \lambda e^{-\lambda t} - \lambda p_{01}^{(t)}.$$

Applying the standard techniques for solving first-order linear differential equations with constant coefficients and using the initial condition $p_{01}^{(0)} = 0$, we find that $p_{01}^{(t)} = \lambda t e^{-\lambda t}$. Continuing this process, we will see that the transition probabilities $p_{0k}^{(t)}$, $k = 0, 1, 2, \ldots$, (and hence also $p_{j,j+k}^{(t)}$, $k = 0, 1, 2, \ldots$, for all $j > 0$) are given by the Poisson distribution with parameter λt.

If we attempted to find the stationary distribution of the Poisson process (does it exist?) solving the system (6.10), which now has the form

$$\begin{aligned} 0 &= -\lambda \pi_1, \\ 0 &= \lambda \pi_1 - \lambda \pi_2, \\ 0 &= \lambda \pi_2 - \lambda \pi_3, \\ &\ldots \end{aligned}$$

we would immediately get $\pi_1 = \pi_2 = \pi_3 = \cdots = 0$ (which is quite natural, for the process clearly drifts away to infinity; moreover, all the states are seen to be non-essential).

Example 6.2. Consider a jump MP $\{X_t\}_{t\geq 0}$ with state space $S = \{1,2\}$ and generator

$$A = \begin{pmatrix} -\lambda & \lambda \\ \lambda & -\lambda \end{pmatrix} \equiv \lambda D, \quad \text{where} \quad D = \begin{pmatrix} -1 & 1 \\ 1 & -1 \end{pmatrix}, \quad \lambda > 0.$$

It follows from our discussion of the meaning of the entries of A that the process evolves as follows: it simply alternates its values from 0 to 1 and the other way around at consecutive jump times in a Poisson process with rate λ. Such a process (or, rather, the process $\{2X_t - 1\}$) is known as the *telegraph signal process.*

In this simple case we can use (6.7) to find a closed form expression for the transition matrix $P^{(t)}$. Indeed, note that

$$D^2 = -2D, \; D^3 = D^2 \times D = -2D^2 = (-2)^2 D, \; \dots \; , \; D^n = (-2)^{n-1} D.$$

Therefore $A^n = -\frac{1}{2}(-2\lambda)^n D$, $n \geq 1$, and hence by (6.7)

$$P^{(t)} = \sum_{n=0}^{\infty} \frac{t^n}{n!} A^n = I - \frac{1}{2} \sum_{n=1}^{\infty} \frac{(-2\lambda t)^n}{n!} D = I - \frac{1}{2}(e^{-2\lambda t} - 1)D,$$

i.e., for $j, k = 1, 2$,

$$p_{jj}^{(t)} = \frac{1}{2}(1 + e^{-2\lambda t}), \quad p_{jk}^{(t)} = \frac{1}{2}(1 - e^{-2\lambda t}), \quad j \neq k.$$

Clearly, as $t \to \infty$, all $p_{jk}^{(t)} \to 1/2$, which means that the process is ergodic with the uniform stationary distribution $(1/2, 1/2)$.

6.2 Inhomogeneous Processes

Example 6.3. *Time-inhomogeneous Poisson processes.* The notion of the Poisson process can readily be extended to describe situations where occurrence of events, at any given (infinitely) small time interval $(t, t+dt)$, is independent of the past, but can have different transition rates at different times t. Assume that, for a (measurable) function $\lambda(t) \geq 0$ on $[0, \infty)$,

$$\mathbf{P}\left(N_{t+h} = N_t + k \,|\, N_t = x\right) = \begin{cases} 1 - \lambda(t)h + o(h), & k = 0, \\ \lambda(t)h + o(h), & k = 1, \end{cases}$$

as $h \searrow 0$.

If $\lambda(t) \equiv \lambda = \text{const}$, $\{N_t\}$ will be an ordinary Poisson process. In the general case, $\{N_t\}$ is called the **time-inhomogeneous Poisson process** with **rate function** $\lambda(t)$. The distribution of the number N_B of jumps occurred at times from a Borel set $B \subset [0, \infty)$ is then $Po(\Lambda(B))$ with

$$\Lambda(B) = \int_B \lambda(t)\, dt \tag{6.13}$$

being the so-called *intensity measure* of the process. Moreover, for disjoint sets B_j, the RVs N_{B_j} are independent. A further generalisation assumes that the measure Λ does not need to be absolutely continuous. If, say, $\Lambda(\{t_0\}) = \lambda_0 > 0$, then the process $\{N_t\}$ will have (with the positive probability $1 - e^{-\lambda_0}$) a jump at the fixed point t_0 (which is impossible for the ordinary Poisson process or for inhomogeneous Poisson process with intensity measure of the form (6.13)), and the size of the jump can be any integer $k \geq 1$ (it will be a Poisson RV with mean λ_0).

The remainder of the section will be devoted to a brief discussion of more general *time-inhomogeneous* jump MPs. For such processes, similarly to (6.2), one has

$$\begin{aligned}\mathbf{P}\left(X_{t_1} = k_1, X_{t_2} = k_2, \ldots, X_{t_n} = k_n\right)& \\ = \sum_j p_j\, p_{jk_1}(0, t_1)\, p_{k_1 k_2}(t_1, t_2) \cdots p_{k_{n-1}k_n}(t_{n-1}, t_n),&\end{aligned}$$

which follows from the Chapman-Kolmogorov equation taking now the form

$$p_{jk}(s,t) = \sum_m p_{jm}(s,u)\, p_{mk}(u,t), \quad 0 < s < u < t,$$

or, in matrix notation, setting $P(s,t) := (p_{jk}(s,t))$, the form

$$P(s,t) = P(s,u)P(u,t). \tag{6.14}$$

Under additional regularity conditions (in particular, transition probabilities $p_{jk}(s,t)$ are supposed to be continuously differentiable), proceeding as in (6.3), we obtain from (6.14) the relation

$$P(s, t+h) - P(s,t) = P(s,t)(P(t, t+h) - I).$$

Dividing both sides by h and letting $h \searrow 0$, we obtain **Kolmogorov forward equations** (differentiating is w.r.t. the "forward" variable t):

$$\frac{\partial}{\partial t} P(s,t) = P(s,t)A(t) \tag{6.15}$$

with

$$A(t) := \frac{\partial}{\partial v} P(t,v)\bigg|_{v=t} = \lim_{h \searrow 0} \frac{1}{h}(P(t, t+h) - I) \tag{6.16}$$

being the matrix of transition rates $a_{jk}(t)$ (now dependent on time t!): as $h \searrow 0$,

$$p_{jk}(t, t+h) = \begin{cases} ha_{jk}(t) + o(h), & j \neq k, \\ 1 + ha_{jj}(t) + o(h), & j = k. \end{cases} \tag{6.17}$$

Similarly, for $0 \leq s < s + h < t$, writing

$$h^{-1}[P(s+h, t) - P(s, t)] = h^{-1}[I - P(s, s+h)]P(s+h, t)$$

and letting $h \searrow 0$ (that is, differentiating w.r.t. the "backward" variable s), we get the **Kolmogorov backward equations**

$$\frac{\partial}{\partial s} P(s, t) = -A(s)P(s, t). \tag{6.18}$$

In the regular case (when, say, $\sup_{j,k} |a_{jk}(t)| < \infty$ for all t), the systems (6.15) and (6.18) are equivalent. Generally, (6.18) is more universal (it may happen that $P(s,t)$ satisfies (6.18), but not (6.15)).

As it was the case for time homogeneous processes, one clearly has $a_{jk}(t) \geq 0$, $j \neq k$, while

$$-a_{jj}(t) = \sum_{k \neq j} a_{jk}(t) \geq 0$$

is the total transition rate from j at time t (or, equivalently, the hazard rate of the residual sojourn time at state j). In contrast to the homogeneous case, it does not need to be constant anymore, and hence the sojourn times for our MP are not exponential in the general case. Instead we have, for the distribution tail $\overline{F}_{s,j}$ of the residual sojourn time

$$\tau_s := \min\{t > s : X_t \neq X_s\} - s,$$

the following relations: for $v, h > 0$,

$$\begin{aligned}
\overline{F}_{s,j}(v+h) &:= \mathbf{P}\left(\tau_s > v + h \,|\, X_s = j\right) \\
&= \mathbf{P}\left(\tau_s > v + h \,|\, \tau_s > v,\, X_s = j\right) \times \mathbf{P}\left(\tau_s > v \,|\, X_s = j\right) \\
&= \mathbf{P}\left(\tau_{s+v} > h \,|\, X_{s+v} = j\right) \overline{F}_{s,j}(v) \\
&\equiv \mathbf{P}\left(\text{no transition in } (s+v, s+v+h] \,|\, X_{s+v} = j\right) \overline{F}_{s,j}(v) \\
&= (1 + a_{jj}(s+v)h + o(h))\, \overline{F}_{s,j}(v)
\end{aligned}$$

as $h \searrow 0$, so that

$$\frac{\partial}{\partial v} \overline{F}_{s,j}(v) = a_{jj}(s+v) \overline{F}_{s,j}(v), \quad \text{or} \quad \frac{\partial}{\partial v} \log \overline{F}_{s,j}(v) = a_{jj}(s+v).$$

Integrating the last relation with the obvious initial condition $\overline{F}_{s,j}(0) = 0$ leads to the DF

$$F_{s,j}(v) = 1 - \exp\left\{\int_s^{s+v} a_{jj}(u)\,du\right\} \tag{6.19}$$

having the density

$$\frac{dF_{s,j}(v)}{dv} = |a_{jj}(s+v)| \exp\left\{\int_s^{s+v} a_{jj}(u)\,du\right\}. \tag{6.20}$$

So we see that the time-inhomogeneous processes are much more flexible models capable of reproducing rather general sojourn time distributions (for example, in an MP modelling the condition of an individual: healthy, sick *etc.*, it is natural to have the times which the individual spends in one or another state dependent on the individual's age).

Note also that, similarly to the homogeneous case, (6.17) implies that, given a transition from state j occurs at time t, the MP moves then to state $k \neq j$ with probability $a_{jk}(t)/|a_{jj}(t)|$.

One more important observation is that, conditioning on the residual sojourn time and the value of our process after the *last transition* prior to time t, we get from the TPF a useful integrated form of the forward equation. To derive it, fix $t > s \geq 0$, set $h := (t-s)/n$ for $n \geq 1$ (later we will let $n \to \infty$), $t_r := s + rh$, $r = 0, 1, \ldots$ (note that both h and t_r depend on n, which we didn't reflect in our notation for simplicity's sake), and let

$$n^* := \min\{r \leq n : X_u = X_t, u \in [t_r, t]\}.$$

Then, for $j \neq k$,

$$\begin{aligned} p_{jk}(s,t) &= \sum_{r=1}^{n} \mathbf{P}\,(X_t = k, n^* = r \,|\, X_s = j) \\ &= \sum_{r=1}^{n} \sum_{m \neq k} \mathbf{P}\,(X_t = k, n^* = r, X_{t_{r-1}} = m \,|\, X_s = j): \end{aligned} \tag{6.21}$$

By the Markov property and (3.10), the term in the last sum is equal to

$$\begin{aligned} &\mathbf{P}\,(X_t = k, n^* = r \,|\, X_{t_{r-1}} = m, X_s = j)\mathbf{P}\,(X_{t_{r-1}} = m \,|\, X_s = j) \\ &\quad = \mathbf{P}\,(X_t = k, n^* = r \,|\, X_{t_{r-1}} = m) p_{jm}(s, t_{r-1}) \\ &\quad = \mathbf{P}\,(n^* = r, X_{t_r} = k \,|\, X_{t_{r-1}} = m) p_{jm}(s, t_{r-1}) \\ &\quad = \mathbf{P}\,(n^* = r \,|\, X_{t_r} = k) p_{mk}(t_{r-1}, t_r) p_{jm}(s, t_{r-1}) \\ &\quad = \overline{F}_{t_r,k}(t - t_r)(a_{mk}(t_{r-1})h + o(h)) p_{jm}(s, t_{r-1}). \end{aligned}$$

Substituting that into (6.21) and using (6.19), one obtains that

$$\begin{aligned} p_{jk}(s,t) &= \sum_{m\neq k}\sum_{r=1}^{n} p_{jm}(s,t_{r-1})e^{\int_{t_r}^t a_{kk}(u)\,du}(a_{mk}(t_{r-1})h + o(h)) \\ &\to \sum_{m\neq k}\int_s^t p_{jm}(s,v)e^{\int_v^t a_{kk}(u)\,du}a_{mk}(v)\,dv \qquad \text{as } n\to\infty. \quad (6.22) \end{aligned}$$

As the left-hand side does not depend on n, the above relation can only hold if $p_{jk}(s,t)$ is equal to the expression in the second line of (6.22).

When $j = k$, all what changes is that one needs to add to the right-hand side the probability of staying at j all the time during $[s,t]$, which leads to

$$p_{jj}(s,t) = \sum_{m\neq j}\int_s^t p_{jm}(s,v)e^{\int_v^t a_{jj}(u)\,du}a_{mk}(v)\,dv + e^{\int_s^t a_{jj}(u)\,du}.$$

As we will see soon, the representation for $p_{jk}(s,t)$ given by the right-hand side in (6.22) can be used for numerical computation of transition probabilities.

Example 6.4. *Sickness and death.* The states of a JMP represent the state of a person as follows:

$$1 = \text{"healthy"}, \quad 2 = \text{"sick"}, \quad 3 = \text{"dead"}.$$

Possible transitions are from 1 to 2 or to 3, and from 2 to 1 or 3 (state 3 is sadly absorbing), with the respective intensities given by known functions: $a_{12}(t) = \sigma(t)$ is the rate indicating the likelihood of getting sick at age t, $a_{21}(t) = \rho(t)$ is the recovery intensity at age t, while $a_{13}(t) = \mu(t)$ and $a_{23}(t) = \nu(t)$ show how likely death is at age t when our person is healthy or sick, respectively. The intensity matrix has the form

$$A(t) = \begin{pmatrix} -\sigma(t)-\mu(t) & \sigma(t) & \mu(t) \\ \rho(t) & -\rho(t)-\nu(t) & \nu(t) \\ 0 & 0 & 0 \end{pmatrix}.$$

From (6.19) we immediately obtain the probability of remaining continuously healthy over the time interval $[s,t]$:

$$\mathbf{P}(\tau_s > t-s \,|\, X_s = 1) = \exp\left\{-\int_s^t (\sigma(u)+\mu(u))\,du\right\},$$

and that of remaining continuously sick over the same time interval:

$$\mathbf{P}(\tau_s > t-s \,|\, X_s = 2) = \exp\left\{-\int_s^t (\rho(u)+\nu(u))\,du\right\}.$$

As we said earlier, even in the homogeneous case, the series representation (6.7) for transition probabilities may prove to be less convenient than numerical integration of differential equations (6.5). In the inhomogeneous case, there is almost no hope to get a closed form representation for them, and one has to mainly rely on numerical methods for solving Kolmogorov equations. We will now outline some of the basic approaches to doing that. We will deal with the forward equations (6.15) only: we fix s and sketch possible algorithms for computing $P(s,t)$.

The initial conditions are simply $P(s,s) = I$, i.e.,

$$p_{jk}(s,s) = \delta_{jk} \equiv \begin{cases} 0, & j \neq k; \\ 1, & j = k. \end{cases}$$

Choose a fixed time increment h and consider the grid $t_m = s + mh$, $m = 0, 1, 2, \ldots$, on which we will be calculating the (approximate) values $p_{jk}(s, t_m)$; we will denote them by $p_{jk}^{[m]}$, with $P^{[m]} = (p_{jk}^{[m]})$. At points other than t_m, we can approximate $P(s,t)$ using linear (or some other) interpolation.

Euler's method. This is the simplest (and oldest) approach, based on replacing the time derivative in (6.15) with the finite difference

$$(P^{[m+1]} - P^{[m]})/h = (P(s, t_m + h) - P(s, t_m))/h,$$

which leads to the following system of finite-difference equations:

$$P^{[m+1]} = P^{[m]} + hP^{[m]} A(t_m). \tag{6.23}$$

This is a rather crude approach which works well only for sufficiently small h and only for a relatively few initial points. Its "local error" (arising at each single step of the scheme due to approximating the derivative with the difference) is of order h^2 (indeed, the right-hand side of (6.23) is a truncated Taylor series, the order of the first discarded term being h^2) .

Runge-Kutta[1] *methods.* This is a family of more sophisticated schemes giving higher order approximations. The most popular of them is the following fourth order (the local error is of order h^5) method:

$$P^{[m+1]} = P^{[m]} + \frac{h}{6}(M_1 + 2M_2 + 2M_3 + M_4),$$

[1] Carl David Tolmé Runge (30.08.1856–03.01.1927) and Martin Wilhelm Kutta (03.11.1867–25.12.1944), German mathematicians who derived methods for numerical solutions of ordinary differential equations (in 1895 and 1901, resp.).

where

$$M_1 := P^{[m]}A(s, s+hm),$$
$$M_2 := \left(P^{[m]} + \frac{h}{2}M_1\right)A\left(s, s+\left(m+\frac{1}{2}\right)h\right),$$
$$M_3 := \left(P^{[m]} + \frac{h}{2}M_2\right)A\left(s, s+\left(m+\frac{1}{2}\right)h\right),$$
$$M_4 := (P^{[m]} + hM_3)A(s, s+(m+1)h).$$

This method, in a sense, attempts to replace the right-hand side in the difference approximation to (6.15) evaluated at the *initial* point of the interval $[t_m, t_{m+1}] = [s+mh, s+(m+1)h]$ with a sort of "weighted average" of the values of the right-hand side of (6.15) inside this interval.

Integral approximation. The idea is to construct a sequence of successive approximations $p_{jk}^{\{n\}}(s,t)$ to $p_{jk}(s,t)$ as follows: we start with

$$p_{jk}^{\{0\}}(s,t) := \delta_{jk}e^{\int_s^t a_{jj}(u)\,du}$$

and then compute recursively, based on (6.22),

$$p_{jk}^{\{n+1\}}(s,t) := \delta_{jk}e^{\int_s^t a_{jj}(u)\,du} + \sum_{m\neq k}\int_s^t p_{jm}^{\{n\}}(s,v)\,a_{mk}(v)\,e^{\int_v^t a_{kk}(u)\,du}dv.$$

The remarkable property of this sequence of functions $p_{jk}^{\{n\}}$, $n = 0, 1, 2, \ldots$, is that it *increases* to the solution of (6.22) as $n \to \infty$ (monotonicity can easily be seen by the induction argument once you notice that one always has $p_{jk}^{\{1\}}(s,t) \geq p_{jk}^{\{0\}}(s,t)$). To do the integration numerically, one can use any approximate integration formula, e.g., the popular Simpson's[2] rule. Namely, if $g(x)$ is a function on $[a, b]$, $h = (b-a)/n$ for some even $n > 0$, then, setting $g_k := g(a+kh)$, one has

$$\int_a^b g(x)\,dx = \frac{h}{3}\left[g_0 + \sum_{j=1}^{n-1}(3+(-1)^{j+1})g_j + g_n\right] + R_n, \tag{6.24}$$

[2]Thomas Simpson (20.08.1710–14.05.1761), an English mathematician whose first job was a silk weaver. Later he became a professor of the Royal Military Academy in Woolwich and was elected to the Royal Society of London (which sponsored the first scientific expedition to the Pacific under James Cook in 1768). He derived (6.24) in 1743 and was one of the founders of Error Theory. In fact, the rule (6.24) was known earlier to James Gregory (11.1638–10.1675), a Scottish mathematician and astronomer who published the first proof of the fundamental theorem of calculus. He also designed the first practical reflecting telescope, derived the numerical integration formula now known as Simpson's rule (1668) and actually knew what is now called Taylor's expansion well before Taylor (1715). By the way, one more elegant and famous Gregory's formula states that $\pi = 4(1 - 1/3 + 1/5 - 1/7 + \cdots)$.

where the remainder term has the form

$$R_n = -\frac{(b-a)^4}{180} g^{(4)}(x_0)h^4 \quad \text{for some } x_0 \in [a,b].$$

6.3 Birth-and-Death Processes

A pure jump MP $\{X_t\}$ with the state space $S = \{0, 1, 2, \dots\}$ is called a **birth-and-death process** (B+DP) if, for its generator $A = (a_{jk})$, one has

$$a_{jk} = 0 \quad \text{if } |j-k| > 1. \tag{6.25}$$

The values

$$\begin{cases} \lambda_j = a_{j,j+1} & \\ & \text{are called} \quad \begin{matrix}\textbf{birth} \\ \\ \textbf{death}\end{matrix} \quad \text{rates,} \\ \mu_j = a_{j,j-1} & \end{cases}$$

respectively; observe that one always has $\mu_0 = 0$. The terminology is due to a biological interpretation of the model, where X_t denotes the number of alive individuals in a population.

Since in any generator all the row sums are zeros, (6.25) yields that the diagonal elements are equal to $a_{jj} = -(\lambda_j + \mu_j)$, so that

$$A = \begin{pmatrix} -\lambda_0 & \lambda_0 & 0 & 0 & 0 & \cdots \\ \mu_1 & -(\lambda_1+\mu_1) & \lambda_1 & 0 & 0 & \cdots \\ 0 & \mu_2 & -(\lambda_2+\mu_2) & \lambda_2 & 0 & \cdots \\ 0 & 0 & \mu_3 & -(\lambda_3+\mu_3) & \lambda_3 & \cdots \\ 0 & 0 & 0 & \mu_4 & -(\lambda_4+\mu_4) & \cdots \\ \cdot & \cdot & \cdot & \cdot & \cdot & \ddots \end{pmatrix}. \tag{6.26}$$

In words, immediate transitions are *only possible to the neighbouring states*, and, as $h \searrow 0$,

$$\begin{aligned} \mathbf{P}(X_{t+h} = k+1 | X_t = k) &= \lambda_k h + o(h), \\ \mathbf{P}(X_{t+h} = k-1 | X_t = k) &= \mu_k h + o(h), \\ \mathbf{P}(X_{t+h} = k | X_t = k) &= 1 - (\lambda_k + \mu_k)h + o(h). \end{aligned}$$

Thus, a B+DP sits at state k for a random time $\sim Exp(\lambda_k + \mu_k)$, and then

$$\text{jumps to} \quad \begin{matrix} \nearrow & k+1 \quad \text{w.p. } \dfrac{\lambda_k}{\lambda_k+\mu_k}, \\ \\ \searrow & k-1 \quad \text{w.p. } \dfrac{\mu_k}{\lambda_k+\mu_k}. \end{matrix}$$

Alternatively, according to our discussion of property E2 in Section 5.1, the B+DP sits at k for the time $\tau = \min\{\tau_+, \tau_-\}$, where $\tau_+ \sim Exp(\lambda_k)$ and $\tau_- \sim Exp(\mu_k)$ are independent RVs, and if $\tau = \tau_+$, then it jumps to $k+1$, otherwise to $k-1$.

Remark 6.1. The Poisson process is a special case of the general B+DP; it is actually a pure birth process with rates $\lambda_k \equiv \lambda$, $\mu_k \equiv 0$, $k \geq 0$.

Example 6.5. *Binary continuous time branching processes.* Assume we have a population of individuals, or particles. Each particle lives for a random time distributed according to $Exp(\lambda)$, $\lambda > 0$, and then either splits into two new particles (w.p. p) or just dies (w.p. $q = 1 - p$). All particles evolve independently.

We will begin with the following question: Given $X_t = k \geq 1$, what is the time till any of the existing particles splits or dies? This time is the minimum of k i.i.d. $Exp(\lambda)$-RVs (which are the residual lifetimes of the k particles alive at time t), and hence it is exponentially distributed with parameter $k\lambda$ (property E2). When an event occurs, it is a transition either to $k+1$ (division, occurs w.p. p) or to $k-1$ (death, occurs w.p. q). This is exactly the behaviour of a B+DP with rates $\lambda_k = p\lambda k$ and $\mu_k = q\lambda k$, $k \geq 0$ (so that $\lambda_0 = 0$, 0 is an absorbing state!). The generator of the process has the form

$$A = \begin{pmatrix} 0 & 0 & 0 & 0 & 0 & \cdots \\ q\lambda & -\lambda & p\lambda & 0 & 0 & \cdots \\ 0 & 2q\lambda & -2\lambda & 2p\lambda & 0 & \cdots \\ 0 & 0 & 3q\lambda & -3\lambda & 3p\lambda & \cdots \\ 0 & 0 & 0 & 4q\lambda & -4\lambda & \cdots \\ \cdot & \cdot & \cdot & \cdot & \cdot & \ddots \end{pmatrix}. \tag{6.27}$$

Note that there exists a trivial stationary distribution $\pi = (1, 0, 0, \ldots)$ (which is a solution to $0 = \boldsymbol{\pi} A$, $\sum_j \pi_j = 1$). When $p \leq q$, the process is ergodic with that stationary distribution (eventual extinction = absorption); when $p > q$, there is *no ergodicity*: with a probability depending on the initial state, the process becomes extinct, while on the complementary event one has an *explosion*: $X_t \to \infty$ exponentially fast as $t \to \infty$.

To understand why the above statement concerning the extinction probabilities holds, note the following. If $T_0 = 0 < T_1 < T_2 < \cdots$ denote the times of successive events (divisions and deaths of the particles) in our process, then $X_{T_n} = X_{T_{n-1}} + Y_n$, where $Y_n = \pm 1$ w.p.'s p and q, respectively, independently of the past values $X_{T_0}, \dots, X_{T_{n-1}}$. That is, when $X_0 = m$,

$$X_{T_n} = m + S_n, \quad \text{where} \quad S_n = Y_1 + \cdots + Y_n$$

is a simple RW "embedded" into the branching process $\{X_t\}$, with the drift $\mathbf{E}\, Y_1 = p - q$ (so that $\mathbf{E}\, X_{T_n} = m + n(p-q)$).

It is obvious that extinction occurs when $W_- = \inf_{n\ge 0} S_n \le -m$. As we already know from Section 3.5, $W_- = -\infty$ a.s. when $\mathbf{E}\, Y_n \le 0$, i.e., $p \le q$. So when the last relation holds, the branching process becomes extinct w.p. 1.

If $p > q$, then W_- is a finite RV. Moreover, as we also noticed in Section 3.5 (cf. (3.39) for W_+; by symmetry, the same assertion holds for $-W_-$ as well), the RV $-W_-$ is geometrically distributed:

$$\mathbf{P}\,(-W_- \ge m) = b^m, \quad m = 0, 1, 2, \dots,$$

where $b = \mathbf{P}\,(W_- < 0)$. So the extinction probability for our $\{X_t\}$ is given by $\mathbf{P}\,(W_- < 0)^m$.

On the other hand, one could also observe that we can view $\{X_t\}$ as a sum of m independent copies of the branching process starting with $X_0 = 1$ particle (corresponding to the progenies of the initial m particles in the original process). The extinction of the original process occurs when all these m processes become extinct, so the probability of this event will by independence be equal to the same b^m—as b gives the extinction probability when $X_0 = 1$.

Denote by

$$\varphi(t, z) := \sum_{k=0}^{\infty} p_{1k}^{(t)} z^k \equiv \mathbf{E}\,(z^{X_t} | X_0 = 1)$$

the GF of X_t given that initially there was only one particle. We have already noted that if $X_0 = m$, the process X_t can be thought of as the sum of m independent realisations of the branching processes starting with *one particle*, so that

$$\sum_{k=0}^{\infty} p_{mk}^{(t)} z^k = \varphi^m(t, z).$$

Now, making use of (6.5) ($p_{1k}^{(t)}$, $k = 0, 1, 2, \dots$, form the second row of the matrix $P^{(t)}$), we get

$$\begin{aligned}
\frac{\partial}{\partial t}\varphi(t, z) &= \sum_{k=0}^{\infty} \frac{dp_{1k}^{(t)}}{dt} z^k = \sum_{k=0}^{\infty} \left(\sum_{m=0}^{\infty} a_{1m} p_{mk}^{(t)} \right) z^k \\
&= \sum_{m=0}^{\infty} a_{1m} \left(\sum_{k=0}^{\infty} p_{mk}^{(t)} z^k \right) = \sum_{m=0}^{\infty} a_{1m} \varphi^m(t, z) \\
&= \lambda(q - \varphi(t, z) + p\varphi^2(t, z)). \qquad (6.28)
\end{aligned}$$

in the special case of the binary branching process from (6.27). [One gets a similar relation for non-binary branching processes as well. In that case, on the right-hand side one has $Q(f(t,z))$, where $Q(z) = \sum_{m=0}^{\infty} z^m Q_m$ is the GF of the intensities Q_m, $m \neq 1$, of producing m particles by any existing particle, $Q_1 = -\sum_{m\neq 1} Q_m$; in the binary case, $Q_0 = \lambda q$, $Q_2 = \lambda p$.] Analyzing the derived differential equation $\varphi(t,z)$, one can extract a lot of information about the behaviour of the branching process (see, e.g., Problem 3 below).

Example 6.6. A rope initially consists of n nylon threads. Each thread, under a load of x kg, breaks at the rate λx, and all threads are independent. We also assume that the total load on the rope is evenly distributed among the threads. If the rope is under a constant load of L kg, what is the distribution of the lifetime T of the rope? (The lifetime is the time till the last thread breaks.)

Let $X_t := \#$ of threads **broken** by the time t. Clearly, $X_0 = 0$ and $T = \min\{t > 0 : X_t = n\}$. We know that when thread j is under a load of x kg, for the time T_j when it breaks we have, for any $t > 0$,

$$\mathbf{P}(T_j \in (t, t+h) | \, T_j > t) = \lambda x h + o(h)$$

as $h \searrow 0$. Now when $X_t = n - k$, each of the k unbroken threads bears a load of L/k kg, so that

$$\mathbf{P}(T_j \in (t, t+h) | \, T_j > t, X_t = n - k) = h\lambda L/k + o(h),$$

i.e., the failure rate for one thread is $\lambda L/k$. Since there are k threads, one can easily see that the total failure rate for the rope will be k times that rate (property E2), or just λL. When one more thread breaks, the load will be re-distributed among the remaining $k - 1$ threads, and the *individual* rate will increase. However, since the total number of the remaining threads will decrease, this will compensate for the said increase, so that the *total* failure rate *remains unchanged.*

Hence the process $\{X_t\}$ sits at each of the states $k = 0, 1, \ldots, n - 1$ for an $Exp(\lambda L)$-distributed random time, just as the Poisson process with rate λL. Absorption at state n occurs at time T equal to the sum of n independent $Exp(\lambda L)$-distributed RVs, so that T has the gamma distribution with parameters $n, \lambda L$, like the time of the nth jump in the Poisson process with parameter λL.

For a B+DP, one can find the (unconditional) probabilities $p_j(t) = \mathbf{P}(X_t = j)$ for a given initial distribution $\boldsymbol{p}(0) = \boldsymbol{p}$ by solving the system of differential equations

$$\frac{d}{dt}\boldsymbol{p}(t) = \boldsymbol{p}(t)A, \qquad \boldsymbol{p}(t) = (p_0(t), p_1(t), \ldots),$$

which is obtained by differentiating the relation $\boldsymbol{p}(t) = \boldsymbol{p}P^{(t)}$ and applying the second equation from (6.5). The system has the form

$$\begin{cases} p_0'(t) = -\lambda_0 p_0(t) + \mu_1 p_1(t), \\ p_k'(t) = \lambda_{k-1} p_{k-1}(t) - (\lambda_k + \mu_k) p_k(t) + \mu_{k+1} p_{k+1}(t), \quad k \geq 1. \end{cases} \tag{6.29}$$

This system, governing the "re-distribution" of the "probability mass" as time passes, can be solved numerically or, for finite state spaces, even explicitly.

In the steady state, the distribution "stops varying" in time, so that the derivatives on the left-hand side of (6.29) turn zero. As we have already noted, this leads to the system $0 = \boldsymbol{\pi} A$, which, in the special case of B+DPs, is just

$$\begin{cases} 0 = -\lambda_0 \pi_0 + \mu_1 \pi_1, \\ 0 = \lambda_{k-1} \pi_{k-1} - (\lambda_k + \mu_k) \pi_k + \mu_{k+1} \pi_{k+1}, \quad k \geq 1. \end{cases} \tag{6.30}$$

We have already encountered a similar system when dealing with RWs whose jumps' absolute values do not exceed one (see Example 3.18). We saw that such a system can be solved recursively.

Indeed, the kth equation can be re-written as

$$-\lambda_{k-1} \pi_{k-1} + \mu_k \pi_k = -\lambda_k \pi_k + \mu_{k+1} \pi_{k+1}, \quad k \geq 1.$$

Note that here the right-hand side can be obtained from the left-hand side by simply replacing k with $k+1$ in the latter. Therefore the equation means that the left-hand side does not depend on k and hence is zero from the first equation in (6.30). Summarising, we get the relation

$$\lambda_k \pi_k = \mu_{k+1} \pi_{k+1}, \quad k \geq 0,$$

which can be thought of as a sort of "balance equation": there is no transfer of the "probability mass" across the "boundary" between the neighbouring states k and $k+1$. Indeed, λ_k represents the rate at which the mass is transferred from k to $k+1$, while μ_{k+1} is the rate at which the mass is transferred from $k+1$ to k, cf. (6.29).

Therefore, assuming that all $\lambda_j > 0$, $\mu_{j+1} > 0$, $j = 0, 1, \dots$ (which means, in particular, that the B+DP is irreducible), we obtain

$$\pi_{k+1} = \frac{\lambda_k}{\mu_{k+1}} \pi_k = K_{k+1} \pi_0, \quad K_{k+1} := \frac{\lambda_0 \lambda_1 \cdots \lambda_k}{\mu_1 \mu_2 \cdots \mu_{k+1}}, \quad k \geq 0; \quad K_0 := 1. \tag{6.31}$$

The last expression is very similar to what we got in Example 3.18. Similarly to that example, the B+DP is *ergodic* **iff** $\sum_{j\geq 0} K_j < \infty$. If this is the case, we have from the usual condition

$$1 = \sum_{j\geq 0} \pi_j = \pi_0 \sum_{j\geq 0} K_j$$

that the stationary distribution has the form

$$\pi_0 = \frac{1}{\sum_{j\geq 0} K_j}, \quad \pi_j = K_j \pi_0, \quad j \geq 1, \tag{6.32}$$

where the coefficients K_j are given by (6.31).

6.4 PASTA

The abbreviation PASTA stands for *Poisson Arrivals See Time Averages.* A system possesses this property if the "average characteristics" (e.g., proportions of the time spent at particular states) from the point of view of arriving "customers" coincide with the respective values from the point of view of an "outside observer".

More formally, suppose we have a stochastic system, and the value X_t describes the state of the system at time t. Suppose further that the system is "fed" by a Poisson process $\{N_t\}$, so that the process is one of the inputs to the system:

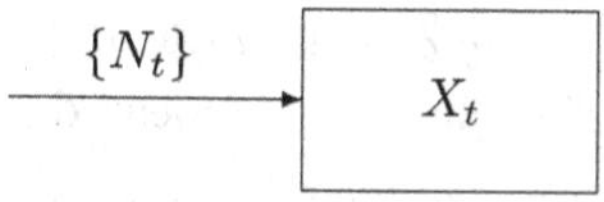

Thus, N_t could be the total number of customers arrived at a supermarket during the time interval $[0, t]$.

If the process $\{X_t\}$ is ergodic, there exists a stationary regime for the system and a stationary distribution; if X_t assumes integer values, we have in the "steady-state" that $X_t = j$ with the stationary probability π_j. On the other hand, we know that the same π_j gives the proportion of the time the system spends in state j in the long run: the "time average"

$$\frac{\text{time spent at } j \text{ during } [0, t]}{t} \to \pi_j \quad \text{as } t \to \infty \tag{6.33}$$

w.p. 1.

Now the natural question is whether arriving customers "see" the same averages. That is, how often do new customers see the system at state j, or what is the proportion of the times T_k (of new arrivals) when the system is at j? Shall we have

$$\frac{\#\{k \le n : X_{T_k - 0} = j\}}{n} \overset{?}{\to} ? \quad \text{as } n \to \infty?$$

We wrote here X_{T_k-0} for the value of the process *immediately before* the time T_k: this is what the arriving customer sees *when s/he arrives* (all our processes are, by definition, right-continuous, so the value X_{T_k} would represent the state of the system *with* the arrived customer).

The **PASTA** property holds under rather wide assumptions and states that when the system's input is an *independent Poisson process* (that is, the "future input" $\{N_s,\ s > t\}$ and the "past/present evolution" $\{X_s,\ s \le t\}$ of the system are independent), the limit exists and coincides with π_j—as it was the case for the time averages in (6.33).

Example 6.7. Suppose we know that, in a queueing system with a Poisson arrival flow, the server is idle w.p. π_0 (we will discuss the notions of queueing systems and servers later, in Chapter 7). Then, by the PASTA property, the proportion of arriving customers which find the server idle is also π_0.

6.5 Recommended Literature

KIJIMA, M. *Markov processes for stochastic modeling.* Chapman & Hall, New York, 1997.

KULKARNI, V.G. *Modeling and Analysis of Stochastic Systems.* Chapman & Hall, New York, 1995 [2nd edn. appeared in 2009].

ROSS, S. M. *Stochastic processes.* Wiley, New York, 1983 [2nd edn. appeared in 1995].

WANG, T.-K. AND YANG, H.-C. *Birth and death processes and Markov chains.* Springer, New York, 1992.

6.6 Problems

1. Prove the implication in (6.4) and also that each of the conditions imply that $\{X_t\}$ is stochastically continuous at t. Show that the two relations in (6.4) are equivalent if $\mathbf{P}(X_t = j) > 0$ for all $j \in S$.
2. Using (6.7), show that if $0 = \boldsymbol{\pi} A$, then $\boldsymbol{\pi} P^{(t)} = \boldsymbol{\pi}$ for any $t \ge 0$.
3. For the branching process from Example 6.5, find the probability of the eventual extinction given $X_0 = 1$ (that is, the value $p_{10}^{(\infty)} = \lim_{t\to\infty} p_{10}^{(t)}$; note that

the limit always exists since $p_{10}^{(t)}$ is increasing—the state 0 is absorbing, and the probability to be in it can only increase as the time increases). What is the answer to this question in the general case when $X_0 = k$, $k \geq 1$?

Hints. Note that $p_{10}^{(t)} = \varphi(t, 0)$ and make use of the relation (6.28); to derive from it a relation for $p_{10}^{(\infty)}$, one should only notice that $\frac{d}{dt}p_{10}^{(t)} \to 0$ as $t \to \infty$ (why?).

4. Consider a population consisting of particles arriving from outside at a constant rate λ (i.e., according to a Poisson process with rate λ). Particles' lifetimes are independent exponentially distributed RVs with mean α, and the particles do not produce any new particles.

   (i) Model the system as a birth-and-death process. Find the birth and death rates. Draw a transition diagram indicating rates for transitions shown on it.

   (ii) Show that this process is ergodic and find its steady-state distribution.

   (iii) What is the expected number of live particles in the population in equilibrium?

5. Customers arrive at a bus stop according to a Poisson process with rate λ, while buses (of capacity $N > 0$) arrive one after another according to an independent Poisson process with rate $\mu > 0$. Each bus departs (almost) immediately having $\min\{x, N\}$ customers aboard, where x is the number of customers waiting at the time when the bus arrived.

   Show that X_t = the number of customers waiting at the bus stop at time t is a (time-homogeneous) jump Markov process and find its generator. Sketch the transition diagram.

6. The (random) lifetime of an atom of a radioactive isotope can be modelled by an exponential RV whose parameter is called the *decay constant.* Suppose we have a specimen containing, at time $t = 0$, $X_0 = n$ atoms of a given radioactive isotope with decay constant λ. Lifetimes $\tau_j \sim Exp(\lambda)$ of all atoms are independent RVs. Denote by T_k the time of the kth disintegration (i.e., T_k is the kth smallest element of the sample $(\tau_1, \ldots, \tau_n)$, a.k.a. the kth *order statistic* of the sample).

   (i) What is the distribution of the time T_1?

   (ii) For $k = 2, \ldots, n$, what is the distribution of the time interval $T_k - T_{k-1}$ between the $(k-1)$st and kth disintegrations? Show that these time intervals are independent RVs.

   (iii) What is the mean number $\mathbf{E}\, X_t$ of the atoms of the isotope which remain by the time t? The isotope's *half-life* is the time interval required for one-half of the atomic nuclei of a radioactive sample to decay. Using your answer to the above question, find an expression for the half-time in terms of the decay constant.

   Hint: You may to wish to use indicators.

7. Martian "amoebae" have the following properties: they live exponentially long, with a mean lifetime of one hour. At the end of its life, an amoeba either simply dies (with probability 0.2) or splits either into two amoebae (with probability 0.4) or into three amoebae (with probability 0.4). The first successful Martian expedition brought to Earth $X_0 = 66$ Martian amoebae, which were put in a specially prepared (according to the amoebae's taste) red culture dish. Denote by X_t the number of alive amoebae in the culture dish at time t (hours after the expedition's arrival back to Earth). Assuming that all amoebae behave independently of each other, find the generator of the process X_t and sketch the transition diagram. Can you find the expected number $\mathbf{E}\, X_t$ of amoebae alive at time t?

8. Consider a workshop with two machines and one repairman. Each machine can be either up (functioning) or down (non-functioning). If the ith machine is up ($i = 1, 2$), it fails after an $Exp(\lambda_i)$-distributed random time. If the ith machine is down, it takes the repairman an $Exp(\mu_i)$-distributed random time to fix it (he can only work on one machine at a time). Once it is fixed, it is as good as new. Assume that all the lifetimes and repair times are independent.

   (i) Construct an appropriate jump Markov process to describe the system (give the state space, transition diagram, generator).

   (ii) Putting $\lambda_i := \mu_i := i$, $i = 1, 2$, find the stationary distribution of the process.

   Hint: The machines *are different!*

9. *A nucleotide substitution model.* A DNA strand is a (quite long!) sequence of nucleotide bases traditionally encoded with the first letters of their names by A, G, T and C. The simplest mutation consists of replacing a particular base with another one (e.g., an A can be replaced by G etc).

   Let X_t be base type at a given location on the DNA strand at time t. The simplest model for $\{X_t\}$ (Jukes–Cantor (1969)) assumes that it is a continuous time MP with constant transition rates which are equal to each other.

   (i) Denoting the common transition rate by α, write down the generator of the process $\{X_t\}$.

   (ii) Find the stationary distribution of $\{X_t\}$.

   (iii) Compute the transition matrix $P^{(t)}$ and verify if it converges, as $t \to \infty$, to a matrix of the form (3.7) with rows equal to the distribution you found in part (ii).

   Hint: Cf. Example 6.2.

10. A simple model of "accident proneness" assumes that, given j accidents have occurred prior to time t, the probability of an accident in the time interval $(t, t+h)$ is $(a + jb)h + o(h)$, $h \searrow 0$, for some $a, b > 0$. Write the generator of the B+DP defined by the above assumption and show that the transition

probabilities $p_{00}^{(t)} = e^{-at}$,

$$p_{0k}^{(t)} = \frac{a(a+b)\cdots(a+(k-1)b)}{k!b^k} e^{-at}(1-e^{-bt})^k, \quad k \geq 1,$$

solve the (forward) Kolmogorov equations for $p_{0k}^{(t)}$.

Hint: Cf. Example 6.1: one has $\boldsymbol{p}_0^{(t)} := (p_{0k}^{(t)}) = \boldsymbol{p}P^{(t)}$ with $\boldsymbol{p} = (1, 0, 0, \ldots)$, and we get the forward differential equations for $p_{0k}^{(t)}$ by multiplying the first of relations (6.5) from the left by $\boldsymbol{p}$.

Chapter 7

Elements of Queueing Theory

7.1 Definitions and Notation

Queueing theory deals with mathematical models for various kinds of real queues, i.e., situations where congestion occurs due to randomness in arrival and service times, and customers have to wait for service. Queues occur when the current demand for service exceeds the capacity of service facilities, and the main purpose of the theory is to provide means for designing/modifying/optimizing service systems in such a way as to reduce the likelihood of queues, the customers' waiting times and so on. Analysis of queueing models may include determining the distributions of the queue length, waiting times (before commencement of the service of a customer) and the durations of the busy/idle periods for servers. In many cases, only finding the long time limits (equilibrium values) for the characteristics of interest is possible.

More formally, a **queue**, or **queueing system** (QS), is a system which includes a random "input stream" of service requests (often referred to as "customers", although they can also be phone calls, information packets to be transferred *etc.*), and a mechanism (server(s), with a prescribed algorithm of its operation) which provides the requested service. Typical examples of queues are telephone exchanges (requests are calls of subscribers, the service mechanism consists of channels of communication, each of which may be busy with the service of a call for a random time), customer queues at checkout counters in a supermarket, an airport (with the flow of aircraft that need to land), and a network of time sharing computers (with a flow of programs which must be processed).

The simplest QS can be schematically depicted as follows:

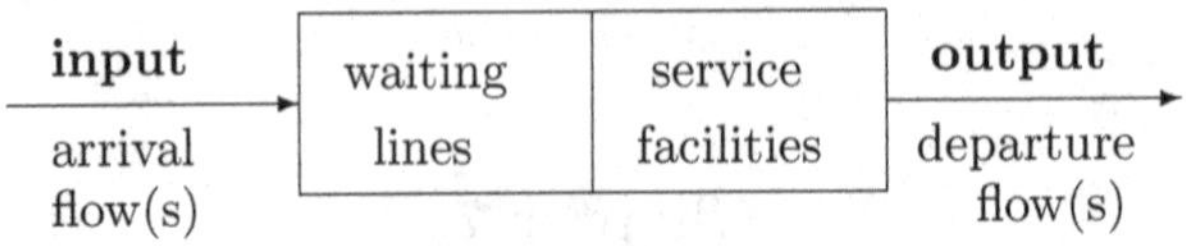

Arrival flow(s): As we said, arriving customers can be real customers in a shop, broken cars, computer jobs *etc.* For mathematical modelling purposes, what really arrives does not matter. In fact, what actually arrives, is "demand for service" which should be met. The process is specified by the *arrival times* $T_1, T_2, \ldots$ at which customers arrive at the QS, or, equivalently, by the *inter-arrival times* $\tau_j := T_j - T_{j-1}$. Alternatively, the input could be specified by the *counting process* $N_t :=$ # of arrivals in $[0, t]$.

In general situations, one can use the machinery of the so-called *marked point process* in which not only the times of arrivals are specified, but also the sizes of the arriving batches (customers may arrive in groups) and particular customers' demands for service.

To satisfy the customers' demand for service, the QS has got certain facilities: $a \geq 1$ **servers**. There exist various algorithms producing diverse forms of queues; we will only be dealing with the simplest one. If there is an idle server at the time of a customer's arrival, the customer goes to it—and will be being served for a random *service time*. We use notation $s_j^{(i)}$ for the time spent by the ith server processing its jth customer. Randomness in $s_j^{(i)}$ is due to variability in both service demands and operation of the service facilities. During that time, the server is **busy** and cannot provide any service to other customers[1]. When all a servers are busy, the arriving customer joins the queue (which is assumed to be common for all servers in the simplest case) if there is enough space for it; otherwise it is rejected and leaves the QS. To specify this situation formally, we introduce

$$m := \# \text{ of spaces for customers in the QS} \leq \infty;$$

we included here both receiving and awaiting service customers, so that $m = a +$ size of the waiting space. If $m < \infty$, then all the customers arriving when the QS is full, will be *lost*.

Next we need to describe the so-called *service discipline* specifying the order in which the customers waiting in the queue will be served. We will only consider the most common case of the so-called FIFO (standing for First-In-First-Out) discipline (a.k.a. FCFS, for First-Come-First-Served). Sometimes one considers different disciplines, e.g., LIFO (for Last-In-First-Out), random order *etc.*, which may be preferable when customers can have different priorities *etc.*

[1] In the general setup, *time sharing protocols* are often used according to which the server can provide its resources to serve several customers at any one time.

One can further use such simple QS's as building blocks to construct series models or networks of QS's, where departing from one QS customers are forwarded (by a random mechanism in the general case) to other QS's.

The standard notation[2] used for QS's is as follows:

$$A\,/\,B\,/\,a\,/\,m,$$

where:

• A denotes the type of the interarrival times distribution. If the symbol at that position is M, it means that the times are i.i.d. RVs with a common exponential distribution (M stands for "Markovian", as the customers arrive according to a Poisson process, which is Markovian). If the symbol is GI ("General Independent"), the τ_j's are just i.i.d. RVs (often denoted by G as well; the latter notation indicates, strictly speaking, that one actually has a *general* arrival stream). When the τ_j's are non-random, one uses D (for "Deterministic");

• B denotes the type of the service times distribution. Again, M means that the s_j's are i.i.d. exponential, while GI indicates that the s_j's are simply i.i.d., and D is used when the service times are non-random;

• a denotes the number of servers in the QS, $1 \le a \le \infty$;

• m is the size of the system. When $m = \infty$, the symbol is usually omitted.

Remark 7.1. It is worth stressing that the behaviour of stochastic models can differ qualitatively from that of deterministic ones. To illustrate the statement, let us compare two "heavy-traffic" (the term refers to situations where the arrival rate equals the service rate, which in the case of the simplest QS means that $\mathbf{E}\,\tau_j = \mathbf{E}\,s_j$):

(i) $D/D/1$ and, say,
(ii) $M/D/1$

(any other truly stochastic QS would also be OK). The two QS's have completely different long-run behaviour!

Putting $X_t :=$ # of customers in the QS, it is easy to see that, in case (i), the only possible values for X_t are X_0 and $X_0 \pm 1$ (depending on the possible "lag" if, at time 0, the remaining service time and the time till the next arrival do not coincide with the deterministic times specified for

[2]Introduced in Kendall, D. (1953), Stochastic processes occurring in the theory of queues and their analysis by the method of embedded Markov chains. *Ann. Math. Statist.* 24, 338–354.

the model). On the other hand, in case (ii) we have an incoming stochastic flow modelled by the Poisson process and, for the time intervals when there is a queue, a sort of constant negative drift (with the same rate as the arrival process) added to the process. As you might have expected from our previous discussion, since the process has a zero trend "everywhere but at the point 0", it will eventually hit the point 0 due to its random fluctuations, which corresponds to the server becoming idle. After that, the value of X_t will stay zero until the next customer arrives, and so on. As we know, for RWs with zero trend, it is typical to have very large and long "excursions" away from the point zero (see Section 3.5), and it is exactly what happens in case (ii) as well. For any fixed $x > 0$, we will have

$$\mathbf{P}(X_t < x) \to 0 \quad \text{as} \quad t \to \infty,$$

which means that X_t tends to infinity *in probability* (but note that it will keep returning to zero!). In fact, under proper scaling (to get a non-trivial picture of what happens during a *long* time interval $[0, T]$, one considers the scaled process $T^{-1/2}X_{sT}$, $s \in [0, 1]$), the process will converge in distribution to the so-called *reflected Brownian motion.*

Typical problems of queuing theory include:

• Find out whether a given QS has a steady-state (= equilibrium) regime, or the queue increases unboundedly ("explodes").

• What is the steady-state distribution of a given QS? What is the average load on the server? What fraction of time is the server idle?

• Optimization problems. There are different types of costs associated with a QS. Thus, too much service available means excessive costs (of providing additional unnecessary servers, their maintenance *etc.*). On the other hand, if there is not enough service, that would also incur extra costs (e.g., "lost business opportunities" when your clients give up waiting and leave looking for faster service). One has to "balance" situation and, say, find the optimal number of servers (e.g., checkouts open in a supermarket).

In conclusion of this section, we will list the basic notation to be used throughout the chapter (all the expectations and probabilities in the remaining part of this chapter **are taken under the steady-state, or**

stationary distribution):

$$L := \mathbf{E}\,X_t = \text{the expected \# of customer in the QS;}$$
$$L_q := \mathbf{E}\,\max\{X_t - a, 0\} = \text{the expected \# of customers in the } \textit{queue};$$
$$T_q := \text{the time spent by a customer in the queue (this is an RV!);}$$
$$W := \mathbf{E}\,T_q \equiv \mathbf{E}\,(\text{the waiting time});$$
$$D := W + \mathbf{E}\,(\text{service time}) = \text{the expected delay,}$$

delay being the total time spent by a customer waiting *and* obtaining service in the system.

7.2 Exponential Queueing Systems

7.2.1 *M/M/1 Systems*

As we know from the notation, for such QS's, the arrival stream is the Poisson process (of a given intensity λ), and there is a single server with the service times being i.i.d. $\sim Exp(\mu)$ for some $\mu > 0$; and an infinite waiting space ($m = \infty$). Although the above assumptions are rarely quite realistic, the system is still used since one can easily derive explicit expressions for many quantities of interest. Experience shows that even when we do not expect the assumptions to hold exactly, the use of the system—just to get an idea of what can happen—could be quite justified.

As usual, we denote by X_t the number of customers in the QS. Clearly,

$$X_t = 0 \iff \text{idle server}$$
$$X_t = k > 0 \iff \text{one customer obtains service, } k-1 \text{ in the queue.}$$

First of all note that $\{X_t\}$ is an MP. Indeed, regardless of the past history, the evolution of the process is as follows. If $X_t = 0$, the process stays at 0 for an $Exp(\lambda)$-distributed random time τ_+ till a new customer arrives (Poisson arrivals!), and then $X_{t+\tau_+} = 1$. Now if $X_t = k > 0$, the process remains at the state k for a random time $\tau = \min\{\tau_+, \tau_-\}$, where $\tau_+ \sim Exp(\lambda)$ is the time till the first arrival after t, and $\tau_- \sim Exp(\mu)$ is the time till the end of the service of the customer obtaining it at time t, the RVs being independent.

Then we have the respective transitions:

$$\text{if } \tau = \tau_+ \quad \text{then} \quad X_{t+\tau} = k+1,$$
$$\text{if } \tau = \tau_- \quad \text{then} \quad X_{t+\tau} = k-1.$$

Recall that this behaviour is exactly what we saw when discussing B+DPs (see Section 6.3)! Therefore $\{X_t\}$ is a B+DP with rates $\lambda_k \equiv \lambda$, $\mu_{k+1} \equiv \mu$, $k = 0, 1, 2, \ldots$

For the steady-state distribution, we have from (6.32) that

$$\pi_n = K_n \pi_0, \quad K_n = \frac{\lambda^n}{\mu^n} = \rho^n, \tag{7.1}$$

where $\rho = \lambda/\mu$ is called the **traffic intensity parameter**. The process is *ergodic* iff

$$\sum_{n=0}^{\infty} K_n = \sum_{n=0}^{\infty} \rho^n < \infty \quad \Longleftrightarrow \quad \rho < 1, \tag{7.2}$$

and then the sum of the (geometric) series is $1/(1-\rho)$, and hence $\pi_0 = \left(\sum_{n\geq 0} K_n\right)^{-1} = 1 - \rho$.

Therefore, if $\rho = \lambda/\mu < 1$ (i.e., $\lambda < \mu$, the arrival rate < service rate), the process $\{X_t\}$ is *ergodic*: there exists the steady-state regime, and from (7.1) the stationary distribution is the **geometric** one:

$$\pi_n = (1-\rho)\rho^n, \quad n = 0, 1, 2, \ldots \tag{7.3}$$

Knowing the stationary distribution, we can answer a lot of important questions:

(i) What fraction of the time is the server idle (in equilibrium)?

This fraction is given by $\pi_0 = 1 - \rho$.

(ii) What is L = the expected number of customers in the QS?

This is just the expectation of the geometric distribution (7.3). The simplest way to compute it is by differentiating the GF

$$g(z) = \sum_{k=0}^{\infty} \pi_k z^k = (1-\rho) \sum_{k=0}^{\infty} \rho^k z^k = \frac{1-\rho}{1-\rho z},$$

so that

$$L = g'(z)\Big|_{z=1} = \frac{\rho(1-\rho)}{(1-\rho z)^2}\Big|_{z=1} = \frac{\rho}{1-\rho}. \tag{7.4}$$

(iii) What is L_q = the expected number of customers in the queue in our QS?

Note that $L_q \neq L - 1$; in fact, $L_q = \mathbf{E} \max\{X_t - 1, 0\}$. Hence,

$$L_q = \sum_{k=1}^{\infty} (k-1)\pi_k = \underbrace{\sum_{k=1}^{\infty} k\pi_k}_{L} - \underbrace{\sum_{k=1}^{\infty} \pi_k}_{1-\pi_0} = \frac{\rho}{1-\rho} - \rho = \frac{\rho^2}{1-\rho}.$$

(iv) If the QS is viewed at the arrival epoches, what is the probability of having k customers in the QS? The expected # of customers in the QS? In the queue itself?

By the PASTA property, the answers will be the same as above: π_k, L, L_q, respectively.

(v) What is W = the expected waiting time for a newly arrived customer (under the FCFS discipline)?

An arriving customer © finds a random number ν of customers in the QS, the distribution of ν being (by PASTA) the geometric law (7.3), so that $\mathbf{E}\,\nu = L = \rho/(1-\rho)$. The *residual* service time s_1 for the customer obtaining service at the arrival epoch of © (if there is one) and the *full* service times $s_2, \ldots, s_\nu$ for those waiting in the queue are all i.i.d. RVs $\sim Exp(\mu)$. Moreover, they are independent of ν (which is determined by the past—prior to the arrival of ©—evolution of the process). Therefore the expectation of the waiting time $w^{©} := \sum_{j\le\nu} s_j$ of our customer © is

$$W = \mathbf{E}\,w^{©} = \mathbf{E}\sum_{j\le\nu} s_j \overset{TPF}{=} \sum_{k=0}^{\infty} \underbrace{\mathbf{E}\left[\sum_{j=1}^{\nu} s_j \middle| \nu = k\right]}_{k\mathbf{E}\,s_j = k/\mu} \underbrace{\mathbf{P}\,(\nu = k)}_{\pi_k}$$

$$= \frac{1}{\mu}\mathbf{E}\,\nu = \frac{\rho}{(1-\rho)\mu} = \frac{\rho}{\mu-\lambda}. \tag{7.5}$$

(vi) What is D = the expected delay time for a newly arrived customer? Clearly,

$$D = W + \frac{1}{\mu} = \frac{\rho}{(1-\rho)\mu} + \frac{1}{\mu} = \frac{1}{(1-\rho)\mu} = \frac{1}{\mu-\lambda}.$$

Remark 7.2. Observe that the following two relations hold for the $M/M/1$ system:

$$L = \lambda D \qquad \text{and} \qquad L_q = \lambda W. \tag{7.6}$$

Both are called **Little's law** and take place under much more general assumptions (it is important that the service discipline is FIFO—why?). They are very simple and yet important relations. Indeed, once we know L, we can immediately get D, and vice versa. If the service rate μ = const (which is *not the case* when we have $a > 1$ servers), then $D = W + 1/\mu$, so that of the four quantities L, D, L_q and W, it suffices to find only one of them to know them all!

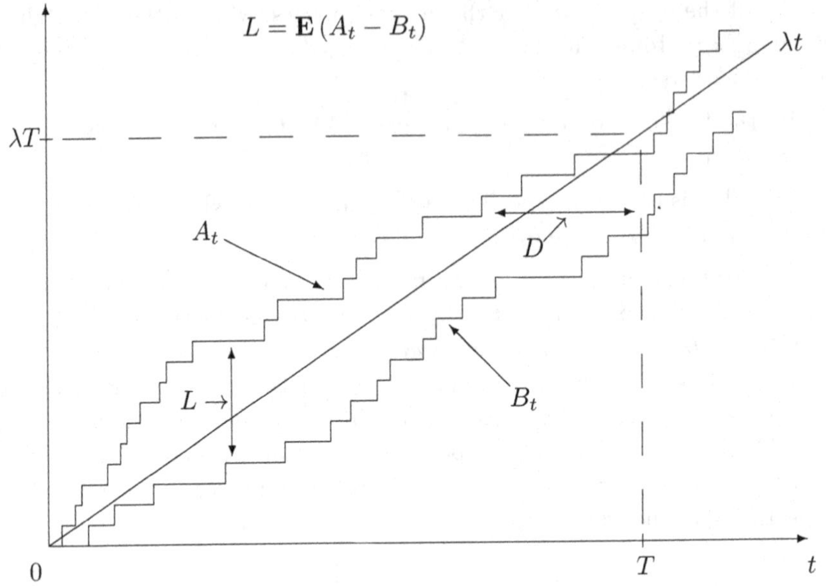

Fig. 7.1 A graphical illustration to Little's law.

Fig. 7.1 illustrates Little's law. It depicts the trajectories of the arrival process A_t (counting arrivals) and departure process B_t (counting departures). Clearly, $B_t - A_t \geq 0$ is the number of customers in the QS at time t, while, for a particular customer, its delay time is the duration of the time interval between the respective jumps in the arrival and departure processes. In the long run, the "typical height" of the strip between the trajectories of the processes A_t and B_t is $L = \mathbf{E}(A_t - B_t)$, while its typical width is D. It remains to note that the "average slope" of both processes A and B is λ (why?), and hence, for large T, the area of the "jagged strip" over the time interval $[0, T]$ will be $\approx L \times T$ and, at the same time, it will also be $\approx D \times \lambda T$. From here we infer the first relation in (7.6).

Example 7.1. A repairman is assigned to service a bank of machines in a shop. Assume that failures occur according to the Poisson process with rate $\lambda = 1/12\,\text{min}^{-1}$, while the repair rate is $\mu = 1/8\,\text{min}^{-1}$. Find the stationary distribution and the basic characteristics of the steady-state regime of the system.

In this example, the "customers" are clearly the machines from the bank. Demand for service arises at the *constant* rate λ which apparently indicates that the bank is pretty large, so that the rate does not depend on the number of non-working machines (which is small relative to the total size of the bank). We have a single "server"—the repairman, who also works at a constant rate. When it is not stated otherwise, mentioning the rate *only* means that the service times are independent and exponentially distributed (for only that distribution has a constant hazard rate, see E1 in Section 5.1). So we have an $M/M/1$ system with the traffic intensity

$$\rho = \frac{\lambda}{\mu} = \frac{8}{12} = \frac{2}{3} < 1,$$

which ensures that the system is stable and ergodic. The stationary distribution is given by (7.3):

$$\pi_k = (1-\rho)\rho^k = \frac{1}{3}\left(\frac{2}{3}\right)^k, \quad k = 0, 1, 2, \ldots$$

The repairman is idle w.p. $\pi_0 = 1 - \rho = 1/3$ (= the fraction of the time he is idle in the long run). Note that this is also the percentage of the machines that do not wait in the queue before repair (PASTA!).

The expected number of machines in the QS (that is, *non-working* machines) is

$$L = \frac{\rho}{1-\rho} = \frac{2/3}{1-2/3} = 2.$$

The expected waiting time is

$$W = \frac{\rho}{(1-\rho)\mu} = 2 \times 8 = 16 \text{ (min)}.$$

Suppose now that the failure rate of the machines increases (e.g., due to their ageing) by 20%: the new rate is

$$\lambda' = 120\% \text{ of } \lambda = 1.2 \times \frac{1}{12} = \frac{1}{10}.$$

The new traffic intensity is $\rho' = 8/10 = 4/5 < 1$, so that the system is still stable. The new mean number of non-working machines is

$$L' = \frac{\rho'}{1-\rho'} = 4, \qquad W' = 4 \times 8 = 32 = 2W.$$

Thus, a relatively moderate 20% increase in the arrival rate has **doubled** the expected number of non-working machines! When ρ is close to one, the effect of small changes in the traffic intensity ρ is profound! (You might have already noticed this phenomenon observing traffic jams.) On the other hand, this also means that if a QS is characterised by long waiting times and lines, rather modest increase in the service rate can bring about dramatic reduction in waiting times.

Suppose now that a new piece of equipment is available which will increase the repair rate from $\mu = 1/8$ to $\mu^* = 1/6$ (so that the expected repair time will drop from 8 to 6 min). The maintenance cost for this new equipment is $C_M = 10$ (\$/min), while the cost incurred due to the lost production when a machine is out of order is $C_D = 5$ (\$/min). Should the new equipment be purchased?

To answer the question, we need to compare the expected costs (in equilibrium, for they would give us long-term time averages). The expected number of non-working machines is equal to the number L of machines in the QS, and hence the delay cost rate is

$$LC_D = \frac{\rho C_D}{1-\rho} \qquad (\$/\text{min}).$$

With the new equipment, the traffic intensity becomes $\rho^* = \lambda/\mu^* = 1/2$, and the expected total cost is

$$L^*C_D + C_M = \frac{\rho^* C_D}{1-\rho^*} + C_M.$$

Therefore we decide to adopt the new equipment iff

$$\frac{\rho^* C_D}{1-\rho^*} + C_M < \frac{\rho C_D}{1-\rho},$$

which is clearly equivalent to

$$\underbrace{\frac{C_M}{C_D}}_{2} < \underbrace{\frac{\rho}{1-\rho}}_{2} - \underbrace{\frac{\rho^*}{1-\rho^*}}_{1}.$$

Thus, in this case, the new equipment does not justify its cost and hence *should not* be adopted.

If, however, the failure rate increases by 20% to $\lambda' = 1/10$, then, with the new traffic intensity $\rho'^* = \lambda'/\mu^* = 6/10 = 3/5$, the above difference becomes

$$\frac{\rho'}{1-\rho'} - \frac{\rho'^*}{1-\rho'^*} = 4 - \frac{3}{2} = 2.5 > 2 \equiv \frac{C_M}{C_D}.$$

Now purchasing the new equipment is justified!

Example 7.2. Arrivals at a public telephone booth form a Poisson process with a rate of 12 per hour. The duration of a phone call made from the booth is an exponential RV with the expected value of 2 min.

What is (i) the traffic intensity? (ii) the probability that an arrival will find the phone occupied? (iii) the average length of the queue (including the person speaking) when it forms? (iv) The telephone company installs additional booths if customers wait on average at least 3 min for the phone. By how much should the flow of arrivals increase to justify a second booth?

Solution. (i) The customer arrival rate is $\lambda = 12$ hour^{-1} $= \frac{1}{5}$ min^{-1}; the service rate is $\mu = 1/(\text{mean service time}) = \frac{1}{2}$ min^{-1} (one has to express all quantities in the same units!). Hence the traffic intensity is

$$\rho = \frac{\lambda}{\mu} = \frac{1/5}{1/2} = 0.4.$$

(ii) By the PASTA property, this is $1 - \pi_0 = \rho = 0.4$.

(iii) This is the conditional expectation (recall that, in the remaining part of this chapter, we assume $\mathbf{E} = \mathbf{E}_{\text{st}}$ and $\mathbf{P} = \mathbf{P}_{\text{st}}$—the process is in the stationary regime)

$$\begin{aligned}
\mathbf{E}\left(X_t | X_t > 0\right) &= \sum_{k=1}^{\infty} k\mathbf{P}\left(X_t = k | X_t > 0\right) \\
&= \frac{1}{\mathbf{P}\left(X_t > 0\right)} \sum_{k=1}^{\infty} k \underbrace{\mathbf{P}\left(X_t = k,\, X_t > 0\right)}_{\mathbf{P}\,(X_t = k) = \pi_k} \\
&= \frac{1}{\rho} \sum_{k=1}^{\infty} k\pi_k = \frac{1}{\rho} L = \frac{1}{\rho} \times \frac{\rho}{1-\rho} = \frac{1}{1-\rho} = \frac{5}{3} \approx 1.67.
\end{aligned}$$

(iv) Denoting the unknown (higher) arrival rate by λ', we have to solve the following inequality for it: the new mean waiting time

$$W = \frac{\rho}{(1-\rho)\mu} = \frac{\lambda'}{\mu(\mu - \lambda')}\bigg|_{\mu=0.5,\lambda'<\mu} = \frac{\lambda'}{0.5(0.5 - \lambda')} \geq 3.$$

The solution is $\lambda' \geq 0.3$ (min^{-1}).

So, when the arrival rate exceeds 0.3 min^{-1} $= 18$ hour^{-1}, an additional booth will be justified. The present arrival rate should increase by 6 arrivals per hour.

Example 7.3. At a car service station, the service rate is μ cars per hour, and the rate of arrivals is λ (hour^{-1}). The cost incurred by the service station due to delaying cars is $\$C_1$ per car per hour, whereas the operating and service costs depend on the service rate and are $\$\mu C_2$ per hour. The rate μ is our *control* parameter: we can vary it.

Determine the value of μ that results in the least expected cost and the value of the latter.

Assume that the system has settled down in equilibrium. Then, on average, there are

$$L = \frac{\rho}{1-\rho} = \frac{\lambda}{\mu - \lambda}$$

cars at the service station. Hence the total (average) cost to be minimized is, in \$/hour,

$$C(\mu) = \frac{\lambda}{\mu - \lambda} C_1 + \mu C_2 \longrightarrow \min_{\mu > \lambda} .$$

The necessary condition for the extremum is

$$0 = \frac{dC(\mu)}{d\mu} = -\frac{\lambda}{(\mu - \lambda)^2} C_1 + C_2.$$

Solving the equivalent quadratic equation $(\mu - \lambda)^2 = \lambda C_1 / C_2$, we get $\mu = \lambda \pm (\lambda C_1/C_2)^{1/2}$. Now since we must have $\rho = \lambda/\mu < 1$, the minus in the expression is impossible, and so the optimum value is

$$\mu^* = \lambda + \sqrt{\frac{\lambda C_1}{C_2}}.$$

Note that this is a minimum indeed (for $C(\mu) \to \infty$ as $\mu \searrow \lambda$ or $\mu \nearrow \infty$; alternatively, you can show that the second derivative of C at $\mu = \mu^*$ is positive). The minimum expected cost is

$$C(\mu^*) = \frac{\lambda}{\sqrt{\lambda C_1 / C_2}} C_1 + \left(\lambda + \sqrt{\frac{\lambda C_1}{C_2}} \right) C_2 = \lambda C_2 + 2\sqrt{\lambda C_1 C_2}.$$

So far we have been dealing with the *analysis of averages* only. In particular, we found the expectation $W = \mathbf{E}\, T_q$ of the waiting time of an arriving customer. It turns out that we can find the *distribution* of T_q as well, which is much more informative than just the mean. We begin by noting that

$$\mathbf{P}\,(T_q = 0) = \mathbf{P}\,(\text{server is idle}) = \pi_0 = 1 - \rho.$$

Further, when $T_q > 0$, one of the events

$$A_k = \{\text{arrival sees } k \text{ customers in QS}\}, \quad k = 1, 2, \ldots,$$

occurs; by PASTA, one has $\mathbf{P}(A_k) = \pi_k = (1-\rho)\rho^k$. Therefore, for $t > 0$, by the TPF and (5.6),

$$\begin{aligned}
\mathbf{P}(T_q > t) &= \sum_{k=1}^{\infty} \underbrace{\mathbf{P}(T_q > t|\, A_k)}_{\mathbf{P}([\text{sum of } k \text{ i.i.d.}\sim Exp(\mu)]>t)} \quad \mathbf{P}(A_k) \\
&= (1-\rho)\sum_{k=1}^{\infty}\left[\int_t^{\infty} \frac{1}{(k-1)!}\mu^k x^{k-1} e^{-\mu x}\,dx\right]\rho^k \\
&= \rho \underbrace{(1-\rho)\mu}_{\mu-\lambda} \int_t^{\infty} \underbrace{\left[\sum_{k=1}^{\infty} \frac{(\mu\rho x)^{k-1}}{(k-1)!}\right]}_{e^{\mu\rho x}=e^{\lambda x}} e^{-\mu x}\,dx \\
&= \rho \int_t^{\infty} \underbrace{(\mu-\lambda)e^{-(\mu-\lambda)x}}_{Exp(\mu-\lambda)\text{-density}}\,dx = \rho e^{-(\mu-\lambda)t}.
\end{aligned}$$

Thus, the waiting time T_q has the *mixture distribution* $(1-\rho)I_0 + \rho Exp(\mu-\lambda)$, I_0 denoting the degenerate at zero law. Note that the *conditional* distribution of T_q given that $T_q > 0$ is just $Exp(\mu-\lambda)$.

Example 7.1 (continued). In our repairman example, what is the probability that a newly broken machine must wait for more than $t = 10$ min before its repair starts?

Recall that we have $\lambda = 1/12$ and $\mu = 1/8$, so that $\rho = 2/3$ and $\mu - \lambda = 1/24$. Therefore,

$$\mathbf{P}(T_q > t) = \rho e^{-(\mu-\lambda)t} = \frac{2}{3}e^{-10/24} \approx 0.44.$$

7.2.2 $M/M/a$ Systems

Now we have $a \geq 1$ servers; all service times are i.i.d. RVs $\sim Exp(\mu)$; arrivals follow the Poisson process with rate λ. Also, the FIFO discipline is adopted, and when an arrival finds several idle servers, the server to be used is chosen at random.

We begin by noting that when k servers are working, the time till the next departure (from one of the servers) is $\min\{s_1, \ldots, s_k\} \sim Exp(k\mu)$, where s_j is the residual service time for the customer currently obtaining service on the jth working server. Hence we have the following transition diagram for the process $\{X_t\}$ whose values are the numbers of customers in the QS at the respective times:

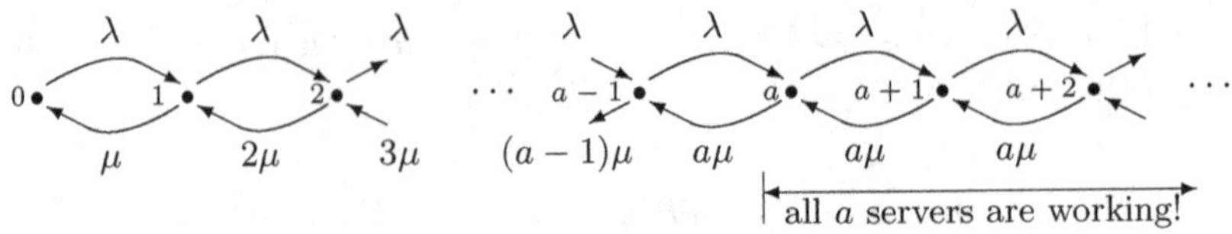

This is easily seen to be a B+DP with the rates

$$\lambda_j = \lambda, \quad j = 0, 1, 2, \ldots,$$
$$\mu_j = j\mu, \quad j = 1, 2, \ldots, a,$$
$$\mu_j = a\mu, \quad j = a+1, a+2, \ldots,$$

so that we have a *variable service rate*.

The first question one can ask is when the system is ergodic. The coefficients (6.31) participating in the expressions for the stationary distribution of B+DPs are in our case

$$K_j = \frac{\lambda_0\lambda_1\cdots\lambda_{j-1}}{\mu_1\mu_2\cdots\mu_j} = \begin{cases} \dfrac{\rho^j}{j!}, & j \le a, \\[2ex] \dfrac{\rho^j}{a!a^{j-a}}, & j > a, \end{cases} \qquad \rho := \frac{\lambda}{\mu}. \tag{7.7}$$

Clearly, $\sum_{j=0}^{\infty} K_j < \infty$ iff $\sum_{j=a}^{\infty} K_j < \infty$, and the latter sum can be found using the formula for the sum of the geometric series:

$$\sum_{j\ge a} \frac{\rho^j}{a!a^{j-a}} = \frac{\rho^a}{a!}\sum_{k\ge 0}\left(\frac{\rho}{a}\right)^k = \frac{\rho^a}{a!} \times \frac{1}{1-\rho/a} < \infty$$

when $\rho/a = \lambda/a\mu < 1$. Thus, as one could expect, an $M/M/a$ system is stable and ergodic iff the arrival rate is less than the **maximum service rate**. If this is the case, the stationary distribution is given by $\pi_k = K_k\pi_0$ which, in view of (7.7), becomes

$$\pi_k = \frac{\rho^k}{k!}\pi_0, \quad k \le a,$$
$$\pi_k = \frac{\rho^k}{a!a^{k-a}}\pi_0, \quad k > a,$$

where

$$\pi_0 = \left[\sum_{k=0}^{\infty} K_k\right]^{-1} = \left[\sum_{k=0}^{a-1} + \sum_{k=a}^{\infty}\right]^{-1} = \left[\sum_{k=0}^{a-1}\frac{\rho^k}{k!} + \frac{\rho^a}{a!(1-\rho/a)}\right]^{-1}. \tag{7.8}$$

Having found the stationary distribution of the QS, we can now answer further questions regarding the system's behaviour. One of the basic ones is, what proportion of the time are all a servers busy? By PASTA, this will also be the probability that an arriving customer will have to wait. The desired probability is clearly

$$P_q := \mathbf{P}\,(\text{all servers busy}) = \sum_{k=a}^{\infty} \pi_k = \sum_{k=a}^{\infty} K_k \pi_0 = \pi_0 \frac{\rho^a}{a!(1-\rho/a)}. \quad (7.9)$$

One can also find the expected queue length

$$L_q = \mathbf{E}\,\max\{X_t - a, 0\} = \sum_{j=a+1}^{\infty} (j-a)\pi_j = \sum_{j=a}^{\infty} (j-a)\frac{\rho^j}{a!a^{j-a}}\pi_0$$
$$= \frac{\rho^a}{a!}\pi_0 \sum_{k=0}^{\infty} k\left(\frac{\rho}{a}\right)^k = \frac{\rho^a}{a!}\pi_0 \frac{\rho/a}{(1-\rho/a)^2} = \frac{\lambda}{a\mu - \lambda} P_q,$$

where we made use of the formula $\sum_{k \geq 0} k b^k (1-b) = b/(1-b)$ for the mean of the geometric distribution with parameter $b = \lambda/a\mu$ (cf. (7.4)).

One more natural average quantity of interest is the expected number of busy servers. It is equal to

$$\mathbf{E}\,\min\{X_t, a\} = \sum_{j=1}^{a} j\pi_j + \sum_{j=a+1}^{\infty} a\pi_j$$
$$= \pi_0 \left[\sum_{j=1}^{a} j\frac{\rho^j}{j!} + \sum_{j=a+1}^{\infty} a \frac{\rho^j}{a!a^{j-a}} \right]$$
$$= \pi_0 \frac{\lambda}{\mu} \underbrace{\left[\sum_{j=1}^{a} \frac{\rho^{j-1}}{(j-1)!} + \sum_{j=a+1}^{\infty} \frac{\rho^{j-1}}{a!a^{(j-1)-a}} \right]}_{\pi_0^{-1}} = \rho, \quad (7.10)$$

cf. (7.8).

Remark 7.3. Note that the answer to the question about the average number of busy servers—somewhat surprisingly from the first glance—does not depend on the number of servers in the system (provided that $\rho/a = \lambda/a\mu < 1$)! This fact is actually nothing else but the balance equation stating that, in equilibrium, the arrival rate λ equals the mean service rate = (average # of working servers)$\times\mu$. The *distribution* of the number of busy servers *does depend* on a (even its "support" does: it is $\{0, 1, \dots, a\}$).

Using the above result, we can easily find the expected number of customers in the QS:

$$L = \mathbf{E}\,(\#\text{ of busy servers}) + L_q = \frac{\lambda}{\mu} + \frac{\lambda}{a\mu - \lambda} P_q. \quad (7.11)$$

From here, the expected delay is, by Little's law, equal to

$$D = \frac{1}{\lambda} L = \frac{1}{\mu} + \frac{1}{a\mu - \lambda} P_q. \tag{7.12}$$

Similarly, from $L_q = \lambda W$ we find that the expected waiting time is

$$W = \frac{1}{\lambda} L_q = \frac{1}{a\mu - \lambda} P_q.$$

Example 7.4. An insurance company has three claim adjusters in its branch office. People with claims against the company are found to arrive according to the Poisson process at an average rate of 20 per eight hour working day. The amount of time an adjuster spends with a claimant is an independent exponential RV with the mean (service) time of 40 min.

(i) How many hours a week can an adjuster expect to spend with the claimants?

(ii) How much time does a claimant spend on average in the branch office?

To answer the above questions, first note that here we are dealing with an $M/M/a$ QS, whose parameters are: the number of servers $a = 3$; the arrival rate $\lambda = 20/8 = 5/2$ (hour^{-1}); the service rate $\mu = 1/(\textit{mean service time}) = 1/(40 \text{ min}) = 3/2$ (hour^{-1}). Since

$$\frac{\lambda}{a\mu} = \frac{5}{2} \times \frac{1}{3} \times \frac{2}{3} = \frac{5}{9} < 1,$$

the steady-state regime exists.

(i) During a 40 hour week, an adjuster (say, the first one) will, on average, spend with the claimants

$$40 \times \underbrace{\mathbf{P}\left(1^{\text{st}} \text{ is busy}\right)}_{\mathbf{E}\,\mathbf{1}_{\{1^{\text{st}} \text{ is busy}\}}} \quad \text{(hours)} \tag{7.13}$$

(recall that we agreed that all the probabilities and expectations in this chapter are under the stationary distribution).

But we know (see (7.10)) that the average number of busy adjusters is

$$\rho = \mathbf{E}\,(\#\text{ of busy adjusters}) = \mathbf{E}\left(\mathbf{1}_{\{1^{\text{st}}\}} + \mathbf{1}_{\{2^{\text{nd}}\}} + \mathbf{1}_{\{3^{\text{rd}}\}}\right) = 3\mathbf{E}\,\mathbf{1}_{\{1^{\text{st}}\}}$$

by symmetry, so that

$$\mathbf{P}\left(1^{\text{st}} \text{ is busy}\right) = \rho/3.$$

Note that, in the general case of a servers, one similarly has

$$\mathbf{P}\left(1^{\text{st}} \text{ is busy}\right) = \rho/a.$$

Therefore the answer to (i) is

$$40 \times \frac{\rho}{3} = \frac{40}{3} \times \frac{5}{2} \times \frac{2}{3} = \frac{200}{9} \approx 22.2 \quad \text{(hours per week).}$$

(ii) This is nothing else but the expected delay (7.12). To find its value, it remains to compute

$$P_q = \frac{\rho^a}{a!(1-\rho/a)}\pi_0 = \frac{1}{3!}\left(\frac{5}{2} \times \frac{2}{3}\right)^3 \times \frac{1}{1 - \frac{1}{3} \times \frac{5}{2} \times \frac{2}{3}}\pi_0 = \frac{125}{72}\pi_0,$$

where we have from (7.8) that

$$\pi_0^{-1} = \sum_{j=0}^{a-1} \frac{\rho^j}{j!} + \frac{\rho^a}{a!(1-\rho/a)} = 1 + \rho + \frac{1}{2}\rho^2 + \frac{125}{72} = \frac{139}{24},$$

and hence $\pi_0 = 24/139$. Therefore expression (7.12) is equal in our case to

$$D = \frac{2}{3} + \frac{1}{2} \times \frac{125}{72} \times \frac{24}{139} \approx 0.817 \text{ hour} \approx 49 \text{ min.}$$

Note: since the mean service time is 40 min, we see that a claimant will spend on average only 9 min in the queue!

Now let us see what would happen if there were only $a = 2$ adjusters in the office. We would still have

$$\frac{\rho}{a} = \frac{5}{6} < 1,$$

so that the QS is stable, the steady-state regime exists. Further,

$$\mathbf{P}\,(1^{\text{st}} \text{ is busy}) = \frac{\rho}{a} = \frac{5}{6},$$

which is 1.5 times the value for $a = 3$. Therefore, due to (7.13), the answer to (i) will be $40 \times \frac{5}{6} \approx 33.3$ hours per week, also a 50% increase. But what about the average claimant delay?

Now we have

$$\pi_0^{-1} = 1 + \rho + \frac{1}{2}\rho^2 \frac{1}{1-\rho/2} = 1 + \frac{5}{3} + \frac{1}{2} \times \frac{25}{9} \times 6 = 11,$$

so that $\pi_0 = 1/11$ and

$$P_q = \frac{\rho^2}{2!}\frac{1}{1-\rho/2}\pi_0 = \frac{25}{33}.$$

Hence the delay time (7.12) is

$$D = \frac{2}{3} + \frac{1}{2 \times \frac{3}{2} - \frac{5}{2}} \times \frac{25}{33} \approx 2.18 \text{ (hour)},$$

more than 1.5 hour in the queue! Compare the result with 9 min in the case $a = 3$! We again see how a (relatively) moderate drop in productivity can cause dramatic changes in the service quality—in this case, an 11-fold increase of the average waiting time.

One more interesting question we can now address is as follows. What is better: to have a single "mighty server" or several smaller ones of the same total productivity? Assume that the arrival process is a Poisson one with rate λ and consider, for an $a > 1$, the two alternatives:

(i) $M/M/1$: a single server with the service rate $a\mu$;
(ii) $M/M/a$: a parallel servers with the service rate μ each.

In which case shall we have a smaller expected delay time and number of customers in the system?

A heuristic argument runs as follows: when $X_t \geq a$, both systems work at the same (maximum) rate; when $X_t = k < a$, the service rate in the multiserver system is $k\mu < a\mu$, while that in the single-server one is still $a\mu$. Therefore the latter is more efficient. (Observe also that, in the extreme case when $\rho = \lambda/\mu \to 0$, there is "almost no queue" and the delay will basically coincide with the service time, which is much longer in the case of $a > 1$ servers.) Note, however, that we did not take into account other (possibly important) aspects such as reliability, maintenance costs *etc.*

Example 7.5. In a police department, only selected clerks are allowed to locate files in the vault. When police officers (POs) want to access a file, they must queue until a clerk becomes available. Suppose interarrival times between the POs with requests for files are exponentially distributed i.i.d. RVs with the mean 5 min, the time for a clerk to locate a file is also an exponential RV with the mean 7.5 min. A clerk's salary is \$10 per hour, a PO's salary is \$15 per hour. The police department can hire up to three clerks. How many should they hire to minimise delay and service costs?

First of all note that here we deal with an $M/M/a$ system that has a servers = clerks, a common infinite waiting space, and exponential times. The rates, being reciprocal to the respective mean times, are:

$$\lambda = \frac{1}{5}\ \text{min}^{-1} = 12\ \text{hour}^{-1}, \quad \mu = \frac{1}{7.5}\ \text{min}^{-1} = 8\ \text{hour}^{-1},$$

so that

$$\rho = \frac{\lambda}{\mu} = \frac{12}{8} = \frac{3}{2} > 1, \quad \frac{\rho}{2} = \frac{3}{4} < 1.$$

Therefore they need to hire $a \geq 2$ clerks to ensure that the queue size does not blow up and hence the delay cost does not blow up, too.

The expected cost per hour is the sum of the expected service cost and expected delay cost: $a \times \$10 + L \times \$15$. From (7.11) and (7.9),

$$L = \frac{3}{2} + \frac{3/2}{a - 3/2} P_q, \qquad P_q = \frac{(3/2)^a}{a!(1 - 3/2a)} \pi_0.$$

Case $a = 2$. From (7.8) one has

$$\pi_0^{-1} = 1 + \rho + \frac{1}{2!}\rho^2 \frac{1}{1 - \rho/2} = 1 + \frac{3}{2} + \frac{9}{2} = 7, \quad \pi_0 = \frac{1}{7},$$

so that $P_q = 9/14$ and $L = 24/7$. Therefore, the expected average cost is $2 \times 10 + (24/7) \times 15 \approx 71.43$ (\$/hour).

Case $a = 3$. Now

$$\pi_0^{-1} = 1 + \rho + \frac{1}{2!}\rho^2 + \frac{1}{3!}\rho^3 \frac{1}{1 - \rho/3} = 1 + \frac{3}{2} + \frac{9}{8} + \frac{9}{2} = \frac{19}{4}, \quad \pi_0 = \frac{4}{19},$$

so that $P_q = 9/38$ and $L = 33/19$. Hence the expected cost is $3 \times 10 + (33/19) \times 15 \approx 56.06$ (\$/hour). One can easily verify that the expected costs will exceed that value for $a \geq 4$.

Therefore the variant $a = 3$ is preferable.

7.2.3 $M/M/a/N$ Systems

In this subsection we will deal (for the first time) with a QS having a finite waiting space. The *state space* is now $\{0, 1, \ldots, N\}$, while the transition rates are the same as for $M/M/a$ when restricted to that state subspace:

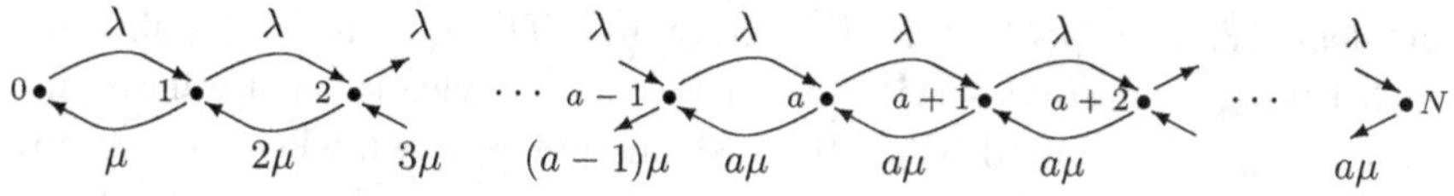

so that

$$K_j = \begin{cases} \dfrac{\rho^j}{j!}, & j = 0, 1, \ldots a - 1, \\[2ex] \dfrac{\rho^j}{a!a^{j-a}}, & j = a, \ldots, N, \end{cases} \qquad \rho = \frac{\lambda}{\mu}. \tag{7.14}$$

The state space is finite, and hence the irreducibility of the process implies ergodicity (the sum $\sum_{j=0}^{N} K_j$ is always finite). Therefore, whatever λ, μ, N and a, the QS is always stable, the stationary distribution exists and is given by the same formulae as for $M/M/a$:

$$\pi_k = \begin{cases} \dfrac{\rho^k}{k!}\pi_0, & k = 0, 1, \ldots a - 1, \\[2ex] \dfrac{\rho^k}{a!a^{k-a}}\pi_0, & k = a, \ldots, N, \end{cases}$$

but with a different π_0, which is now determined from[3]

$$\pi_0^{-1} = \sum_{k=0}^{N} K_k = \sum_{k=0}^{a-1} + \sum_{k=a}^{N} = \sum_{k=0}^{a-1} \frac{\rho^k}{k!} + \frac{\rho^a(1-(\rho/a)^{N-a+1})}{a!(1-\rho/a)}.$$

All other characteristics of the QS can be computed in the same way as we did it for the $M/M/a$ QS's.

7.3 The Machine Repair Problem

In all the models we have been dealing with so far there was an "infinite pool" of potential customers. This assumption is justified in the real-life situations when the number of potential customers is much bigger than the typical number of customers in the system. If this is not the case, one should take into account the variation of the arrival rate caused by the variation in the number of customers in the QS—if they are in the queue or are being served, they cannot contribute to the arrival flow at that time!

In the **machine repair problem** (a.k.a. the **finite source queue**), one has a finite number M of machines each having a constant hazard rate λ (so that the times till the breakdowns of particular machines are independent RVs $\sim Exp(\lambda)$, cf. E1 in Section 5.1). There are also a repairmen, each having a constant service rate μ (i.e., the service times are independent $Exp(\mu)$-distributed RVs). If a machine breaks down when there is an idle repairman, the latter starts servicing the broken machine immediately. Otherwise, the broken machine joins a (common for all repairmen) queue. When one of the repairmen finishes servicing a machine, he at once starts working on the first (to arrive) machine from the queue.

Denoting, as usual, by X_t the number of customers in the QS at time t, we see that $\{X_t\}$ is a B+DP with the state space $\{0, 1, \ldots, M\}$. To find the *rates* of the process, note that when there are j customers in the QS, there remain only $M - j$ "potential customers" outside, so that the arrival rate is

$$\lambda_j = (M-j)\lambda, \quad j = 0, 1, \ldots, M.$$

[3]Here we used the formula $\sum_{k=0}^{n} b^k = (1-b^{n+1})/(1-b)$ for partial sums of the geometric series; note that the sum is just $n+1$ when $b = 1$, which is consistent with the first formula—just apply to it the l'Hospital rule. The rule was in fact discovered by Johann Bernoulli who communicated it (not without compensation) to Guillaume François Antoine, Marquis de l'Hospital (1661–02.02.1704), and the latter published the rule in his textbook on calculus (1696); the whole book was actually based on the Johann Bernoulli's lectures.

For the service rates, we have the same expressions as in the case of an $M/M/a$ QS:

$$\mu_j = \begin{cases} j\mu, & j = 1, 2, \ldots, a-1, \\ a\mu, & j = a, \ldots, M. \end{cases}$$

Since the process is irreducible with a finite state space, it is always ergodic. The stationary distribution of X_t will then be given by $\pi_k = K_k \pi_0$ with the K_k's specified by (6.31), which in this particular case becomes (we set $\rho := \lambda/\mu$)

$$K_k = \frac{M\lambda \cdot (M-1)\lambda \cdots (M-k+1)\lambda}{\mu \cdot 2\mu \cdots k\mu} = \binom{M}{k}\rho^k, \quad k = 0, 1, \ldots, a-1,$$

and

$$K_k = \frac{M\lambda \cdot (M-1)\lambda \cdots (M-k+1)\lambda}{\mu \cdot 2\mu \cdots (a-1)\mu \cdot a\mu \cdots a\mu} = \binom{M}{k}\frac{k!}{a!a^{k-a}}\rho^k, \quad k = a, \ldots, M.$$

Therefore we have

$$\pi_k = \pi_0 \binom{M}{k}\rho^k \times \begin{cases} 1, & k = 0, 1, \ldots, a-1, \\ \\ \dfrac{k!}{a!a^{k-a}}, & k = a, \ldots, M, \end{cases} \tag{7.15}$$

where, of course,

$$\pi_0^{-1} = \sum_{k=0}^{a-1}\binom{M}{k}\rho^k + \sum_{k=a}^{M}\frac{k!}{a!a^{k-a}}\binom{M}{k}\rho^k. \tag{7.16}$$

The expected number of customers in the QS is given by the sum

$$L = \sum_{k=0}^{M} k\pi_k.$$

One is tempted to use Little's law to get the expected delay from the known value of L; note, however, that we cannot use (7.6) directly, since the arrival rate is now *variable*. However, the law still holds in the form

$$L = \Lambda D,$$

where Λ is the **average arrival rate** (in equilibrium):

$$\Lambda = \sum_{k=0}^{M}\lambda_k\pi_k = \sum_{k=0}^{M}\lambda(M-k)\pi_k = \lambda\Bigg(M\underbrace{\sum_{k=0}^{M}\pi_k}_{1} - \underbrace{\sum_{k=0}^{M}k\pi_k}_{L}\Bigg) = \lambda(M-L),$$

so that

$$D = \frac{L}{\Lambda} = \frac{L}{\lambda(M-L)}. \tag{7.17}$$

Example 7.6. A laundry has got five old tumble dryers. A typical dryer breaks down once every five days. A repairman can fix a machine in an average of 2.5 days. Currently there are three workers on duty. The owner can replace them with a "superworker" who works three times faster—and is to be paid three times the salary of an ordinary worker. Assuming that breakdown and repair times are all independent exponential RVs, is it profitable for the owner to employ the superworker instead of the other three?

The salary cost is the same in both cases. Therefore we only need to compare the average numbers of working machines.

Three normal workers. The QS is clearly a finite source queue—there are only $M = 5$ potential customers (= machines), $a = 3$, while the rates are $\lambda = 1/5$ (day^{-1}) and $\mu = 2/5$ (day^{-1}); $\rho = \lambda/\mu = 1/2$. To evaluate L, we need to find the stationary distribution (7.15) (recall that it always exists):

$$k < a: \begin{cases} \pi_1 = \binom{5}{1}\left(\frac{1}{2}\right)^1 \pi_0 = \frac{5}{2}\pi_0, \\ \pi_2 = \binom{5}{2}\left(\frac{1}{2}\right)^2 \pi_0 = \frac{5}{2}\pi_0, \end{cases}$$

and

$$k \geq a: \begin{cases} \pi_3 = \binom{5}{3}\frac{3!}{3!3^0}\left(\frac{1}{2}\right)^3 \pi_0 = \frac{5}{4}\pi_0, \\ \pi_4 = \binom{5}{4}\frac{4!}{3!3^1}\left(\frac{1}{2}\right)^4 \pi_0 = \frac{5}{12}\pi_0, \\ \pi_5 = \binom{5}{5}\frac{5!}{3!3^2}\left(\frac{1}{2}\right)^5 \pi_0 = \frac{5}{72}\pi_0. \end{cases}$$

Now to find π_0, we equate, as usual,

$$1 = \pi_0 + \pi_1 + \cdots + \pi_5 = \left(1 + \frac{5}{2} + \frac{5}{2} + \frac{5}{4} + \frac{5}{12} + \frac{5}{72}\right)\pi_0 \approx 7.736\pi_0,$$

so that $\pi_0 \approx 0.129$ and

$$L = \sum_{j=0}^{5} j\pi_j = \pi_0\left(0\times 1 + 1\times\frac{5}{2} + 2\times\frac{5}{2} + 3\times\frac{5}{4} + 4\times\frac{5}{12} + 5\times\frac{5}{72}\right) \approx 1.715.$$

Therefore $\mathbf{E}$ (the number of working machines) $= 5 - L \approx 3.285$.

One superworker. This is again a finite source queue model, with $M = 5$, $a = 1$, $\lambda = 1/5$, $\mu = 6/5$; $\rho = 1/6$. To find the stationary distribution, we only need the second of the formulae (7.15), which yields

$$\pi_1 = \binom{5}{1} 1! \left(\frac{1}{6}\right)^1 \pi_0 = \frac{5}{6}\pi_0,$$

$$\pi_2 = \binom{5}{2} 2! \left(\frac{1}{6}\right)^2 \pi_0 = \frac{5}{9}\pi_0,$$

$$\pi_3 = \binom{5}{3} 3! \left(\frac{1}{6}\right)^3 \pi_0 = \frac{5}{18}\pi_0,$$

$$\pi_4 = \binom{5}{4} 4! \left(\frac{1}{6}\right)^4 \pi_0 = \frac{5}{54}\pi_0,$$

$$\pi_5 = \binom{5}{5} 5! \left(\frac{1}{6}\right)^5 \pi_0 = \frac{5}{324}\pi_0.$$

Further, $\pi_0 \approx 0.360$, for

$$1 = \pi_0 + \pi_1 + \cdots + \pi_5 \approx 2.775\pi_0,$$

and hence

$$L = \sum_{j=0}^{5} j\pi_j = \pi_0 \left(0 \times 1 + 1 \times \frac{5}{6} + \cdots + 5 \times \frac{5}{324}\right) \approx 1.161.$$

Thus, in that case, $\mathbf{E}$ (the number of working machines) $= 5 - L \approx 3.839$, and the owner will possibly prefer hiring the superworker. Again note that we did not take into account several important aspects (e.g., the reliability of the system).

7.4 Exponential Queueing Networks

All the multiserver QS's we have been considering so far have a "parallel structure". A more complex and general scheme is that of the **network of queues** where customers may queue in front of each of the servers they need to meet their service demand and can "travel" within the network: once the current service of a customer is completed on one of the servers constituting the network, the customer is forwarded (at random in the general case) to another server or leaves the network.

First we will discuss the so-called **open networks**, where customers can arrive to the system from outside and depart from it as well. A fragment of an example of the general scheme is depicted in Fig. 7.2.

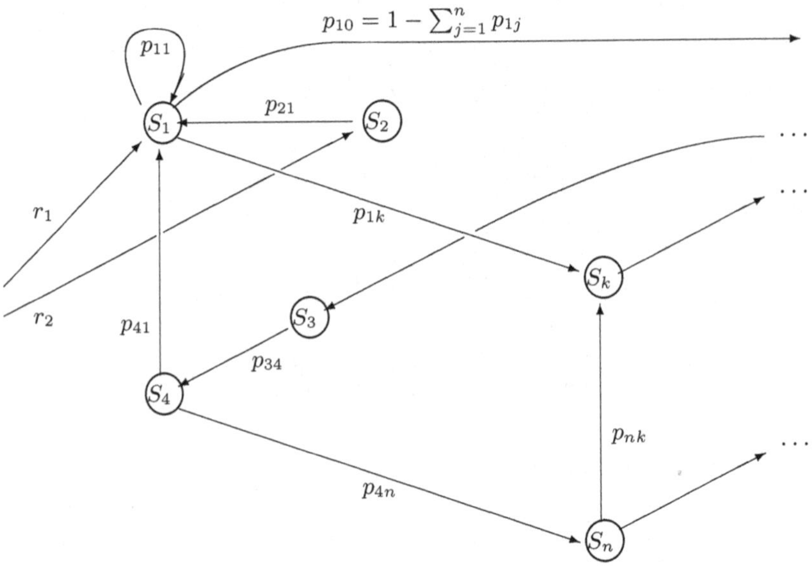

Fig. 7.2 A fragment of an open network.

We assume that:

• there are n servers (nodes) $S_1, \ldots, S_n$, service times at node S_j being $Exp(\mu_j)$-RVs, $j = 1, 2, \ldots, n$ (all the service times are jointly independent);

• for each server S_j (working under the FIFO discipline), there is a separate infinite waiting space;

• customers arrive from that outside at node S_j according to a Poisson process with rate $r_j \geq 0$, $j = 1, 2, \ldots, n$. Hence the total arrival rate is $\lambda := \sum_{j=1}^{n} r_j$; the value $p_{0j} := r_j/\lambda$ can be interpreted as the probability that an arriving in the network customer will first go to node S_j;

• once a customer has completed his service at node S_j, with probability $p_{jk} \geq 0$ he proceeds to node S_k, $k = 1, 2, \ldots, n$, where $\sum_{k=1}^{n} p_{jk} \leq 1$, or departs from the network with probability $p_{j0} := 1 - \sum_{k=1}^{n} p_{jk} \geq 0$. The matrix $P := (p_{jk})_{j,k=0,\dots,n}$ is clearly a stochastic one. That matrix—or its (substochastic) submatrix $\widetilde{P} := (p_{jk})_{j,k=1,\dots,n}$ (with positive indices)—is called the *routing matrix.*

All the random variables/processes used in the definition are supposed to be independent. A queueing network of that type is referred to as a *Jackson network.* Denoting by $X_j(t)$ the number of customers at node S_j at time t, one can see from the exponentiality assumptions that the vector of the X_j's is a Markov process.

Assume that P is irreducible, i.e., an arriving from outside customer can reach (directly or via a sequence of nodes) any node of the network, and any customer can eventually leave the network. The assumption ensures that the system of the *balance equations*

$$\lambda_k = r_k + \sum_{j=1}^{n} \lambda_j p_{jk}, \quad j = 1, \dots, n \tag{7.18}$$

(in words: in equilibrium, the total arrival rate at a node equals the arrival rate from outside plus the total arrival rate *from other nodes*), or, in matrix notation (setting $\boldsymbol{r} := (r_1, \dots, r_n)$),

$$\boldsymbol{\lambda} = \boldsymbol{r} + \boldsymbol{\lambda}\widetilde{P},$$

has a unique solution $\boldsymbol{\lambda} = (\lambda_1, \dots, \lambda_n)$ which can actually be written as

$$\boldsymbol{\lambda} = \boldsymbol{r}(I - \widetilde{P})^{-1} = \boldsymbol{r}\sum_{k=0}^{\infty} \widetilde{P}^k$$

(note that $\widetilde{P}^k \to 0$ exponentially fast as $k \to \infty$: it is a *substochastic* matrix, and hence the absolute value of its largest eigenvalue—which in fact is positive by the *Perron*[4]*–Frobenius*[5] *theorem*[6]—is less than one, so that the matrix series converges).

The component λ_j of the solution is called the *equilibrium arrival rate* at node S_j (due to stationarity, it is also the equilibrium departure rate—why?). But when does the equilibrium actually exist? The answer to this question is given by the following

[4]Oskar Perron (07.05.1880–22.02.1975), a German mathematician who worked on a wide range of mathematical topics.

[5]Ferdinand Georg Frobenius (26.10.1849-03.08.1917), a German mathematician who, in particular, made contributions to the theory of elliptic functions, differential equations and group theory.

[6]See e.g., Gantmakher, F.R. *The theory of matrices.* Vol. 2. Chelsea, New York, 1989.

Theorem 7.1. *If*

$$\rho_j = \frac{\lambda_j}{\mu_j} < 1, \quad j = 1, \ldots, n,$$

where λ_j *form the unique solution to the balance equations* (7.18), *then the vector-valued MP* $\{(X_1(t), \ldots, X_n(t))\}$ *is ergodic and has the stationary distribution*

$$\mathbf{P}(X_1 = k_1, \ldots, X_n = k_n) = \prod_{j=1}^{n} (1 - \rho_j)\rho_j^{k_j}, \quad k_j = 0, 1, 2, \ldots \tag{7.19}$$

Therefore, if there are no *bottlenecks* (nodes with the ratio $\lambda_j/\mu_j \geq 1$), then there exists the steady state and, in equilibrium, $X_j(t)$, $j = 1, \ldots, n$, are independent geometric RVs—as if they were describing n independent $M/M/1$ QS's! But the nodes, in the general case, are by no means independent; that is, the *processes* $\{X_j(t), t \geq 0\}$, $j = 1, \ldots, n$, will not be independent. Indeed, arrival flows to $S_1, \ldots, S_n$ are *dependent* and *not* Poissonian, for there can be such things as *feedback*. For example, consider the simplest case depicted on the diagram below:

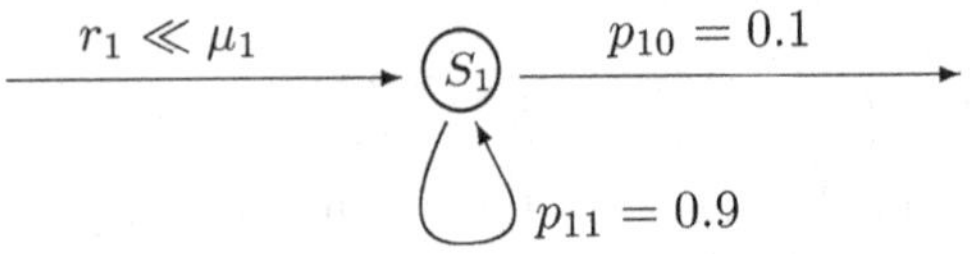

At an arrival epoch, we have a high probability of another arrival in a short time (it is likely that the customer, upon completing its service at S_1, will immediately be back), whereas at an arbitrary time point, that probability is relatively low. Therefore the arrival process at S_1 is definitely not Poissonian. Solving the balance equation (7.18) in this case yields the total arrival rate of $\lambda_1 = 10r_1$, so that the system will be stable iff $r_1 < 0.1\mu_1$. If this is the case, the stationary distribution of the number of customers in it will be the same geometric distribution as one would have for an $M/M/1$ system with a ten times more intensive arrival stream.

Knowing the stationary distribution (7.19), we can answer further questions about the long-run behaviour of the network. For example, the average number of customers in it will just be the sum of the expectations of the marginal geometric distributions:

$$L = \sum_{j=1}^{n} (\text{average \# at node } S_j) = \sum_{j=1}^{n} \frac{\lambda_j}{\mu_j - \lambda_j}.$$

The average delay in the network will be given, by Little's law, by the relation $L = \lambda D$ with $\lambda = \sum_{j=1}^{n} r_j$ being the total arrival rate at the network:

$$D = \frac{1}{\lambda} L = \frac{\sum_{j=1}^{n} \lambda_j/(\mu_j - \lambda_j)}{\sum_{j=1}^{n} r_j}.$$

Example 7.7. *A tandem queueing system.* Suppose we have a tandem network of two nodes connected as shown below:

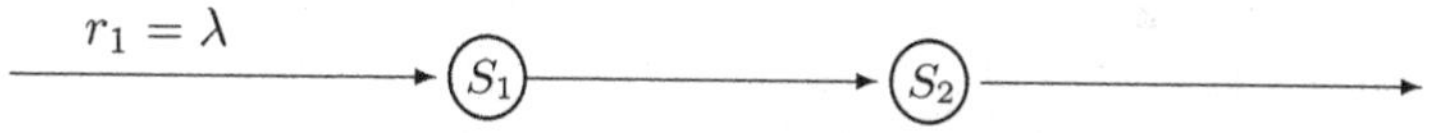

Clearly, the probabilities of transitions along the arrows from S_1 to S_2 and from S_2 are ones. So, we have here the following numeric values:

$$n = 2,\ r_1 = \lambda,\ r_2 = 0,\ p_{01} = 1,\ p_{12} = 1,\ p_{20} = 1.$$

The routing matrix

$$P = \begin{pmatrix} 0 & 1 & 0 \\ 0 & 0 & 1 \\ 1 & 0 & 0 \end{pmatrix}$$

is irreducible. The balance equations (7.18) take the form

$$\begin{cases} \lambda_1 = r_1 \\ \lambda_2 = \lambda_1 \end{cases} \implies \lambda_1 = \lambda_2 = r_1 = \lambda,$$

so that the stationary distribution exists if $\lambda < \mu_j$, $j = 1, 2$, and is then given by

$$\mathbf{P}\left(X_1 = k_1, X_2 = k_2\right) = \left(1 - \frac{\lambda}{\mu_1}\right)\left(\frac{\lambda}{\mu_1}\right)^{k_1}\left(1 - \frac{\lambda}{\mu_2}\right)\left(\frac{\lambda}{\mu_2}\right)^{k_2}, \quad k_1, k_2 \geq 0.$$

Now the average number of customers in the tandem network and average delay are, respectively,

$$L = \frac{\lambda}{\mu_1 - \lambda} + \frac{\lambda}{\mu_2 - \lambda}, \quad D = \frac{L}{\lambda} = \frac{1}{\mu_1 - \lambda} + \frac{1}{\mu_2 - \lambda}.$$

In *closed networks* we have the same situation as in Fig. 7.2 with the only difference that there is no in- or outgoing streams of customers (no arrows leading from the outside to the network, and from the network to the outside), so that there is a fixed population of m customers who are being served and who travel within the network. We again assume exponentiality and independence of the service times. Then it turns out that when the routing matrix

$$P = (p_{jk})_{j,k=1,\ldots,n}$$

is *irreducible*, the process $(X_1(t), \ldots, X_n(t))$ is always ergodic. The balance equations, which now take the form

$$\lambda_k = \sum_{j=1}^{n} \lambda_j p_{jk}, \quad j = 1, \ldots, n, \qquad \text{or} \qquad \boldsymbol{\lambda} = \boldsymbol{\lambda} P$$

in matrix notation, have in that case a unique (up to a common constant multiplier) solution $\boldsymbol{\lambda} = \boldsymbol{\pi} =$ the stationary distribution of P, and the stationary distribution of the process $(X_1(t), \ldots, X_n(t))$ has a simple closed product form:

$$\mathbf{P}\,(X_1 = k_1, \ldots, X_n = k_n) = \begin{cases} C_m \prod_{j=1}^{n} \rho_j^{k_j} & \text{if } k_1 + \cdots + k_n = m, \\ 0 & \text{otherwise.} \end{cases} \tag{7.20}$$

Note that, despite the product form of the distribution, X_j are *not independent* (which was the case for open networks); even the support of the distribution is not a product set (which is a necessary condition for independence), but a "discrete simplex" consisting of the points from $\mathbf{R}^m$ with non-negative integer-valued components whose sum is m.

One more remark is that, despite its simple closed form, there is a "nasty" parameter in that product form distribution. Namely, computing the value of the (normalising) constant C_m in (7.20) can be very hard even for moderate n. It requires finding the sum of a (very large) number of the products of the ρ's.

7.5 Recommended Literature

KLEINROCK, L. *Queueing Systems.* Vols. 1,2. Wiley, New York, 1975–1976.

LEE, A.M. *Applied Queueing Theory.* St. Martin's Press, New York, 1966. [*Interesting case studies using elementary theory.*]

MEDHI, J. *Stochastic Models in Queueing Theory.* Academic Press, Boston, 1991.

7.6 Problems

1. The first five jobs in the queue to enter a job shop on a day have the following service times:

$$1.3,\ 0.7,\ 4.1,\ 2.9,\ 3.1 \quad \text{(hours)}.$$

   (i) Suppose the jobs are processed in the strict arrival order (i.e., according to FIFO). Find the waiting time and delay time (waiting time plus service time) for each job. Find the average delay time per job for these jobs.

   (ii) In order to minimise the waiting time per customer, the priority rule "shortest-in, first-out" (SIFO) may be established. This means that the next customer from the queue to be served will be the one with the shortest service time. Repeat the calculations of (i) for this service discipline.

   (iii) Suppose the cost of one hour delay for a customer is \$10. Compare the total cost of delay for FIFO and SIFO rules.

   Hint. There is no randomness in this problem! No probability calculus!

2. What is the expected number of customers being served (i.e., obtaining service *now*) in an $M/M/1$ queue?

3. Show that, in an $M/M/1$ queue, the expected lengths of the idle and busy periods are $1/\lambda$ and $1/(\mu - \lambda)$, respectively.

   Hint. The idle time is the time between the departure of a customer when there is nobody left in the system and the arrival of the next customer. Recall also that the proportion of the time the server is idle is equal to the stationary probability of having no customers in the system.

4. A rent-a-car maintenance facility has capabilities for routine maintenance for only one car at a time. Cars arrive for this routine maintenance according to a Poisson process at the mean rate of 3 per day, and the service time to perform this maintenance has the exponential distribution with the mean of 7/24 days. It costs the company a fixed \$150 a day to operate the facility. The company estimates a loss in profit on a car of \$10 per day for every day the car is tied up in the shop. The company, by changing certain procedures and hiring faster mechanics, can decrease the mean service time to 1/4 day. This also increases their operating costs. Up to what value can the operating cost increase before it is no longer economically attractive to make the change?

5. Suppose that there were an unlimited number of servers ($a = \infty$). This can be an appropriate model for a self-service situation.

   (i) Draw the transition rate diagram and calculate the steady state probability distribution.

   (ii) Find the expected number of customers in the system and the expected number of customers waiting in the (common) queue. What is the expected delay time D?

Hint. To model the system by a B+DP, note that customers (= particles, individuals) arrive at a constant rate (not depending on the current state of the system), and then "live" (= stay in the system) for an exponential random time. What is the death rate when there are k "alive" individuals?

6. Draw the transition rates diagram for an $M/M/N/N$ queue. Prove that the equilibrium distribution $\{\pi_j\}$ (of the number of customers in the system) for this system is given by

$$\pi_j = \frac{\rho^j}{j!}\pi_0, \quad j = 1, \ldots, N, \quad \pi_0 = \left(\sum_{j=0}^{N} \frac{\rho^j}{j!}\right)^{-1},$$

regardless of the value of $\rho = \lambda/\mu$.

7. Plot both stationary distributions in Example 7.6 and discuss differences in their shape and reasons for them.

8. Damdam Pty Ltd is building a dam. A total of 10,000,000 cu ft of dirt is needed to construct the dam. A bulldozer is used to collect dirt for the dam. Then the dirt is moved via dumpers to the dam site. Only one bulldozer is available and it rents for \$100 per hour. Damdam can rent, at \$40 per hour, as many dumpers as desired. Each dumper can hold 1,000 cu ft of dirt. It takes an average of 12 minutes for the bulldozer to load a dumper with dirt and takes each dumper an average of five minutes to deliver the dirt to the dam and return to the bulldozer. Making appropriate assumptions about exponentiality, determine how Damdam can minimise the total expected cost of moving the dirt needed to build the dam.

   Hints. The objective function (which is to be minimised in M, the number of rented dumpers) is

$$\mathbf{E}\{\text{\# of hours to work}\} \times \big(\{\text{cost of bulldozer}\} + \{\text{cost of } M \text{ dumpers}\}\big). \quad (7.21)$$

   The bulldozer is clearly the server, the dumpers being the "customers". What model suits the situation?

   How to find the first factor in (7.21)? It will suffice to consider the values $M = 1, 2$ only; show that to rent $M \geq 3$ dumpers will definitely be worse than any of these first two variants.

9. Solve the previous problem under the assumption that there are two bulldozers (all other parameters remain unchanged).

10. Customers arrive at a bank according to a Poisson process with rate λ. The service times are i.i.d. exponential random variables with rate μ. The bank follows this policy: when there are fewer than four customers in the bank, only one teller is active; for four to nine customers, there are two tellers, and beyond nine customers there are three tellers. Model the number of customers in the bank as a birth-and-death process. When is this system stable? Assuming stability, compute the steady state distribution.

Hint. You will get a queueing model which we did not discuss in the text.

11. New Melbourne City has 10,000 street lights. A street light burns out after an average of 100 days use. The city has hired Small Company Pty Ltd to replace the burnt out lamps. The contract states that Small Company Pty Ltd is supposed to replace a burnt out lamp in an average of 7 days.

    City investigators have determined that, at any given time, an average of 1,000 lights are burnt out. Do you think that Small Company Pty Ltd is keeping to the contract?

    Hints. Assume that the lifetimes are exponentially distributed. Small Company Pty Ltd is the server, while street lights are the "customers".

12. For a finite source queue (machine repair problem), express the probability that an arrival finds the system in state k in terms of the stationary probabilities π_k for the system. (This model does not have the PASTA property!) Give a verbal explanation for the result.

    Hint. The desired probability is actually the conditional probability of the form $\mathbf{P}\left(X_t = k \,|\, \text{arrival during } (t, t+dt)\right)$. Use the Bayes formula and the interpretation of the birth rates for the modelling birth-and-death process.

13. A small company has two old trucks. On average, a truck breaks down once in 40 days. When a truck breaks down, its repair starts *immediately* and takes an average of 4 days. All times are supposed to be independent and exponentially distributed.

    (i) Formulate a three state birth-and-death process for modelling this situation. Draw the transition diagram for this process and find all the birth and death rates. How is such a queueing model called?

    (ii) In the steady state, find the fraction of the time that both trucks are running.

    (iii) Find the fraction of the time that no trucks are running.

Chapter 8

Elements of Renewal Theory

8.1 Definitions and Notation. Renewal Theorems

A continuous time process $\{N_t\}_{t\geq 0}$ is said to be a **renewal process** (RP) when it is a *counting process* for which the times between consecutive events/jumps (called "renewals") are non-negative i.i.d. RVs τ_j, $j = 1, 2, \ldots$, with a common DF F. We assume w.l.o.g. that $F(0) < 1$ (thus excluding the case $\tau_j \equiv 0$) and set $\mu := \mathbf{E}\,\tau_j \in (0, \infty]$.

Sometimes it is convenient to assume that the time τ_1 of the first event in the process follows a *different* distribution, and then the RP is called a **delayed RP**.

So the trajectory of an RP is a step function with unit jumps, with the position of the kth jump being given by the so-called *renewal epoch* $T_k := \tau_1 + \cdots + \tau_k$, $k = 1, 2, \ldots$ (note that $\{T_k\}_{k\geq 0}$, $T_0 := 0$, itself is an RW with jumps τ_j):

$$N_t := \max\{k \geq 0 : T_k \leq t\}, \tag{8.1}$$

see Fig. 8.1.

Example 8.1. Suppose we have a lamp which should always be switched on and an (at least theoretically) infinite supply of light bulbs whose lifetimes are i.i.d. RVs. Once the bulb currently on burns out, it is immediately replaced by a new one. Then $N_t = \#\{\text{bulbs failed by the time } t \geq 0\}$ will be an RP.

In fact, we have already encountered such processes. Thus,

(i) in discrete time, due to the Markov property, the "cycles" between consequent visits by a MC to a fixed recurrent state j_0 are i.i.d. Therefore, the time instances when the MC is at this state are renewal epoches and

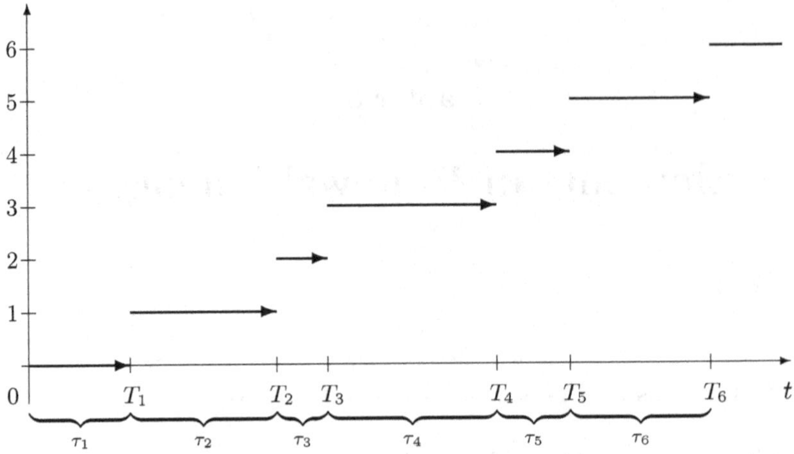

Fig. 8.1 A typical trajectory of a renewal process

form an RP. If the MC does not start at j_0, the time of its first visit to that state will, generally speaking, have a different distribution—and then we will get a delayed RP.

(ii) In continuous time, the Poisson process is actually an RP, with $\tau_j \sim Exp(\lambda)$ for a common $\lambda > 0$. Note also that the Poisson process is an MP, and, as we saw in Section 5.1, when $F \neq Exp(\cdot)$, a general RP $\{N_t\}$ is not Markovian!

Quite often, stochastic systems have what is called "regenerations": roughly speaking, there are (random) instances when the system starts, in a sense, its evolution anew (like the times when a MC visits a fixed selected state; an example of a non-Markovian process having such regeneration property is an $M/GI/1$ system at times when the server becomes idle). In all such cases, renewal theory provides powerful means for studying the behaviour of the modelled systems: once we know the laws governing the long-run regeneration occurrence, we can also infer what happens, in the long term, to the system itself.

But first of all note that RPs are *regular processes* in the sense that there can be no such things as *explosions* (i.e., infinite numbers of renewals in a finite time) in an RP. Indeed, the following events are equivalent:

$$\{N_t \geq n\} = \{T_n \leq t\}, \quad t \geq 0, \quad n = 0, 1, 2 \ldots \tag{8.2}$$

Therefore, for any fixed $t < \infty$, from (2.9) (where we set $B_n := \{N_t \geq n\}$;

clearly, $B_n \supset B_{n+1}$) we have

$$\mathbf{P}(N_t = \infty) = \lim_{n\to\infty} \mathbf{P}(N_t \ge n) = \lim_{n\to\infty} \mathbf{P}(T_n \le t)$$
$$\equiv \lim_{n\to\infty} \mathbf{P}\left(\frac{T_n}{n} \le \frac{t}{n}\right) = 0,$$

because, as $n \to \infty$, we have $T_n/n \to \mu > 0$ by the LLN, whereas $t/n \to 0$.

On the other hand, $\lim_{t\to\infty} N_t = \infty$ w.p. 1, since

$$\left\{\lim_{t\to\infty} N_t < \infty\right\} = \bigcup_{j=1}^{\infty} \underbrace{\{\tau_j = \infty\}}_{\mathbf{P}(\,\cdot\,)=0},$$

and the union of a countable number of events of zero probability also has zero probability.

In fact, one can write a simple explicit formula for the distribution of N_t in terms of the convolutions of F. Indeed, from (8.2), for $n = 0, 1, \ldots,$

$$\begin{aligned}\mathbf{P}(N_t = n) &= \mathbf{P}(N_t \ge n) - \mathbf{P}(N_t \ge n+1)\\ &= \mathbf{P}(T_n \le t) - \mathbf{P}(T_{n+1} \le t) = F_n(t) - F_{n+1}(t),\end{aligned}$$

where $F_n(t) = F^{*n}(t)$ is the DF of T_n, i.e., the n-fold convolution of F with itself. However, as we already said, computation of such convolutions is a very tedious task even for moderate values of n, while the most interesting question is what happens to the RP in the long term. The first result in this direction we cite here is actually a simple consequence of the strong LLN.

Theorem 8.1. *With probability* 1,

$$\frac{N_t}{t} \to \frac{1}{\mu} \quad as \quad t \to \infty;$$

the case $\mu = \infty$ *is not excluded.*

Indeed, relation (8.1) actually means that, as a function of t, the RP $\{N_t\}$ is the *generalised* inverse of the (continuous time) process $\{T_{\lfloor t\rfloor}\}_{t\ge 0}$. Since under identical scaling in both time and space variables (i.e., when considering the process $\{n^{-1}T_{\lfloor nt\rfloor}\}$ as $n \to \infty$) that process is close, by the LLN, to the straight line going through the origin with slope μ (see Section 2.9), the "inverse" process $\{n^{-1}N_{nt}\}$ will also be close to a straight line—but with the reciprocal slope $1/\mu$. Theorem 8.1 immediately follows from this observation. Note also that the above proposition holds under more general assumptions as well (when the τ's are "weakly dependent").

We will illustrate this theorem by a simple example.

Example 8.2. Ann has a radio which works on a single battery. When the battery fails, Ann immediately replaces it (regardless of the time of the day). Supposing that a battery's lifetime is an RV $\sim U(30,60)$, at what rate does Ann have to change batteries? (Note that the assumption on uniformity does not necessarily mean low quality and unreliability of the batteries; it could rather reflect variable load on them.)

Since the expectation equals $\mu = \frac{1}{30}\int_{30}^{60} x\,dx = \frac{1}{2}(30+60) = 45$, the desired long-run rate is $\lim_{t\to\infty} t^{-1}N_t = 1/\mu = 1/45$ (hour^{-1}). If, in addition, Ann must go shopping each time to buy a new battery which takes her a random time $\xi \sim U(0,2)$, then the rate will drop to $1/(\mu + \mathbf{E}\,\xi) = 1/46$.

Example 8.3. *An $M/G/1/1$ queueing system.* There is actually no queue in this QS: when an arriving customer finds the server busy, he does not enter the system. Arrival stream is given by a Poisson process with rate λ, while the service times are positive i.i.d. RVs following a common DF G with mean m_G. What is the rate at which new customers enter the system? What proportion of potential customers actually enter the QS?

Due to the memoryless property of the exponential distribution, the times between successive instances when the customers successfully enter the system are i.i.d. RVs which can be represented as sums of independent $Exp(\lambda)$-distributed and G-distributed RVs. (Indeed, the time from the moment when a customer leaves the server till the arrival of the next customer is independent of the "past" and $Exp(\lambda)$-distributed.) Therefore the desired entrance rate is

$$\frac{1}{\text{mean}} = \frac{1}{1/\lambda + m_G} = \frac{\lambda}{1+\lambda m_G}.$$

On the other hand, since the potential customers arrive at the rate λ, the proportion of them entering the QS is

$$\frac{\text{entrance rate}}{\text{arrival rate}} = \frac{\lambda/(1+\lambda m_G)}{\lambda} = \frac{1}{1+\lambda m_G}.$$

Thus, for example, if $\lambda = 7.5$ hour^{-1} and $m_G = 15$ (min) $= 0.25$ (hour), so that the service rate is 53.3% of the arrival one, then

$$\frac{1}{1+\lambda m_G} = \frac{1}{1+7.5\times 0.25} = \frac{8}{23},$$

i.e., on the average, only about one customer out of three will actually enter the QS. Note that in the deterministic system $D/D/1/1$ with zero initial delay and the same arrival/service rates as in the $M/G/1/1$ system, the proportion will be 1/2—much higher!

As it was the case for RWs, the LLN can be "refined" by the CLT when the second moments are finite.

Theorem 8.2. (CLT for RPs) *If* $\mathbf{E}\,\tau_j = \mu$ *and* $\mathrm{Var}\,(\tau_j) = \sigma^2 < \infty$, *then*

$$\xi_t := \frac{N_t - t/\mu}{\sigma\sqrt{t/\mu^3}} \xrightarrow{distr} N(0,1).$$

That is, for any x, $\mathbf{P}\,(\xi_t \le x) \to \Phi(x)$ *as* $t \to \infty$.

Convergence here is actually uniform in x:

$$\sup_x |\mathbf{P}\,(\xi_t \le x) - \Phi(x)| \to 0 \quad \text{as} \quad t \to \infty.$$

Theorem 8.2 can be easily proved using representation (8.2) and the standard CLT.

Along with the number of renewals by the time t, one is often interested in another important characteristic of the process, namely, the "residual lifetime"

$$Y_t := T_{N_t+1} - t > 0,$$

i.e., the time from the current t till the first renewal to occur in the "future".

In discrete time, we have actually already dealt with that characteristic, recall Example 3.13. Moreover, we showed later that if $\mu < \infty$ and $\mathrm{GCD}\{k : \mathbf{P}\,(\tau_j = k) > 0\} = 1$, the respective MC is ergodic and has the limiting distribution (3.32), which is equivalent to

$$\lim_{t\to\infty} \mathbf{P}\,(Y_t = j) = \frac{1}{\mu}(1 - F(j-0)), \quad j = 1, 2, \dots \tag{8.3}$$

Similarly, when the τ_j's are "non-arithmetic" (that is, there is no such number a that all possible values of τ_j are multiples of a), the limiting DF also exists and is given by the so-called *integrated tail distribution.*

Theorem 8.3. *If the distribution of* τ_j *is non-arithmetic, then, for* $x > 0$,

$$\lim_{t\to\infty} \mathbf{P}\,(Y_t \le x) = \frac{1}{\mu}\int_0^x \overline{F}(y)\,dy, \quad \overline{F}(y) := 1 - F(y). \tag{8.4}$$

That is, as $t \to \infty$, the limiting for Y_t distribution is absolutely continuous with density $\mu^{-1}\overline{F}(y)$, $y > 0$. Not that if $F = Exp(\lambda)$ (so that $\mu = 1/\lambda$), this expression is equal to $f(y) = \lambda e^{-\lambda y}$, $y > 0$, which is nothing else but the density of F itself (recall that, in that case, $f(y)$ is the density not only of the limiting distribution, but also of the distribution of Y_t for any finite t as well). At the end of this section we will sketch the proof of a more general result based on the renewal Theorem 8.4 below.

Remark 8.1. It is interesting to note the following apparent paradox: the limiting distribution of the residual lifetime has a finite expectation iff the lifetimes themselves have a finite second moment: $\mathbf{E}\,\tau_j^2 < \infty$ (cf. (2.55)). On the other hand, for any fixed t, Y_t is always *less* than the value of the length τ_j of the renewal interval covering the time point t. The explanation is as follows: a *long* random interval $[T_{j-1}, T_j)$ is *more likely* to cover the given value t. Hence the "covering interval" (its number is also random, of course) will tend to be *longer* (in distribution) than an "ordinary" $[T_{j-1}, T_j)$ (cf. Problem 3 in Section 5.3).

Remark 8.2. Note also that if τ_1 (which, as we said, can be distributed differently from $\tau_2, \tau_3, \ldots$, and then $\{N_t\}$ is called a *delayed RP*) follows the DF (8.4), then $\{N_t\}$ will be *stationary* (see p. 63).

One more very important notion is that of the **renewal function**[1] (for the distribution F) defined by

$$H(t) := \mathbf{E}\,N_t = \int_0^\infty (1 - F_{N_t}(x))\,dx = \sum_{n=1}^\infty \mathbf{P}\,(N_t \geq n)$$
$$= \sum_{n=1}^\infty \mathbf{P}\,(T_n \leq t) = \sum_{n=1}^\infty F^{*n}(t). \quad (8.5)$$

One can show that there exists a one-to-one correspondence between the DF F and the respective renewal function H.

Example 8.4. For a Poisson process $\{N_t\}$ with rate λ, the renewal function is clearly $H(t) = \lambda t$.

We saw above, in Theorem 8.1, that $N_t/t \to 1/\mu$ a.s. It turns out that the expectations of the left-hand side converge to the same limit. This assertion is called the *key renewal theorem*. Moreover, it admits refinements in both non-arithmetic and arithmetic cases (it is clear that the latter can be reduced to the case of integer-valued RVs).

Theorem 8.4. (i) *There exists the limit*

$$\lim_{t\to\infty} \frac{H(t)}{t} = \frac{1}{\mu};$$

we do not exclude the case $\mu = \infty$ (then we set $1/\infty = 0$).

[1] The renewal function is sometimes defined as $H(t) := 1 + \mathbf{E}\,N_t$.

(ii) *If τ_j are non-arithmetic, then, for any fixed $u > 0$,*

$$H(t) - H(t-u) \to \frac{u}{\mu} \quad as \quad t \to \infty.$$

(iii) *If τ_j are integer-valued and* $\mathrm{GCD}\{k > 0 : \mathbf{P}(\tau_j = k) > 0\} = 1$, *then, as $n \to \infty$,*

$$h(n) := H(n) - H(n-1) \to \frac{1}{\mu}, \quad \sum_{l=1}^{n} h(l)\, g(n-l) \to \frac{1}{\mu} \sum_{m=1}^{\infty} g(m) \quad (8.6)$$

for any sequence $g(m)$ such that $\sum_{m=1}^{\infty} |g(m)| < \infty$.

Thus, in case (ii), for any fixed (integer) k, the mean number of events in the interval $(n-k, n]$ will converge to k/μ as $n \to \infty$. Note that the standard proof of the ergodic theorem for MCs (Theorem 3.4) is based on part (iii) of the above theorem, where the τ_j's are just times between successive visits of the MC to a fixed recurrent state j_0 (with a finite mean recurrence time). The mean number of renewals (= visits to j_0) on a one-point interval $\{n\}$ is just the probability of being at j_0 at the time n, and (iii) means that this value converges to 1/(mean recurrence time to j_0) as time increases.

The proof of Theorem 8.4 can be found in any advanced textbook on probability theory. Here we will only give an idea how the elegant *coupling* method can be used to prove part (iii).

Introduce an independent of $\{\tau_j\}$ sequence of independent RVs $\{\tau_j'\}$, with τ_1' being distributed according to (8.4) and τ_j', $j = 2, 3, \ldots$, having the same distribution as τ_1. Then the renewal function $H'(n)$ for the delayed RP $\{N_n'\}$ defined by the sequence $\{\tau_j'\}$ will simply be n/μ for integer n (Problem 7; cf. Remark 8.2: the RP $\{N_n'\}$ is stationary!). On the other hand, one can show that, sooner or later, we will have renewal epochs for both sequences at a *common time*. Moreover, one has

$$\nu := \min\{m \ge 1 : T_m = T_m'\} < \infty \quad \text{a.s.}, \qquad T_m' = \tau_1' + \cdots + \tau_m'.$$

Indeed, that follows from the fact that an RW with symmetrically distributed jumps $\tau_j - \tau_j'$ will a.s. visit all integer points (it is the proof of that fact where we will need the assumption that the GCD of the possible values of τ_1 is one).

Now note that we can *couple* ("glue together") the processes $\{N_n'\}$ and $\{N_n\}$ *after* the time T_ν—say, by putting $N_n := N_n'$ for all $n \ge T_\nu$. This will not change the distribution of any of the RPs! Therefore we can write

$$\begin{aligned} H(n) - H(n-1) &= \mathbf{E}\,(N_n - N_{n-1}) \\ &= \mathbf{E}\,(N'_n - N'_{n-1};\, T_\nu < n) + \mathbf{E}\,(N_n - N_{n-1};\, T_\nu \ge n) \\ &= \frac{1}{\mu} - \mathbf{E}\,(N'_n - N'_{n-1} - (N_n - N_{n-1});\, T_\nu \ge n). \end{aligned}$$

But $0 \le N_n - N_{n-1} \le 1$ always (the same is true of $\{N'_n\}$), so that the absolute value of the expression in the last expectation cannot exceed one and hence

$$|H(n) - H(n-1) - 1/\mu| \le \mathbf{P}\,(T_\nu \ge n) \to 0 \qquad \text{as} \quad n \to \infty$$

since $T_\nu < \infty$. The second part of (iii) is an elementary consequence of the first assertion.

As promised, in conclusion of this section we will show how to derive the assertion of Theorem 8.3 (and more!) from Theorem 8.4 (ii), (iii). Consider, in addition to the residual life-time Y_t, the "current age" of the currently working device:

$$Z_t := t - T_{N_t}.$$

In the RW terminology, Z_t and Y_t are called, respectively, the *defect* and *excess* of the level t for the RW $\{T_n\}$. Then, in the case of integer-valued τ_j's, for any $i \ge 0$ and $j \ge 1$, one has

$$\begin{aligned} \mathbf{P}\,(Z_t = i, Y_t = j) &= \sum_{n=0}^{\infty} \mathbf{P}\,(Z_t = i, Y_t = j, N_t = n) \\ &= \sum_{n=0}^{\infty} \mathbf{P}\,(T_n = t - i, \tau_{n+1} = i + j) \\ &= \sum_{n=0}^{\infty} \mathbf{P}\,(T_n = t - i)\, \mathbf{P}\,(\tau_{n+1} = i + j) \qquad \text{[by independence]} \\ &= p_{i+j} \sum_{n=0}^{\infty} \mathbf{P}\,(T_n = t - i) \\ &= p_{i+j} \Bigl(\sum_{n=0}^{\infty} \mathbf{P}\,(T_n \le t - i) - \sum_{n=0}^{\infty} \mathbf{P}\,(T_n \le t - i - 1) \Bigr) \\ &= p_{i+j} (H(t-i) - H(t-i-1)) = p_{i+j} h(t-i) \to \frac{p_{i+j}}{\mu} \end{aligned}$$

as $t \to \infty$ by (8.6). Since $\mathbf{P}\,(Y_t = j) = \sum_{i \ge 0} \mathbf{P}\,(Z_t = i, Y_t = j)$, the above result immediately implies (8.3).

Likewise, in the non-arithmetic case, one obtains from Theorem 8.4 that, for $u, v > 0$,

$$\lim_{t \to \infty} \mathbf{P}\,(Z_t > u, Y_t > v) = \frac{1}{\mu} \int_{u+v}^{\infty} \overline{F}(y)\, dy.$$

Letting here $u := 0$ yields (8.4).

One more important class of objects from renewal theory is the so-called renewal and renewal-type equations which often arise in various areas of mathematics. We discuss them to some extent in the problem section below.

8.2 Problems

1. Prove Theorem 8.2.

2. A faculty course advisor is giving advice to students doing either course A or course B. Course A is well structured, there is almost no space for electives, so advice is simple. Course B students need much more time to select subjects; moreover, in particularly complex situations, when advising time reaches 15 minutes, the student is sent to talk to a faculty officer. Accordingly, we assume that the distributions of the time the advisor spends talking to a student have the densities

   (A) $f_A(x) = C_A(1/4 - x)$, $x \in [0, 1/4]$, and

   (B) $f_B(x) = C_B x$, $x \in [0, 1/4]$,

   for courses A and B, respectively (the time unit is hour). On each day, either only course A students or only course B students can come to the course advice session. We assume that there are always some students waiting for advice, so the advisor works without breaks.

   Denote by N_t the number of students who have already finished talking to the advisor by the time t (hours after the beginning of the advice session on a given day).

   For both cases (A) and (B):

   (i) Find the constants C_A and C_B in the expressions for the densities and plot the densities.

   (ii) Find and plot the density of the stationary residual lifetime distribution (i.e., the limiting law for $Y_t = T_{N_t+1} - t$ as $t \to \infty$). Compute the mean of this distribution.

   (iii) Find an approximate value of the number N_8 of students the advisor can talk to during an eight hour long working day.

   (iv) Give (approximate) intervals to which the RV N_8 belongs with probability 90%. How could you explain the difference in the sizes of the intervals for courses A and B?

   Hint: You can greatly reduce calculations by making use of the symmetry between the densities: $f_A(x) = f_B(1/4 - x)$.

3. Show that the renewal function $H(t)$ satisfies the *renewal equation*

$$H(t) = F(t) + \int_0^t H(t-s)\,dF(s) \equiv F(t) + (H * F)(t), \quad t \geq 0.$$

4. Show that the *renewal-type equation*

$$M(t) = D(t) + \int_0^t M(t-s)\,dF(s) \equiv D(t) + (M * F)(t), \quad t \geq 0,$$

for unknown M and given D admits solution $M(t) = D(t) + (D * H)(t)$, where H is the renewal function for F. (This solution is unique.)

5. Set $H_2(t) := \mathbf{E}\, N_t^2$. Show that

$$H_2(t) = H(t) + 2\int_0^t H(t-s)\, dH(s).$$

Find $H_2(t)$ when $\{N_t\}$ is a Poisson process.

6. Prove that, for a stationary delayed RP (i.e., when τ_1 follows the DF (8.4)), the renewal function

$$H_S(t) := \sum_{n=1}^{\infty} F_1 * F^{*(n-1)}(t) = \frac{t}{\mu}, \quad t \geq 0.$$

Hint. You may wish to show first that $F_1 = \mu^{-1}(I - F) * J$, where we put $J(x) := xI(x)$, $I(x) := \mathbf{1}_{\{x \geq 0\}}$. Alternatively, you can compute the Laplace-Stieltjes transform of F_1 (integrating by parts), and then find the transform of H_S and that of t/μ (cf. (2.71)).

7. Verify that, for the delayed RP, when the τ_j's are integer-valued and τ_1 follows the distribution

$$\mathbf{P}(\tau_1 = k) = \frac{1}{\mu}(1 - F_{\tau_2}(k-1)) = \frac{1}{\mu}\sum_{j=k}^{\infty} \mathbf{P}(\tau_2 = j), \tag{8.7}$$

the renewal function has the form

$$H_S(k) \equiv \sum_{n=1}^{\infty} F_1 * F^{*(n-1)}(k) = \frac{k}{\mu}, \quad k = 1, 2, \ldots$$

Hint. You may wish to use an argument similar to the one from Problem 6. Alternatively, use GFs.

Chapter 9

Elements of Time Series

A **time series** (TS) is a sequence of observations $\{x_t\}$ recorded at (usually) regular time intervals. For convenience sake, we often assume that the time values are $t = 1, 2, \ldots, T$. The difference from the notion of a *random sample* is that the *order* of observations is now important. In other words, a TS is a realisation of an SP in discrete time. The term is often used to mean both the data $\{x_t\}$ (the x_t's are *numbers*) *and* the process $\{X_t\}$ itself (the X_t's are *RVs*!), of which the former is a realisation.

The main objective of TS analysis is to draw inferences from observed segments of TS's. To do this, we need to choose a (hypothetical) model or, rather, a family of models for $\{X_t\}$, estimate its parameters and check the goodness of fit. A satisfactory model can help to better understand how the TS is generated and, importantly, to predict the future values of the TS. The present chapter is only a brief introduction to the theory of TS's.[1] In particular, we will mostly be discussing modelling aspects rather than "more statistical" problems (fitting models, forecasting *etc.*) of the TS analysis.

Example 9.1. *White noise.* The simplest model for a TS is an i.i.d. noise (e.g., the Gaussian white noise from Example 2.7), where X_t are i.i.d. RVs with a common mean m and variance σ^2. Since the conditional distribution of X_{t+1} given $X_1, \ldots, X_t$ coincides with the unconditional one, the best (in mean quadratic) predictor for X_{t+1} from $X_1, \ldots, X_t$ is simply its expected value $m = \mathbf{E}\, X_{t+1}$ (Problem 27 on p.73).

[1]Very good much more detailed texts (at an introductory and a more advanced level, respectively) are Brockwell and Davis (1995) and Brockwell and Davis (1991) listed in Section 9.5 below. The now-classical references are Anderson (1976) and Box and Jenkins (1976). An interesting modern source on various practical aspects of time series analysis is Akaike and Kitagawa (1999).

Such **white noise** (WN) sequences are used as basic building blocks for more complicated models. Quite often, however, the independence assumption is relaxed so that X_t are only assumed to be *uncorrelated*:

$$\mathbf{E}\,(X_t - m)(X_{t+h} - m) = 0, \quad h \neq 0,$$

and such a sequence is also called a white noise. In what follows, we will be dealing with that type of WN sequences, always assuming that they have zero means: $\mathbf{E}\,X_t = 0$ and finite variances $\mathrm{Var}\,(X_t) =: \sigma^2 < \infty$; for such a sequence, we will be using notation $\mathrm{WN}(0, \sigma^2)$. In that case, the *best linear predictor* for X_{t+1} from $X_1, \ldots, X_t$ will still be its expectation, see Problem 1 below (although there may exist a much better non-trivial *non-linear predictor* which will depend on the joint distribution of $(X_1, \ldots, X_t, X_{t+1})$, see Problem 2).

Most popular TS models are build from WNs by applying to them certain linear transformations and also adding deterministic components. Namely, one deals with the *classical decomposition model*

$$X_t = m_t + s_t + Y_t, \tag{9.1}$$

where m_t is a slowly changing function called a **trend component**, s_t is a function with a period $d > 1$ called a **seasonal component**, and Y_t is (weakly) stationary (see Section 2.10 for the definition); it is the last process which is a result of transforming WNs.

Decomposition (9.1) may be suggested from inspecting the plot of the data (observed values vs times), where the presence of trend (which can usually be well approximated by a linear or polynomial function) or seasonal oscillations is noticeable. When data exhibits increasing or decreasing fluctuations, it is advisable to first transform it (often applying the logarithmic transformation) and then proceed to further analysis. The point of getting a representation involving a stationary process (which, in turn, can be expressed as a linear transformation of WNs) is that one can easily estimate the parameters of such a process (using LLN-type results) and hence fit the model for the original data as well.

One of the main tasks of the analysis of TS's is to reconstruct the transformation yielding the stationary TS $\{Y_t\}$ in (9.1) from WNs and to estimate the trend and seasonal components in the model. We will discuss some popular approaches to these problems in this chapter. But first of all we need to look more closely at the nature of weakly stationary sequences (called in what follows simply stationary).

9.1 Stationary Sequences

Recall that an SP $\{X_t\}$ is called (weakly, or wide-sense) stationary if its mean is constant, while the covariance of X_{t+h} and X_t depends on the time **lag** h only. That is, for all $t = 0, \pm 1, \pm 2, \dots,$ (in this chapter, it will often be convenient to consider infinite in both directions sequences of RVs)

$$\mathbf{E}\, X_t =: m = \text{const}, \quad \text{Cov}\,(X_{t+h}, X_t) =: \gamma(h).$$

The function $\gamma(h)$ is the key tool for analysing, understanding and describing the stationary process; it is called the **autocovariance function** (ACVF) of $\{X_t\}$ ($\gamma(h)$ is said to be the value of the function "at lag h"), while

$$\rho(h) := \frac{\gamma(h)}{\gamma(0)} = \text{Corr}(X_{t+h}, X_t)$$

is referred to as the **autocorrelation function** (ACF) of the TS. Note that the common variance of the X_t's is equal to

$$\sigma^2 = \text{Var}\,(X_t) = \text{Cov}\,(X_t, X_t) = \gamma(0) \geq \gamma(h),$$

the last relation following from the Cauchy–Bunyakovskii inequality (2.52).

The ACVF (and ACF) of a TS is a measure of (linear) dependence between the values X_t and X_{t+h} for different lags h. The faster $\gamma(h)$ vanishes as $|h| \to 0$, the sooner decays the dependence between these values as the lag increases. More specific information about the shape of the function $\gamma(h)$—as we will see below—can tell one a lot about the nature of the TS.

Example 9.1 (continued). For a $\text{WN}(0, \sigma^2)$,

$$\gamma(h) = \sigma^2 \delta_{h0} \equiv \begin{cases} \sigma^2, & h = 0; \\ 0, & h \neq 0. \end{cases}$$

Note that since $e^{ih\pi} = e^{-ih\pi} = \pm 1$ for any integer $h \neq 0$, this ACVF can be represented for any $h = 0, \pm 1, \pm 2, \dots$ by the integral

$$\gamma(h) = \frac{\sigma^2}{2\pi} \int_{(-\pi,\pi]} e^{ih\lambda} d\lambda = \frac{\sigma^2}{2\pi} \times \begin{cases} 2\pi, & h = 0; \\ (e^{ih\pi} - e^{-ih\pi})/ih = 0, & h \neq 0. \end{cases} \tag{9.2}$$

We will see that the existence of such an integral representation for $\gamma(h)$ is a general and important fact.

Example 9.2. Let

$$X_t := Y_1 \cos(\Lambda t) + Y_2 \sin(\Lambda t), \tag{9.3}$$

where the RVs Y_j are uncorrelated and have common mean $\mathbf{E}\,Y_j = 0$ and variance $\sigma^2 < \infty$, and Λ is a random frequency (radians per unit time) independent of them and having a DF F_Λ. Since both sin and cos have a period of 2π, and t is integer, we can assume w.l.o.g. that $\Lambda \in (-\pi, \pi]$. Note that X_t has a (random) period of $2\pi/\Lambda$. Moreover, it is just a sinusoidal function with a random phase. Indeed, let $Y := \sqrt{Y_1^2 + Y_2^2}$ and Φ be the angle between the vectors $(1, 0)$ and (Y_1, Y_2) on the plane. Then

$$X_t = Y\big(\cos(\Lambda t)\cos\Phi + \sin(\Lambda t)\sin\Phi\big) = Y\sin(\Lambda t - \Phi).$$

Clearly, $\mathbf{E}\,X_t = 0$ is constant, while since $\mathbf{E}\,Y_j Y_k = \sigma^2\delta_{jk}$ by independence, one has

$$\begin{aligned}
\operatorname{Cov}(X_{t+h}, X_t) &= \mathbf{E}\,X_{t+h}X_t \\
&= \sigma^2\mathbf{E}\left[\cos(\Lambda(t+h))\cos(\Lambda t) + \sin(\Lambda(t+h))\sin(\Lambda t)\right] \\
&= \sigma^2\mathbf{E}\,\cos(\Lambda h) = \sigma^2\int_{(-\pi,\pi]}\cos(\lambda h)\,dF_\Lambda(\lambda) =: \gamma(h) \\
&= \frac{1}{2}\sigma^2\mathbf{E}\left(e^{i\Lambda h} + e^{-i\Lambda h}\right) = \sigma^2\int_{(-\pi,\pi]} e^{i\lambda h} d\left[\sigma^2(F_\Lambda(\lambda) + F_{(-\Lambda)}(\lambda))/2\right],
\end{aligned}$$

which is a function of h only. So our TS (9.3) is stationary.

In particular, if Λ can only take finitely many values $\lambda_1, \ldots, \lambda_n$ with respective probabilities $q_1, \ldots, q_n$, we have

$$\gamma(h) = \sigma^2\sum_{k=1}^{n} q_k\cos(\lambda_k h). \tag{9.4}$$

Now observe that, along with our process (9.3), the same ACVF describes a very different process. Recall that, in model (9.3), any fixed realisation of the process (i.e., the sequence of values $X_t(\omega)$ for a fixed "chance" value ω) is a sinusoidal oscillation with a *fixed frequency* $\Lambda(\omega)$. On the other hand, consider the model

$$\widetilde{X}_t = \sum_{k=1}^{n}[Y_k'\cos(\lambda_k h) + Y_k''\sin(\lambda_k h)],$$

where all Y_k', Y_k'' are uncorrelated with each other, and

$$\mathbf{E}\,Y_k' = \mathbf{E}\,Y_k'' = 0, \quad \operatorname{Var}(Y_k') = \operatorname{Var}(Y_k'') = \sigma_k^2 = \sigma^2 q_k, \quad k = 1, \ldots, n.$$

Realisations of the process $\{\widetilde{X}_t\}$ are sums of sinusoids with *different frequencies*, which can be functions of a very complex form. But the ACVF of $\{\widetilde{X}_t\}$

will be the same $\gamma(h)$: using relations $\mathbf{E}\,Y_j'Y_k' = \sigma_j^2\delta_{jk}$ and $\mathbf{E}\,Y_j'Y_k'' = 0$, we get

$$\begin{aligned}\gamma(h) &= \sum_{k=1}^{n}\sigma_k^2[\cos(\lambda_k(t+h))\cos(\lambda_k t) + \sin(\lambda_k(t+h))\sin(\lambda_k t)]\\ &= \sigma^2\sum_{k=1}^{n} q_k\cos(\lambda_k h).\end{aligned}$$

Note also that

$$\gamma(h) = \frac{1}{2}\sum_{k=1}^{n}\sigma_k^2[e^{i\lambda_k h} + e^{-i\lambda_k h}] = \int_{(-\pi,\pi]} e^{i\lambda h}\,dF(\lambda),$$

where $F(\lambda)$ has jumps $\sigma_k^2/2$ at the points $\pm\lambda_k$ (we assume that all $|\lambda_k|$ are different).

The fact that two *very different* processes can have one and the same ACVF should be no surprise: in Problem 32 on p.74 we saw that the standard Brownian motion process and the Poisson process with rate one have one and the same ACVF! After all, when dealing with ACVFs, we are looking at the second moments only. And if, say, two RVs X_1 and X_2 have the same means and variances, this by no means implies that the X_j's should have the same distribution! Likewise here: $\gamma(h)$ only summarises the linear dependence structure within a TS, and it may be the same for quite different processes.

Remark 9.1. Observe that, in the last example, we would obtain exactly the same ACVF if we added a random *phase shift* to the model. That is, we could consider $\cos(\Lambda t + \varphi)$ and $\sin(\Lambda t + \varphi)$ instead of simply $\cos(\Lambda t)$ and $\sin(\Lambda t)$ *etc.*, where φ is an RV independent of the Y_j's.

Example 9.3. Assume that $Y_1, Y_2, \ldots$ are i.i.d. RVs with zero means and $\sigma^2 := \mathbf{E}\,Y_1^2 < \infty$ (or one could just consider a WN$(0,\sigma^2)$-sequence). Then, for the RW

$$X_0 := 0, \quad X_t := Y_1 + \cdots + Y_t, \quad t = 1, 2, \ldots,$$

we have $\mathbf{E}\,X_t = 0$ which is independent of t, but, for $h > 0$,

$$\begin{aligned}\mathrm{Cov}\,(X_{t+h}, X_t) &= \mathrm{Cov}\,(X_t + Y_{t+1} + \cdots + Y_{t+h}, X_t)\\ &= \mathrm{Cov}\,(X_t, X_t) = \sigma^2 t, \quad t \ge 0.\end{aligned}$$

Clearly, $\{X_t\}$ is not stationary.

However, if we only look at the *increments* of this TS over time intervals of a fixed length n:

$$V_t := X_t - X_{t-n} = Y_{t-n+1} + \cdots + Y_t, \quad t \geq n$$

(such a sequence is called a *moving average*), we will have $\mathbf{E}\, V_t = 0$ and, for $h > 0$,

$$\mathrm{Cov}\,(V_{t+h}, V_t) = \begin{cases} 0, & h \geq n, \\ \mathbf{E}\,(Y_{t+h-n+1} + \cdots + Y_t)^2 = \sigma^2(n-h), & h < n. \end{cases}$$

The last relation holds since V_t and $V_{t+h} = Y_{t+h-n+1} + \cdots + Y_{t+h}$ have no common Y's in the former case and only $n - h$ common Y's in the latter one.

The case $h < 0$ is considered similarly yielding that $\{V_t\}$ is stationary with the ACVF

$$\gamma(h) = \sigma^2 \max\{n - |h|, 0\}.$$

One can show that in this case

$$\gamma(h) = \int_{(-\pi,\pi]} e^{i\lambda h} \frac{\sigma^2}{2\pi} \left(\frac{\sin(n\lambda/2)}{\sin(\lambda/2)} \right)^2 d\lambda. \tag{9.5}$$

The proof is left to the reader as an exercise.

Now note that any ACVF $\gamma(h)$ is obviously

(i) *even*: $\gamma(h) = \gamma(-h)$, and

(ii) *non-negative definite*: for any n, $a_1, \ldots, a_n$ and integer $h_1, \ldots, h_n$,

$$\sum_{j,k=1}^{n} a_j a_k \gamma(h_j - h_k) \equiv \mathbf{E} \left| \sum_{j=1}^{n} a_j (X_{h_j} - m) \right|^2 \geq 0.$$

That is, the ACVF has the same properties as the ChF of a symmetric RV (see Problem 15 on p.73). This has profound consequences.

Theorem 9.1.[2] (Herglotz[3]) *Any function* γ *satisfying* (i) *and* (ii) *is, up to a constant factor (equal to* $\sigma^2 := \gamma(0)$*), the ChF of a symmetric RV:*

$$\gamma(h) = \int_{(\pi,\pi]} e^{i\lambda h} dF(\lambda), \tag{9.6}$$

where the **spectral function** F *equals* $\sigma^2 F_\Lambda$ *for some symmetric RV* Λ.

[2]For the proof of the theorem see, e.g., Shiryaev (1984), ref. in Section 2.11.

[3]Gustav Herglotz (02.02.1881–22.03.1953), a German mathematician known for his work in a number of fields, including celestial mechanics, special and general relativity theory, hydrodynamics, and mathematical seismology.

It is often more convenient and natural to consider complex-valued stationary processes $\{X_t\}$, with the ACVF defined as

$$\gamma(h) = \mathbf{E}\left[(X_{t+h} - m)\overline{(X_t - m)}\right] \tag{9.7}$$

(which is not a true covariance anymore, of course; here $\overline{z} = x - iy$ stands for the complex conjugate of $z = x + iy$). Instead of property (i) above, we will in this case have $\gamma(h) = \overline{\gamma(-h)}$, while (ii) will be true for any complex a_j; we will need to consider the sums $\sum_{j,k} a_j \overline{a}_k \gamma(h_j - h_k)$. Representation (9.6) will also be valid—but the spectral function F can now be arbitrary (no symmetry is required). For example, setting $X_t := Ye^{i\lambda_0 t}$ for some fixed $\lambda_0 \in (-\pi, \pi]$ and an RV Y with zero mean yields a stationary sequence whose spectral function is the step function $F(\lambda) := \mathbf{E}\,|Y|^2 \mathbf{1}(\lambda \ge \lambda_0)$:

$$\mathbf{E}\left[(X_{t+h} - m)\overline{(X_t - m)}\right] = \mathbf{E}\,|Y|^2 e^{i\lambda_0 (t+h)} e^{-i\lambda_0 t} = \mathbf{E}\,|Y|^2 e^{i\lambda_0 h} =: \gamma(h).$$

Most of our statements below hold true for complex-valued TS's as well.

It turns out that this representation has a very important counterpart for the SP itself that clarifies the meaning of the spectral function. Namely, the following **spectral representation** for stationary process takes place.

Theorem 9.2. *Let $\{X_t\}$ be a stationary sequence with zero mean and spectral function F. Then there always exists a (complex-valued)* random process $\{Z(\lambda),\ \lambda \in (-\pi, \pi]\}$, *with zero mean and uncorrelated increments:* $\mathbf{E}\, Z(\lambda) = 0$ *for any λ and, for any $-\pi < a_1 \le b_1 \le a_2 \le b_2 \le \pi$,*

$$\mathbf{E}\,(Z(b_1) - Z(a_1))\overline{(Z(b_2) - Z(a_2))} = 0,$$

such that

$$\mathbf{E}\,|Z(b) - Z(a)|^2 = F(b) - F(a) \quad \textit{for any} \quad -\pi < a < b \le \pi,$$

which is usually written as the equality

$$\mathbf{E}\,|dZ(\lambda)|^2 = dF(\lambda), \tag{9.8}$$

and the following spectral representation[4] *holds: for any t,*

$$X_t = \int_{(-\pi,\pi]} e^{it\lambda} dZ(\lambda). \tag{9.9}$$

Relation (9.9) can be viewed as a *decomposition of the process in the frequency domain.*

That is, the stationary process $\{X_t\}$ can be thought of as a *superposition of sinusoids* at different frequencies with uncorrelated random amplitudes, and, for a given frequency λ, the "power" of the respective component

[4]The *stochastic* integral in (9.9) is actually not a conventional Riemann (or even Lebesgue) integral. It is, in fact, defined as the so-called *mean-quadratic limit* of approximating it random sums (that is, integrals in which the integrands are simple functions). A more technical discussion of this object is beyond the scope of the present text.

(the second moment of its random amplitude) is given by (9.8). So the increments of the spectral function *tell us how large the contributions of these sinusoids at the respective frequency intervals are.* (More precisely, they give the contributions' variances.)

A large jump in the spectral function F (or a large peak in the spectral density $f = F'$) at points $\pm\lambda_0$ indicates that there is a *strong sinusoidal component* at (or near) the frequency λ_0. (More precisely, the amplitude of the respective sinusoidal component has a large variance.)

Thus, in Example 9.2, the ACVF (9.4) corresponds to the case of a finite sum of sinusoids of frequencies $\lambda_1, \ldots, \lambda_n$ to which the spectral function assigns positive weights, the weights being the variances of the respective amplitudes. That is, the spectral function is piece-wise constant, having jumps (equal to $\frac{1}{2}\sigma^2 q_j$) at points λ_j (and hence, by symmetry, at $-\lambda_j$, too). Such a TS is said to have a **discrete spectrum**. When this is the case, and one has observed a long segment of the TS, it is possible to get very good estimates of the frequencies present in the TS and the amplitudes of sinusoids at those frequencies. One can obtain arbitrary precise estimates of them given a sufficiently long series of observations, and hence in that case *very good forecasting* even for the very distant future is possible. Such a stationary process is called *singular*. A process is singular iff its spectral function has no absolutely continuous component (recall the so-called Lebesgue decomposition mentioned at the end of Section 2.2).

On the other hand, the ACVFs (9.2) and (9.5) have absolutely continuous spectral functions with the densities

$$f(\lambda) = \frac{\sigma^2}{2\pi} \quad \text{and} \quad f(\lambda) = \frac{\sigma^2}{2\pi}\left(\frac{\sin(n\lambda/2)}{\sin(\lambda/2)}\right)^2,$$

respectively. In the cases where **spectral density** f exists, i.e., when one has $F(\lambda) = \int_{(-\pi,\lambda]} f(u)\,du$, $\lambda \in (-\pi, \pi]$, and hence

$$\gamma(h) = \int_{(-\pi,\pi]} e^{i\lambda h} f(\lambda)\,d\lambda, \tag{9.10}$$

one says that the TS has a **continuous spectrum**. In this case a (nearly) perfect prediction is impossible even if you have at your disposal an infinitely long series of observations of the past values of the stationary process (it is often convenient to extend time backwards assuming that $t \in \{\ldots, -1, 0, 1, \ldots\}$).

More specifically, if condition

$$\int_{(-\pi,\pi]} \log f(\lambda)\,d\lambda > -\infty \tag{9.11}$$

holds (which means, in particular, that $f(\lambda) > 0$ almost everywhere on $(-\pi, \pi]$), then the *best forecast* for the distant future will simply be the mean value of the process (in which case the stationary process is called *regular*). Condition (9.11) is necessary and sufficient for regularity.

Note that when $\{\gamma(h)\}$ is absolutely summable (so that the correlation decays fast enough as the lag $h \to \infty$), the spectral density always exists.

Theorem 9.3. *If $\sum_{h=-\infty}^{\infty} |\gamma(h)| < \infty$, the spectral density exists and is given by*

$$f(\lambda) = \frac{1}{2\pi} \sum_{k=-\infty}^{\infty} e^{-i\lambda k} \gamma(k). \tag{9.12}$$

Indeed, swapping the order of summation/integration, we see that with this choice of f,

$$\int_{(-\pi,\pi]} e^{i\lambda h} f(\lambda)\, d\lambda = \frac{1}{2\pi} \sum_{k=-\infty}^{\infty} \gamma(k) \int_{(-\pi,\pi]} e^{i\lambda(h-k)} d\lambda = \gamma(h), \tag{9.13}$$

since clearly $\int_{(-\pi,\pi]} e^{i\lambda m} d\lambda = 2\pi\delta_{m0}$, cf. (9.2). [For those with a strong mathematical background: thus $\{\gamma(h)\}$ is nothing else but the Fourier series for f.]

Example 9.4. *Moving averages* MA(q), $1 \le q \le \infty$. It turns out that a (zero-mean) stationary process $\{X_t\}$ is regular iff it can be represented as a *one-sided moving average* of a WN$(0, \sigma^2)$ sequence $\{Y_t\}$: for a sequence $\{a_k\}_{k \ge 0}$ satisfying the *absolute summability condition*:

$$\sum_{k=0}^{\infty} |a_k| \le C < \infty, \tag{9.14}$$

we have

$$X_t = \sum_{k=0}^{\infty} a_k Y_{t-k}. \tag{9.15}$$

Such processes are denoted by MA(∞) and are referred to as **causal** (or future-independent) processes.

Assumption (9.14) implies that the process X_t exists: the series in (9.15) is absolutely convergent w.p. 1. To see why this is true, observe first that the sequence of RVs $S_n := \sum_{k=0}^{n} |a_k Y_{t-k}|$ *increases* and hence has a limit $\lim_{n\to\infty} S_n =: S \le \infty$ a.s. The sequence will *converge* a.s. (and then the series (9.15) will be absolutely convergent) if the limit $S < \infty$ w.p. 1. The last relation will certainly hold if $\mathbf{E}\, S < \infty$. To verify the last condition, recall that, by the monotone convergence theorem (p.38), $\mathbf{E}\, S_n \nearrow \mathbf{E}\, S$ as $n \to \infty$ and note that the sequence on the left-hand side is bounded:

$$\mathbf{E}\, S_n \le \sum_{k=0}^{n} |a_k|\, \mathbf{E}\, |Y_{t-k}| \le \sigma \sum_{k=0}^{n} |a_k| \le \sigma \sum_{k=0}^{\infty} |a_k| \le \sigma C < \infty$$

due to (2.53) and (9.14). Therefore the limit also satisfies $\mathbf{E}\, S \le \sigma C$, and the desired convergence follows.

If $a_q \ne 0$ for some $q \ge 1$, but $a_k = 0$ for all $k > q$, the process is called a **moving average of order** q (denoted by MA(q)).

The ACVF of a moving average process $\{X_t\}$ is given, for $h \ge 0$, by

$$\gamma(h) = \mathrm{Cov}\left(\sum_{j=-\infty}^{t} a_{t+h-j} Y_j + \sum_{j=t+1}^{t+h} a_{t+h-j} Y_j,\ \sum_{j=-\infty}^{t} a_{t-j} Y_j \right)$$
$$= \sigma^2 \sum_{k=0}^{\infty} a_{k+h} a_k$$

(we again used the fact that $\mathrm{Cov}\,(Y_j, Y_k) = \sigma^2 \delta_{jk}$).

Since $\gamma(h)$ is an even function, one has

$$\gamma(h) = \sigma^2 \sum_{k=0}^{\infty} a_{k+|h|} a_k = \sigma^2 \sum_{k=0}^{\infty} a_{k+h} a_k, \tag{9.16}$$

the last relation being true if we put, for convenience sake, $a_k = 0$ for all $k < 0$.

Note that the ACVF of an MA(q) process has a "cut-off" at lag q: $\gamma(h) = 0$ once $|h| > q$. Indeed, in this case all the products $a_{k+h} a_k = 0$ since we can only have $a_k \ne 0$ for $k = 0, 1, \dots, q$. The cut-off means that there is a "finite-range" dependence in the process.

Due to the assumption (9.14), the sequence $\{\gamma(h)\}$ is absolutely summable. Indeed,

$$\sum_{h=-\infty}^{\infty} |\gamma(h)| = \sum_{h=-\infty}^{\infty} \left| \sum_{k=0}^{\infty} a_{k+h} a_k \right| \le \sum_{h=-\infty}^{\infty} \sum_{k=0}^{\infty} |a_{k+h} a_k|$$
$$= \sum_{k=0}^{\infty} |a_k| \sum_{h=-\infty}^{\infty} |a_{k+h}| \le C \sum_{k=0}^{\infty} |a_k| \le C^2 < \infty.$$

So we can apply (9.12) to see that $\gamma(h)$ has a spectral density given by

$$\begin{aligned} f(\lambda) &= \frac{\sigma^2}{2\pi} \sum_{h=-\infty}^{\infty} e^{-i\lambda h} \sum_{k=0}^{\infty} a_{k+h} a_k \\ &= \frac{\sigma^2}{2\pi} \sum_{h=-\infty}^{\infty} \sum_{k=0}^{\infty} (a_{k+h} e^{-i\lambda(k+h)}) \overline{(a_k e^{-i\lambda k})} \\ &= \frac{\sigma^2}{2\pi} \sum_{m=0}^{\infty} \sum_{k=0}^{\infty} (a_m e^{-i\lambda m}) \overline{(a_k e^{-i\lambda k})} = \frac{\sigma^2}{2\pi} |a(e^{-i\lambda})|^2, \qquad (9.17) \end{aligned}$$

where $a(z) := \sum_{k=0}^{\infty} a_k z^k$ is the generating function of the sequence $\{a_k\}$. Note that (9.5) is a special case of this result: in that case, $a_0 = a_1 = \cdots = a_{n-1} = 1$, whereas all the other a_k's are zeros, cf. Problem 3 below).

Transformation (9.15) of the white noise process is a special case of *linear filtering*. This very important notion will be discussed in the next section. In particular, we will see below that (9.17) follows from a general result for linear filters.

9.2 Linear Filters and Linear Processes

Operations similar to the one applied to $\{Y_t\}$ in Example 9.4 are very useful tools for analysing TS's. In particular, they can be used to detect, isolate and remove deterministic trend in a TS. Before discussing them, we will introduce a few important notions.

The **backward shift operator** B is defined on TS's $\boldsymbol{X} = \{X_t\}$, and the result of its application is another TS $\boldsymbol{Y} = \{Y_t\}$ denoted by $B\boldsymbol{X} = \{BX_t\}$ such that $Y_t = X_{t-1}$. That is,

$$\begin{aligned} B\boldsymbol{X} &= B\{\dots, X_{-2}, X_{-1}, X_0, X_1, X_2, \dots\} \\ &= \{\dots, X_{-3}, X_{-2}, X_{-1}, X_0, X_1, \dots\}. \end{aligned}$$

Note that BX_t is **not** the result of an operation applied to the value (or RV) X_t, but the tth element of the TS obtained by applying a certain operation to the **whole** TS $\boldsymbol{X}$.

The operator B is clearly *linear*, i.e., for any TS's $\boldsymbol{X}$ and $\boldsymbol{Y}$ and constants a and b,

$$B(a\boldsymbol{X} + b\boldsymbol{Y}) = aB\boldsymbol{X} + bB\boldsymbol{Y}$$

(addition of TS's and multiplication of a TS by a constant are understood component-wise: $a\boldsymbol{X} = \{aX_t\}$ *etc.*).

The **backward difference operator** ∇ (read "nabla"[5]) is defined by

$$\nabla X_t = (1-B)X_t = X_t - X_{t-1} \tag{9.18}$$

(we again stress that ∇ is an *operator* applied to the *whole* TS; it is a discrete analog of differentiation). The operator is, of course, also linear.

Powers of the operators B and ∇ are defined in the standard way:

$$B^k\boldsymbol{X} = B(B^{k-1}\boldsymbol{X}), \text{ so that } B^kX_t = X_{t-k},$$
$$\nabla^k\boldsymbol{X} = \nabla(\nabla^{k-1}\boldsymbol{X}),$$

with both $B^0 = \nabla^0 = 1$ (so that $B^0X_t = \nabla^0X_t = X_t$). Note also that B^{-k}, $k>0$, is well-defined, too: $B^{-k}X_t = X_{t+k}$.

A negative power ∇^{-k} of the difference operator is a trickier thing. Formally, it is tempting to write

$$\nabla^{-1} \equiv (1-B)^{-1} = \sum_{k=0}^{\infty} B^k.$$

The right-hand side here is, however, not defined in the general case. But there is no problem with defining it when applying this operator to a TS $\{Y_t\}$ with $Y_t = 0$ for all $t \le 0$ (or, more generally, $t \le s$ for some s). The result is clearly an RW:

$$X_t := \left(\sum_{k=0}^{\infty} B^k\right)Y_t = \sum_{k=0}^{\infty} B^kY_t = \sum_{k=0}^{\infty} Y_{t-k} = Y_1 + \cdots + Y_t, \qquad t \ge 0.$$

As we saw in Example 9.3, this is *not a stationary process* (when $\{Y_t\}$ is). Nevertheless, such processes can also be quite useful for modelling TS's. In a slightly more general situation (allowing for non-zero initial values X_0), a TS $\{X_t\}$ is called **integrated of order one** (denoted by I(1)) if

$$Y_t = \nabla X_t, \quad t \ge 1,$$

is a stationary process (NB: we start at $t = 1$!). This is clearly equivalent to

$$X_t = X_0 + \nabla^{-1}Y_t = X_0 + Y_1 + \cdots + Y_t, \qquad t \ge 1 \tag{9.19}$$

(again stipulating that $Y_t = 0$ for $t \le 0$). Similarly, $\{X_t\}$ is called **integrated of order** d (denoted by I(d)) if $Y_t = \nabla^d X_t$, $t \ge d$, is a stationary process. Accordingly, such models are used when one can establish that the TS of interest can be reduced by differencing to a stationary process (and the latter can be analysed using the standard techniques for such processes).

[5] *Nabla* is the (Greek) name of an Egyptian or Assyrian harp (the name for the symbol was suggested because of its similarity to a harp).

Notation I(0) is sometimes used for stationary processes.

Returning to positive powers of ∇, we see that

$$\begin{aligned}\nabla^2 X_t &= \nabla(X_t - X_{t-1}) = (X_t - X_{t-1}) - (X_{t-1} - X_{t-2}) \\ &= X_t - 2X_{t-1} + X_{t-2} \end{aligned} \tag{9.20}$$

and

$$\nabla^3 X_t = \nabla(\nabla^2 X_t) = X_t - 3X_{t-1} + 3X_{t-2} - X_{t-3}. \tag{9.21}$$

We can introduce and manipulate polynomials in B and ∇ in the same way as polynomials in real variables. And the polynomials will also be linear operators. For example, (9.20) and (9.21) follow immediately from the binomial formula

$$\nabla^n = (1-B)^n = \sum_{k=0}^{n} \binom{n}{k} (-1)^{n-k} B^{n-k}.$$

Operators ∇^n can be used to remove trends from TS's. First note that if we apply the operator ∇ to a linear trend function $m_t = c_0 + c_1 t$, the result

$$\nabla m_t = m_t - m_{t-1} = c_0 + c_1 t - (c_0 + c_1(t-1)) = c_1$$

is constant. Similarly, any *polynomial trend*

$$m_t = \sum_{j=0}^{k} c_j \, t^j \tag{9.22}$$

of degree k is reduced to a constant by applying the operator ∇^k, and, for a TS $X_t = m_t + Y_t$ with such a trend $\{m_t\}$ and stationary $\{Y_t\}$,

$$\nabla^k X_t = \nabla^k m_t + \nabla^k Y_t = k!\, c_k + \nabla^k Y_t. \tag{9.23}$$

Moreover, as one could expect, $\{\nabla^k Y_t\}$ is *again a stationary* TS. This is a special case of a general fact which holds for the so-called *linear filters* (LF) and will be demonstrated below.

A **linear filter** is a transformation of the *input* TS $\{X_t\}$ into the *output*

$$V_t := \sum_{k=-\infty}^{\infty} a_k X_{t-k} \equiv a(B) X_t, \tag{9.24}$$

where $\{a_k\}$ is a real sequence and

$$a(z) := \sum_{k=-\infty}^{\infty} a_k z^k$$

its generating function. It will usually be assumed that

$$\sum_{k=-\infty}^{\infty} |a_k| < \infty. \tag{9.25}$$

The sequence $\{a_k\}$ itself is also often called a linear filter.

The same argument as in Example 9.4 shows that when condition (9.25) is satisfied and $\{X_t\}$ is a stationary process (or even merely a process with $\sup_t \mathbf{E}\,|X_t| < \infty$), the output process V_t is well-defined: the series in (9.24) is absolutely convergent w.p. 1.

Note that if the input consists of a unit pulse at zero:

$$X_t = \delta_{0t} := \begin{cases} 1 & \text{if } t = 0, \\ 0 & \text{otherwise,} \end{cases}$$

then the output will simply be $V_t = a_t$. That is why $\{a_k\}$ is also referred to as the *unit impulse response*.

When the input $\{X_t\}$ of an LF is a $\mathrm{WN}(0, \sigma^2)$, the output process is said to be a **linear process**. If this is the case and, moreover, all $a_k = 0$, $k < 0$, so that

$$V_t = \sum_{k=0}^{\infty} a_k X_{t-k},$$

then $\{V_t\}$ is called a **causal function** of $\{X_t\}$, or simply a **causal** (or future-independent) **process**, since its values can be calculated from the *past and present* values of $\{X_t\}$ only. Recall that such processes are denoted by $\mathrm{MA}(\infty)$, see Example 9.4. Causality is a desirable and convenient property of stochastic processes.

Observe that (9.24) has the convolution form (2.63) and hence, as we already know, for manipulating with LFs it might be convenient to consider the Fourier transforms (or generating functions) of the respective sequences (to convolutions of sequences there correspond just the products of the respective transforms!). The Fourier transform of an LF $\{a_k\}$

$$A(\lambda) := \sum_{k=-\infty}^{\infty} a_k e^{-i\lambda k} \equiv a(e^{-i\lambda}) \tag{9.26}$$

is called the **transfer function** of the filter.

The transfer function $A(\lambda)$ of an LF $\{a_k\}$ completely characterises the filter. Indeed, we can directly compute the weights a_j of the LF as follows:

$$\begin{aligned} \frac{1}{2\pi}\int_{(-\pi,\pi]} e^{i\lambda j} A(\lambda)\, d\lambda &= \frac{1}{2\pi}\int_{(-\pi,\pi]} e^{i\lambda j} \sum_k a_k e^{-i\lambda k}\, d\lambda \\ &= \frac{1}{2\pi}\sum_k \int_{(-\pi,\pi]} e^{i\lambda(j-k)}\, d\lambda = a_j \end{aligned} \tag{9.27}$$

similarly to (9.13). Moreover, the transfer function describes how the input signal is amplified/dampened by the LF. Indeed, first we note that, for any integer k,

$$B^k e^{i\lambda t} = e^{i\lambda(t-k)} = e^{i\lambda t} e^{-i\lambda k}.$$

Therefore, if we apply an LF with transfer function (9.26) to the input $X_t = e^{i\lambda t}$ (which is just a sequence of observations of a harmonic oscillation of unit amplitude with a fixed frequency λ), the output is

$$V_t = a(B)X_t = a(B)e^{i\lambda t} = \sum_k a_k B^k e^{i\lambda t} = e^{i\lambda t} \sum_k a_k e^{-i\lambda k} = A(\lambda)e^{i\lambda t}. \tag{9.28}$$

That is, all what happens is that the original "monochromatic" signal is multiplied by $A(\lambda)$ (and that is why $A(\lambda)$ is called the transfer function).

In the general case, applying the LF to a stationary process $\{X_t\}$ with the spectral representation (9.9) yields, as $a(B)$ is linear, the output

$$\begin{aligned} V_t = a(B)X_t = a(B) \int e^{i\lambda t} dZ(\lambda) &= \int \left(a(B)e^{i\lambda t}\right) dZ(\lambda) \\ = \int \left(\sum_k a_k B^k e^{i\lambda t}\right) dZ(\lambda) &= \int e^{i\lambda t} \left(\sum_k a_k e^{-i\lambda k}\right) dZ(\lambda) \\ &= \int e^{i\lambda t} A(\lambda)\, dZ(\lambda), \end{aligned}$$

which is again of the form (9.9), but with a different random process Z_V given by

$$dZ_V(\lambda) = A(\lambda)\, dZ(\lambda), \quad \text{or} \quad Z_V(\lambda) = \int_{(-\pi,\lambda]} A(u)\, dZ(u). \tag{9.29}$$

The process Z_V is easily seen to also have *zero mean* and *uncorrelated increments*. To understand this better, assume for a moment that the original Z was a step function having random (zero-mean and uncorrelated) jumps $Z_1, \ldots, Z_n$ at points $\lambda_1, \ldots, \lambda_n$, respectively. Then

$$Z(\lambda) = \sum_{j:\, \lambda_j \le \lambda} Z_j \quad \text{and} \quad Z_V(\lambda) = \sum_{j:\, \lambda_j \le \lambda} A(\lambda_j) Z_j$$

from (9.29). So the trajectories of the process $Z_V(\lambda)$ are also step functions, with jumps at the same points λ_j, but of the sizes $A(\lambda_j)Z_j$ instead of Z_j. Since $A(\lambda_j)$ are simply constant coefficients, the jumps in the new process are also zero-mean and uncorrelated. Note that the variances of these jumps are equal to $|A(\lambda)|^2 \mathrm{Var}\,(Z_j)$.

The next theorem follows from the observation that any process having a spectral representation of the form (9.9) with a random process Z with zero mean and uncorrelated increments is always stationary. To understand this, one can again use the above simplifying assumption that Z is a step function having jumps at finitely many fixed points. In this case the integral becomes a sum of sinusoids with fixed frequencies but random amplitudes given by the jumps of Z at the respective frequency values.

Theorem 9.4. *The output $V_t = a(B)X_t$ of an LF with a stationary input $\{X_t\}$ having the spectral representation* (9.9) *is also stationary and has the spectral process* (9.29).

From (9.29) *and* (9.8) *it follows that the spectral function G of $\{V_t\}$ is given by*

$$dG(\lambda) = |A(\lambda)|^2 dF(\lambda). \tag{9.30}$$

In particular, when the original TS $\{X_t\}$ has a spectral density f, the output $\{V_t\}$ will also have a spectral density, which is given by

$$g(\lambda) = |A(\lambda)|^2 f(\lambda). \tag{9.31}$$

Since the power of a signal at a given frequency λ is proportional to the square of the amplitude of the sinusoidal component at that frequency, and the amplitudes of the signal's components passing an LF are transformed according to (9.28), the function $|A(\lambda)|^2$ is often called the **power transfer function**.

Relation (9.30) can be verified by a direct computation as well:

$$\begin{aligned}
\mathbf{E}\left(V_{t+h}V_t\right) &= \mathbf{E}\left(\sum_{j=-\infty}^{\infty} a_j X_{t+h-j} \times \sum_{k=-\infty}^{\infty} a_k X_{t-k}\right) \\
&= \sum_{j,k} a_j a_k \mathbf{E}\left(X_{t+h-j}X_{t-k}\right) \\
&= \sum_{j,k} a_j a_k \gamma(h-j+k) \qquad\qquad (9.32) \\
&= \sum_{j,k} a_j a_k \int e^{i\lambda(h-j+k)} dF(\lambda) \qquad \text{by (9.6)} \\
&= \int e^{i\lambda h}\left(\sum_{j,k} a_j a_k e^{-i\lambda j} e^{i\lambda k}\right) dF(\lambda) \\
&= \int e^{i\lambda h}|A(\lambda)|^2 dF(\lambda) =: \gamma_V(h)
\end{aligned}$$

depends on h only, which confirms the above observation that $V_t = a(B)X_t$ is stationary.

Example 9.5. *The perfect delay.* A delayed (or lagged) by (a fixed number) τ periods TS $\{X_t\}$ is the TS $V_t = B^\tau X_t = X_{t-\tau}$. This is clearly an LF with $a_k = \delta_{k\tau}$, the transfer function being $A(\lambda) = e^{-i\lambda\tau}$. Further, $|A(\lambda)|^2 = |e^{-i\lambda\tau}|^2 = 1$, and hence by (9.30) the spectral function of the delayed process is equal to that of the original process. Hence the ACVF of the delayed process equals

$$\gamma_V(h) \equiv \gamma(h)$$

(no surprise: shifting the *whole* TS does not change the covariance structure of the stationary process). Of course, the last relation can be proved by a direct computation as well.

Example 9.6. An example of the so-called *low-pass* filter (removing high frequency components and hence leaving the slowly varying trend, so that the LF can be used to estimate the latter) is a *two-sided moving average* with

$$a_k = \begin{cases} \dfrac{1}{2q+1} & \text{for } \; -q \le k \le q, \\ 0 & \text{otherwise,} \end{cases}$$

q is a fixed positive integer.

The LF transforms a TS $\{X_t\}$ into

$$V_t = \frac{1}{2q+1} \sum_{k=-q}^{q} X_{t-k}.$$

The transfer function of the LF equals

$$A(\lambda) = \frac{1}{2q+1} \sum_{k=-q}^{q} e^{-i\lambda k} = \frac{1}{2q+1} \frac{e^{i\lambda q} - e^{-i\lambda(q+1)}}{1 - e^{-i\lambda}} = \frac{\sin(\lambda(q+1/2))}{(2q+1)\sin(\lambda/2)}.$$

The value of $A(\lambda)$ is close to one in vicinity of zero and, even for moderate values of q, is quite small elsewhere, which confirms our expectations that, due to averaging, the low frequency components will pass the LF almost unchanged, while the rapidly fluctuating ones will basically be removed.

Note also that, for large enough q, if the correlation between the terms in the TS $\{X_t\}$ is small, the LF will not only alleviate noise but also pass the linear trend without distortion. Indeed, assume for illustration purposes that

$$X_t = c_0 + c_1 t + Y_t,$$

where c_j are constants and $\{Y_t\}$ is a WN$(0, \sigma^2)$. Then filtering yields

$$V_t = \frac{1}{2q+1} \sum_{k=-q}^{q} (c_0 + c_1(t-k)) + \underbrace{\frac{1}{2q+1} \sum_{k=-q}^{q} Y_{t-k}}_{\widetilde{Y}_t} = c_0 + c_1 t + \widetilde{Y}_t,$$

where clearly

$$\mathrm{Var}\,(\widetilde{Y}_t) = \mathrm{Var}\left(\frac{1}{2q+1} \sum_{k=-q}^{q} Y_{t-k} \right) = \frac{\sigma^2}{2q+1},$$

so that the magnitude of the "noise" is reduced by the factor $(2q+1)^{-1/2}$.

The two-sided moving average from the above example is not the only smoothing LF. One can design a filter that will effectively remove noise and allow a larger class of trend functions (e.g., all polynomials of degree ≤ 3) to pass through without distortion; for more detail on smoothing filters see, e.g., Chapter 46 in Kendall and Stuart (1976).

An LF is said to be **recursive** if its output $\{V_t\}$ depends linearly on a fixed number of the "past values" of the *output* and input TS's: for some integers $p \geq 1$ and $q \geq 0$,

$$V_t = \sum_{k=1}^{p} \beta_k V_{t-k} + \sum_{k=0}^{q} \alpha_k X_{t-k}.$$

Example 9.7. An example of a recursive filter is given by

$$V_t = aV_{t-1} + (1-a)X_t.$$

When $|a| < 1$, this is equivalent to the so-called exponential smoothing:

$$\begin{aligned} V_t &= a(aV_{t-2} + (1-a)X_{t-1}) + (1-a)X_t \\ &= a^2 V_{t-2} + (1-a)(X_t + aX_{t-1}) \\ &= a^2(aV_{t-3} + (1-a)X_{t-2}) + (1-a)(X_t + aX_{t-1}) \\ &= a^3 V_{t-3} + (1-a)(X_t + aX_{t-1} + a^2 X_{t-2}) = \cdots \\ &= (1-a) \sum_{k=0}^{\infty} a^k X_{t-k}, \qquad (9.33) \end{aligned}$$

which is an LF with the weights $a_k = (1-a)a^k$, $k \geq 0$, decreasing exponentially fast as $k \to \infty$. The transfer and power transfer functions of the filter are, respectively,

$$A(\lambda) = (1-a)\sum_{k=0}^{\infty} a^k e^{-i\lambda k} = \frac{1-a}{1-ae^{-i\lambda}}$$

and

$$|A(\lambda)|^2 = \left|\frac{1-a}{1-ae^{-i\lambda}}\right|^2 = \frac{(1-a)^2}{1-2a\cos\lambda + a^2}.$$

Thus, for small values of a, the transfer function is relatively "flat", meaning that the LF passes all frequencies quite well (the weights a_k decay too fast to be able to smooth anything), while for a close to one, there is a sharp peak at zero and rather small values outside the vicinity of zero, which means that high frequencies are significantly damped.

Example 9.8. *The first order autoregression* model (denoted by AR(1)) is very similar to the exponentially smoothed white noise. An AR(1) process $\{X_t\}$ is defined by

$$X_t = \beta X_{t-1} + Y_t, \tag{9.34}$$

where $\{Y_t\}$ is a WN$(0, \sigma^2)$ and β is a constant.

If $|\beta| < 1$ and (9.34) holds for all $t = 0, \pm 1, \ldots$, then the series

$$X_t = \sum_{k=0}^{\infty} \beta^k Y_{t-k} \tag{9.35}$$

converges (see the argument in Example 9.4) and is easily seen to be a solution to (9.34), cf. (9.33). Being a special case of an MA(∞) with $a_k = \beta^k$, $k = 0, 1, 2, \ldots$, the process (9.35) is stationary itself with the ACVF given, by virtue of (9.16), by

$$\gamma_X(h) = \sigma^2 \sum_{k\geq 0} \beta^{k+|h|}\beta^k = \sigma^2\beta^{|h|}\sum_{k\geq 0}\beta^{2k} = \frac{\sigma^2\beta^{|h|}}{1-\beta^2}, \tag{9.36}$$

and hence the ACF of $\{X_t\}$ is simply $\rho(h) = \beta^{|h|}$. Note that the ACVF displays a very simple pattern: it decreases exponentially fast (alternating its sign when $\beta < 0$) as the absolute value of the lag h goes to infinity.

Since $\{X_t\}$ is an output of an exponential LF with $a_k = \beta^k$, of which the transfer function is

$$A(\lambda) = a(e^{-i\lambda}) = \sum_{k\geq 0} e^{-i\lambda k}\beta^k = \frac{1}{1-\beta e^{-i\lambda}},$$

and the input WN process has the constant spectral density $\sigma^2/2\pi$, the output process will have, by (9.31), the spectral density

$$f(\lambda) = |A(\lambda)|^2 \times \frac{\sigma^2}{2\pi} = \frac{\sigma^2}{2\pi|b(e^{-i\lambda})|^2}, \quad \text{where } b(z) = 1 - \beta z. \tag{9.37}$$

It is important to note that representation (9.35) can formally be obtained using the following observation as well. Using the backward shift operator, we can re-write (9.34) as $\boldsymbol{X} = \beta B \boldsymbol{X} + \boldsymbol{Y}$, or

$$b(B)\boldsymbol{X} \equiv (1 - \beta B)\boldsymbol{X} = \boldsymbol{Y}. \tag{9.38}$$

The operator $b(B)$ has a formal inverse given by the series

$$(1 - \beta B)^{-1} = \sum_{k \geq 0} \beta^k B^k,$$

and we get from (9.38) that

$$\boldsymbol{X} = (1 - \beta B)^{-1}\boldsymbol{Y} = \sum_{k \geq 0} \beta^k B^k \boldsymbol{Y}, \tag{9.39}$$

which coincides with (9.35). We will see below that a similar argument can be used to show the existence of stationary solutions in more general situations as well.

In conclusion note that (9.35) is the *only stationary solution* of (9.34). Indeed, if $\{X'_t\}$ is another stationary solution, then, similarly to (9.33), for any $n > 1$,

$$X'_t = \beta^n X'_{t-n} + \sum_{k=0}^{n-1} \beta^k Y_{t-k}, \tag{9.40}$$

which differs from (9.35) by

$$R_n := \beta^n X'_{t-n} - \sum_{k=n}^{\infty} \beta^k Y_{t-k},$$

which vanishes in probability as $n \to \infty$. To see that, it only remains to use the fact that, for any RVs ξ_1 and ξ_2,

$$\mathrm{Var}\,(\xi_1 + \xi_2) \leq 2(\mathrm{Var}\,(\xi_1) + \mathrm{Var}\,(\xi_2)) \tag{9.41}$$

(Problem 4), and hence, since $\{X'_t\}$ is stationary (and has therefore a constant variance) and $\{Y_t\} \sim \mathrm{WN}(0, \sigma^2)$,

$$\mathrm{Var}\,(R_n) \leq 2\beta^{2n}\mathrm{Var}\,(X'_1) + 2\sigma^2 \sum_{k=n}^{\infty} \beta^{2k} = \text{const} \times \beta^{2n} \to 0 \quad \text{as } n \to \infty.$$

Thus, we showed that $X_t - X'_t = R_n$ holds with $\mathrm{Var}\,(R_n)$ becoming arbitrary small as $n \to \infty$. This can only hold if $X_t - X'_t = 0$ a.s.

Example 9.9. To illustrate how AR(1) models can arise in real-life situations, we will now construct a model for the water level in a reservoir.

Denoting by L_t the level in the tth year, we can write the following *balance equation:*

$$L_t = L_{t-1} - \mu S(L_{t-1}) + D_t, \tag{9.42}$$

where μ is the evaporation rate, $S(l)$ the surface area of the reservoir when the water is at level l, and D_t is the total drainage to the reservoir during the tth year.

Now put $X_t := L_t - \overline{L}$, where $\overline{L}$ is the long-term average level, and assume that $S(l) = S(\overline{L}) + s(l - \overline{L})$ for some $s > 0$ (which is the first-order linear approximation to the function $S(l)$ in the vicinity of the point $\overline{L}$; it is well justified when the oscillations of the level from year to year are not too large and $S(l)$ is a smooth function). Then (9.42) becomes

$$X_t = (1 - s\mu)X_{t-1} + Y_t, \quad Y_t := D_t - \mu S(\overline{L}).$$

It is natural to assume that the RVs Y_t have zero means ("long-term balance" between drainage and evaporation) and, at least as the first approximation, that they are uncorrelated for different t's. Then, as we showed, if $|1 - s\mu| < 1$, this equation has a unique stationary solution describing the oscillation of the water level.

A more general class of processes often admitting a causal representation are **autoregressive of order** p processes (denoted by AR(p)). An AR(p) process is defined as the solution of the equation

$$X_t = \beta_1 X_{t-1} + \beta_2 X_{t-2} + \cdots + \beta_p X_{t-p} + Y_t, \quad \{Y_t\} \sim \mathrm{WN}(0, \sigma^2). \tag{9.43}$$

Similarly to Example 9.8, this can be re-written as

$$b(B)X_t = Y_t, \quad b(z) = 1 - \beta_1 z - \beta_2 z^2 - \cdots - \beta_p z^p. \tag{9.44}$$

The polynomial $b(z)$ is called the **characteristic polynomial** of the TS given by (9.43). As is well known, any polynomial of degree p can be factorised into a product of p monomials. So if $z_1, \ldots, z_p$ are the roots of the **characteristic equation** $b(z) = 0$ (among them there can be complex and equal values), then

$$\begin{aligned} b(z) &= -\beta_p(z - z_1)\cdots(z - z_p) \\ &= (-1)^{p+1}\beta_p z_1 \cdots z_p\left(1 - \frac{z}{z_1}\right)\cdots\left(1 - \frac{z}{z_p}\right) \\ &= (1 - z_1^{-1}z)\cdots(1 - z_p^{-1}z) \end{aligned} \tag{9.45}$$

(the last relation is due to the fact that $(-1)^{p+1}\beta_p z_1 \cdots z_p = 1$ since the constant terms in both $b(z)$ and $(1 - z/z_1)\cdots(1 - z/z_p)$ are equal to one).

Therefore our (9.44) has the form

$$(1 - z_1^{-1}B)\cdots(1 - z_p^{-1}B)X_t = Y_t.$$

Now, if $|z_1^{-1}| < 1$, we can proceed as in Example 9.8 (see (9.38)–(9.39)) using the existence of the inverse operator

$$(1 - z_1^{-1}B)^{-1} = \sum_{k=0}^{\infty} z_1^{-k}B^k$$

to get

$$(1 - z_2^{-1}B)\cdots(1 - z_p^{-1}B)X_t = \widetilde{Y}_t \quad \text{with} \quad \widetilde{Y}_t := \sum_{k=0}^{\infty} z_1^{-k}Y_{t-k}.$$

Next, if $|z_2^{-1}| < 1$, we can repeat the trick to obtain

$$(1 - z_3^{-1}B)\cdots(1 - z_p^{-1}B)X_t = \widetilde{\widetilde{Y}}_t$$

with

$$\widetilde{\widetilde{Y}}_t := \sum_{j=0}^{\infty} z_2^{-j} \sum_{k=0}^{\infty} z_1^{-k}Y_{t-k-j}. \tag{9.46}$$

Observe that the process $\{\widetilde{\widetilde{Y}}_t\}$ is again an MA(∞) whose mth coefficient $\tilde{\tilde{a}}_m$ can be found by grouping together all terms on the right-hand side of (9.46) with $k + j = m$:

$$\tilde{\tilde{a}}_m = \sum_{j,k\ge 0:\, j+k=m} z_2^{-j}z_1^{-k} = \sum_{j=0}^{m}\left(\frac{1}{z_1}\right)^{m-j}\left(\frac{1}{z_2}\right)^{j}$$

which is clearly a sum of the form (2.63). This means that the sequence $\tilde{\tilde{a}}_m$ is a convolution of the sequences $\{z_1^{-m}\}_{m\ge 0}$ and $\{z_2^{-m}\}_{m\ge 0}$ and hence has the GF $\tilde{\tilde{a}}(z)$ given by the product

$$\tilde{\tilde{a}}(z) = \sum_{m\ge 0}(z/z_1)^m \times \sum_{m\ge 0}(z/z_2)^m = \frac{1}{(1 - z_1^{-1}z)(1 - z_2^{-1}z)}. \tag{9.47}$$

We can continue in the same way, and after p such steps we will arrive at the following conclusion:

When all $|z_k^{-1}| < 1$, $k = 1, \ldots, p$, the operator $b(B)$ has an inverse which is a one-sided moving average:

$$X_t = (b(B))^{-1}Y_t = \sum_{k=0}^{\infty} a_k Y_{t-k} = \left(\sum_{k=0}^{\infty} a_k B^k\right) Y_t,$$

where the coefficients a_k can be found from the Taylor expansion

$$\frac{1}{(1-z_1^{-1}z)\cdots(1-z_p^{-1}z)} = \frac{1}{b(z)} = \sum_{k=0}^{\infty} a_k z^k \tag{9.48}$$

of the function $1/b(z)$ about zero.

The last statement can also be seen from the fact that when $\{a_k\}$ is defined by (9.48), one has

$$1 = b(z) \times (b(z))^{-1} = b(z) \times \sum_{k=0}^{\infty} a_k z^k$$

and hence $b(B)(\sum a_k B^k) = 1$ (the identity operator), so that $(b(B))^{-1} = \sum_{k=0}^{\infty} a_k B^k$ indeed.

From that we infer that the transfer function of the LF $(b(B))^{-1}$ is $1/b(e^{-i\lambda})$, and hence the spectral density of $\{X_t\}$ is given, due to (9.31), by

$$f(\lambda) = \frac{\sigma^2}{2\pi|b(e^{-i\lambda})|^2}. \tag{9.49}$$

This means that the following result is true:

Theorem 9.5. *If all the roots of the characteristic equation lie outside the unit disk: $|z_j| > 1$, $j = 1, \ldots, p$, then the* AR(p) *process* (9.43) *is stationary and can be represented as a one-sided moving average of the form* (9.15) *of the white noise $\{Y_t\}$. The process has the spectral density* (9.49).

Example 9.10. *Second order autoregression* AR(2). When $p = 2$, we have

$$X_t - \beta_1 X_{t-1} - \beta_2 X_{t-2} = (1 - z_1^{-1}B)(1 - z_2^{-1}B)X_t = Y_t,$$

where

$$z_{1,2} = -\frac{\beta_1}{2\beta_2} \pm \sqrt{\left(\frac{\beta_1}{2\beta_2}\right)^2 + \frac{1}{\beta_2}}$$

are the roots of the characteristic equation $1 - \beta_1 z - \beta_2 z^2 = 0$. Relations (9.46) and (9.47) yield

$$X_t = \sum_{m=0}^{\infty} z_1^{-m} \left(\sum_{j=0}^{m} \left(\frac{z_1}{z_2}\right)^j\right) Y_{t-m}$$
$$= \begin{cases} \dfrac{1}{z_1^{-1} - z_2^{-1}} \displaystyle\sum_{m=0}^{\infty} (z_1^{-(m+1)} - z_2^{-(m+1)}) Y_{t-m} & \text{if } z_1 \neq z_2, \\ \displaystyle\sum_{m=0}^{\infty} (m+1) z_1^{-m} Y_{t-m} & \text{if } z_1 = z_2. \end{cases}$$

Now we can use (9.16) to find the ACVF of our AR(2) process. Thus, when $z_1 \neq z_2$, putting for brevity's sake $\xi_i := z_i^{-1}$, we have

$$a_j = \frac{\xi_1^{j+1} - \xi_2^{j+1}}{\xi_1 - \xi_2},$$

and hence, for $h \geq 0$,

$$\gamma(h) = \sigma^2 \sum_{j=0}^{\infty} a_{j+h} a_j = \frac{\sigma^2}{(\xi_1 - \xi_2)^2} \sum_{j=0}^{\infty} (\xi_1^{j+h+1} - \xi_2^{j+h+1})(\xi_1^{j+1} - \xi_2^{j+1})$$

$$= \frac{\sigma^2}{(\xi_1 - \xi_2)^2} \sum_{j=0}^{\infty} \left(\xi_1^{h+2} \xi_1^{2j} + \xi_2^{h+2} \xi_2^{2j} - (\xi_1^{h+1} \xi_2 + \xi_1 \xi_2^{h+1})(\xi_1 \xi_2)^j \right)$$

$$= \frac{\sigma^2}{(\xi_1 - \xi_2)^2} \left[\xi_1^{h+1} \left(\frac{\xi_1}{1 - \xi_1^2} - \frac{\xi_2}{1 - \xi_1 \xi_2} \right) + \xi_2^{h+1} \left(\frac{\xi_2}{1 - \xi_2^2} - \frac{\xi_1}{1 - \xi_1 \xi_2} \right) \right].$$

For $h < 0$, as we know, $\gamma(h) = \gamma(|h|)$.

This ACVF can have, for different values of z_1 and z_2, quite different behaviour (cf. exponential decay for AR(1)), see Problem 6.

To get the ACVF for a general AR(p) process, one can multiply both sides of (9.43) by X_{t-h}, $h \geq 0$, and take expectations to get

$$\gamma(h) = \beta_1 \gamma(h-1) + \beta_2 \gamma(h-2) + \cdots + \beta_p \gamma(h-p), \tag{9.50}$$

or

$$b(B)\gamma(t) = 0.$$

This is a linear difference equation (we have already encountered such equations in Examples 3.21 and 4.4), and it can be shown (see p.120 for references) that, when the stability condition is met, its general solution is given by a sum of vanishing exponents and damped sine waves (perhaps multiplied by polynomials—in case of multiple roots z_j).

The next general class of processes is obtained by combining MA(q) and AR(p) constructions. Namely, an ARMA(p, q) process (autoregression of order p with a noise which is a moving average of order q) is defined by

$$X_t = \beta_1 X_{t-1} + \cdots + \beta_p X_{t-p} + \alpha_0 Y_t + \cdots + \alpha_q Y_{t-q}, \quad \{Y_t\} \sim \text{WN}(0, \sigma^2), \tag{9.51}$$

or, equivalently,

$$b(B) X_t = \alpha(B) Y_t, \quad \alpha(z) = \alpha_0 + \alpha_1 z + \cdots + \alpha_q z^q, \tag{9.52}$$

with the same $b(z)$ as in (9.44); w.l.o.g. we might assume that $\alpha_0 = 1$ (all the α's can be multiplied by one and the same quantity, this only results in changing the variance of the WN input process). The same argument as for AR(p) processes, combined with the properties we derived for MA(q) processes, yields the following result.

Theorem 9.6. *If all the roots of the characteristic equation $b(z) = 0$ satisfy $|z_j| > 1$, $j = 1, \ldots, p$, then the* ARMA(p, q) *process* (9.51) *is stationary and can be represented as a one-sided moving average of the form* (9.15) *of the white noise* $\{Y_t\}$. *The process has the spectral density*

$$f(\lambda) = \frac{\sigma^2|\alpha(e^{-i\lambda}|^2}{2\pi|b(e^{-i\lambda})|^2}. \tag{9.53}$$

The stability condition that all $|z_j| > 1$ is actually quite natural. Indeed, imagine that the TS $\{X_t\}$ is given by a version of (9.51) *without* any random noise:

$$X_t = \beta_1 X_{t-1} + \cdots + \beta_p X_{t-p}, \quad t \geq p, \tag{9.54}$$

with some initial conditions $X_0, X_1, \ldots, X_{p-1}$. This is a linear difference equation of the type we have already seen earlier (see, e.g., (4.11); in fact, (9.54) has the same form as (9.50)). To solve it, we substitute $X_t := z^{-t}$ for some fixed $z \neq 0$ into (9.54). This clearly yields the equation $z^{-t}b(z) = 0$ for z, which is obviously equivalent (as $z \neq 0$) to the characteristic equation $b(z) = 0$. So if z_j is a root of the characteristic equation, then $X_t = z_j^{-t}$ solves (9.54). The general solution to (9.54) is then of the form

$$X_t = \sum_{j=1}^{p} C_j z_j^{-t}$$

(given all z_j's are different), where the coefficients C_j can be found from the initial conditions.

So if all $|z_j| > 1$, then $X_t \to 0$ as $t \to \infty$, i.e., $\{X_t\}$ is stable. Adding a stationary random (MA) noise does *perturb* the trajectory of $\{X_t\}$, but as the "AR-part" (9.54) is "inherently stable", its "influence" will eventually dampen the perturbations—as a car's suspension dampens shocks from bumps on the road. So the new (ARMA) process will also be stable.

The ACVF of the ARMA(p, q) process can be found by substituting (9.53) into the spectral representation (9.10). An alternative approach is to use the causal LF representation

$$X_t = \sum_{k=0}^{\infty} a_j Y_{t-j} \tag{9.55}$$

and formula (9.16) for the ACVF of a linear process. To find the coefficients a_k in (9.55), one can proceed as follows. Multiplying both sides of the relation (9.51) by Y_{t-k} and taking expectations, we get from (9.55) that, for $m = \max\{p, q+1\}$,

$$a_k - \beta_1 a_{k-1} - \cdots - \beta_p a_{k-p} = \alpha_k, \qquad k = 0, 1, \ldots, m-1, \tag{9.56}$$

and

$$a_k - \beta_1 a_{k-1} - \cdots - \beta_p a_{k-p} = 0, \qquad k \geq m. \tag{9.57}$$

Relation (9.57) is a homogeneous linear difference equation (for unknowns a_k, $k \geq m-p$) with constant coefficients. Its general solution is well known to be of the form

$$a_k = C_1 z_1^{-k} + \cdots + C_p z_p^{-k}, \quad k \geq m-p, \tag{9.58}$$

where z_j are distinct roots of the characteristic equation (when not all the roots are distinct, we will get terms of the form $Ck^l z_j^{-k}$ as well, for more detail see texts referred to in Example 3.21).

To find the p constants C_j from (9.58) and the first $m-p$ values $a_0, a_1, \ldots, a_{m-p-1}$, we substitute the representation (9.58) into (9.56) and get a linear system of m equations for $p + (m-p) = m$ unknowns.

9.3 A General Approach to Time Series Modelling

So far in our discussion of AR, MA and ARMA processes we have been basically talking about zero-mean stationary sequences $\boldsymbol{X}$ satisfying relations of the form

$$b(B)X_t = a(B)Y_t, \quad t = \ldots, -1, 0, 1, \ldots; \quad \boldsymbol{Y} \sim \mathrm{WN}(0, \sigma^2), \tag{9.59}$$

with both $b(B)$ and $a(B)$ being some polynomials in B.

Models based on such processes can actually refer to finite time intervals, have non-zero means and be non-stationary.

Firstly, if our process $\boldsymbol{X}$ is given by (9.59), but starts at time $t = 0$ only (in which case to use the relation we formally need p "initial values" $X_{-1}, \ldots, X_{-p}$ of the sequence $\boldsymbol{X}$ if the polynomial b is of degree p), it will not be stationary. But, if the condition of Theorem 9.6 is met, then the process $\boldsymbol{X}$ will be stable and will approach the stationary regime rather quickly (cf. relation (9.40) for AR(1) processes: the contribution of the "distant" past values of $\boldsymbol{X}$ to the newly formed term X_t is vanishing, the principal contribution is that from the causal function of the white noise obtained when one inverts $b(B)$). In particular, $\mathrm{Cov}\,(X_{t+h}, X_t) \to \gamma(h)$, the ACVF of the stationary ARMA, as $t \to \infty$.

Secondly, note that if $\{X_t - m\}$ is a zero-mean stationary ARMA(p,q), then $\{X_t\}$ will also be stationary with the same ACVF (and hence the same spectral function) as $\{X_t - m\}$, but with mean m. Hence the general form of the relation giving ARMA processes:

$$b(B)(X_t - m) = a(B)Y_t.$$

Non-stationary processes can be obtained either by adding a *mean function* $\{m_t\}$ or forming an integrated process (I(d), see p.248). The last approach leads to the following most general standard model:

A process $\{X_t\}$ is said to be ARIMA(p, k, q) ("autoregressive integrated process with moving average residuals") if

$$\nabla^k X_t = V_t,$$

where V_t is a (causal) ARMA(p, q) process. In other words,

$$(1 - B)^k b(B) X_t = \alpha(B) Y_t, \tag{9.60}$$

where $b(z) = 1 - \beta_1 z - \cdots - \beta_p z^p$ is a polynomial of degree p whose zeros satisfy $|z_j| > 1$, $\alpha(z)$ is a polynomial of degree q, and $\boldsymbol{Y}$ is a WN process. Such an $\boldsymbol{X}$ will be non-stationary (unless $k = 0$), and moreover, adding an arbitrary polynomial trend m_t of degree $< k$ will not violate equation (9.60), which means that the model is quite good for modelling TS with trends!

Now suppose we are given a TS $\{x_t\}_{t=1}^n$ (observed values). How could one model such a TS? Here modelling means selecting a theoretical model (a random process) such that the observed sequence would be a likely realisation of that process.

A popular *Box-Jenkins approach* has a formal objective of finding an LF which would "satisfactory" reduce the original TS to a residual WN process with a small variance. The LF is chosen from a certain parametric class (including ARIMA—and hence all the simpler versions thereof, i.e., AR, MA and ARMA processes), and one attempts to keep the number of parameters employed as small as reasonably possible. We can achieve a small estimated white noise variance by fitting a model of a high order. The excessive number of parameters means that one can "bend" the model forcing it to follow in detail the "truly random" oscillations which in fact cannot be explained at that level of modelling. Such overparametrization, as usual, leads to poor models: forecasts based on such models can be bad for the above reason and also due to errors in estimates of the parameters of the model.

As it is often the case in statistical modelling, one begins with postulating a plausible class of parametric models for initial investigation. Next one identifies a likely member of the class, estimates its parameters and assesses the success of the fit. The fitted model is either accepted at the verification stage, or one should suggest a sensible modification and repeat the identification/estimation cycle until a satisfactory model is obtained.

For a detailed description of model fitting procedures (including, in particular, criteria for model selection penalising for fitting models with too many parameters and more sophisticated models as well) see, e.g., Brockwell and Davis (1991, 1996) in the list of recommended literature.

9.4 Forecasting of Time Series

In conclusion we will briefly touch upon the problem of forecasting the future values of a TS X_{t+s}, $s > 0$, from the "past values" $X_1, \ldots, X_t$. As we know, the best (in the mean-quadratic sense) predictor is the conditional expectation $\mathbf{E}(X_{t+s}|X_1, \ldots, X_t)$. To calculate it, the joint distribution of the X's *should be known*, which is rarely the case. Recall that, in this chapter, we in fact operate with much less restrictive assumptions only regarding the first two moments of the processes (means and ACVFs). A natural approach in that case is to use a *linear predictor*—a linear combination of $1, X_1, \ldots, X_t$ which would predict X_{t+s} with the minimum possible mean squared error.[6] The coefficients in the best linear combination will only depend on the means and ACVF (and hence do not require more detailed information about the joint distributions).

For those with a wider mathematical background: note that one can think about the X_t's as elements of the linear space of square integrable random variables (given on a common probability space). Then $\mathbf{E}(XY)$ can be viewed as an *inner* (or *scalar*) product in this space, and two (zero-mean) RVs X and Y are uncorrelated iff they are *orthogonal* (w.r.t. this inner product). The best linear prediction of X_{t+s} from $1, X_1, \ldots, X_t$ is nothing else but the *projection* of X_{t+s} onto the linear space spanned by the "vectors" $1, X_1, \ldots, X_t$.

The basic task here is to predict the value X_{t+s}, $s > 0$, of a stationary TS with zero mean and a known ACVF γ in terms of the values $X_1, \ldots, X_t$ ("prediction made at origin t for lead time s"). More sophisticated problems can, in a sense, be reduced to this one. (We again refer the interested reader to the texts listed in Section 9.5.) That is, one has to solve the following minimisation problem:

$$\Sigma(a_1, \ldots, a_t) := \mathbf{E}(X_{t+s} - a_1 X_t - \cdots - a_t X_1)^2 \longrightarrow \min_{\{a_j\}}.$$

The function Σ is a non-negative quadratic form in variables a_j, hence it has at least one global minimum. To find it, we have to solve the system

[6] We should also mention the well-known fact that, in the important special case when the WN is Gaussian (i.e., the Y_t's are i.i.d. $N(0, \sigma^2)$ RVs), the best linear predictor coincides with the conditional expectation (and hence is the best predictor as well).

of equations

$$\frac{\partial}{\partial a_j}\Sigma(a_1,\dots,a_t) = 0, \quad j = 1,\dots,t,$$

which is easily seen (by evaluating the derivatives) to be equivalent to

$$\mathbf{E}\left[X_{t+1-j}\left(X_{t+s} - \sum_{k=1}^{t} a_k X_{t+1-k}\right)\right]$$
$$\equiv \gamma(s+j-1) - \sum_{k=1}^{t} a_k\gamma(k-j) = 0, \quad j = 1,\dots,t.$$

The last system can, in its turn, be re-written in matrix form as

$$\boldsymbol{a}_t\Gamma_t = \boldsymbol{\gamma}_t(s), \tag{9.61}$$

where $\boldsymbol{a}_t = (a_1, a_2, \dots, a_t)$, $\boldsymbol{\gamma}_t(s) = (\gamma(s), \gamma(s+1), \dots, \gamma(s+t-1)) \in \mathbf{R}^t$ and the $t \times t$ matrix $\Gamma_t := (\gamma(j-k))_{j,k=1,\dots,t}$ has the form

$$\Gamma_t = \begin{pmatrix} \gamma(0) & \gamma(1) & \gamma(2) & \cdots & \gamma(t-1) \\ \gamma(1) & \gamma(0) & \gamma(1) & \cdots & \gamma(t-2) \\ \gamma(2) & \gamma(1) & \gamma(0) & \cdots & \gamma(t-3) \\ \cdots & \cdots & \cdots & \cdots & \cdots \\ \gamma(t-1) & \gamma(t-2) & \gamma(t-3) & \cdots & \gamma(0) \end{pmatrix}. \tag{9.62}$$

This Γ_t is nothing else but the covariance matrix of the random vector

$$\boldsymbol{X}_t = (X_t, X_{t-1}, \dots, X_1).$$

If the matrix Γ_t is non-degenerate, the linear system (9.61) will have a unique solution given by

$$\boldsymbol{a}_t(s) := \boldsymbol{\gamma}_t(s)\Gamma_t^{-1},$$

and we will get the *best linear predictor* for X_{t+s} as

$$\Pi_t X_{t+s} := \boldsymbol{a}_t(s)\boldsymbol{X}_t^T,$$

where ^{T} stands for transposition. One can similarly solve the *interpolation* problem when one has to estimate missing values in a TS.

The case where the matrix Γ_t is degenerate (i.e., $\det\Gamma_t = 0$) is trivial. Indeed, a well-know fact of linear algebra is that then there exists a vector $\boldsymbol{v} = (v_1, \dots, v_t) \neq 0$ such that

$$\boldsymbol{v}\Gamma_t = 0.$$

But then

$$\mathrm{Var}\,(\boldsymbol{v}\boldsymbol{X}_t^T) = \mathbf{E}\left(\boldsymbol{v}\boldsymbol{X}_t^T(\boldsymbol{v}\boldsymbol{X}_t^T)^T\right) = \mathbf{E}\left(\boldsymbol{v}\boldsymbol{X}_t^T\boldsymbol{X}_t\boldsymbol{v}^T\right) = \boldsymbol{v}\Gamma_t\boldsymbol{v}^T = 0,$$

so that

$$vX_t^T \equiv v_1X_t + v_2X_{t-1} + \cdots + v_tX_1 = 0 \quad \text{a.s.}$$

Denoting by $k \geq 1$ the minimum value such that $v_k \neq 0$ (note that always $k < t$—why?) and setting $c_j := -v_{k+j}/v_k$, $j = 1, \dots, t-k-1$, we get

$$X_{t-k+1} = c_1X_{t-k} + \cdots + c_{t-k}X_1 \quad \text{a.s.}$$

By stationarity this implies that, for any $u > m := t - k > 0$,

$$X_u = c_1X_{u-1} + \cdots + c_mX_{u-m} \quad \text{a.s.}$$

That is, we have a *perfect prediction* for X_{t+1} from $X_t, \dots, X_{t-m+1}$! Moreover, since we can also get a perfect prediction for X_{t+2} from $X_{t+1}, \dots, X_{t-m+2}$, this means that we will have a perfect prediction for X_{t+2} from the same $X_t, \dots, X_{t-m+1}$ as well, and so on. So all the exact future values of the TS can be obtained from $X_t, \dots, X_{t-m+1}$!

Observe that the mean squared prediction error is

$$\begin{aligned}\mathbf{E}\,(X_{t+s} - \Pi_tX_{t+s})^2 &= \mathbf{E}\,X_{t+s}^2 - 2\mathbf{E}\,X_{t+s}(a_t(s)X_t^T) + \mathbf{E}\,(a_t(s)X_t^T)^2\\ &= \gamma(0) - 2a_t(s)\gamma_t(s)^T + \mathbf{E}\,(a_t(s)X_t^T)(a_t(s)X_t^T)^T\\ &= \gamma(0) - a_t(s)\gamma_t(s)^T = \gamma(0) - \gamma_t(s)\Gamma_t^{-1}\gamma_t(s)^T\end{aligned}$$

since

$$\begin{aligned}\mathbf{E}\,(a_t(s)X_t^T)(a_t(s)X_t^T)^T &= \mathbf{E}\,(a_t(s)X_t^TX_ta_t(s)^T)\\ &= a_t(s)\,\mathbf{E}\,(X_t^TX_t)\,a_t(s)^T = a_t(s)\Gamma_ta_t(s)^T = a_t(s)\gamma_t(s)^T\end{aligned}$$

from (9.61).

There exist popular simple recursive algorithms for computing one-step predictors and then using them to derive the s-step ones (the so-called Durbin-Levinson and Innovations algorithms), but discussing them is beyond the scope of the present text. The interested reader is again referred to the books listed in Section 9.5.

Example 9.11. Forecasting X_{t+1} from X_t is particularly simple for AR(p) processes. Indeed, let X be a TS given by (9.43). Then, for $t > p$,

$$\begin{aligned}&\mathbf{E}\,[X_{t+1} - a_tX_t^T]^2\\ &= \mathbf{E}\left[Y_{t+1} + (\beta_1 - a_1)X_t + \cdots + (\beta_p - a_p)X_{t+1-p} - \sum_{k=p+1}^{t} a_kX_{t+1-k}\right]^2\\ &= \sigma^2 + \mathbf{E}\left[(\beta_1 - a_1)X_t + \cdots + (\beta_p - a_p)X_{t+1-p} - \sum_{k=p+1}^{t} a_kX_{t+1-k}\right]^2\end{aligned}$$

since Y_{t+1} and X_j are uncorrelated, $j = 1, \dots, t$. Now the last expectation is clearly non-negative (as that of a square) and is equal to zero when we choose $a_1 = \beta_1, \dots, a_p = \beta_p$ and $a_k = 0$ for all $k > p$. So the minimum error is attained for that choice of the a_j's, and hence the best linear predictor is

$$\Pi_t X_{t+1} = \beta_1 X_t + \cdots + \beta_p X_{t+1-p}$$

with the mean squared error $\mathbf{E}\,(X_{t+1} - \Pi_t X_{t+1})^2 = \sigma^2$.

Now observe that the forecast we have just derived is exactly what the general formula would suggest: the forecast corresponds to the vector

$$\boldsymbol{a}_t(1) := (\beta_1, \dots, \beta_p, 0, \dots, 0),$$

which is the solution to (9.61) for $s = 1$. Indeed, if we substitute it into the system, we will get nothing else but the collection of our equations (9.50) with $h = 1, \dots, t$.

How could one get forecasts for leads $s \geq 2$?

Consider the case $s = 2$ first. We have to solve (9.61), i.e., find an $\boldsymbol{a}_t(2)$ such that

$$\boldsymbol{a}_t(2)\Gamma_t = (\gamma(2), \dots, \gamma(t+1)). \tag{9.63}$$

This can be done in two steps. First, again appealing to (9.50), note that the vector on the right-hand side of the above relation can be obtained as the product

$$\boldsymbol{a}_t(1) \begin{pmatrix} \gamma(1) & \gamma(2) & \gamma(3) & \cdots & \gamma(t) \\ \gamma(0) & \gamma(1) & \gamma(2) & \cdots & \gamma(t-1) \\ \gamma(1) & \gamma(0) & \gamma(1) & \cdots & \gamma(t-2) \\ \cdots & \cdots & \cdots & \cdots & \cdots \\ \gamma(t-2) & \gamma(t-3) & \gamma(t-4) & \cdots & \gamma(1) \end{pmatrix}.$$

Second, observe that the matrix above can, in its turn, be obtained as the matrix product of the form $A\Gamma_t$ with

$$A = \begin{pmatrix} \beta_1 & \beta_2 & \beta_3 & \cdots \\ 1 & 0 & 0 & \cdots \\ 0 & 1 & 0 & \cdots \\ \cdots & \cdots & \cdots & \cdots \end{pmatrix}.$$

Thus equation (9.63) is equivalent to

$$\boldsymbol{a}_t(2)\Gamma_t = \boldsymbol{a}_t(1)A\Gamma_t.$$

As we saw earlier, in all non-trivial cases the matrix Γ_t is non-degenerate, so multiplying both sides of the above equation by Γ_t^{-1} from the right yields

$$\boldsymbol{a}_t(2) = \boldsymbol{a}_t(1)A = (\beta_1^2 + \beta_2, \beta_1\beta_2 + \beta_3, \beta_1\beta_3 + \beta_4, \dots, \beta_1\beta_p, 0, \dots, 0). \tag{9.64}$$

This solution has the following very simple and graphical interpretation: for an AR(p) model, to get a forecast for X_{t+2} from $X_t, X_{t-1}, \ldots, X_1$ ($t > p$), we first predict

$$X_{t+1} \quad \text{by} \quad \Pi_t X_{t+1} = \beta_1 X_t + \cdots + \beta_p X_{t-p+1},$$

and then use the predicted value instead of the unknown X_{t+1} in the prediction formula $\Pi_{t+1}X_{t+2}$ with lead $s = 1$ for X_{t+2}. This way we get

$$\Pi_t X_{t+2} = \beta_1 \Pi_t X_{t+1} + \beta_2 X_t + \cdots + \beta_p X_{t-p+2},$$

which can easily be seen to be the same as (9.64).

Forecasts $\Pi_t X_{t+s}$, $s > 2$, can be obtained in a similar way.

9.5 Recommended Literature

AKAIKE, H. AND KITAGAWA, G. *The practice of time series analysis.* Springer, New York, 1999.

ANDERSON, O.D. *Time series analysis and forecasting: the Box-Jenkins approach.* London, Butterworth, 1976.

BOX, G.E.P. AND JENKINS, G.M. *Time series analysis: Forecasting and control.* Holden-Day, San Francisco, 1976.

BROCKWELL, P.J. AND DAVIS, R.A. *Introduction to time series and forecasting.* Springer, New York, 1996.

BROCKWELL, P.J. AND DAVIS, R.A. *Time series: Theory and methods.* 2nd ed. Springer, New York, 1991.

D'AGOSTINO, R.B. Tests for the normal distribution. In: D'AGOSTINO, R.B. AND STEPHENS, M., EDS., *Handbook of goodness-of-fit techniques.* Dekker, New York, 1986, pp.367–419.

KENDALL, M.G. AND STUART, A. *The advanced theory of statistics. Vol. 3.* Griffin, London, 1976.

9.6 Problems

1. (i) Let X_0 and X_1 be RVs, $\mathbf{E}\, X_j = m_j$, $\mathrm{Var}\,(X_j) = \sigma_j < \infty$, $\mathrm{Cov}\,(X_0, X_1) = C$. Find the best (in mean quadratic) linear predictor $\widehat{X}_0 = aX_1 + b$ for X_0 from X_1 and the error of the predictor. That is, find $\min_{a,b} \mathbf{E}\,|X_0 - (aX_1 + b)|^2$ and the values of a and b for which the minimum is attained.

(ii) Extend the result to the situation when one has to predict X_0 using linear combinations of $X_1, \ldots, X_n$. We know all the expectations $m_j = \mathbf{E}\, X_j$ and covariances $C_{jk} = \mathrm{Cov}\,(X_j, X_k)$, $j, k = 0, 1, \ldots, n$.

2. Let (X_1, X_2) be uniformly distributed over the triangle with vertices at the points $(1, 0)$, $(-1, 0)$ and $(0, 1)$. Find (i) the best linear predictor for X_2 from X_1 and (ii) the best predictor for X_2 from X_1 (note that it is non-linear indeed!).

   Hint. Recall that the best predictor is given by the respective conditional expectation (cf. Problem 27 on p.73).

3. Prove (9.5).

   Hint. Verify that

$$\left(\frac{\sin(n\lambda/2)}{\sin(\lambda/2)}\right)^2 = (1 + e^{i\lambda} + \cdots + e^{i\lambda(n-1)})(1 + e^{-i\lambda} + \cdots + e^{-i\lambda(n-1)})$$

   (cf. Example 9.4).

4. Prove (9.41).

5. Let Y and $\varphi \sim U(0, 2\pi)$ be independent RVs, $\mathbf{E}\, Y = 0$, $\mathrm{Var}\,(Y) = \sigma^2 < \infty$, $\lambda_0 \in (-\pi, \pi]$ a constant. Show that $X_t = Y\cos(\lambda_0 t + \varphi)$ is a stationary process with the ACVF $\gamma(h) = \sigma^2 \cos(\lambda_0 h)$. What happens when $\mathbf{E}\, Y \neq 0$?

   Hint. You may wish to use the fact that, for a fixed s, the value of the integral $\int_x^{x+2\pi} \cos(u+s)\cos(u)\, du$ does not depend on x (why?). Also, $\sin(u) = \cos(u - \pi/2)$ for any u, so that replacing cos by sin in the above integral will not change its value (for this will in essence be equivalent to replacing x with $x - \pi/2$).

6. For the four AR(2) models below ($\{Y_t\} \sim$WN(0,1)), write down and solve the characteristic equations, and plot the ACVF $\gamma(h)$ for $h = 0, 1, \ldots, 20$:

   (i) $X_t = 0.7X_{t-1} - 0.1X_{t-2} + Y_t$;
   (ii) $X_t = 1.4X_{t-1} - 0.45X_{t-2} + Y_t$;
   (iii) $X_t = -0.4X_{t-1} + 0.45X_{t-2} + Y_t$;
   (iv) $X_t = \frac{3}{4}X_{t-1} - \frac{9}{16}X_{t-2} + Y_t$.

7. Suppose the spectral density of a stationary process $\{X_t\}$ is given by

$$f(\lambda) = \frac{1}{2\pi}(5 + 4\cos(\lambda)).$$

   (i) Identify an MA process which would have this spectral density. [Such an MA process is not unique.]

   (ii) Find the ACVF of this process.

   (iii) For the process you suggested in (i), find the best linear prediction for X_{t+s}, $s \geq 1$, from $X_t, X_{t-1}, \ldots$

Hints: (i) Using the relation $\cos(\lambda) = (e^{i\lambda} + e^{-i\lambda})/2$, you can derive for the spectral function a representation of the form $c|c_1 + c_2 e^{-\lambda}|^2$. Thus the spectral density coincides with the product of the power transfer function of a simple linear filter and the spectral density of a white noise. (ii) You can directly compute the ACVF using the integral representation (9.10). Alternatively, you can derive it from your answer to part (i). (iii) What is the covariance of X_{t+s} and X_t for $s > 1$? For $s = 1$, you can directly minimise the mean squared error.

8. For the AR(1) process

$$X_t = \beta X_{t-1} + Y_t, \quad \{Y_t\} \sim \mathrm{WN}(0, 2.25), \quad \beta = 0.8, \tag{9.65}$$

(i) find the ACVF using direct computation;

(ii) find the spectral density and plot it;

(iii) comment on the shape of the spectral density and how it relates to the representation (9.65).

(iv) How will your answers to (i)–(iii) change if we put $\beta = -0.8$ in (9.65)?

(v) Find the best (linear) predictor for X_{t+s}, $s \geq 1$, from $X_1, \ldots, X_t$.

9. Verify that, for a stationary ARMA(1,1) process $X_t = \beta X_{t-1} + Y_t + \alpha Y_{t-1}$, $|\beta| < 1$, $\{Y_t\} \sim \mathrm{WN}(0, \sigma^2)$, the ACVF $\gamma(h)$ is equal to

$$\gamma(h) = \sigma^2 \times \begin{cases} \dfrac{1 + 2\alpha\beta + \alpha^2}{1 - \beta^2}, & h = 0, \\[2ex] \dfrac{(1 + \alpha\beta)(\alpha + \beta)}{1 - \beta^2}\beta^{|h|-1}, & h \neq 0. \end{cases}$$

Hint: You may wish to use the relation $(1 - \beta B)^{-1} = \sum_{k \geq 0} \beta^k B^k$ to represent X_t as a one-sided moving average $\sum_{k \geq 0} a_k Y_{t-k}$ (of infinite order), find the coefficients a_k and then apply (9.16) to compute $\gamma(h)$.

10. (i) Write down the characteristic equation for the AR(2) process

$$X_t = -X_{t-1} - 0.5 X_{t-2} + Y_t, \qquad \{Y_t\} \sim \mathrm{WN}(0, 4).$$

Solve the equation and verify the condition for the process to be causal.

(ii) Say how your answer to part (i) will change if, instead of the AR(2) process above, we have an ARMA(2,2) process

$$X_t = -X_{t-1} - 0.5 X_{t-2} + Y_t + 1.7 Y_{t-1} - 7 Y_{t-2}, \qquad \{Y_t\} \sim \mathrm{WN}(0, 0.01).$$

(iii) Find the spectral density $f(\lambda)$ of a stationary process $\{X_t\}$ described by the AR(2) model in (i). Plot the density and comment on its shape (what it means in terms of the spectral decomposition of the process).

(iv) Suppose that we apply the linear filter

$$V_t = X_{t-1} - 2X_t + X_{t+1}$$

to our stationary process $\{X_t\}$ from part (iii). Find the transfer function of this filter and the spectral density of the new process $\{V_t\}$.

(v) Using the formula for the transfer function of the linear filter, explain what the result of applying the filter to a constant sequence $Z_t \equiv c = \text{const}$ will be.

(vi) Using a representation for the filter in terms of the operators B and ∇, say what the result of applying the filter to a sequence of the form

$$Z_t = c_0 + c_1 t + c_2 t^2$$

will be. Verify your result by directly computing V_1 and V_2.

Hints: For parts (i) and (iii), make use of the following relations: $|z|^2 = z\bar{z}$ for a complex number z; $\overline{e^{-it}} = e^{it}$; $e^{-it} + e^{it} = 2\cos(t)$.

Chapter 10

Elements of Simulation

10.1 Basics. Random Number Generators

Simulation and extensive use of (pseudo) random numbers have become feasible with the emergence of electronic computers. Over the past decades, the fast progress of technology in this area has made it possible for the simulation of complex systems with inherent randomness to turn into a very popular and important component of such systems' study.

In its essence, a system is a set of interacting components and processes; the interactions may be internal or may be linked to the external environment. Simulating a system basically means studying and analysing a *model of the system* on a computer. Such a model can be thought of as a representation of the system in which the processes or interactions bear a close resemblance to those in the specific system being studied, and which has a form suitable for implementation on a computer. Computer models used in simulation are rarely "strictly mathematical" and highly abstract, and therefore will not be discussed in the present course in any detail. Designing such models is rather a question of programming skills and knowledge of a particular system to be modelled than a mathematical problem; there even exist special programming languages created exclusively for that task, the most famous of them being *Simula*.[1] However, there are crucial elements of the models (first of all, the ones responsible for "imitating" the randomness in the system's behaviour) that are very interesting and important from the mathematical point of view, and we will consider some of them.

[1] *Simula 1* was a simulation programming language designed (on the basis of *Algol 60*) to model discrete-event systems, like the queueing networks discussed in Section 7.4. In fact, it was the first *object-oriented language*; it provided objects, classes and inheritance (dynamic typing was added in 1967). What is called *Simula* nowadays is actually the later general-purpose language *Simula 67*.

In this part of the chapter, we will briefly discuss a few basic ideas of random number generation, simulation of RVs with prescribed distributions, and some techniques aimed to improve the performance of algorithms based on the use of random numbers. For further reading, we recommend the books from the list at the end of the chapter.

First of all, we will make the following comment: One attempts to simulate a system when it is hard/impossible to find explicit formulae (in the framework of the *mathematical model* of the system) or reasonable approximations to them (using the limit theorems of probability theory such as the LLN, CLT *etc.*) for the characteristics of interest—which is typically the case once the systems of interest become more or less complicated. The main stages can be described—rather loosely—as follows:

1. Formulate a mathematical model for the system/phenomenon of interest (usually, in the form of a stochastic process).

2. Simulate (a large number of independent) realisations of the process constructed at Step 1.

3. Statistically analyse the data from observing these realisations.

Example 10.1. How could one evaluate the performance of a statistical procedure (estimating a parameter or testing a hypothesis)? It is not very often that one can say which procedure is optimal in this or that sense. Even when it is possible, such an assertion is usually based on specific assumptions on the distribution of the observed samples *etc.* When the sample size is *large*, one could rely on the *large sample theory* based on the use/derivation of specific probabilistic limit theorems. However, in the most interesting cases, the sample size is not that large—statistics is especially important when there is no large data arrays, and one needs to derive conclusions from a very limited amount of information.

So what one does is producing a long sequence of "artificial samples" $(X_1, \dots, X_n)$ of desired length n from known distributions and then testing the statistical procedures of interest on the simulated data. If a particular procedure performs better on that data set, one may agree that it would most likely be better on the real-life data as well. It is important, of course, to ensure that the simulated data have the "randomness" properties similar to those of the real world data.

Example 10.2. *Modelling a $GI/GI/1$ queue.* There are ways to analyse this non-Markovian system theoretically, so there is little sense in simulating it on computer. However, this example allows one to illustrate the important idea that a good (e.g., "economical" in terms of its memory/processor time requirements) model of a system is not necessarily a sort of "direct translation" of the system to a programming language, but rather a result of certain preliminary (and sometimes complex) analysis.

Our system is fed by an RP $\{N_t\}$ with i.i.d. interarrival times $\tau_j \sim F_\tau$. Suppose we want to compare the performance of two different servers:

- server A, with service times $s_j \sim F_A$, and
- server B, with service times $s_j \sim F_B$.

We want to compare "typical" values (say, means or medians) of the main characteristics for the QS with different servers and, on the basis of this information, to decide which server would be better for our purposes. Should we model the entire process $X_t :=$ # of customers in the QS at time t? Of course, our approach will depend on what exactly we want to compare. It may turn out that, say, modelling the whole trajectory of $\{X_t\}$ is meaningless.

Anyway, we will need to simulate independent sequences of i.i.d. RVs $\{\tau_j\} \sim F_\tau$ and $\{s_j\} \sim F_A$ (or F_B). Later we will discuss to some extent how to do that. For the time being, assume we have already got them.

Now the trajectory of $\{X_t\}$ is a step-function with jumps

$$+1 \text{ at times } T_j = \tau_1 + \cdots + \tau_j,$$

$$-1 \text{ at times } S_j = \begin{cases} S_{j-1} + s_j & \text{if } X_{S_{j-1}} > 0; \\ \\ \min\{T_k,\, k \geq 1 : T_k > S_{j-1}\} + s_j & \text{if } X_{S_{j-1}} = 0. \end{cases}$$

If we need to know, say, only the mean waiting or delay time or the distribution of the waiting times, we can avoid modelling the process $\{X_t\}$ itself by observing that, for the waiting time w_n of the nth customer, one has the following recurrence:

$$w_{n+1} = \max\{0, w_n + \underbrace{s_n - \tau_{n+1}}_{Y_n}\}, \tag{10.1}$$

where Y_n are i.i.d. RVs (see Example 3.4).

Set $a := \mathbf{E}\, Y_n = \mathbf{E}\, s_1 - \mathbf{E}\, \tau_1$. Then one can show that, as $n \to \infty$,

$$w_n \overset{\text{distr}}{\longrightarrow} W := \max_{k \geq 0} \sum_{j=1}^{k} Y_j \begin{cases} = \infty & \text{if } a \geq 0 \\ < \infty & \text{if } a < 0 \end{cases} \quad \text{a.s.} \tag{10.2}$$

(cf. Section 3.5; the condition $a < 0$ is clearly equivalent to the stability condition on traffic intensity: $\rho = \mathbf{E}\, s_j / \mathbf{E}\, \tau_j < 1$). The RV W is often much easier to simulate than the whole process $\{X_t\}$ or the sequence $\{w_n\}$.

The moral is that one should not model any redundant aspects of the behaviour of the stochastic system of interest.[2]

In the above example, we said that we had to simulate sequences of i.i.d. RVs and "run" a series of independent realisations of the system. *How could one do that?*

First of all, what does it mean to "simulate a random sequence"? Any such simulation produces a *deterministic number sequence* $x_1, x_2, \dots$ (of the so-called *pseudo-random numbers*). To use such a sequence to model stochastic systems, it should have properties similar to those of the real-life random sequences.[3]

Assume we want to simulate a sequence of i.i.d. RVs $\{X_j\}$ following a prescribed DF F. As we will see later, it actually suffices to know how to simulate a sequence of $U(0,1)$-distributed RVs. The most important properties such a simulated sequence $\{x_j\}$ of "uniform variates" should display are, loosely speaking, as follows:

(i) *frequency stabilisation*: for any $a \in [0,1]$,

$$\frac{1}{n}\#\{j \le n : x_j \le a\} \to a \quad \text{as} \quad n \to \infty;$$

(ii) *statistical independence*: stated in terms of the relative frequencies, we would like to have that, for any fixed $a, b \in (0,1)$ and $u \ge 1$,

$$\frac{1}{n}\#\{j \le n : x_j \le a, x_{j+u} \le b\} \to ab \quad \text{as} \quad n \to \infty;$$

(iii) $\{x_j\}$ must be rather "irregular".

If the simulated sequences do not have these properties, using them for modelling stochastic systems can (and most likely will) produce unreliable

[2]The principle is important not only for simulation, of course; it is applicable to any kind of modelling and is closely related to the famous *Ockham's razor* (called after William of Ockham, also spelled Occam (c. 1285–1349), one of the most influential philosophers of the 14th century). The rule says that the fewest possible assumptions are to be made in explaining a thing.

[3]It is interesting to note that pseudo-random sequences are used not only for the purposes of simulating stochastic systems or Monte Carlo computations (discussed to some extent in our Section 10.4 below). Since the 1950s, astronomers used radar pulses emitted by radio telescopes at times following such a sequence to examine other planets. The travel time (delay in the weak radar reflections from the surface of the planet) is found by measuring when the transmitted and reflected signals were most correlated. To have a reliable method of doing so, the transmitted sequence should be very "irregular", and a (deterministic) pseudo-random sequence is a good choice here. A similar idea was implemented in the Global Positioning System (GPS) enabling one to work out one's whereabouts on our planet to within about a centimetre.

and even surprising results. For more detailed discussion of pseudo-random sequences see, e.g., Gentle (1998), Moeschlin et al. (1998) and Chapter 7 in Fishman (1996) from the list of recommended literature in Section 10.6.

Once we know the desired properties, the next question is how to produce such sequences. At early times, one would sometimes use *analogous devices*: thus, in a specimen of monoatomic radioactive material, the times between consequently registered decays of the atoms form a "nearly perfect" realisation of a sequence of independent exponentially distributed RVs. So, if you can connect a radioactive particle counter to your computer, you get a ready (although relatively slow) random number generator. The RAND corporation published in 1955 a table of a million random digits obtained using such a device. For many modern applications, such a table would be too short!

With the development of numerical algorithms and computers, the so-called **arithmetic generators** became dominating and eventually the only ones used in practice.

An arithmetic generator produces a deterministic sequence of numbers from a finite subset A of non-negative integers, which is then transformed into values from $[0, 1]$ to get $U(0, 1)$-variates. Such generators are usually defined by a recursive relation

$$x_{k+1} = f(x_k), \quad k = 0, 1, 2, \ldots,$$

where $x_0 \in A$ (the *seed* of the generator) and $f : A \to A$ is a specific mapping. Clearly, any such sequence is *periodic*. The shorter the period, the less useful the generator.

A remarkable feature of arithmetic generators is that they can *reproduce* exactly the same sequence of (pseudo) random numbers (by using one and the same seed), which can be very important, say, for

(i) *comparative simulations* (performance evaluation), when one performs two sets of simulations in two different situations, or two different configurations of the model;

(ii) *sensitivity analysis,* when one has to determine if the value of a characteristic of interest of our model changes significantly when the value of a parameter of the system is changed, or for

(iii) *numerical evaluation* of a derivative of a function whose values can only be found as averages using Monte Carlo techniques.

In two words, employing statistical terminology, this feature allows one to do "paired comparisons", avoiding confounding the effect of changing the parameter value with that of changing random variates used for computations.

Example 10.3. *Congruential generators.*[4] This is a very popular family of simple arithmetic algorithms producing (when their parameters are well chosen) rather good pseudo-random sequences. The sequence is defined by the following recursive relation[5] starting with some initial value (seed) x_0:

$$x_{j+1} = ax_j + c \pmod{M}, \tag{10.3}$$

where a, c and M are fixed given parameters. The quality of the generator strongly depends upon the choice of these constants. There are many theoretical and empirical results on selecting "good" values of the parameters. If, for example, you take $a = 383$, $c = 263$ and $M = 10^4$, the period of $\{x_j\}$ will be the maximum possible (for that M) one: 10^4. The choice[6] of $a = 7^5 = 16,807$, $c = 0$ and $M = 2^{31} - 1 = 2,147,483,647$ is a very good set of parameters. This variant of the generator is sometimes called the *minimal standard random number generator* and is often built into software packages.

Example 10.4. *Shift-register generators* are of the form

$$x_{n+1} = \sum_{j=0}^{p-1} a_j x_{n-j} \pmod{2},$$

where the x_n's and a_j's are all either 0 or 1 (so we actually generate random bits). One can produce sequences of the maximum period (equal to $2^p - 1$) even when among the coefficients $a_0, \ldots, a_{p-1}$ there are only two a_j's equal to one, and the resulting generators are very fast.

Thus, a very popular shift-register generator R250[7] has $a_j = 0$ for all $j \neq 102$ or 249:

$$x_{n+1} = x_{n-102} + x_{n-249} \pmod{2},$$

with its period being equal to $2^{250} - 1 \approx 1.80925 \times 10^{75}$. Note that addition modulo 2 is equivalent to the so-called "exclusive-or" (XOR) operation, so

[4]Suggested in: Thompson, W.E. (1958), A modified congruence method of generating pseudo random numbers. *Comp. J.* 1, 83–86.

[5]Notation $x+y \pmod{M}$ means addition modulo M; for the definition of this operation, see p. 111.

[6]These parameters were published in: Park, S.K. and Miller, K.W. (1988), Random number generators: Good ones are hard to find. *Comm. ACM*, 31, no. 10, 1192–1201.

[7]Suggested in: Kirkpatrick, S. and Stoll, E. (1981), A very fast shift-register sequence random number generator. *J. Comp. Phys.*, 40, 517–526. The idea of such a generator goes back to Tausworthe, R.C. (1965) Random numbers generated by linear recurrence modulo two. *Mathem. Comp.*, 19, 201–209.

that all the additions can be made in parallel (which is much faster than multiplication used in the linear congruential method). Note also that, to start the generator, one needs p initial x_n's (which can be produced, say, using a congruential generator).

Developing and choosing good arithmetic methods are very hard and interesting problems. We will not discuss them any further, referring the interested reader to the above-mentioned books for more information and bibliography. In practice, if for the application one is interested in, it is not of utmost importance to ensure that certain specific criteria of the "randomness" of the used pseudo-random sequences are met, there is no need to look for a new random number generator or try to choose something special. One typically uses generators provided by various software packages (including realisations of popular computer languages) producing *uniform* random numbers. So, in what follows, we will consider in more detail how to transform simulated uniform RVs into ones following a prescribed law. For more detailed exposition, see, e.g., Devroye (1986) and Gentle (1998).

Remark 10.1. It turns out that for some applications pseudo-random numbers are *too* random, and having a "more uniform" density of coverage by the sampled points of the domain may be preferable. Such random number generators (avoiding clusters by "giving up" independence) are called *quasi-random number generators.*

10.2 The Inverse Function Method

This very simple and nice method uses the *generalised inverse function* of the DF F (we want to simulate from) defined as

$$Q(u) := \min\{x : F(x) \geq u\}, \quad u \in (0,1), \tag{10.4}$$

and denoted sometimes by $F^{(-1)}$ or simply F^{-1}; it is just the inverse of F when the latter is continuous and strictly increasing. In mathematical statistics, $Q(u)$ is called the *quantile function* of the distribution F: its values are just the respective *quantiles* of F (thus, $Q(1/2)$ is the median, $Q(3/4)$ is the upper quartile of F *etc.*).

Note that an equivalent to (10.4) definition of Q is that

$$Q(u) \leq x \quad \text{iff} \quad F(x) \geq u. \tag{10.5}$$

The *inverse function method* is based on the following observation.

Theorem 10.1. *If $U \sim U(0,1)$, then, for any DF F, $X := Q(U) \sim F$.*

Proof Indeed, the DF of the thus defined RV X is

$$F_X(x) \equiv \mathbf{P}(Q(U) \le x) \overset{\text{by (10.5)}}{=} \mathbf{P}(F(x) \ge U) = F_U(F(x)) \equiv F(x).$$

□

Remark 10.2. For continuous DFs F, the assertion of Theorem 10.1 has a simple "converse": if $X \sim F$ then $F(X) \sim U(0,1)$. In the general case, there is also a version of such a relationship — but that one requires an additional "source of randomness" to "fill the holes" in the image $F(\mathbf{R})$ that exist when F has jumps (for more detail, see Problems 2 and 3 in Section 10.7 below).

Example 10.5. Let $F := Exp(\lambda)$, $\lambda > 0$. Then

$$F(x) = \begin{cases} 1 - e^{-\lambda x}, & x \ge 0, \\ 0, & x < 0, \end{cases}$$

whose restriction to $(0, \infty)$ has an inverse function. Solving the equation $u = 1 - e^{-\lambda x}$ for x, we find that the inverse function equals

$$Q(u) = -\frac{1}{\lambda}\log(1-u).$$

Observing that, for an RV $U \sim U(0,1)$, one has $1 - U \sim U(0,1)$ as well, we conclude that

$$X := -\frac{1}{\lambda}\log U \sim Exp(\lambda).$$

When one *can compute* the quantile function Q for a given F (there is a closed form expression or a relatively fast algorithm for calculating the values of Q), one has the following simple

Algorithm for simulating an RV $X \sim F$ when you know Q:

1. Generate a $U \sim U(0,1)$.
2. Put $X := Q(U)$. Stop.

Unfortunately, it is rarely the case that one can use the inversion method in a straightforward way. However, when there is no explicit formula for Q, there may still be a way of transforming the problem into one admitting solution based on the inversion method.

Example 10.6. Suppose we want to simulate an RV following the normal law $N(\mu, \sigma^2)$. First note that if $Y \sim N(0,1)$, then $X := \mu + \sigma Y \sim N(\mu, \sigma^2)$.

Hence it suffices to simulate a standard normal RV. Unfortunately, there is no simple formula for $Q = F^{-1}$. Yet one can still find several rather simple (exact or approximate) methods for simulating normal RVs.

(i) *The CLT.* As we know, for i.i.d. RVs $Y_1, \ldots, Y_n$ with $\mathbf{E}\, Y_i = m$ and finite $\mathrm{Var}\,(Y_i) = s^2 < \infty$, the distribution of the normalised sum

$$S_n := (Y_1 + \cdots + Y_n - mn)/\sqrt{ns^2}$$

will, for large n, be close to $N(0,1)$ (cf. (2.88)). Moreover, for Y's with "smooth distributions" (especially symmetric unimodal ones), the value of n for which the approximation will be very good, can be quite small, cf. our illustration in Example 2.3. In practice, many calculators used to employ just this approach to produce standard normal RVs: using standard algorithms for generating i.i.d. $U(0,1)$-RVs Y_i, one simply takes $n = 12$ and claims that

$$S_{12} = Y_1 + \cdots + Y_{12} - 6$$

is (approximately) normally distributed. Indeed, we know that $m = \mathbf{E}\, Y_i = 1/2$, while $s^2 = \mathrm{Var}\,(Y_i) = 1/12$ (see (2.56)), so that $\mathbf{E}\, S_{12} = nm - 6 = 0$ and $\mathrm{Var}\,(S_{12}) = n \times s^2 = 1$. For a graphical illustration of the approximation rate for the distribution densities, we again refer to Example 2.3. For most practical purposes, sequences produced using this approach are quite satisfactory.

(ii) *An exact method* (the Box-Muller algorithm). This method is based on the following observation: if X_1 and X_2 are i.i.d. $N(0,1)$-RVs, then, for the polar coordinates (R, Θ) of the vector (X_1, X_2), the following holds true:

$$R^2 := X_1^2 + X_2^2 \sim Exp(1/2) \quad \text{and} \quad \Theta \sim U(0, 2\pi) \tag{10.6}$$

are independent RVs.

That R are Θ are independent RVs and Θ is uniformly distributed on $[0, 2\pi]$ immediately follows from the symmetry of the density of (X_1, X_2). More formally, since this density is equal to

$$f(x_1, x_2) = \frac{1}{2\pi} e^{-(x_1^2 + x_2^2)/2}, \quad -\infty < x_1, x_2 < \infty$$

(see (2.37)), changing to the polar coordinates (ρ, θ) we see (e.g., from the change of variables formula (2.33)) that our normal vector will have the following density in the (ρ, θ)-plane:

$$\varphi(\rho, \theta) = \frac{\rho}{2\pi} e^{-\rho^2/2}, \quad \rho > 0, \quad \theta \in [0, 2\pi].$$

This is obviously also a product of two densities: $\varphi(\rho, \theta) = \varphi_\Theta(\theta)\varphi_R(\rho)$, where

$$\varphi_\Theta(\theta) = \frac{1}{2\pi}, \quad \theta \in [0, 2\pi], \quad \text{and}$$

$$\varphi_R(\rho) = \rho e^{-\rho^2/2}, \quad \rho > 0,$$

so that the RVs R and Θ are independent indeed, and $\Theta \sim U(0, 2\pi)$. The assertion about the distribution of R^2 is almost obvious as

$$\mathbf{P}(R^2 > t) = \mathbf{P}(R > \sqrt{t}) = \int_{\sqrt{t}}^{\infty} \rho e^{-\rho^2/2} d\rho = -e^{-\rho^2/2}\Big|_{\sqrt{t}}^{\infty} = e^{-t/2}, \quad t > 0$$

(alternatively, we could merely refer to the standard formula (2.27) for the density of a transformed RV).

Hence the

Algorithm.

1. Simulate independent $U_j \sim U(0, 1)$, $j = 1, 2$.
2. Put $\Theta := 2\pi U_1$ ($\sim U(0, 2\pi)$).
3. Put $R := \sqrt{-2\log U_2}$ (cf. Example 10.5).
4. Return

$$X_1 := R\cos\Theta, \qquad X_2 := R\sin\Theta,$$

the components of the random vector with the polar coordinates (R, Θ). They will be independent $N(0, 1)$-RVs. Stop.

A problem with this algorithm when a lot of simulations are required is that computing trigonometric functions is time consuming. There is another way (called the *polar algorithm*) of making use of the same observation (10.6), which also exploits the idea of the rejection method to be discussed in the next section. Instead of using $(\cos\Theta, \sin\Theta)$ to get a point uniformly distributed on the unit circle, we first get a random point uniformly distributed *inside a disk* and then normalize it to get the same result:

Algorithm.

1. Simulate independent $U_j \sim U(0, 1)$, $j = 1, 2$.
2. Put $V_j = 2U_j - 1 \sim U(-1, 1)$ and $S = V_1^2 + V_2^2$.
3. If $S > 1$, go to Step 1. [Given $S \leq 1$, the point (V_1, V_2) is uniform inside the unit disk, while $S \sim U(0, 1)$. (Why? Verify this!)]
4. Return

$$X_j := V_j\sqrt{\frac{-2\log S}{S}}, \qquad j = 1, 2.$$

They will be independent $N(0, 1)$-RVs. Stop.

We will only comment on Step 4. Since S is $(0, 1)$-uniform, the RV $\sqrt{-2\log S}$ will have distribution $Exp(1/2)$ (like the R—the length of the standard normal vector (X_1, X_2)—from the previous algorithm). Given the value $S = s$, the point (V_1, V_2) is uniformly distributed on the circle of radius $\sqrt{s}$. Therefore by taking $(V_1/\sqrt{S}, V_2/\sqrt{S})$ we will get an independent of S (and hence of $\sqrt{-2\log S}$) point uniformly distributed on the *unit circle* (like $(\cos\Theta, \sin\Theta)$ in the previous algorithm).

In conclusion of this example, note that once we can simulate a vector $\boldsymbol{X} = (X_1, \ldots, X_k)$ of independent $N(0,1)$-RVs, we can generate an arbitrary multivariate normal vector $\boldsymbol{Y} = (Y_1, \ldots, Y_k) \in \mathbf{R}^k$ as well. Assume w.l.o.g. that $\mathbf{E}\,\boldsymbol{Y} = 0$ and let $C_{\boldsymbol{Y}} := \mathbf{E}\,\boldsymbol{Y}^T\boldsymbol{Y}$ be the covariance matrix of the vector $\boldsymbol{Y}$. Now if K is a $k \times k$-matrix such that $K^T K = C_{\boldsymbol{Y}}$, then

$$\mathbf{E}\,(\boldsymbol{X}K)^T(\boldsymbol{X}K) = K^T(\mathbf{E}\,\boldsymbol{X}^T\boldsymbol{X})K = K^T K = C_{\boldsymbol{Y}},$$

so that $\boldsymbol{X}K$ has the same covariance matrix as $\boldsymbol{Y}$ and hence the desired distribution of $\boldsymbol{Y}$. The methods of extracting square roots of (positive definite) matrices can be found in any advanced text on matrices.[8]

In the case of **discrete distributions**, the inversion method works as follows. Let F be a discrete DF and $\cdots < x_{k-1} < x_k < x_{k+1} < \cdots$ the set of its jump points (if necessary, we may assume that the index runs from $-\infty$ to $+\infty$). Denote by

$$p_k = \mathbf{P}\,(X = x_k) = \mathbf{P}\,(X \le x_k) - \mathbf{P}\,(X < x_k) = F(x_k) - F(x_k - 0)$$

the jump of F at the point x_k, and put

$$u_k := F(x_k) = \sum_{j \le k} p_j. \tag{10.7}$$

Then the quantile function of F is given by

$$Q(u) = x_k \quad \text{for} \quad u \in (u_{k-1}, u_k].$$

Let $U \sim U(0,1)$. By Theorem 10.1, the RV X such that $X = x_k$ when $U \in (u_{k-1}, u_k]$, will have the desired DF F. We could easily verify this without referring to the theorem as well: indeed, since U is uniform on $[0,1]$, the probability of the last event is exactly the length $u_k - u_{k-1} = p_k$ of the interval, which is the desired probability for the value x_k.

Unfortunately, simple explicit formulae for Q are usually not available in the discrete case. It is often easier to exploit the same principle, but working with F instead of Q. One could use the following algorithm when X takes possible values $x_1, x_2, \ldots$ with probabilities $p_1, p_2, \ldots$, respectively.

---

[8]See, e.g., Gantmakher, F.R. *The theory of matrices.* Chelsea, New York, 1989.

Algorithm for discrete DFs.

1. Generate a $U \sim U(0,1)$.
2. Set $k := 1$.
3. While $F(x_k) \leq U$, put $k := k+1$.
4. Return $X := x_k$. Stop.

The algorithm can be impractical when there are (infinitely) many different values x_k. For specific DFs, there may exist much better solutions.

Example 10.7. *Simulating a Poisson RV $X \sim Po(\lambda)$.* In this important special case, it turns out that it is sometimes simpler to simulate not a single value X, but the whole trajectory of a Poisson process $\{X_t\}$ (with rate λ), and then put $X := X_1$. Such a simulation requires generating a sequence of i.i.d. $Exp(\lambda)$-RVs, and we have a simple exact method for doing that. (But there exist other—often faster—approaches as well.)

Hence one can use the following

Algorithm for simulating a Poisson process $\{X_s\}_{s\in[0,t]}$ with rate λ.

1. Put $k := 0$, $T_0 := 0$.
2. Set $k := k+1$ and simulate an independent $U(0,1)$-distributed RV U_k.
3. Put $\tau_k := -\frac{1}{\lambda}\log U_k$.
4. Put $T_k := T_{k-1} + \tau_k$.
5. If $T_k < t$, then go to Step 2.
6. To get the value of X_s for a given $s \in [0,t]$, find k such that $T_k \leq s < T_{k+1}$ and put $X_s := k$ (such a k always exists, cf. Step 5). Stop.

One can use a modified version of the inverse function method to simulate random vectors $\boldsymbol{X} = (X_1, \ldots, X_k) \in \mathbf{R}^k$ as well. A possible approach can be outlined as follows. First generate a sample of k i.i.d. RVs $U_j \sim U(0,1)$. Then choose one of the components (the choice may depend on the specific distribution of $\boldsymbol{X}$), say X_1, and simulate it by setting

$$X_1 := Q_1(U_1), \tag{10.8}$$

where Q_1 is the quantile function of the distribution of X_1. The latter is given by the marginal distribution

$$F_1(x) := F(x, \infty, \ldots, \infty),$$

where F is the DF of $\boldsymbol{X}$.

Now we cannot simply set $X_2 := Q_2(U_2)$ in a similar way, for X_2 is, generally speaking, *dependent* of X_1. What we can do is to find the *conditional DF* $F_2(x_2|x_1)$ of X_2 given $X_1 = x_1$, invert it as a function of x_2 to get the conditional quantile function $Q_2(u|x_1)$ and then set

$$X_2 := Q_2(U_2|X_1) \tag{10.9}$$

with X_1 defined by (10.8). Next we find the conditional DF $F_3(x_3|x_1, x_2)$ of X_3 given that $X_1 = x_1$ and $X_2 = x_2$, invert it as a function of x_3 to get the conditional quantile function $Q_3(u|x_1, x_2)$ and then set

$$X_3 := Q_3(U_3|X_1, X_2)$$

with X_1 and X_2 already set by (10.8) and (10.9), and so on.

Example 10.8. *Simulating a Markov chain.* We will illustrate the above approach by showing how to simulate an MC. Let $\{X_n\}$ be a (homogeneous) MC with state space $\{1, 2, \dots\}$ and transition matrix $P = (p_{jk})$. Similarly to (10.7), put

$$u_{jk} := \sum_{m \le k} p_{jm}, \qquad Q_2(u|\, j) := k \quad \text{for} \quad u \in (u_{j,k-1}, u_{jk}].$$

We have to start with simulating the initial value X_0—and one can do that by simply following the algorithm for discrete distributions. Next note that, due to the Markov property,

$$F_n(x_n|\, x_{n-1}, x_{n-2}, \dots, x_1) = F_2(x_n|\, x_{n-1}),$$

and hence we only need to know Q_2 to simulate our MC step by step. We simulate a sequence of i.i.d. $U_j \sim U(0, 1)$ and then set

$$X_1 := Q_2(U_1|\, X_0), \quad X_2 := Q_2(U_2|\, X_1), \quad \dots.$$

In the general case, when using the inverse function method is impractical, one of the popular alternatives is the so-called *rejection method.*

10.3 The Rejection Method

When F is an absolutely continuous DF but cannot be easily inverted, one can use the (acceptance-) **rejection method**, which is based on the following

Theorem 10.2. *Let f be the density of a DF F,*

$$A := \{(x, y) \in \mathbf{R}^2 : -\infty < x < \infty,\, 0 \le y \le f(x)\} \tag{10.10}$$

and (X, Y) be a random vector which is uniformly distributed over A. Then the first component of the vector (X, Y) follows the desired distribution F, while the conditional distribution of the second component given the first one equals x is uniform on $(0, f(x))$:

$$X \sim F, \qquad \mathbf{P}(Y \leq y | X = x) = y/f(x), \quad y \in (0, f(x)).$$

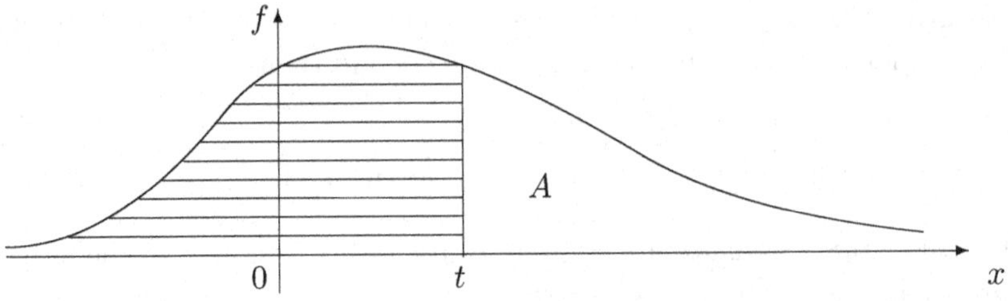

Fig. 10.1 Illustration to Theorem 10.2.

Proof First note that the expression "uniform distribution on A" does make sense, for the total area of A is finite: it is clearly given by the integral $\int_{-\infty}^{\infty} f(x)\,dx = 1$. Therefore, this condition actually means that the probability of the vector to be in a given (measurable) subset $B \subset A$ is nothing else but the area of B. Hence the DF of the first component is

$$\mathbf{P}(X \leq t) = \text{area of } (A \cap \{(x, y) : x \leq t\}) = \int_{-\infty}^{t} f(x)\,dx = F(t),$$

see Fig. 10.1 ($F(t)$ gives the area of the dashed part of A).

The second half of the assertion is almost obvious. Indeed, recall that, for any fixed set $C \subset A$, the conditional distribution of our vector (X, Y) given $(X, Y) \in C$ will be uniform on C: for $B \subseteq C$,

$$\begin{aligned}\mathbf{P}((X, Y) \in B \,|\, (X, Y) \in C) &= \frac{\mathbf{P}((X, Y) \in B)}{\mathbf{P}((X, Y) \in C)} \\ &= \frac{\text{area of } B/\text{area of } A}{\text{area of } C/\text{area of } A} = \frac{\text{area of } B}{\text{area of } C}. \end{aligned} \tag{10.11}$$

Assuming for simplicity's sake that f is continuous at x and $f(x) > 0$, we note that $\sup_{u \in [x-\varepsilon, x+\varepsilon]} |f(u)/f(x) - 1| \to 0$ as $\varepsilon \to 0$. That implies that the area of the figure

$$C := \{(u, v) \in \mathbf{R}^2 : |u - x| \leq \varepsilon, 0 \leq v \leq f(u)\}$$

is equal to $(1+o(1))f(x)\times 2\varepsilon$. Hence, for $y < f(x)$ and $B := \{(u,v) \in \mathbf{R}^2 : |u-x| \le \varepsilon,\ 0 \le v \le y\}$, relations (2.79) and (10.11) yield

$$\mathbf{P}(Y \le y \,|\, X = x) = \lim_{\varepsilon\to 0} \frac{\mathbf{P}(Y \le y, |X-x| \le \varepsilon)}{\mathbf{P}(|X-x| \le \varepsilon)} = \lim_{\varepsilon\to 0} \frac{y \times 2\varepsilon}{(1+o(1))f(x)\times 2\varepsilon} = \frac{y}{f(x)}.$$

□

The above observation also indicates how to generate a random vector (X,Y) **uniformly distributed** over A. Indeed, if we had a random vector (V,W) uniformly distributed over some $B \supset A$, its conditional distribution given $(V,W) \in A$ would be uniform on A. The idea of the *rejection method* is to somehow produce a sample of random vectors uniformly distributed over B and accept only those of them that hit A (i.e., reject all the simulated points falling outside A).

So one first looks for an "envelope" $B \supset A$—a set B which would be "nice" in the sense that there exists an effective algorithm for simulating independent random vectors $(V_1,W_1),(V_2,W_2),\ldots$ uniformly distributed over the set B (often it is a rectangle or a combination of rectangles and/or triangles).

The **rejection algorithm** for generating an RV $X \sim F$ works then as follows:

Algorithm (acceptance-rejection).

1. Simulate an independent copy of (V,W) uniformly distributed over B.
2. If $(V,W) \notin A$ [i.e., $W \notin (0, f(V))$], then go to Step 1.
3. Set $(X,Y) := (V,W)$ and return X. Stop.

The efficiency of the method is determined by the proportion of accepted points in the original sample of the (V_j,W_j)'s, which is equal to the ratio

$$\frac{\text{area of } A}{\text{area of } B} \equiv \frac{1}{\text{area of } B}.$$

So one always tries to choose the set B as small (i.e., as close to A) as possible to maximize the efficiency of the algorithm.

An important popular variant of the rejection method is illustrated by the following

Example 10.9. Let $g(x)$ be a probability density function such that, for some $c > 0$,

$$f(x) \le cg(x) \quad \text{for all } x$$

(note that always $c \geq 1$—why?), and we know how to simulate an RV Z having the density g. We can take

$$B := \{(x, y) : -\infty < x < \infty, 0 \leq y \leq cg(x)\};$$

clearly, $A \subset B$. Now to get a random vector (V, W) uniformly distributed over B we proceed as follows:

1. Simulate an RV $Z \sim$ density g and an independent RV $U \sim U(0, 1)$.
2. Put $V := Z$, $W := cg(V)U$ [so that $W \sim U(0, cg(v))$ given $V = v$].

Then it follows from Theorem 10.2 that (V, W) will have the desired uniform distribution over B, and we can use the above rejection algorithm to generate $X \sim F$.

The efficiency of the rejection algorithm based on this approach is clearly determined by the value of c: the proportion of accepted points equals $1/c$.

To conclude this section, note that the rejection method can work in the multidimensional case $\boldsymbol{X} \in \mathbf{R}^k$ as well. However, the main technical problem of simulating a random vector uniformly distributed in the $(k+1)$-dimensional analog of (10.10)—the set

$$A = \{(\boldsymbol{x}, y) \in \mathbf{R}^{k+1} : 0 \leq y \leq f(\boldsymbol{x})\},$$

f being the density of $\boldsymbol{X}$—is usually a rather difficult task. The higher the dimensionality k, the lower the efficiency of the rejection method: typically, the proportion of the volume occupied by A in the respective "envelope" B (e.g., a parallelepiped) is extremely low even for moderate k, and when one generates points inside that envelope, almost all of them will be rejected. Yet there are alternative more efficient methods, e.g., based on running (for sufficiently long time) a specially constructed MC (or a continuous time MP) whose stationary distribution is uniform over A or B (the so-called *Markov chain Monte Carlo*, to be discussed briefly in Section 10.5).

10.4 Monte Carlo. Variance Reduction Methods

The purpose of Monte Carlo methods is to estimate unknown quantities of interest—integrals (in particular, expectations of RVs and volumes), parameters of complex systems *etc.*—by simulating random processes.

Suppose we want to find (or, at least, estimate) the value of an unknown quantity (parameter) θ. For that purpose, one constructs an estimator $\hat{\theta}$ which is a function of (simulated) RVs and hence an RV itself.

One says that $\hat{\theta}$ is an *unbiased* estimator of θ if $\mathbf{E}\,\hat{\theta} = \theta$. This is a highly desirable property. The precision of the estimator $\hat{\theta}$ is usually measured by its variance $\mathrm{Var}\,(\hat{\theta})$. The lower the variance, the better the estimator.

One of the most common and important problems requiring the use of Monte Carlo methods is the evaluation of (multidimensional) integrals

$$\theta = \int H(\boldsymbol{x})\, d\boldsymbol{x}$$

of some real-valued functions $H(\boldsymbol{x})$, $\boldsymbol{x} \in \mathbf{R}^m$, the integral being over the whole space $\mathbf{R}^m$. On can always choose a density $f(\boldsymbol{x})$ on $\mathbf{R}^m$ such that, for $h(\boldsymbol{x}) := H(\boldsymbol{x})/f(\boldsymbol{x})$, the parameter of interest can be expressed as the expectation

$$\theta = \mathbf{E}\, h(\boldsymbol{X}) = \int h(\boldsymbol{x}) f(\boldsymbol{x})\, d\boldsymbol{x} \tag{10.12}$$

for an RV $\boldsymbol{X} \in \mathbf{R}^m$ following the density f (one must only ensure that if $|H(\boldsymbol{x})| > 0$ then also $f(\boldsymbol{x}) > 0$).

10.4.1 *The Crude Monte Carlo*

The simplest way to estimate θ is to simulate n i.i.d. copies $\boldsymbol{X}_1, \ldots, \boldsymbol{X}_n$ of $\boldsymbol{X}$ and take

$$\hat{\theta} := \frac{1}{n} \sum_{j=1}^{n} h(\boldsymbol{X}_j). \tag{10.13}$$

Clearly, $\mathbf{E}\,\hat{\theta} = \theta$, while by independence

$$\mathrm{Var}\,(\hat{\theta}) = \frac{1}{n^2} \sum_{j=1}^{n} \mathrm{Var}\,(h(\boldsymbol{X}_j)) = \frac{1}{n} \mathrm{Var}\,(h(\boldsymbol{X})), \tag{10.14}$$

so that a $\hat{\theta}$ is an unbiased estimator of θ and its "typical error" is given by its standard deviation $\sqrt{\mathrm{Var}\,(\hat{\theta})} = \sqrt{\mathrm{Var}\,(h(\boldsymbol{X}))}/\sqrt{n}$.

Due to the square root, the accuracy of such an estimator is rather low, and to get an acceptable estimate, one may need to generate an extremely large number of the $\boldsymbol{X}$'s. However, there exist several ways of reducing the error and hence the required number of simulated RVs to achieve prescribed precision.

10.4.2 *The Stratified Sample Method*

Suppose that, for some partition $A_1, \dots, A_k$ of the domain of integration, the values of the integrand $h(\boldsymbol{x})$ *vary insignificantly* within each A_i, but can be substantially different for different A_i's. In this case one can reduce the variance of $\hat{\theta}$ by *stratifying* the sample:

Instead of n independent copies of $\boldsymbol{X}$, we take another collection of n independent RVs. For each $i = 1, \dots, k$, we get n_i (with $n_1 + \cdots + n_k = n$) independent realisations $\boldsymbol{X}_j^{(i)}$, $j = 1, \dots, n_i$, of the RV $\boldsymbol{X}^{(i)}$ distributed according to the conditional distribution of $\boldsymbol{X}$ given that $\boldsymbol{X} \in A_i$. Then, putting $p_i := \mathbf{P}(\boldsymbol{X} \in A_i)$, we construct the estimator

$$\hat{\theta}_s = \sum_{i=1}^{k} p_i \left[\frac{1}{n_i} \sum_{j=1}^{n_i} h(\boldsymbol{X}_j^{(i)}) \right]. \qquad (10.15)$$

The expectations of the expressions in the brackets are clearly equal to $\mathbf{E}(h(\boldsymbol{X})|\boldsymbol{X} \in A_i)$, $i = 1, \dots, k$, respectively, so that, by the TPF,

$$\mathbf{E}\,\hat{\theta}_s = \sum_{i=1}^{k} p_i \mathbf{E}(h(\boldsymbol{X})|\boldsymbol{X} \in A_i) = \mathbf{E}\,h(\boldsymbol{X}) = \theta,$$

which means that $\hat{\theta}_s$ is unbiased for θ.

The variance of the estimator $\hat{\theta}_s$, due to the independence of all $\boldsymbol{X}_j^{(i)}$'s, is equal to

$$\mathrm{Var}(\hat{\theta}_s) = \sum_{i=1}^{k} p_i^2 \mathrm{Var}\left[\frac{1}{n_i} \sum_{j=1}^{n_i} h(\boldsymbol{X}_j^{(i)}) \right] = \sum_{i=1}^{k} \frac{p_i^2}{n_i} \mathrm{Var}(h(\boldsymbol{X}^{(i)})), \qquad (10.16)$$

which can be much *lower* than that of the crude Monte Carlo estimator due to the small variances of the conditional distributions (this will be the case when the variation of $h(\boldsymbol{x})$ is small on each A_i). If we choose $n_i := p_i n$ then it is not hard to show that it is always the case that

$$\mathrm{Var}(\hat{\theta}_s) = \frac{1}{n} \sum_{i=1}^{k} p_i \mathrm{Var}(h(\boldsymbol{X}^{(i)})) \le \frac{1}{n} \mathrm{Var}(h(\boldsymbol{X})) = \mathrm{Var}(\hat{\theta}),$$

the last relation is due to (10.14). This assertion follows from Problem 17.

Example 10.10. When the variation of $h(\boldsymbol{x})$ on each of the A_i's is nil, the stratified sample estimator gives the exact value of θ. This is obvious from (10.16), but can also be directly seen from (10.15). Indeed, if $h(\boldsymbol{x}) \equiv h_i =$ const for $\boldsymbol{x} \in A_i$, $i = 1, \dots, k$, then, for any choice of $n_i \ge 1$,

$$\hat{\theta}_s = \sum_{i=1}^{k} p_i h_i = \mathbf{E}\,h(\boldsymbol{X}) = \theta.$$

Example 10.11. Suppose $X \sim U(0,1)$ and $h(x) \equiv x$, so that

$$\theta = \mathbf{E}\, h(X) = \int_0^1 x\, dx = 1/2.$$

The variance of the crude Monte Carlo estimator $\hat{\theta} = \frac{1}{n}\sum_{j=1}^n X_j$ is equal to

$$\mathrm{Var}\,(\hat{\theta}) = \frac{1}{n}\mathrm{Var}\,(h(X)) = \frac{1}{12n}.$$

Now take $A_i := ((i-1)/k, i/k]$ (so that the variation of $h(x)$ on A_i is $1/k$ only), $p_i := 1/k$ and $n_i := m$, $i = 1, \ldots, k$. Then $X_j^{(i)} \sim U((i-1)/k, i/k)$ with $\mathrm{Var}\,(X_j^{(i)}) = 1/12k^2$, and the variance of the stratified sample estimator for the total sample size $n = mk$ is

$$\mathrm{Var}\,(\hat{\theta}_s) = \sum_{i=1}^k \frac{(1/k)^2}{m} \times \frac{1}{12k^2} = \frac{1}{12k^2 n}.$$

That is, the standard error of $\hat{\theta}_s$ is k times less than that of $\hat{\theta}$!

10.4.3 *The Antithetic Variables Method*

The idea is to use negatively correlated (*antithetic*[9]) copies of $\boldsymbol{X}$ instead of independent ones. This leads to a sort of "more balanced" filling the integration domain with sampled points. As a consequence, the variation of the estimator value tend to be lower.

Before giving a more meaningful example, consider the following simple situation. Let, as in Example 10.11, $h(x) \equiv x$ and $X \sim U(0,1)$, so that $\theta = 1/2$. Take $n = 2$. Then the crude Monte Carlo estimator is

$$\hat{\theta} := \frac{1}{2}(X_1 + X_2) \quad \text{with} \quad \mathrm{Var}\,(\hat{\theta}) = 1/24.$$

However, one could expect that the RVs X_1 and $1 - X_1$ would be "more evenly spread" over $(0,1)$ than the original independent copies of X; note that

$$\mathrm{Cov}(X_1, 1 - X_1) = \mathbf{E}\,(X_1 - 1/2)(1/2 - X_1) = -\mathrm{Var}\,(X_1) = -1/12 < 0$$

(in fact, the correlation between these RVs equals -1). And if we use the "sample" $(X_1, 1 - X_1)$ instead of (X_1, X_2) in our $\hat{\theta}$, the new estimate will simply be

$$\hat{\theta}_a := \frac{1}{2}(X_1 + (1 - X_1)) = \frac{1}{2} = \theta$$

[9]Being in diametrical opposition, from Greek *antitithenai*, to set against, to oppose.

(with zero variance!).

Example 10.12. Estimate the value of

$$\theta = \mathbf{E}\, e^X = \int_0^1 e^x\, dx, \quad X \sim U(0,1)$$

(of course, one can easily compute the value of the integral; this is just a model example).

The crude Monte Carlo, $n = 2$: simulate two independent RVs $X_1, X_2 \sim U(0,1)$, and then set $\hat{\theta} := (e^{X_1} + e^{X_2})/2$. Clearly, $\mathbf{E}\,\hat{\theta} = \theta$,

$$\begin{aligned}\mathrm{Var}\,(\hat{\theta}) &= \frac{1}{2}\mathrm{Var}\,(e^X) = \frac{1}{2}\Big[\mathbf{E}\, e^{2X} - (\mathbf{E}\, e^X)^2\Big] \\ &= \frac{1}{2}\left[\int_0^1 e^{2x} dx - \left(\int_0^1 e^x dx\right)^2\right] = \frac{1}{2}\left[\frac{e^2-1}{2} - (e-1)^2\right] \approx 0.1210.\end{aligned}$$

The antithetic variables method: take X and $1 - X$ (both are $U(0,1)$-distributed and clearly negatively correlated). Since e^x is a monotone function, e^X and e^{1-X} will also be negatively correlated. In fact,

$$\mathrm{Cov}(e^X, e^{1-X}) = e - (\mathbf{E}\, e^X)^2 = e - (e-1)^2 \approx -0.2342.$$

Clearly, for $\hat{\theta}_a := (e^X + e^{1-X})/2$ we also have $\mathbf{E}\,\hat{\theta}_a = \theta$, but

$$\mathrm{Var}\,(\hat{\theta}_a) = \frac{1}{2}\left[\mathrm{Var}\,(e^X) + \mathrm{Cov}(e^X, e^{1-X})\right] \approx 0.0039.$$

More generally, if $n = 2m$ and $\hat{\theta}$ is the crude Monte Carlo estimator constructed from $X_1, \ldots, X_n$, while

$$\hat{\theta}_a := \frac{1}{n}\sum_{j=1}^{m}\left(e^{X_j} + e^{1-X_j}\right)$$

uses the (same number n of) antithetic variables, we have the same ratio for the variances as in the case $n = 2$:

$$\mathrm{Var}\,(\hat{\theta}_a) \approx \frac{1}{31}\mathrm{Var}\,(\hat{\theta}).$$

A simple idea which, nevertheless, can be very efficient!

10.4.4 *The Importance Sampling Method*

We want to estimate integral (10.12). Suppose that we can simulate RVs $\boldsymbol{Y}$ following another density $g(\boldsymbol{x})$ with the property[10] that $g(\boldsymbol{x}) > 0$ whenever

[10] Recall that if the property holds, one says that the distribution of $\boldsymbol{X}$ is *absolutely continuous* w.r.t. that of $\boldsymbol{Y}$, see the footnote on p. 25.

$f(\boldsymbol{x}) > 0$ (that is, if $\mathbf{P}(\boldsymbol{X} \in A) > 0$ for some A, then $\mathbf{P}(\boldsymbol{Y} \in A) > 0$ also; in other words, $\boldsymbol{Y}$ can "visit" the same sets as $\boldsymbol{X}$).

Then one can estimate the value

$$\theta \equiv \mathbf{E}\, h(X) = \int h(\boldsymbol{x}) f(\boldsymbol{x})\, d\boldsymbol{x} = \int \frac{hf}{g} g\, d\boldsymbol{x} = \mathbf{E}\, \frac{hf}{g}(\boldsymbol{Y})$$

using a sample of i.i.d. RVs $\boldsymbol{Y}_1, \ldots, \boldsymbol{Y}_n$ having the density g:

$$\hat{\theta}_I := \frac{1}{n} \sum_{j=1}^{n} \frac{hf}{g}(\boldsymbol{Y}_j), \tag{10.17}$$

the estimator being clearly unbiased for θ.

When is using this estimator better than just using the crude Monte Carlo estimator (10.13)?

The variance of $\hat{\theta}_I$ is

$$\mathrm{Var}\,(\hat{\theta}_I) = \frac{1}{n} \mathrm{Var}\left(\frac{hf}{g}(\boldsymbol{Y}) \right).$$

Suppose that you can choose g so as to "mimic" the function hf, i.e., hf/g is almost *constant*. Then the variance of $\hat{\theta}_I$ will be small! The "best choice" would be to take

$$g(\boldsymbol{x}) := C h(\boldsymbol{x}) f(\boldsymbol{x}) \tag{10.18}$$

for some normalizing constant C. This, however, would not work, for the value of that C is just $1/\theta$ since

$$1 = \int g(\boldsymbol{x})\, d\boldsymbol{x} = \int C h(\boldsymbol{x}) f(\boldsymbol{x})\, d\boldsymbol{x} = C\theta,$$

so that the summands in the estimator $\hat{\theta}_I$ from (10.17) are all equal to $(hf/g)(\boldsymbol{Y}) \equiv \theta$. But it is our task to find θ, we cannot use its unknown value!

Anyway, when $g(\boldsymbol{x})$ is small where $H(\boldsymbol{x}) = h(\boldsymbol{x}) f(\boldsymbol{x})$ is small, and large where the latter is large, there is a substantial gain in the precision of the estimates.

Example 10.13. In risk theory, one often deals with the tail probabilities of the form

$$\theta := \mathbf{P}\,(S_m \geq a) \equiv \mathbf{E}\, h(\boldsymbol{X}), \quad h(\boldsymbol{x}) := \mathbf{1}_{\{x_1 + \cdots + x_m \geq a\}},$$

for the sums $S_m = X_1 + \cdots + X_m$ of i.i.d. RVs X_j forming the vector $\boldsymbol{X} = (X_1, \ldots, X_m)$, when m and a are large. If the value $a - \mathbf{E}\, S_m = a - m\mathbf{E}\, X_1$ is much greater than the standard deviation $\sqrt{\mathrm{Var}\,(S_m)} = m^{1/2}\sqrt{\mathrm{Var}\,(X_1)}$, the CLT only says that the value $\theta = \mathbf{P}(S_m \geq a)$ is **small**, but usually cannot be used to estimate θ (in that case one has to use the so-called *large deviations* theory).

The crude Monte Carlo is extremely inefficient here: since $h(\boldsymbol{X})$ is an indicator RV, the variance of the estimator (10.13) constructed using a sample of n independent copies of $\boldsymbol{X}$ is

$$\mathrm{Var}\,(\hat{\theta}) = \frac{1}{n}\mathbf{E}\,(h(\boldsymbol{X}) - \theta)^2 = \frac{\theta(1-\theta)}{n},$$

so that the *relative error* (which is of importance now since the value of θ is small)

$$\frac{\mathrm{Var}\,(\hat{\theta})}{(\mathbf{E}\,\hat{\theta})^2} = \frac{1-\theta}{n\theta}$$

can be rather large (when $n\theta$ is small).

In that case one can use the importance sampling method and take a sample of $\boldsymbol{Y}$'s for which the event of interest will be rather likely. Suppose that $X = X_1$ has a density $p(x)$ such that

$$\varphi_X(t) := \mathbf{E}\,e^{tX} < \infty$$

for all values t from a sufficiently large interval. Then we can define the density

$$p_t(x) := \frac{e^{tx}}{\varphi_X(t)}p(x), \qquad -\infty < x < \infty$$

(note that since $\int p_t(x)\,dx = \mathbf{E}\,e^{tX}/\varphi_X(t) = 1$, this is a density indeed).

The random vector $\boldsymbol{Y} = (Y_1, \ldots, Y_m)$ of i.i.d. Y's having the density p_t has, by independence, a density which is equal to the product of the components' densities (see (2.36)):

$$g_t(\boldsymbol{x}) := \prod_{i=1}^{m} p_t(x_i) = \frac{e^{t(x_1+\cdots+x_m)}}{\varphi_X^m(t)}\prod_{i=1}^{m} p(x_i) = \frac{e^{ts}}{\varphi_X^m(t)}f(\boldsymbol{x}),$$

where $s \equiv s(\boldsymbol{x}) := x_1 + \cdots + x_m$ and $f(\boldsymbol{x}) = \prod_{i=1}^m p(x_i)$ is the density of the vector $\boldsymbol{X}$. This g_t tends to be large when our $h = 1$ (s is large and hence e^{ts} is very large) and small when $h = 0$ (then s is small).

Now the function appearing in the sum in (10.17) satisfies the inequality

$$\frac{hf}{g_t}(\boldsymbol{x}) = \frac{\mathbf{1}_{\{s\geq a\}}\varphi_X^m(t)}{e^{ts}} \leq \frac{\varphi_X^m(t)}{e^{ta}}$$

(and the bound on the right-hand side is attained at $s = a$). To get a small value of the variance of $(hf/g_t)(\boldsymbol{Y})$, one has to keep this maximum value of that bound as small as possible. This can be done by choosing an appropriate value of t. One can show that this minimum value of the bound is attained at the point $t = t^*$ which is the solution to the equation

$$m\mathbf{E}_t Y \equiv m \int y p_t(y)\, dy = a, \tag{10.19}$$

see Problem 18.

Numerical illustration. Let $m := 20$, X_j be i.i.d. Bernoulli RVs with success probability $p = 0.4$, and $a = 16$, $\theta = \mathbf{P}(S_m \geq a)$. Compare the variances of the crude Monte Carlo and importance sampling estimators for $n = 1$. (In the general case of sample size n, the variances of the respective estimators will simply be n times smaller.)

The crude Monte Carlo: the estimator $\hat{\theta} := \mathbf{1}_{\{S_m \geq a\}}$ can only take two values (0 and 1), the true value being $\theta \approx 3.17 \times 10^{-4}$, so that the variance equals

$$\mathrm{Var}(\hat{\theta}) \equiv \theta(1-\theta) \approx 3.17 \times 10^{-4}. \tag{10.20}$$

Importance sampling: the Y_j's have the distribution

$$\mathbf{P}(Y_j = x) = \frac{e^{tx}}{\varphi_X(t)} \mathbf{P}(X_j = x), \quad x = 0, 1,$$

where $\varphi_X(t) = pe^t + 1 - p$, so that the Y_j's are also Bernoulli RVs, but with success probabilities

$$\frac{e^t p}{e^t p + 1 - p}.$$

Now equality (10.19) becomes

$$m\mathbf{E}_t Y = 20 \frac{e^t p}{e^t p + 1 - p} = 20 \frac{0.4e^t}{0.4e^t + 0.6} = 16.$$

Solving for e^t, we get $e^{t^*} = 6$ and $\varphi_X(t^*) = 3$, so that the Y_j's are Bernoulli RVs with success probabilities

$$\frac{0.4e^{t^*}}{0.4e^{t^*} + 0.6} = 0.8.$$

The importance sampling estimator is now

$$\hat{\theta}_I = \mathbf{1}_{\{S_m \geq a\}} e^{-t^* S_m} \varphi_X^m(t^*) = \mathbf{1}_{\{S_m \geq a\}} 3^{20} 6^{-S_m} \leq 3^{20} 6^{-16} \approx 0.0012.$$

So the values of this estimator are between 0 and 0.0012 only (whereas the crude Monte Carlo assumes the values 0 and 1!), and hence the variance must be much lower. One can show that $\mathrm{Var}(\hat{\theta}_I) < 2.92 \times 10^{-7}$, which is much smaller than (10.20).

10.5 Markov Chain Monte Carlo

Quite often it is very hard to simulate a random vector whose components are dependent. There is no satisfactory exact method (producing random vectors distributed *exactly* according to the desired distribution) in the general case. To solve this problem, one can use a general powerful approach for generating random vectors whose distributions would be approximately of the desired form. The idea is to construct and run for some time an ergodic MC (or a continuous time MP) whose stationary distribution coincides with the required one. These techniques have greatly expanded the use of simulation over the recent years.

It can be best explained in the discrete case. Suppose we want to simulate an RV X with distribution

$$\mathbf{P}(X = j) = \pi_j, \quad j = 1, \ldots, N.$$

Suppose further that we have at our disposal an irreducible aperiodic $N \times N$ transition matrix (q_{ij}) with the following property: if $q_{ij} > 0$, then $q_{ji} > 0$, too. The so-called *Metropolis algorithm* can modify this matrix in such a way that the MC evolving according to the modified transition matrix will have the desired stationary distribution (π_j).

To explain how the Metropolis algorithm works, we will begin by noting the following important fact: if, for some stochastic matrix (p_{ij}),

$$\pi_i p_{ij} = \pi_j p_{ji}, \quad i, j = 1, \ldots, N, \tag{10.21}$$

then the MC with the transition matrix (p_{ij}) has the stationary distribution π_j, $j = 1, \ldots, N$. Indeed, summing up equations (10.21) over all i values, we get

$$\sum_i \pi_i p_{ij} = \sum_i \pi_j p_{ji} = \pi_j \sum_i p_{ji} = \pi_j,$$

since the row sums of (p_{ij}) are all ones.

Now recall that we have got an irreducible stochastic matrix (q_{ij}). Assume we have also chosen a collection of numbers $a_{ij} \in [0, 1]$, $i, j = 1, \ldots, N$, (how to choose them will be specified below). The modified MC $\{X_n\}$ evolves as follows: if $X_n = i$, we move to state j with probability q_{ij} and then decide to stay there with probability a_{ij}. With the complementary probability $1 - a_{ij}$ we immediately return back to i. So the new one-step transition probabilities are:

$$p_{ij} = q_{ij} a_{ij}, \quad j \neq i;$$
$$p_{ii} = q_{ii} + \sum_j q_{ij}(1 - a_{ij}).$$

As we saw above, this new MC has the desired stationary distribution (π_j) if (10.21) holds, which is equivalent to

$$\pi_i q_{ij} a_{ij} = \pi_j q_{ji} a_{ji},$$

which, in its turn, can be re-written as

$$\frac{a_{ij}}{a_{ji}} = \frac{\pi_j q_{ji}}{\pi_i q_{ij}}. \tag{10.22}$$

Recall that always $a_{ij} \le 1$ (they are probabilities!). So if the right-hand side (RHS) of (10.22) is > 1, we set

$$a_{ij} := 1, \quad a_{ji} := 1/\text{RHS};$$

otherwise

$$a_{ij} := \text{RHS}, \quad a_{ji} := 1.$$

In other words, we have to put

$$a_{ij} := \min\left\{1, \frac{\pi_j q_{ji}}{\pi_i q_{ij}}\right\}, \quad i, j = 1, \ldots, N. \tag{10.23}$$

Since the original transition matrix was irreducible, so is the modified one. This implies that the modified MC $\{X_n\}$ is ergodic by Corollary 3.2 (we cannot have periodicity in it—why?) and the distribution of X_n will converge to (π_j) as $n \to \infty$.

Remark 10.3. Observe that, to get an MC whose stationary probabilities are π_j, we do not even need to know the π_j's *exactly*! From (10.23) it follows that it would suffice to know the values c_j such that, for some normalizing constant C,

$$\pi_j = \frac{c_j}{C}, \quad j = 1, \ldots, N$$

(recall the formula for the stationary distribution of a closed queueing network: the c_j's are of a known product form, while computing the value of the normalising factor C is a hard problem).

Now once we know that the modified MC is ergodic and has the desired stationary distribution (π_j), we can estimate $\theta = \mathbf{E}\, h(X)$ by running our MC for some (sufficiently long) time and taking the average

$$\hat{\theta} = \frac{1}{n} \sum_{j=1}^{n} h(X_j).$$

Since the early states of the MC can be strongly influenced by the initial state, it is common practice to disregard several first steps.

An important special case of the MC Monte Carlo approach is the **simulated annealing** algorithm of which the main purpose is to find maxima (or minima) of functions. Here is how it works.

Let S be a finite set, $V(\boldsymbol{x})$ a function on it, and suppose that we want to maximise $V(\boldsymbol{x})$, $\boldsymbol{x} \in S$. Choose a $\lambda > 0$ and consider the following distribution on S:

$$p_\lambda(\boldsymbol{x}) := \frac{e^{\lambda V(\boldsymbol{x})}}{\sum_{\boldsymbol{y} \in S} e^{\lambda V(\boldsymbol{y})}}, \quad \boldsymbol{x} \in S. \tag{10.24}$$

For large λ, this distribution will assign almost all the probability to the points $\boldsymbol{x}$ where $V(\boldsymbol{x})$ is close to its maximum value! So if we run an MC with the stationary distribution (10.24) (for sufficiently large λ) using the Metropolis algorithm, it will eventually enter (and stay for a long time in) the class of sets $\boldsymbol{x}$ with the values of $V(\boldsymbol{x})$ close to the maximum one.

Annealing means that one actually runs a time-inhomogeneous MC letting $\lambda \to \infty$ (slowly enough to escape "traps" = local maxima) as time passes.

Example 10.14. *The travelling salesman problem.* Suppose there are r cities in a state, connected with each other by a road network with known distances for all the links. A salesman has to visit all of the cities enumerated by $1, 2, \dots, r$. The order in which the cities are visited is given by a *permutation* $\boldsymbol{x} = (x_1, x_2, \dots, x_r)$ of $(1, \dots, r)$, and we can take S to be the set of all such permutations (which is a **huge** set even for moderate r: $|S| = r!$). The problem is to maximise (in $\boldsymbol{x}$) the total "return"

$$V(\boldsymbol{x}) := \sum_{i=1}^{r} v(x_{i-1}, x_i)$$

for some function $v(\cdot, \cdot)$ (in particular, if $v(i, j) = -$ distance between the cities i and j, then one minimizes the total distance covered by the travelling salesman).

This famous (NP-hard) problem is very hard to solve.[11] Roughly speaking, exact solution can only be found using enumerative algorithms which is impractical. Simulated annealing can give a "reasonable" solution in reasonable time.

[11] For more thrill, see `http://www.travellingsalesmanmovie.com/`.

10.6 Recommended Literature

DEVROYE, L. *Non-uniform random variate generation.* Springer, New York. 1986. [*A voluminous text on various techniques of generating RVs with various distributions.*]

FISHMAN, G.S. *Monte Carlo: concepts, algorithms, and applications.* Springer, New York, 1996.

GENTLE, J.E. *Random number generation and Monte Carlo methods.* Springer, New York, 1998. [*A good text on random number generators, with discussion of methods for simulating from selected popular distributions.*]

MOESCHLIN, O., GRYCKO, E., POHL, C., AND STEINERT, F. *Experimental stochastics.* Springer, New York, 1998. [*Includes a CD-ROM with programs illustrating random number generators etc.* Version 2.0 edition appeared in 2003.]

ROSS, S. *Simulation.* 5th edn. Academic Press, San Diego, 2012.

10.7 Problems

1. Verify (10.1).

2. Prove that if an RV X has a continuous DF F, then $F(X) \sim U(0,1)$ (an assertion converse to Theorem 10.1). Explain why this is not true when F is not continuous.

3. Let X be an RV following an arbitrary DF F and $U \sim U(0,1)$ be independent of X. Show that one always has $V := (1-U)F(X-0) + UF(X) \sim U(0,1)$. [Note that if F is continuous then $F(X-0) \equiv F(X)$ and so $V \equiv F(X)$, we are back to the conditions of Problem 2.]

4. Give at least two different proofs of the fact that if $X \sim N(0,1)$, then $Y = \mu + \sigma X \sim N(\mu, \sigma^2)$. [Possible options include the use of ChFs and a density transformation for functions of RVs.]

5. The *Pareto distribution* with parameters $a, b > 0$ has the density

$$f_{a,b}(x) = \begin{cases} C_{a,b}x^{-a-1} & \text{if } x \geq b, \\ 0 & \text{otherwise.} \end{cases} \tag{10.25}$$

This distribution has been used to model the distribution of income, city population size, size of firms *etc.*

(i) Find the value of the normalising constant $C_{a,b}$ (for which (10.25) defines a probability density function). Sketch the graphs of $f_{1,1}$ and $f_{2,1}$.

(ii) Find the DF $F_{a,b}$ having the density $f_{a,b}$. Sketch the graphs of the DFs $F_{1,1}$ and $F_{2,1}$.

(iii) For an RV $X \sim F_{a,b}$, obtain an explicit expression for $\mathbf{P}(x \le X \le y)$ when a) $x < b < y$ and b) $b < x < y$.

(iv) Give an explicit description of the inversion method algorithm for simulating a random sample from $F_{a,b}$ (containing an explicit formula for the quantile function Q *etc.*).

(v) We know that $T_1 = 0.6578$ and $T_2 = 2.0137$ are the times of the first two events in a Poisson process with rate $\lambda = 1$. Using these RVs, simulate two independent copies of $X \sim F_{2,3}$.

(vi) Somebody suggested that using the rejection method for simulating RVs $X \sim F_{a,b}$ would be more appropriate than using the inverse function method. Comment on this statement.

6. Let X be an RV distributed according to the *Cauchy density*

$$f(x) = \frac{1}{\pi(1+x^2)}, \qquad -\infty < x < \infty.$$

(i) Find the DF F of X.

(ii) Give an explicit description of the inversion method algorithm for simulating a random sample from F (containing an explicit formula for the quantile function Q *etc.*).

7. Let $U_1, U_2, \dots$ be i.i.d. $U(0,1)$-RVs, X the smallest integer such that

$$\prod_{j=1}^{X+1} U_j < e^{-\lambda}.$$

Find the distribution of X and describe an algorithm for simulating an RV following that distribution which is based on the above fact.

8. Suggest and describe in detail a method for generating a random vector (N_{t_1}, N_{t_2}) of which the components are the values of the Poisson process $\{N_t,\, t \ge 0\}$ with rate λ at times $0 < t_1 < t_2$, respectively.

9. Suppose you have already simulated the values N_{t_j} from the previous problem. For a given $s \in (t_1, t_2)$, describe in detail how to simulate the value of N_s (for the *same* trajectory of the process).

Hint. Cf. Example 2.5.

10. (i) Give an algorithm for simulating the values of the Brownian motion process W_t at times from a "grid" $0 < t_1 < t_2 < \cdots < t_m$.

(ii) Now suppose that, after having done the above simulation, you need to simulate the value W_s (for the same realisation of W!) for some $s > 0$ not belonging to the grid $\{t_j\}$. Describe how to do that.

Hint. Cf. Problem 37 on p.74.

11. In Operations Research, when analysing multistage projects' completion times, one often assumes that the completion time T of a particular stage of a project is an RV having the *beta distribution.*[12] Consider the special case when, for some $0 < A < B$, the density of T is of the form

$$f(t) = 12\left(\frac{t-A}{B-A}\right)^2\left(1-\frac{t-A}{B-A}\right)\frac{1}{B-A}, \qquad t \in (A, B).$$

Describe a rejection method procedure for simulating a sample of n independent identically distributed random variates $T_1, \ldots, T_n$ having the density f.

Hints. First you may wish to note that T can be obtained as a linear transformation $T = A + (B - A)X$, where X has a simpler density g (not depending of A and B). Find this density. It is a bounded function supported by a finite interval (in fact, $g(x) > 0$ iff $x \in (0,1)$). Find an upper bound C for $\max_x g(x)$. Make use of the rejection method's variant based on simulating random points uniformly distributed over rectangles (in our case, over the rectangular $[0,1] \times [0, C]$).

12. Suggest and describe in detail a version of the rejection method for generating a random vector (X_1, X_2) distributed according to a bounded density $f(x_1, x_2) \leq C < \infty$ on the disk $D = \{(x_1, x_2) : x_1^2 + x_2^2 \leq r^2\}$.

13. The *Weibull distribution*[13]

$$F(x) = 1 - e^{-((x-x_0)/a)^c}, \quad x > x_0,$$

where $a, c > 0$ and $x_0 \in \mathbf{R}$ are parameters, is a good model for material strength. Suggest and describe in detail the most suitable (from your point of view) algorithm for simulating Weibull RVs.

14. (i) Apply the inverse function method to generate an observation from the density

$$f_1(x) = \frac{1}{(1+x)^2}, \qquad x > 0,$$

using a (0,1)-uniform (pseudo) random number U. Explain how this can be extended to generate an observation from the symmetrized form of the same

[12]The beta distribution $B(\alpha, \beta)$, $\alpha, \beta > 0$, has density $C_{\alpha,\beta}x^{\alpha-1}(1-x)^{\beta-1}$, $x \in (0,1)$, where $C_{\alpha,\beta} = \frac{\Gamma(\alpha+\beta)}{\Gamma(\alpha)\Gamma(\beta)}$ is the reciprocal of the beta function.

[13]After Ernst Hjalmar Waloddi Weibull (18.08.1887–12.10.1979), a Swedish engineer and scientist. Weibull actually derived the distribution assuming that there is a random Poisson number of cracks in a material specimen, the lengths of the cracks being i.i.d. RVs following a power/Pareto type distribution. He showed, in particular, that the agreement between the empirical data such as yield strength of steel and fiber strength of cotton and the fitted Weibull distributions is very impressing.

density

$$f_2(x) = \frac{1}{2(1+|x|)^2}, \qquad -\infty < x < \infty.$$

(ii) Consider the Cauchy distribution with density

$$f(x|\theta) = \frac{\theta}{\pi(x^2+\theta^2)}, \qquad -\infty < x < \infty,$$

where θ is a positive parameter. Show that

$$f(x|\theta) \le Cf_2(x), \qquad -\infty < x < \infty,$$

so long as $C \ge \frac{2}{\pi}(\theta + \theta^{-1})$. Hence devise a method based on acceptance-rejection sampling for generating observations from the Cauchy distribution.

15. The following is a sample of seven (0,1)-uniform random numbers:

$$0.4537,\ 0.3038,\ 0.5591,\ 0.6154,\ 0.5832,\ 0.8501,\ 0.7700.$$

(i) Use the above random sample and the Box-Muller algorithm to simulate three independent normally distributed random variables with the common mean $\mu = 1$ and standard deviation $\sigma = 2$.

(ii) Use the same sample to simulate an initial segment of the trajectory of an MC $\{X_n\}$ with the state space $\{1,2,3\}$, initial distribution $(0.2, 0.7, 0.1)$ and transition matrix

$$P = \begin{pmatrix} 0.3 & 0.2 & 0.5 \\ 0.4 & 0 & 0.6 \\ 0.3 & 0.5 & 0.2 \end{pmatrix}.$$

(iii) Use the same sample to simulate an initial segment of the trajectory of a birth-and-death process $\{X_t\}$ starting at $X_0 = 4$ and having the birth rates $\lambda_j = 4/(1+j)$, $j = 0, 1, 2, \ldots$, and death rates $\mu_j = 1$, $j = 1, 2, \ldots$ You have enough uniform random numbers to simulate its behaviour till the time of the fourth transition.

16. Suppose that $T \ge 0$ is an RV modelling lifetime, following one of the two following densities: [a] $f_a(x) = 2(1-x)$, $x \in [0,1]$, or [b] $f_b(x) = 0.5\pi \sin(\pi x)$, $x \in [0,1]$. In both cases, using the $U(0,1)$-sample 0.1101, 0.8523, 0.5442, 0.2145, simulate as many independent copies of T as you can employing (i) the inverse function method and (ii) the rejection method.

17. Prove that for any RVs Y and Z, one has $\mathbf{E}\,\mathrm{Var}\,(Z|Y) \le \mathrm{Var}\,(Z)$, where

$$\mathrm{Var}\,(Z|Y) = \mathbf{E}\,((Z - \mathbf{E}\,(Z|Y))^2|Y) = \mathbf{E}\,(Z^2|Y) - (\mathbf{E}\,(Z|Y))^2$$

is the conditional variance of Z given Y.

18. In Example 10.13, show that the minimum $\min_t e^{-ta}\varphi_X^m(t)$ is attained at the value t^* giving the solution to equation (10.19).

Chapter 11

Martingales and Stochastic Calculus

In this chapter we will present the basics of stochastic calculus[1], a part of Probability Theory that deals with constructing integrals with respect to random processes that, generally speaking, would make no "path-wise" sense. For more advanced treatment of stochastic calculus, the reader is referred to the books from the list of recommended literature in Section 11.6.

11.1 Martingales

The *martingale property* is one of the most important concepts in stochastic calculus. Moreover, it proved to be central in the modern theory of mathematical finance to be discussed in Chapter 13. In this section, we will introduce and discuss martingales in the simple case of discrete time processes.

Recall that by a discrete time process we mean a collection of RVs $\{X_t\}_{t\in \boldsymbol{T}}$ on a common probability space $(\Omega, \mathcal{F}, \mathbf{P})$, where the set $\boldsymbol{T}$ of "time" values t is usually either $\{0, 1, 2, \ldots, T\}$ for a $T < \infty$ or $\{0, 1, 2, \ldots\}$ (or even $\{\ldots, -1, 0, 1, \ldots\}$).

It turned out that the most natural in the probabilistic context way of formalising the idea of information available (say, to the observer) by a given time t is to specify the collection $\mathcal{F}_t$ of events "observable" by that time. This leads to the concept of *filtration* $\mathbf{F}$ which is defined as an increasing sequence of sub-σ-fields $\mathcal{F}_0 \subset \mathcal{F}_1 \subset \mathcal{F}_2 \subset \cdots \subset \mathcal{F}$ (which one can interpret as the "history" of the modelled random phenomenon). We

[1]The Latin word *calculus* actually means *pebble* (you might hear it when visiting your dentist: in his/her trade, the word means a form of hardened dental plaque). The word's mathematical meaning is due to use of pebbles in ancient counting mechanisms, such as abacus and Vitruvius' odometer.

say that an SP $\{X_t\}_{t\geq 0}$ is *adapted to filtration* $\mathbf{F}$ if, for any $t = 0, 1, 2, \ldots$, the RV X_t is $\mathcal{F}_t$-measurable, i.e.,

$$\{X_t \in B\} \in \mathcal{F}_t \quad \text{for any } B \in \mathcal{B}.$$

In words, the property means that we know the value of X_t if we know the "history" up to time t (i.e., we know about each of the events from $\mathcal{F}_t$ if it has occurred or not). Observe that, for any (measurable) function f, the process $\{Y_t := f(X_t)\}_{t\geq 0}$ will also be adapted and that, since $\mathcal{F}_t \subset \mathcal{F}_s$, $t \leq s$, it is obvious that X_t will be $\mathcal{F}_s$-measurable for any $s \geq t$ as well.

The quadruplet $(\Omega, \mathcal{F}, \mathbf{F}, \mathbf{P})$ is referred to as a *filtered probability space*, or *stochastic basis.*

Example 11.1. In the so-called *binomial market model*, one assumes that the price S_t of a given asset, of which the initial value S_0 is assumed to be fixed, has the following dynamics: for fixed positive values $d < u$, at each time step, the previous price value is multiplied by d (a "down movement") or by u (an "up movement"):

$$S_{t-1} \begin{array}{l} \nearrow S_t = uS_{t-1}, \\ \\ \searrow S_t = dS_{t-1}, \end{array} \qquad t = 1, 2, \ldots, T.$$

Note that we do not say anything neither about the dependence of the moves nor on their probabilities here.

The natural sample space for that model is the set

$$\Omega := \{\boldsymbol{v} = (v_1 v_2 \ldots v_T) : v_j = d \text{ or } u\}$$

of all strings of length T that consist of letters d and u. We endow Ω with the σ-field 2^Ω of all its subsets.

The "history" of the process is then described by the sequence of sub-σ-fields $\mathcal{F}_t$ generated by the first t moves in the process, respectively. First, we set $\mathcal{F}_0 := \{\emptyset, \Omega\}$ (the trivial σ-field, as no information is available at time $t = 0$). Next, we put

$$\mathcal{F}_1 := \sigma(A_d, A_u) \equiv \{\emptyset, A_d, A_u, \Omega\},$$

the σ-field generated by $A_d := \{\boldsymbol{v} \in \Omega : v_1 = d\}$, $A_u := \{\boldsymbol{v} \in \Omega : v_1 = u\}$ (note that $A_u = A_d^c$). Once we know about each of the events from $\mathcal{F}_1$ whether it occurred or not (which is clearly the same as knowing if A_d occurred), we know the value of S_1. Actually, the converse is also true: once we know the value of S_1, for each event from $\mathcal{F}_1$ we know if it occurred or not.

Then we set

$$\mathcal{F}_2 := \sigma(A_{dd}, A_{du}, A_{ud}, A_{uu}),$$

where $A_{dd} := \{\boldsymbol{v} \in \Omega : v_1 = d, v_2 = d\}$, $A_{du} := \{\boldsymbol{v} \in \Omega : v_1 = d, v_2 = u\}$, *etc.*, so that $\mathcal{F}_2$ will have $2^4 = 16$ elements including, in particular, the unions $A_d = A_{dd} \cup A_{du}$ and $A_u = A_{ud} \cup A_{uu}$ (so that $\mathcal{F}_0 \subset \mathcal{F}_1 \subset \mathcal{F}_2$). It is clear that, once we know about each of the events from $\mathcal{F}_2$ whether it occurred or not, we know the value of S_2 (and also that of S_1). And the other way around, the values of S_1 and S_2 will tell us which of the events from $\mathcal{F}_2$ occurred and which didn't. The σ-fields $\mathcal{F}_3, \mathcal{F}_4, \ldots$ are defined in a similar way, using events A_{ddd}, A_{ddu} *etc.*

It should be obvious that the process $\{S_t\}$ is adapted to the filtration $\mathbf{F} := \{\mathcal{F}_0, \mathcal{F}_1, \ldots, \mathcal{F}_T\}$. Moreover, the information contained in $\mathcal{F}_t$ is exactly the past (up to time t) history of the price process, as $\mathcal{F}_t$ is generated by $(S_0, S_1, \ldots, S_t)$. In such a case, one says that $\mathbf{F}$ is the *natural filtration* of the process $\{S_t\}$.

Note that, in this example, $\{S_t^2\}$ is also adapted to $\mathbf{F}$, but the processes $Y_t := S_{t+1} - S_t$ and $Z_t :=$ the time from t to the first epoch $s > t$ when $S_s > 2S_t$, are not.

Now suppose that, on a filtered probability space $(\Omega, \mathcal{F}, \mathbf{F}, \mathbf{P})$, one is given a discrete time SP $\{X_t\}_{t\geq 0}$ that is adapted to the filtration $\mathbf{F}$. One says that $\{X_t\}$ is a *martingale*[2] (MG) if, for any $t = 0, 1, 2, \ldots$, $\mathbf{E}\,|X_t| < \infty$ and

$$\mathbf{E}\,(X_{t+1}|\mathcal{F}_t) = X_t. \tag{11.1}$$

Sometimes no filtration is explicitly specified. In such cases, it is tacitly assumed that the natural filtration $\mathcal{F}_t = \sigma(X_s, s \leq t)$, $t \geq 0$, is used.

[2]For the origins of the word, see *The Origins of the Word "Martingale"* by R. Mansuy, `http://www.emis.de/journals/JEHPS/juin2009/Mansuy.pdf` (*El. J. Hist. Probab. Stat.*, 5, no. 1, 2009). In gambling, the term means a strategy that can be described as "to always bet all that was lost" (the explanation of the entry *martingale* that first appeared in *Dictionnaire de l'Académie Françoise*, Paris, Veuve Brunet, 4th edn., 1762), although the word itself has a much longer history and appears, e.g., in F. Rabelais' *Gargantua* (1534), where Pangure wore a "martingale garment". The reader will also meet it in V. Hugo's *Les Misérables* (V. 5, Book 1, Ch. 6), where Inspector Javert was tied by the insurgents in "that sort of bond which is called in prisons a martingale, which, starting at the neck, forks on the stomach, and meets the hands, after passing between the legs". In the context of random processes, the term *martingale* was introduced by J. Ville (who was a student of É. Borel) in the late 1930s. In the English language literature it was introduced by Joseph Leo Doob (27.02.1910-07.06.2004, an outstanding American mathematician who made important contributions to martingale theory and potential theory), who met the word in J. Ville's thesis that he had to review.

Closely related notions are that of a *submartingale* (which refers to adapted processes such that $\mathbf{E}(X_{t+1}|\mathcal{F}_t) \geq X_t, t = 0, 1, 2, \ldots$, holds instead of (11.1)) and of a *supermartingale* (when $\mathbf{E}(X_{t+1}|\mathcal{F}_t) \leq X_t$)).

The expression on the left-hand side of (11.1) is, of course, the conditional expectation (CE) of X_{t+1} given the σ-field $\mathcal{F}_t$. We introduced and briefly discussed that concept on p. 56. For convenience of referencing, we will just re-state here the main properties of CEs given a sub-σ-field $\mathcal{G} \subset \mathcal{F}$ (basically paraphrasing (i)–(v) of CEs given an RV that we listed in Section 2.8). We assume that all the RVs appearing in CE1–CE4 below are integrable and recall that all the (in)equalities involving CEs should be understood as holding a.s. For brevity, we will omit "a.s." in all such relations below (but won't forget that it actually must be there!).

CE1. Linearity: $\mathbf{E}(aX + bY|\mathcal{G}) = a\mathbf{E}(X|\mathcal{G}) + b\mathbf{E}(Y|\mathcal{G})$ for any constants a and b.

CE2. If Z is $\mathcal{G}$-measurable, then $\mathbf{E}(ZX|\mathcal{G}) = Z\mathbf{E}(X|\mathcal{G})$. That is, $\mathcal{G}$-measurable RVs behave like constants when computing CEs $\mathbf{E}(\,\cdot\,|\mathcal{G})$.

CE3. If X is independent of $\mathcal{G}$ (meaning: for any $B \in \mathcal{B}$ and $A \in \mathcal{G}$, the events $\{X \in B\}$ and A are independent), then $\mathbf{E}(X|\mathcal{G}) = \mathbf{E}\,X$.

In particular, if $X \equiv c = \text{const}$ or $\mathcal{G} = \{\emptyset, \Omega\}$ is trivial, then one has $\mathbf{E}(X|\mathcal{G}) = X$.

CE4. If $\mathcal{G}_0 \subset \mathcal{G}_1 \subset \mathcal{F}$ are σ-fields, then $\mathbf{E}[\mathbf{E}(X|\mathcal{G}_1)|\mathcal{G}_0] = \mathbf{E}(X|\mathcal{G}_0)$.

In particular, when the σ-field $\mathcal{G}_0 = \{\emptyset, \Omega\}$ is trivial, we obtain the TPF: $\mathbf{E}\,\mathbf{E}(X|\mathcal{G}_1) = \mathbf{E}\,X$ (cf. CE3).

We will not assign any label to yet another important property of CE: monotonicity (meaning that $\mathbf{E}(X_1|\mathcal{G}) \leq \mathbf{E}(X_2|\mathcal{G})$ provided that $X_1 \leq X_2$ a.s.), as we will not explicitly refer to it in what follows.

Recursively using CE4, we immediately obtain that, for an MG $\{X_t\}$, for any $s \geq 1$, one has

$$\begin{aligned} \mathbf{E}(X_{t+s}|\mathcal{F}_t) &= \mathbf{E}\big[\underbrace{\mathbf{E}(X_{t+s}|\mathcal{F}_{t+s-1})}_{=X_{t+s-1}}\,|\,\mathcal{F}_t\big] \\ &= \mathbf{E}(X_{t+s-1}|\mathcal{F}_t) = \cdots = \mathbf{E}(X_{t+1}|\mathcal{F}_t) = X_t. \end{aligned} \tag{11.2}$$

Thus, given the past (represented by $\mathcal{F}_t$), an MG stays, on average, at a constant level. Its mean values just remain constant since, taking expectation on both sides of (11.2) and using CE4, one obtains

$$\mathbf{E}\,X_{t+s} = \mathbf{E}\,\mathbf{E}(X_{t+s}|\mathcal{F}_t) = \mathbf{E}\,X_t = \mathbf{E}\,X_0, \tag{11.3}$$

the last equality holding as the previous ones are true for any $t, s \geq 0$.

In the *continuous time* case, relation (11.1) cannot be used to define an MG, as then there is no such thing as "the next value of the process" (as there is no next time value!). Instead, taking into account (11.2), a continuous time MG is defined as an adapted (in that case, to a continuous time filtration, i.e., a family of σ-fields $\{\mathcal{F}_t\}_{t\geq 0}$ such that, for any $s, t \geq 0$, one has $\mathcal{F}_t \subset \mathcal{F}_{t+s} \subset \mathcal{F}$) process $\{X_t\}_{t\geq 0}$ such that, for any $s, t \geq 0$,

$$\mathbf{E}\left(X_{t+s}|\mathcal{F}_t\right) = X_t. \tag{11.4}$$

Of course, the above relation can also be used as the definition of a discrete time MG as well.

Example 11.2. *Random walks.* Assume that $X_0 := 0$, $X_n := Y_1 + \cdots + Y_n$, $n \geq 1$, where the Y_j's are i.i.d. RVs with $\mathbf{E}|Y_1| < \infty$. When is the SP $\{X_n\}_{n\geq 0}$ an MG?

As no filtration was explicitly mentioned in conditions, we consider the natural filtration $\{\mathcal{F}_n\}_{n\geq 0}$ of $\{X_n\}_{n\geq 0}$. First verify integrability: by the triangle inequality,

$$\mathbf{E}|X_n| \leq \mathbf{E}\left(|Y_1| + \cdots + |Y_n|\right) = \mathbf{E}|Y_1| + \cdots + \mathbf{E}|Y_n| = n\mathbf{E}|Y_1| < \infty.$$

Now to the martingale property: for any $n \geq 0$, using CE1–CE3 one has

$$\begin{aligned}\mathbf{E}\left(X_{n+1}|\mathcal{F}_n\right) &= \mathbf{E}\left(X_n + Y_{n+1}|\mathcal{F}_n\right)\\ &= \mathbf{E}\left(X_n|\mathcal{F}_n\right) + \mathbf{E}\left(Y_{n+1}|\mathcal{F}_n\right) = X_n + \mathbf{E}\,Y_1.\end{aligned}$$

Therefore $\{X_n\}$ is an MG iff $\mathbf{E}\,Y_1 = 0$, as one could expect. For instance, if $\mathbf{P}\left(Y_j = 1\right) = 1 - \mathbf{P}\left(Y_j = -1\right) = p$ then $\{X_n\}$ is an MG iff $p = \frac{1}{2}$.

The argument actually shows that any process with zero mean independent increments will be an MG. The converse is not true, as we will see from examples below: the martingale property is much weaker than independence of increments. The property will imply, though, the *orthogonality of increments* (i.e., that the increments are uncorrelated; that claim makes sense for processes with finite second moments only, of course): for any MG $\{X_t\}_{t\geq 0}$ such that $\mathbf{E}\,X_t^2 < \infty$ for all $t \geq 0$, one has

$$\mathbf{E}\left(X_{t_2} - X_{t_1}\right)\left(X_{t_4} - X_{t_3}\right) = 0, \quad 0 \leq t_1 \leq t_2 \leq t_3 \leq t_4.$$

To prove that, just use property CE4 (Problem 5 on p. 345). Note that orthogonality of increments does not imply the (much stronger) martingale property (can you give a counterexample?).

The above property is closely related to the so-called *Doob decomposition*: for any (discrete time) integrable adapted process $\{X_t\}_{t\geq}$ on a filtered probability space $(\Omega, \mathcal{F}, \mathbf{F}, \mathbf{P})$, there exists an MG $\{M_t\}_{t\geq 0}$ with $M_0 = 0$ and a *predictable* (or *previsible*) process $\{A_t\}_{t\geq 0}$ (meaning that A_t is $\mathcal{F}_{t-1}$-measurable for all $t = 1, 2, \dots$) with $A_0 = 0$ such that

$$X_t = X_0 + M_t + A_t, \quad t = 0, 1, \dots$$

The processes $\{M_t\}$ and $\{A_t\}$ with the above property are a.s. unique. One corollary of the decomposition is that an adapted process $\{X_t\}_{t\geq}$ is a submartingale iff the process $\{A_t\}$ is non-decreasing. (There exists a version of the decomposition for continuous time processes as well.) One can actually give a simple explicit formula for $\{A_t\}$ (and hence for $\{M_t\}$): $A_0 := 0$,

$$A_t := \sum_{n=1}^{t} \mathbf{E}\,(X_n - X_{n-1}|\mathcal{F}_{n-1}) \equiv \mathbf{E}\,(X_t - X_{t-1}|\mathcal{F}_{t-1}) + A_{t-1}, \quad t = 1, 2, \ldots$$

That $\{A_t\}$ is predictable follows from the definition of CE, and it is easy to verify that $\{X_t - A_t\}$ is an MG (which is left to the reader, as an exercise).

Example 11.3. *A geometric random walk.* Here we assume that

$$X_n := X_0 e^{Y_1+\cdots+Y_n}, \quad n \geq 1,$$

where $X_0 :=$ const > 0, and the Y_j's are again i.i.d. RVs. When is $\{X_n\}_{n\geq 0}$ an MG w.r.t. the filtration $\mathcal{F}_t := \sigma(Y_1, \ldots, Y_t)$ ($\mathcal{F}_0$ being trivial)?

First verify integrability: by independence and (2.57),

$$\mathbf{E}\,|X_n| = \mathbf{E}\,X_n = X_0 \mathbf{E}\,e^{Y_1+\cdots+Y_n} = X_0 \big(\mathbf{E}\,e^{Y_1}\big)^n < \infty$$

iff $\varphi_Y(1) \equiv \mathbf{E}\,e^{Y_1} < \infty$. Now to the martingale property: for any $n \geq 0$,

$$\mathbf{E}\,(X_{n+1}|\mathcal{F}_n) = \mathbf{E}\,\big(\underbrace{X_0 e^{Y_1+\cdots+Y_n}}_{=X_n} e^{Y_{n+1}} \big|\, \mathcal{F}_n\big)$$

$$\overset{\text{CE2}}{=} X_n \mathbf{E}\,\big(e^{Y_{n+1}} \big|\, \mathcal{F}_n\big) \overset{\text{CE3}}{=} X_n \mathbf{E}\,e^{Y_{n+1}} = X_n \varphi_Y(1),$$

which equals X_n iff $\varphi_Y(1) = 1$.

From here it is clear that, setting $S_0 := 0$, $S_n := Y_1 + \cdots + Y_n$, $n \geq 1$, for any $v \in \mathbf{R}$ such that $\varphi_Y(v) \equiv \mathbf{E}\,e^{vY_1} < \infty$, the process

$$X'_n := e^{vS_n - n\ln\varphi_Y(v)} \equiv \frac{e^{vS_n}}{\varphi_Y^n(v)}, \quad n = 0, 1, 2, \ldots,$$

is an MG. Note that for both $\{X_n\}$ and $\{X'_n\}$ we could, instead of $\mathcal{F}_t := \sigma(Y_1, \ldots, Y_t)$, consider their natural filtrations (they are the same!).

In both theoretical and applied problems, one often needs to deal with the values of an MG $\{X_t\}$ (or another process) at random times. In that context, the concept of stopping (or Markov) time proved to be very useful.

Consider a filtered probability space $(\Omega, \mathcal{F}, \mathbf{F}, \mathbf{P})$ (assuming discrete time for the moment). An RV τ on that space is called a *stopping time*[3] (ST) if one has

$$\{\tau \leq t\} \in \mathcal{F}_t \quad \text{for each } t = 0, 1, 2, \ldots \tag{11.5}$$

[3]As one sometimes encounters *improper random times* (i.e., when $\mathbf{P}\,(\tau = \infty) > 0$), one often reserves the term "stopping time" to the cases where $\tau < \infty$ a.s., whereas in the general case random times satisfying (11.5) are referred to as *Markov times*.

Prior to justifying the use of "stopping" in the term, note that, for an ST τ,

$$\{\tau = t\} = \underbrace{\{\tau \le t\}}_{\in \mathcal{F}_t} \cap \underbrace{\{\tau \le t-1\}^c}_{\in \mathcal{F}_{t-1} \subset \mathcal{F}_t} \in \mathcal{F}_t \quad \text{for each } t = 0, 1, 2, \ldots, \tag{11.6}$$

where we used the observation that $\{\tau > t\} = \{\tau \le t\}^c \in \mathcal{F}_t$. Moreover, (11.6) implies (11.5) since

$$\{\tau \le t\} = \bigcup_{s=0}^{t} \underbrace{\{\tau = s\}}_{\in \mathcal{F}_s \subset \mathcal{F}_t}.$$

Now about the name. Suppose τ it the (random) time when we decide to stop doing something (for instance, to stop gambling or to sell a block of shares at a stock exchange, which is basically the same). When you do that at time t (that is, when the event $\{\tau = t\}$ occurs), you act on the basis of what you *already know by that time*, which can be mathematically expressed as $\{\tau = t\} \in \mathcal{F}_t$, as it is $\mathcal{F}_t$ that represents all the information available to us at time t.

Example 11.4. *First hitting times.* Stopping times of this kind appear in optimal control problems, risk models, sequential statistical analysis *etc.*

Suppose we are given an adapted process $\{X_t\}$ and a (boundary) function u_t, $t = 0, 1, 2, \ldots$ (the boundary is often just a fixed level $u_t \equiv u$). Show that the first hitting (or crossing) time

$$\tau := \inf\{t \ge 0 : X_t \ge u_t\} \tag{11.7}$$

is an ST.

Indeed, for any $t = 0, 1, 2, \ldots$, one has

$$\{\tau \le t\} = \bigcup_{s=0}^{t} \underbrace{\{X_s \ge u_s\}}_{\in \mathcal{F}_s \subset \mathcal{F}_t} \in \mathcal{F}_t.$$

Note the use of inf rather than min in (11.7): when $X_t < u_t$ for all t, the set of t values appearing on the right-hand side of (11.7) is empty, and since $\inf \emptyset = \infty$, we get $\tau = \infty$ then.

Example 11.5. Let $\{S_t\}_{t \le T}$ be an asset price process, $S_0 \equiv x = \text{const}$, $\{\mathcal{F}_t\}$ the natural filtration of the process (so that $\mathcal{F}_0 = \{\emptyset, \Omega\}$ is trivial). Is the RV

$$\tau := \min\Big\{t \ge 0 : S_t = \max_{0 \le s \le T} S_s\Big\},$$

i.e., the first time in $[0, T]$ when the maximum price on that time period is attained, an ST? Note that always $\tau \le T$.

The answer, in the general case, is negative. Indeed, already

$$\{\tau = 0\} = \bigcap_{t=1}^{T} \{S_t \le x\},$$

and, in non-trivial cases, one has $\{S_t \le x\} \notin \mathcal{F}_0$.

Example 11.6. Suppose τ is an RV and $\{X_t\}$ an SP on a common probability space. Take $\mathcal{F}_t := \sigma(\tau, X_s, s \le t)$, $t = 0, 1, 2, \ldots$ Then $\{\mathcal{F}_t\}$ is clearly a filtration and τ an ST w.r.t. it: for any $t \ge 0$,

$$\{\tau \le t\} \in \mathcal{F}_0 \subset \mathcal{F}_1 \subset \cdots \subset \mathcal{F}_t.$$

In this example, τ can be independent of $\{X_t\}$. That would model the following situation: one decides to stop at a random time, independently of the process' evolution (e.g., prior to running the SP, one tosses a coin till it lands heads up, and then stops the process after the number of steps equal to the number of tosses it took to get heads for the first time).

Example 11.7. *American derivative securities.* We already discussed American options in Example 4.2 (p. 134). More generally, an American derivative security (ADS) is a contract of which the owner can exercise it at any time prior to expiry T. Here is a mathematical definition thereof:

Suppose that we have a financial market (i.e., a family of SPs modelling the dynamics of assets' prices) on a probability space $(\Omega, \mathcal{F}, \mathbf{P})$, and let $\mathbf{F}$ denote the natural filtration for all the price processes on the time interval $t \in \{0, 1, \ldots, T\}$. Any adapted SP $X_t \ge 0$, $t = 0, 1, \ldots, T$, on $(\Omega, \mathcal{F}, \mathbf{F}, \mathbf{P})$ is called an ADS.

The layman's interpretation of the above is that the value X_t is the amount the owner of the ADS receives at time t should (s)he decide to exercise the security at that time (more precisely, X_T will be the payoff in the case when the owner exercises the ADS at time T or does not exercise it at all, letting it expire). That $\{X_t\}$ is adapted means that the payoff value X_t will be known at time t if the history of the prices in the market up to that time is known. The *exercise time* τ itself will, generally speaking, be an RV that has to be an ST (unless the owner of the ADS can foresee the future). Indeed, when the owner decides to exercise the ADS at time t, all (s)he knows is $\mathcal{F}_t$, so $\{\tau = t\} \in \mathcal{F}_t$, and hence (11.5) holds as well.

Martingales serve as mathematical models of the so-called "fair games", in which the expected net gain is zero for each gamble. Letting X_t denote our gambler's fortune at time t (cf. Example 3.21), one can write the condition that, given all what happened in the past, the gambler's net gain in the next gamble is zero, as

$$\mathbf{E}\left(X_{t+1} - X_t|\mathcal{F}_t\right) = 0.$$

Using CE2, this is the same as $\mathbf{E}\left(X_{t+1}|\mathcal{F}_t\right) = X_t$, i.e., $\{X_t\}$ is an MG. And hence $\mathbf{E}\,X_t = \mathbf{E}\,X_0$, the mean fortune value remains constant.

Here is a tricky question: can one win with certainty or, at least, on average using a cleverly chosen ST? Paraphrasing the latter, less ambitious goal, can one find an ST τ such that $\mathbf{E}\,X_\tau > \mathbf{E}\,X_0$?

To answer this question, we first establish the useful fact that a *stopped martingale* is an MG again. Recall the notation $a \wedge b = \min\{a, b\}$.

Theorem 11.1. *Let $\{X_t\}_{t\geq 0}$ be an MG and τ an ST on a common filtered probability space. Then $Z_t := X_{t\wedge\tau}$, $t = 0, 1, 2, \ldots$, is also an MG on that space.*

Proof First verify integrability. This is obvious for $Z_0 \equiv X_0$, while for $t \geq 0$ one has

$$Z_{t+1} = \sum_{k=0}^{t} \underbrace{X_k \mathbf{1}_{\{\tau=k\}}}_{|\cdot|\leq|X_k|} + \underbrace{X_{t+1}\mathbf{1}_{\{\tau>t\}}}_{|\cdot|\leq|X_{t+1}|}, \tag{11.8}$$

which yields

$$\mathbf{E}\,|Z_{t+1}| \leq \mathbf{E}\sum_{k=0}^{t+1}|X_k| = \sum_{k=0}^{t+1}\mathbf{E}\,|X_k| < \infty.$$

Now to prove the martingale property, we use representation (11.8) and the fact that X_k and $\mathbf{1}_{\{\tau=k\}}$ are $\mathcal{F}_k$-measurable (as $\{X_t\}$ is adapted and (11.6) holds true) and CE2 to write

$$\begin{aligned}
\mathbf{E}\left(Z_{t+1}|\mathcal{F}_t\right) &= \mathbf{E}\left[\sum_{k=0}^{t} X_k\mathbf{1}_{\{\tau=k\}} + X_{t+1}\mathbf{1}_{\{\tau>t\}}\middle|\mathcal{F}_t\right] \\
&= \sum_{k=0}^{t} X_k\mathbf{1}_{\{\tau=k\}} + \mathbf{E}\left(X_{t+1}\mathbf{1}_{\{\tau>t\}}|\mathcal{F}_t\right) \\
&= \sum_{k=0}^{t} X_k\mathbf{1}_{\{\tau=k\}} + \mathbf{1}_{\{\tau>t\}}\mathbf{E}\left(X_{t+1}|\mathcal{F}_t\right) \\
&= \sum_{k=0}^{t} X_k\mathbf{1}_{\{\tau=k\}} + \mathbf{1}_{\{\tau>t\}}X_t = Z_t,
\end{aligned}$$

where the third last equality holds since $\mathbf{1}_{\{\tau>t\}} = 1 - \mathbf{1}_{\{\tau\le t\}}$, the second last one holds because $\{X_t\}$ is an MG, while the last equality is equivalent to (11.8) (with t instead of $t+1$).

□

Theorem 11.2. (The Optional Stopping Theorem.) *Let $\{X_t\}_{t\ge 0}$ be an MG and τ a bounded ST (i.e., for a constant $C<\infty$, one has $\tau < C$ a.s.). Then*

$$\mathbf{E}\,X_\tau = \mathbf{E}\,X_0. \tag{11.9}$$

Thus, in a fair game, one cannot invent a rule for quitting the game that would "beat the system": the game will remain fair!

Proof The assertion is obvious from the Theorem 11.1: as $\{Z_t = X_{t\wedge\tau}\}$ is an MG, (11.3) implies that

$$\mathbf{E}\,X_{t\wedge\tau} = \mathbf{E}\,X_0.$$

Setting here $t := C$ yields (11.9) (as $C\wedge\tau \equiv \tau$).

□

There are weaker (than the boundedness of τ) sufficient conditions for (11.9) to hold for a martingale sequence. In the general case, when τ can be an improper RV, (11.9) takes the form

$$\mathbf{E}\,(X_\tau;\tau<\infty) = \mathbf{E}\,X_0. \tag{11.10}$$

Provided that X_τ is integrable, a necessary and sufficient condition for (11.10) is that

$$\mathbf{E}\,(X_t;\tau>t)\to 0 \quad\text{as}\quad t\to\infty. \tag{11.11}$$

Indeed, setting $\xi_t := X_t - X_{t-1}$, note that $\mathbf{E}\,(\xi_t|\mathcal{F}_{t-1}) = 0$, $t=1,2,\ldots$ (such a sequence $\{\xi_t\}$ is referred to as a *martingale difference*). This implies that $\mathbf{E}\,(\xi_t;\tau\ge t) = 0$, since $\{\tau\ge t\}\equiv\{\tau\le t-1\}^c\in\mathcal{F}_{t-1}$ by the definition of ST and that of conditional expectation (see p. 56). From here and (11.11) one has

$$\begin{aligned}
\mathbf{E}\,X_0 &= \lim_{t\to\infty}\left[\mathbf{E}\,X_0 + \sum_{k=1}^{t}\mathbf{E}\,(\xi_k;\tau\ge k) - \mathbf{E}\,(X_t;\tau\ge t+1)\right]\\
&= \lim_{t\to\infty}\left[\mathbf{E}\,X_0 + \sum_{k=1}^{t}\mathbf{E}\,(X_k;\tau\ge k) - \sum_{k=1}^{t}\mathbf{E}\,(X_{k-1};\tau\ge k) - \mathbf{E}\,(X_t;\tau\ge t+1)\right]\\
&= \lim_{t\to\infty}\left[\mathbf{E}\,X_0 + \sum_{k=1}^{t}\mathbf{E}\,(X_k;\tau\ge k) - \sum_{k=0}^{t-1}\mathbf{E}\,(X_k;\tau\ge k+1) - \mathbf{E}\,(X_t;\tau\ge t+1)\right]\\
&= \lim_{t\to\infty}\sum_{k=0}^{t}\left[\mathbf{E}\,(X_k;\tau\ge k) - \mathbf{E}\,(X_k;\tau\ge k+1)\right]\\
&= \lim_{t\to\infty}\sum_{k=0}^{t}\mathbf{E}\,(X_k;\tau=k) = \lim_{t\to\infty}\mathbf{E}\,(X_\tau;\tau\le t) = \mathbf{E}\,(X_\tau;\tau<\infty).
\end{aligned}$$

We proved sufficiency of (11.11) for (11.10). The necessity of (11.11) also follows from the above chain of equalities (just follow it in the opposite direction).

Example 11.2 (continued). Let $S_0 := 0$, $S_n := Y_1 + \cdots + Y_n$, $n = 1, 2, \ldots$, be an RW with i.i.d. jumps Y_j, $\mathbf{P}(Y_j = 1) = 1 - \mathbf{P}(Y_j = -1) = p$. Fix an integer $u \geq 1$ and put

$$\tau_u := \inf\{n \geq 0 : S_n \geq u\}. \tag{11.12}$$

This is an ST (see Example 11.4). Note that the RV τ_u will be improper when $p < \frac{1}{2}$, as the RW will then have a negative trend, and so $\mathbf{P}(\max_{n\geq 0} S_n = 0) > 0$, cf. Section 3.5.

Here we assume that $p = \frac{1}{2}$, making $\{S_n\}$ an MG (p. 307). As in that case $\sup_{n\geq 0} S_n = \infty$ a.s. (see Section 3.5), one has $\mathbf{P}(\tau_u < \infty) = 1$.

Now if the Optional Stopping Theorem held for the MG $\{S_n\}$, observing that $S_{\tau_u} \equiv u$, we would have

$$u = \mathbf{E}\, S_{\tau_u} = \mathbf{E}\, S_0 = 0,$$

which is impossible as $u > 0$. So Theorem 11.2 is not applicable. What is wrong?

Of course, τ_u is not bounded (even worse: $\mathbf{E}\,\tau_u = \infty$, as you can see from the GF of τ_u given below in (11.14): just compute its derivative at 0). As condition (11.11) is necessary and sufficient for (11.9) in this example (as $S_{\tau_u} \equiv u$ is integrable, being a constant), it must fail here. Since $S_t < u$ on the event $\{\tau_u > t\}$, the condition "fails on the negative side": very large negative values of S_t are likely enough to occur prior to the RW hitting $u > 0$ for the first time.

Summarising, one can say that, in a fair game (i.e., in the gambling situation of Example 3.21 with $p = \frac{1}{2}$), one can use our stopping rule (11.12) to win *with certainty* the desired u dollars. However, in order to survive to see that objective achieved, one needs to have access to unlimited funds!

Examples 11.2, 11.3 (continued). Consider our simple RW $\{S_n\}$ in the case when $p \geq \frac{1}{2}$ (then, as we know, $\mathbf{P}(\tau_u < \infty) = 1$ for the ST τ_u defined by (11.12)). We will now show how to compute the distribution of τ_u using the *exponential martingale* from Example 11.3.

Indeed, we know from that example that, for any $v \geq 0$ and $\varphi(v) = \varphi_Y(v) = \mathbf{E}\, e^{vY_1} = pe^v + qe^{-v}$, $q := 1 - p$,

$$X_n := e^{vS_n - n \ln \varphi(v)}, \quad n = 0, 1, \ldots, \tag{11.13}$$

is an MG. One can show that the Optional Stopping Theorem holds true in this case (indeed, condition (11.11) is satisfied as $\mathbf{E}(X_t; \tau_u > t) \le e^{vu}\mathbf{P}(\tau_u > t) \to 0$, since $\varphi(v) \ge 1$ when $v \ge 0$ and τ_u is a proper RV), and hence

$$1 = \mathbf{E}\, X_0 = \mathbf{E}\, X_{\tau_u} = \mathbf{E}\, e^{vS_{\tau_u} - \tau_u \ln \varphi(v)} = e^{vu}\mathbf{E}\, z^{\tau_u}$$

where we put $z := 1/\varphi(v)$. Therefore the GF of τ_u is

$$\mathbf{E}\, z^{\tau_u} = e^{-vu} = (e^v)^{-u},$$

where it remains to express the e^v on the right-hand side in terms of z, which can be done by solving the equation

$$z^{-1} \equiv \varphi(v) = pe^v + qe^{-v}.$$

This is clearly a quadratic equation in e^v, with solutions

$$e^v = \frac{1}{2p}\big(z^{-1} \pm \sqrt{z^{-2} - 4pq}\big).$$

One should choose "+" here (note that $z = 1/\varphi(v)$ decreases when $v > 0$ increases, as we assumed that $p \ge \frac{1}{2}$), which yields the following formula for the GF of τ_u:

$$g_{\tau_u}(z) := \mathbf{E}\, z^{\tau_u} = \left(\frac{2pz}{1 + \sqrt{1 - 4pqz^2}}\right)^u, \quad z \in (0, 1]. \tag{11.14}$$

The right-hand side here is clearly the uth power of the GF of τ_1 (you will obtain the latter by just letting $u := 1$ in the above formula), which means that τ_u has the same distribution as the sum of u independent copies of τ_1 (see Section 2.7). That we could have noticed much earlier, as $\tau_u = \tau_1^{(1)} + \tau_1^{(2)} + \cdots + \tau_1^{(u)}$, where

$$\tau_1^{(k)} := \tau_k - \tau_{k-1} = \min\{n > \tau_{k-1} : S_n = S_{\tau_{k-1}} + 1\} - \tau_{k-1}, \quad k = 1, 2, \ldots,$$

are i.i.d. RVs (by the Markov property).

From that formula one can easily recover the distribution of τ_u: just expand $g_{\tau_u}(z)$ into the Taylor series at $z = 0$. It is also easy to compute the moments of τ_u (that will be finite when $p > \frac{1}{2}$), by computing $\frac{d}{dz} g_{\tau_u}(z)$ at $z = 1$. In particular, we will find that way that $\mathbf{E}\, \tau_u = u/(2p-1)$, $p > \frac{1}{2}$, but that could also be done in an easier way, using another MG, as we will see below.

The same exponential MG can be used to derive the famous *Lundberg*[4] *bound* for ruin probabilities. In the simplest setup, suppose we start an insurance company with a reserve capital $u > 0$, assume that the quantities $Y_j :=$ (aggregated claim amount) − (added premium) during the jth time period, $j = 1, 2, \ldots$, are i.i.d. RVs, and let

[4] Ernst Filip Oskar Lundberg (02.06.1876–31.12.1965), a Swedish actuary, the founder of mathematical risk theory.

$S_n := Y_1 + \cdots + Y_n$ represent the total claims paid minus the total premiums received after n time periods. Then the RV τ_u defined by (11.12) will be the time of our ruin.

Now suppose there exists a $v_0 > 0$ such that $\varphi(v_0) = 1$ (in actuarial literature, such a value v_0 is referred to as the *adjustment coefficient*). Note that as $\varphi(0) = 1$ and, as can easily be shown, $\varphi(v)$ is a convex function, the above condition can only be met when $\varphi'(0) < 0$ (meaning that $\mathbf{E}\, Y_1 < 0$, i.e., the mean aggregate claim amount is less than the mean premium paid, which makes the insurance business viable), and if such a v_0 exists then it is unique.

Applying the Optional Stopping Theorem to the exponential MG (11.13) with $v = v_0$ (cf. Example 11.3) and ST $\tau_u \wedge n \le n$ (this is a bounded ST, see Problem 3 on p. 344), and observing that $S_{\tau_u} \ge u$ on the event $\{\tau_u \le n\}$,

$$\begin{aligned} 1 &= \mathbf{E}\, e^{v_0 S_{\tau_u \wedge n} - (\tau_u \wedge n) \ln \varphi(v_0)} = \mathbf{E}\, e^{v_0 S_{\tau_u \wedge n}} \ge \mathbf{E}\, e^{v_0 S_{\tau_u \wedge n}} \mathbf{1}_{\{\tau_u \le n\}} \\ &= \mathbf{E}\, e^{v_0 S_{\tau_u}} \mathbf{1}_{\{\tau_u \le n\}} \ge \mathbf{E}\, e^{v_0 u} \mathbf{1}_{\{\tau_u \le n\}} = e^{v_0 u} \mathbf{P}\,(\tau_u \le n). \end{aligned}$$

Thus we obtained the following bound for the ruin probability: $\mathbf{P}\,(\tau_u \le n) \le e^{-v_0 u}$. As the right-hand side here does not depend on n, we conclude that the ultimate ruin probability admits the same upper bound:

$$\mathbf{P}\,(\tau_u < \infty) = \lim_{n \to \infty} \mathbf{P}\,(\tau_u \le n) \le e^{-v_0 u}.$$

This result is rather precise: it turns out that $\mathbf{P}\,(\tau_u < \infty) = (c + o(1))e^{-v_0 u}$ as $u \to \infty$ for some constant $c \in (1, \infty)$ (provided that $\varphi(v) < \infty$ for some $v > v_0$).

We will conclude this section with the following important observation: the property (11.9) is actually characteristic of MGs! This is an equivalent definition of the martingale property: the value of the process at an ST has the same mean as the initial value.

Theorem 11.3. *Let $\{X_t\}_{t \ge 0}$ be an adapted integrable process on a filtered probability space. If $\mathbf{E}\, X_\tau = \mathbf{E}\, X_0$ for all STs τ on that space assuming at most two different values, then $\{X_t\}$ is an MG.*

Clearly, if (11.9) holds for constant STs τ the process does not need to be an MG. So the condition of the theorem basically says that if (11.9) holds for the simplest non-trivial STs, we are dealing with an MG here.

Proof The proof works for both discrete- and continuous time processes. Fix $0 \le s < t$. First note that $\tau \equiv s$ is an ST, and so $\mathbf{E}\, X_s = \mathbf{E}\, X_0$. Next take an arbitrary event $A \in \mathcal{F}_s$ and let

$$\tau(\omega) := \begin{cases} s & \text{if } \omega \in A^c, \\ t & \text{if } \omega \in A. \end{cases}$$

This is clearly an ST as

$$\{\tau \le u\} = \begin{cases} \emptyset & \text{if } u < s, \\ A^c & \text{if } u \in [s, t), \\ \Omega & \text{if } u \ge t, \end{cases}$$

the events on the right-hand side being elements of the respective $\mathcal{F}_u$ in all three cases. Since by assumption

$$\mathbf{E}\,X_\tau = \mathbf{E}\,X_0 = \mathbf{E}\,X_s \equiv \mathbf{E}\,(X_s; A) + \mathbf{E}\,(X_s; A^c),$$

and

$$\mathbf{E}\,X_\tau = \mathbf{E}\,(X_\tau; A) + \mathbf{E}\,(X_\tau; A^c) = \mathbf{E}\,(X_t; A) + \mathbf{E}\,(X_s; A^c)$$

we conclude that $\mathbf{E}\,(X_t; A) = \mathbf{E}\,(X_s; A)$ for any $A \in \mathcal{F}_s$, which means that $\mathbf{E}\,(X_t|\mathcal{F}_s) = X_s$ (see p. 56).

□

11.2 The Brownian Motion Process

11.2.1 *The Main Properties of the BM Process*

We gave a formal definition of the standard Brownian motion (BM) process (which is also known as the Wiener process) in Section 2.10, see properties W1–W3 on p. 65. Recall that this is a continuous time process $\{W_t\}_{t\geq 0}$ with stationary independent increments such that $W_t \sim N(0,t)$, and continuous trajectories, which appears as the limit in the Functional Central Limit Theorem (thus approximating in distribution the scaled trajectories of RWs with jumps having finite second moments).

Demonstrating that such a process exists at all is actually a non-trivial mathematical problem. It is remarkable, though, that one can yet give infinite series representations for the BM process. One of them is the following *Paley–Wiener construction*: if $\{Z_n\}_{n\geq 0}$ is a sequence of i.i.d. standard normal RVs, then the series

$$W_t := Z_0 \frac{t}{\sqrt{2\pi}} + \frac{2}{\sqrt{\pi}} \sum_{n=1}^{\infty} Z_n \frac{\sin(nt/2)}{n}, \quad t \in [0, 2\pi],$$

converges a.s. uniformly in t, and the resulting process will satisfy W1–W3 (on $[0, 2\pi]$).

Note also that if an SP $\{X_t\}_{t\geq 0}$ satisfies properties W1 and W2 then there exists its *equivalent modification*, i.e., an SP $\{Y_t\}_{t\geq 0}$ such that $\mathbf{P}\,(X_t = Y_t) = 1$ for any $t \geq 0$, that has continuous trajectories (cf. p. 66). Path-wise, equivalent processes can substantially differ from each other. For example, let $U_1, U_2, \ldots \sim U(0,1)$ be i.i.d. RVs, and set $X_t(\omega) := 0$ for all $t \geq 0$, while $Y_t(\omega) := 1$ if $t = U_j(\omega)$ for some $j \geq 1$, otherwise $Y_t(\omega) := 0$. Then, for any given $t \geq 0$, one has

$$\mathbf{P}\,(X_t \neq Y_t) = \mathbf{P}\left(\bigcup_{j\geq 1}\{U_j = t\}\right) \leq \sum_{j\geq 1} \underbrace{\mathbf{P}\,(U_j = t)}_{=0} = 0,$$

so that the processes are equivalent. However, the values of Y differ by 1 from those of X on an everywhere dense in $[0,1]$ set of t-values $\{U_1, U_2, \ldots\}$ (see Problem 12 on p. 346). Anyway, W1 and W2 are the key properties in the definition of the BM process, while

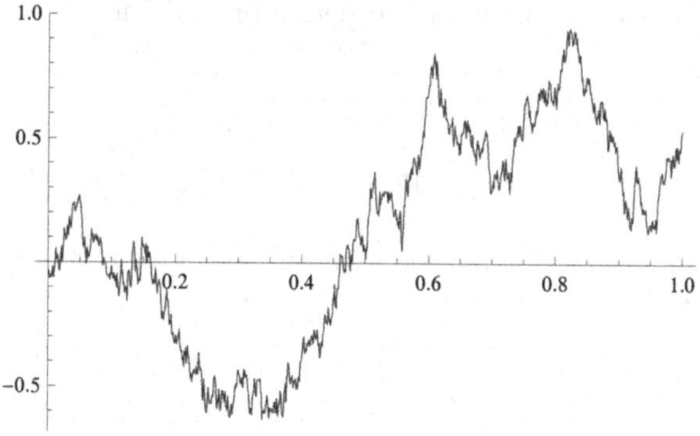

Fig. 11.1 A typical (simulated) trajectory of the standard BM process on $[0, 1]$.

W3 more or less "comes for free". For simplicity, one can always assume it satisfied once W1 and W2 are, which usually causes no problems.

It is clear from W1–W3 that the BM process has the following important *self-similarity property*: for any $a \neq 0$,

$$\{\widetilde{W}_t := a^{-1} W_{a^2 t}\}_{t \geq 0} \quad \text{is a standard BM process as well.} \tag{11.15}$$

We will also often deal with two important related SPs, the so-called *arithmetic Brownian motion* defined as

$$X_t := X_0 + \mu t + \sigma W_t, \quad t \geq 0, \tag{11.16}$$

and the *geometric Brownian motion* defined by

$$Z_t := e^{X_t} = Z_0 \exp\{\mu t + \sigma W_t\}, \quad t \geq 0, \tag{11.17}$$

$Z_0 = e^{X_0}$ (in both cases, $\mu, \sigma \in \mathbf{R}$ are constant).

It is clear that, like the BM itself, the arithmetic BM $\{X_t\}$ is also a Gaussian process with stationary independent increments, but the geometric BM is not.

There is actually a large class of non-Gaussian process with stationary independent increments that are often referred to as *Lévy processes* and that also play an important role in numerous applications. For instance, a Poisson process (with constant intensity) $\{N_t\}_{t\geq 0}$ is a Lévy process, and the linear combination $\{a_1 N_t^{(1)} + \cdots + a_k N_t^{(k)}\}_{t\geq 0}$ of k independent Poisson processes $\{N_t^{(j)}\}$ with constant coefficients $a_j \in \mathbf{R}$, $j = 1, \ldots, k$, is a Lévy process (this will actually be a compound Poisson process, see p. 164). In fact, any Lévy process can be obtained as a limit of such linear combinations (as the number k of terms in them goes to infinity). One can also mention here a simple "jump-diffusion" Lévy process $\{W_t + N_t\}_{t\geq 0}$, where the summands are independent BM and Poisson processes, and the so-called *gamma process* defined as a Lévy process $\{X_t\}_{t\geq 0}$ with the property that $X_t - X_s \sim \gamma(t - s, 1)$ for $0 \leq s < t$. Note that the only Lévy processes that have continuous trajectories are arithmetic BMs.

It immediately follows from the definition that, for $0 \le t_1 < t_2$, the random vector

$$\begin{aligned}(W_{t_1}, W_{t_2}) &= (W_{t_1}, W_{t_1} + (W_{t_2} - W_{t_1})) \\ &= (W_{t_1}, W_{t_2} - W_{t_1}) \begin{pmatrix} 1 & 1 \\ 0 & 1 \end{pmatrix} \end{aligned} \tag{11.18}$$

has a bivariate normal distribution with zero mean (as $\mathbf{E}\, W_t = 0$), variances $\mathrm{Var}\,(W_{t_j}) = t_j$, $j = 1, 2$, and covariance (using independence of increments)

$$\begin{aligned}\mathrm{Cov}\,(W_{t_1}, W_{t_2}) &= \mathbf{E}\, W_{t_1} W_{t_2} = \mathbf{E}\, W_{t_1}(W_{t_1} + (W_{t_2} - W_{t_1})) \\ &= \mathbf{E}\, W_{t_1}^2 + \mathbf{E}\, W_{t_1} \mathbf{E}\,(W_{t_2} - W_{t_1}) = t_1 = t_1 \wedge t_2. \end{aligned} \tag{11.19}$$

Likewise, it is clear that, for any $0 = t_0 \le t_1 < t_2 < \cdots < t_n < \infty$, the distribution of the vector $(W_{t_1}, W_{t_2}, \ldots, W_{t_n})$ is normal (as the vector is a linear transformation of the vector $(W_{t_1} - W_{t_0}, W_{t_2} - W_{t_1}, \ldots, W_{t_n} - W_{t_{n-1}})$ with independent normal components, cf. p. 33 and (11.18)), with zero mean and covariance matrix

$$\begin{pmatrix} t_1 & t_1 & t_1 & \cdots & t_1 \\ t_1 & t_2 & t_2 & \cdots & t_2 \\ t_1 & t_2 & t_3 & \cdots & t_3 \\ \cdots & \cdots & \cdots & \cdots & \cdots \\ t_1 & t_2 & t_3 & \cdots & t_n \end{pmatrix}.$$

Recall that any Gaussian distribution is uniquely specified by its mean vector and covariance matrix (cf. p. 44). Therefore, to uniquely specify the distribution of a general Gaussian SP (i.e., an SP $\{X_t\}$ such that, for any $t_1, t_2, \ldots, t_n$, the vector $(X_{t_1}, X_{t_2}, \ldots, X_{t_n})$ is normal), one just needs to specify the process' *mean function* $m_X(t) := \mathbf{E}\, X_t$ and *covariance function* $\gamma_X(s,t) := \mathrm{Cov}\,(X_s, X_t)$. Thus we arrived at the following result.

Theorem 11.4. *A Gaussian SP* $\{X_t\}_{t \ge 0}$ *is a BM iff* $m_X(t) \equiv 0$ *and* $\gamma_X(s,t) = s \wedge t$, $s, t \ge 0$.

Formally, we also had to require here that the SP would have continuous trajectories a.s. (property W3). However, as we agreed earlier, we would take that for granted once W1 and W2 are satisfied (which is the case in Theorem 11.4) as then there always exists a continuous modification of the SP.

Knowing the mean and covariance matrix of the Gaussian vector $(W_{t_1}, \ldots, W_{t_n})$ one can compute the density of the vector's distribution on $\mathbf{R}^n$ (see (2.61)) or, if the covariance matrix is singular, on the lower dimensionality subspace of $\mathbf{R}^n$ on which the distribution of the vector will be concentrated, which will enable one to compute any probability of the form $\mathbf{P}\left((W_{t_1}, \ldots, W_{t_n}) \in B\right)$, $B \in \mathcal{B}(\mathbf{R}^n)$, by integrating the density over B.

Instead of inverting the covariance matrix that is necessary to use (2.61) to find the density of a BM's FDD, one can immediately compute the density using the Markov property of the BM. Recall that an SP $\{X_t\}$ is said to be Markovian if, given that the exact *present* (say, at time s) position of the SP (i.e., the value X_s) is known, then the future evolution $\{X_t\}_{t \geq s}$ of the SP will not depend on *how* it arrived at X_s by the time s (i.e., on the values X_t with $t < s$; for a formal definition of a continuous time Markov process, see Section 6.1). Now it is easy to see that any SP with independent increments is Markovian. Indeed, given that $X_s = x$, the "future values" of the SP will be of the form $X_t = x + (X_t - X_s)$, where the values of $X_t - X_s$ are independent of the SP's values prior to time s. Alternatively, denoting by $\{\mathcal{F}_t\}$ the natural filtration of $\{X_t\}$, one can compute the conditional ChF of X_t given $\mathcal{F}_s$, $s < t$: using CE2 and CE3, one has

$$\begin{aligned}\mathbf{E}\left(e^{iuX_t} | \mathcal{F}_s\right) &= \mathbf{E}\left(e^{iuX_s} e^{iu(X_t - X_s)} | \mathcal{F}_s\right) \\ &= e^{iuX_s} \mathbf{E}\left(e^{iu(X_t - X_s)} | \mathcal{F}_s\right) = e^{iuX_s} \mathbf{E}\, e^{iu(X_t - X_s)}.\end{aligned}$$

Repeating the above computation with $\mathcal{F}_s$ replaced by X_s will clearly yield the same answer. Since the conditional ChF of X_t given $\mathcal{F}_s$ coincides with that given X_s only, we conclude that the respective conditional distributions of X_t must coincide, too (as distributions are uniquely specified by their ChFs).

Thus, the BM process is a Markov process. As such, it can be described by its initial distribution and *transition probabilities*. Recall that the conditional (given the initial state) FDDs of a time-homogeneous Markov chain are given in terms of the transition probabilities by relation (3.15). In the case of continuous time and space Markov processes, instead of $p_{jk}^{(n)}$ one has to deal with transition probabilities of the form

$$\mathbf{P}\left(X_t \in B \,|\, X_s = x\right), \quad 0 \leq s < t, \ B \in \mathcal{B}(\mathbf{R}), \ x \in \mathbf{R}. \tag{11.20}$$

When $\{X_t\}$ is the BM process $\{W_t\}$, these transition probabilities are just the distributions of $x + W_t - W_s \sim N(x, t - s)$ and hence have densities

$$p(s, x; t, y) = \frac{1}{\sqrt{2\pi(t-s)}} \exp\left\{-\frac{(y-x)^2}{2(t-s)}\right\} =: p_{t-s}(x, y) \tag{11.21}$$

(note that the last notation makes sense as the times s and t appear in the density $p(s, x; t, y)$ in the form of the time lag $t - s$ only). Therefore, similarly to (3.15), for $0 = t_0 < t_1 < \cdots < t_n < \infty$, $x_j \in \Delta_j \subset \mathbf{R}$, Δ_j being intervals of lengths $|\Delta_j| \to 0$, $j = 0, 1, \ldots, k$, $x_0 = 0$, one has

$$\mathbf{P}(W_{t_1} \in \Delta_1, \ldots, W_{t_n} \in \Delta_n) = (1 + o(1)) \underbrace{\prod_{j=1}^{n} p_{t_j - t_{j-1}}(x_{j-1}, x_j)|\Delta_j|}_{=: f_{t_1,\ldots,t_n}(x_1,\ldots,x_n)}. \quad (11.22)$$

That shows that $f_{t_1,\ldots,t_k}(x_1, \ldots, x_n)$ is the density of the random vector $(W_{t_1}, \ldots, W_{t_n})$, cf. (2.32) (formally, one can prove that fact by expressing densities via conditional ones using formula (2.80) and induction). See also Problem 15 on p. 346.

As we said before, to compute a (finite-dimensional) probability of the form $\mathbf{P}\big((W_{t_1}, \ldots, W_{t_n}) \in B\big)$, one can now simply integrate the density $f_{t_1,\ldots,t_k}(x_1, \ldots, x_n)$ over the set B. That, however, can be a challenging computational task even for small values of n and simple regions B, e.g., orthants, in which case one just computes the value of a multivariate normal DF. An example of a real-life situation where this is needed is the computation of the so-called fair price of a barrier option in the Black–Scholes setting, see Section 13.8.

Alternatively, evaluation of such probabilities can be done using the Monte Carlo method by generating a large i.i.d. sample of realisations of $(W_{t_1}, \ldots, W_{t_n})$ and then taking the relative frequency of hitting the set B by the elements of the sample as an estimator of the value of $\mathbf{P}\big((W_{t_1}, \ldots, W_{t_n}) \in B\big)$. Here is an algorithm for simulating a sample point with the desired distribution:

1. Simulate independent standard normal RVs $Z_1, \ldots, Z_n$ (using, say, the Box–Muller algorithm, see p. 281) and set $W_{t_1} := {t_1}^{1/2} Z_1$, $W_{t_2} - W_{t_1} := (t_2 - t_1)^{1/2} Z_2$, …, $W_{t_n} - W_{t_{n-1}} := (t_n - t_{n-1})^{1/2} Z_n$.
2. Take the partial sums
$$\begin{aligned} W_{t_2} &:= W_{t_1} + (W_{t_2} - W_{t_1}), \\ W_{t_3} &:= W_{t_2} + (W_{t_3} - W_{t_2}), \\ &\cdots \\ W_{t_n} &:= W_{t_{n-1}} + (W_{t_n} - W_{t_{n-1}}). \end{aligned}$$
3. Stop.

Of course, to improve the performance of the Monte Carlo estimators of the desired probabilities (or expectations), one can (and should) employ variance reduction methods (see Section 10.4).

11.2.2 *The Path Properties*

By definition, for (almost) all $\omega \in \Omega$, the *trajectory* $W_t(\omega)$, $t \geq 0$, is a continuous function. Most of the continuous functions that we encounter in calculus courses are actually differentiable. Will this be the case for the BM trajectories as well?

Note that, for any fixed $t \geq 0$ and $Z \sim N(0,1)$,

$$\frac{1}{h}\underbrace{(W_{t+h} - W_t)}_{\sim N(0,h)} \stackrel{d}{=} \frac{1}{h} \times \sqrt{h}Z = \frac{Z}{\sqrt{h}}, \qquad h > 0, \tag{11.23}$$

where $\stackrel{d}{=}$ stands for "equality in distribution": $X \stackrel{d}{=} Y$ means that the RVs X and Y have the same distribution. Clearly, if $Z \gtrless 0$ then $Z/\sqrt{h} \to \pm\infty$ as $h \searrow 0$, so that the difference quotient on the left-hand side of (11.23) cannot have any finite limit (even in distribution). In fact, with probability one, the trajectory of the BM process is nowhere differentiable (cf. Fig. 11.1)! Try to construct a deterministic function with that property.[5]

Another important analytic characteristic of a function of real variable is its *variation*[6] which, in particular, is crucial for the so-called Lebesgue-Stieltjes integrals, see (2.44). Without going into any fine detail here, we just observe that, similarly to (11.23), for a fixed $t > 0$ and any $n \geq 1$,

$$\sum_{j=1}^{n} |W_{tj/n} - W_{t(j-1)/n}| \stackrel{d}{=} \sum_{j=1}^{n} \sqrt{\frac{t}{n}}\,|Z_j| = \underbrace{\sqrt{nt}}_{\to\infty} \times \underbrace{\frac{1}{n}\sum_{j=1}^{n} |Z_j|}_{\stackrel{\text{distr}}{\longrightarrow} \mathbf{E}\,|Z| = \sqrt{2\pi}}\ ,$$

where $Z_1, Z_2, \ldots$ are i.i.d. $N(0,1)$–RVs and the last convergence follows from the LLN. That means that the path $W_t(\omega)$ is not a function of bounded variation on finite intervals, and so one cannot use the trajectories of the BM process as "integrators" in a Lebesgue-Stieltjes integral, so that expressions of the form $\int_a^b f(t)\,dW_t$ would be undefined (see p. 37)

In contrast to the above, for any continuously differentiable function $F(s)$ on $[0,t]$, by the mean value theorem one has, for some $\theta_{n,j} \in [0,1]$,

$$\sum_{j=1}^{n} \left| F\left(\frac{tj}{n}\right) - F\left(\frac{t(j-1)}{n}\right) \right| = \sum_{j=1}^{n} \left| F'\left(\frac{t(j-\theta_{n,j})}{n}\right) \right| \frac{t}{n} \to \int_0^t |F'(s)|\,ds$$

---

[5]One example is the *Weierstrass function* (1872) given by $f(x) = \sum_{n=0}^{\infty} a^n \cos(b^n \pi x)$, $x \in \mathbf{R}$, where $b > 0$ is odd, $a \in (0,1)$ is such that $ab > 1 + 3\pi/2$.

[6]Formally, the total variation of a function $f(x)$ on interval $[a,b]$ from its domain is defined as $\sup_{\{t_0,t_1,\ldots,t_n\}} \sum_{j=1}^{n} |f(t_{j+1} - f(t_j)|$, where the supremum is taken over all finite partitions $a = t_0 < t_1 < \cdots < t_n = b$ of $[a,b]$. Any function having a bounded derivative on $[a,b]$ will (by the mean value theorem) have finite total variation on $[a,b]$. In fact, any function of bounded variation can be represented as a difference of two increasing functions.

as $n \to \infty$ (of course, this does not need to be uniform partitions of $[0, t]$; any sequence of partitions with vanishing mesh will lead to the same result). It is not hard to see that in such a case the total variation of F on $[0, t]$ will be equal to the finite quantity $\int_0^t |F'(s)|\, ds$.

An important role in the theory of the BM process (and, more generally, in stochastic calculus) is played by the so-called *quadratic variation* where we sum up the squares of the increments of the function rather than their absolute values. Again, the trajectories of the BM display a rather bizarre behaviour here. Indeed, for any continuously differentiable function $F(s)$ on $[0, t]$,

$$\sum_{j=1}^{n}\left(F\left(\frac{tj}{n}\right) - F\left(\frac{t(j-1)}{n}\right)\right)^2 = \frac{t}{n}\underbrace{\sum_{j=1}^{n}\left(F'\left(\frac{t(j-\theta_{n,j})}{n}\right)\right)^2\frac{t}{n}}_{\to\int_0^t (F'(s))^2\, ds<\infty} \to 0$$

as $n \to \infty$. However, for the BM trajectory, by the LLN one has

$$\sum_{j=1}^{n}(W_{tj/n} - W_{t(j-1)/n})^2 \overset{d}{=} \sum_{j=1}^{n}\left(\sqrt{\frac{t}{n}}\,Z_j\right)^2 = t \times \frac{1}{n}\sum_{j=1}^{n} Z_j^2 \overset{\text{distr}}{\longrightarrow} t. \quad (11.24)$$

As was said earlier, our exposition in this book may sometimes be somewhat sloppy, and deliberately so, as we want to avoid some technicalities of the theory. This is one of such places. Just note that, if one defines the total quadratic variation of a function on $[0, t]$ as the supremum of the sums of its squared increments over all finite partitions of $[0, t]$ (similarly to the definition of the total variation from the footnote on the previous page), then the quadratic variation of the BM will be infinite. If, however, we define it as a limit of such sums along a sequence of (deterministic) partitions of $[0, t]$ with vanishing mesh, then the limit will exists in the sense of L^2 norm (and hence also in probability and in distribution, see p. 43) and will be equal to t. If, moreover, we consider sequences of *nested* partitions with vanishing mesh, then convergence to t will take place with probability one.

Thus the paths of the BM are very irregular (nowhere differentiable) random functions that nevertheless have a very special property that their quadratic variation on $[0, t]$ increases exactly as the deterministic function t. Symbolically, one can write that as

$$(dW_t)^2 = dt. \quad (11.25)$$

It turns out that one can describe in detail the local behaviour of the BM trajectories. As the process has homogeneous independent increments, it suffices to understand how the process behaves as $t \searrow 0$. The first thing to observe is that since $W_t \sim N(0, t)$, one can expect the "typical" values of W_t to be of the order of magnitude of its standard deviation $\sqrt{t}$ (and likewise, for any fixed $s > 0$, the magnitude of the increments $W_{s+t} - W_s$ is of the order of $\sqrt{t}$; no wonder that the trajectory is nowhere differentiable!).

To get further insight, one can actually analyse the process' behaviour as $t \to \infty$. Indeed, using Theorem 11.4 it is easy to show that $\widetilde{W}_t := tW_{1/t}$, $t \geq 0$, is also a standard BM process (Problem 18). Here we put $\widetilde{W}_0 := 0$ "by continuity" as $s^{-1}W_s \to 0$ a.s. as $s \to \infty$ by the strong LLN (cf. (2.84); note that, for any fixed $h > 0$, $\{W_{nh}\}_{n\geq 0}$ is actually a sequence of partial sums of i.i.d. $N(0,h)$–RVs).

Recalling that the BM is the limit of an appropriately scaled simple random walk, one can conclude that, like the latter, it will also be recurrent (see (3.21)). In particular, almost all trajectories $W_s(\omega)$, $s \geq 0$, of the BM visit 0 infinitely many times as $s \to \infty$. From that it immediately follows that $\widetilde{W}_t(\omega)$ (which is also a BM) visits 0 infinitely often as $t \searrow 0$. Further fine detail of its behaviour is given by the law of the iterated logarithm that implies that, roughly speaking, almost any trajectory of W_t "fills out" the interior of the figure

$$\left\{(t,x) : t > 0, |x| \leq \sqrt{2t \ln|\ln t|}\,\right\} \tag{11.26}$$

in the vicinity of its vertex $(0,0)$. And the same applies to any point (s, W_s) on the graph of the trajectory: the values W_{s+t} for small $t > 0$ will fill-out the "near-vertex area" of the translated (so that its vertex is at (s, W_s) now) version of the figure (11.26). For more interesting results of that kind and beyond, see, e.g., Mörters and Peres (2010).

11.2.3 *The Distributions of Some RVs Related to the BM*

One can derive closed-from expressions for the distributions (or their intergal transforms) of amazingly many functionals of the BM process (again, we refer the interested reader to Mörters and Peres (2010)). Here we will only deal with a couple of the most important such functionals, namely, the maximum of the BM on a finite interval and the first hitting time of a level by the BM process.

For $x \in \mathbf{R}$, $t > 0$, put

$$M_t := \max_{0\leq s\leq t} W_s, \quad \tau_x := \min\{t \geq 0 : W_t = x\} \tag{11.27}$$

(note that, for $x > 0$, the definition of τ_x will not change if we replace the equality sign "=" in it with "≥"; also, why can we use here max and min instead of the apparently more appropriate sup and inf?).

Theorem 11.5. (i) *For a fixed* $t > 0$,

$$\mathbf{P}(M_t > x) = 2\mathbf{P}(W_t > x) = 2(1 - \Phi(x/\sqrt{t})), \quad x \geq 0, \tag{11.28}$$

where Φ *is the standard normal DF.*

(ii) *For a fixed* $x > 0$, *the RV* τ_x *has density*

$$f_{\tau_x}(t) := \frac{x}{\sqrt{2\pi}} t^{-3/2} e^{-x^2/(2t)}, \quad t > 0. \tag{11.29}$$

Note that the form of $f_{\tau_x}(t)$ is basically of the same as for the probabilities of first hitting time (3.40) for skip-free RWs (but now expressed in terms of densities rather than probabilities). Both are actually special cases of a general relationship for skip-free processes with independent increments.

Proof (i) By the TPF, using the fact that $W_s = x$ on $\{\tau_x = s\}$ and that $W_t - W_s$ is independent of $\{\tau_x = s\} \in \sigma(W_u, u \leq s)$, one has

$$\begin{aligned}
\mathbf{P}(W_t \geq x) &= \int_0^t \mathbf{P}(W_t \geq x \,|\, \tau_x = s)\mathbf{P}(\tau_x \in ds) \\
&= \int_0^t \mathbf{P}(W_t - W_s \geq 0 \,|\, \tau_x = s)\mathbf{P}(\tau_x \in ds) \\
&= \int_0^t \underbrace{\mathbf{P}(W_t - W_s \geq 0)}_{=1/2 \text{ as } W_t - W_s \sim N(0, t-s)} \mathbf{P}(\tau_x \in ds) \\
&= \frac{1}{2}\int_0^t \mathbf{P}(\tau_x \in ds) = \frac{1}{2}\mathbf{P}(\tau_x \leq t) = \frac{1}{2}\mathbf{P}(M_t \geq x),
\end{aligned}$$

which is equivalent to (11.28) (do you see why?).

(ii) Clearly, $\{\tau_x \leq t\} = \{M_t \geq x\}$, and so, from (i), the DF of τ_x equals

$$\begin{aligned}
\mathbf{P}(\tau_x \leq t) &= 2\mathbf{P}(W_t \geq x) \\
&= \frac{2}{\sqrt{2\pi t}} \int_x^\infty e^{-y^2/(2t)} dy = \sqrt{\frac{2}{\pi}} \int_{x/\sqrt{t}}^\infty e^{-u^2/2} du.
\end{aligned}$$

By the fundamental theorem of calculus and the chain rule, for a continuous function f and differentiable g, one has $\frac{d}{dt}\int_{g(t)}^\infty f(u)\,du = -g'(t)f(g(t))$. Applying this to the above formula yields (11.29).

□

It immediately follows from the self-similarity property (11.15) (put there $a := x$) that $\tau_x \stackrel{d}{=} x^2 \tau_1$, $x > 0$. That, of course, is also obvious from the density transformation formula (2.27). Note that the distribution of the time τ_1 (that is sometimes referred to as the *Lévy distribution*) is one of the so-called *stable distributions* that play a key role in the theory of summation of independent random variables: they are the only distributions that can emerge in the limit when one considers sequences of cumulative sums of i.i.d. RVs. Note that, due to independence of increments, the time since our BM hits level $x = 1$ till it first hits level $x = 2$ is independent of and has the same distribution as τ_1 *etc.* Hence, for any $n \geq 1$,

$$\tau_n = (\tau_1 - \tau_0) + (\tau_2 - \tau_1) + \cdots + (\tau_n - \tau_{n-1}) \stackrel{d}{=} \sum_{k=1}^{n} \tau_1^{(k)},$$

where $\tau_1^{(k)} := \tau_k - \tau_{k-1}$, $k \geq 1$, are i.i.d. copies of τ_1 (cf. our discussion of the simple RW on p. 314). Therefore, the self-similarity implies that

$$n^{-2} \sum_{k=1}^{n} \tau_1^{(k)} \stackrel{d}{=} \tau_1. \tag{11.30}$$

That means, in particular, that the LLN does not hold for τ_1. Moreover, the average of n RVs $\tau_1^{(k)}$ has the same distribution as $n\tau_1$, so averaging these RVs does not lead to averaging anything out! Of course, that the LLN fails should not be surprising since the density (11.29) decays at infinity as $ct^{-3/2}$ implying that $\mathbf{E}\,\tau_x = \int_0^\infty t f_{\tau_x}(t)\,dt = \infty$.

Other examples of stable distributions include the normal and Cauchy distributions. All the stable distributions have infinitely many times differentiable densities, but only these two distribution types, together with the Lévy distribution, have densities that admit closed-form analytic representations. However, the ChFs of all stable distributions have very simple representations. Thus, *symmetric* stable RVs X have ChFs of the form $\mathbf{E}\,e^{iuX} = e^{-c|u|^\alpha}$, where $c > 0$ and $\alpha \in (0, 2]$ are constant, the parameter α being known as the index (or exponent) of the stable distribution ($\alpha = 1$ yields the Cauchy distribution, while $\alpha = 2$ corresponds to the normal one). As $x \to \pm\infty$, the density $f_X(x)$ of a stable RV of index $\alpha < 2$ decays as $c_\pm |x|^{-\alpha-1}$ (except for the so-called one-sided cases, like our τ_1), meaning that the stable distributions with $\alpha < 2$ are *heavy-tailed.*

Since $\{-W_t\}_{t\geq 0}$ is also a BM process (set $a := -1$ in (11.15)), we see that, for $x < 0$,

$$\begin{aligned}\mathbf{P}\Big(\min_{0\leq s\leq t} W_s \leq x\Big) &= \mathbf{P}\Big(-\max_{0\leq s\leq t}(-W_s) \leq x\Big) \\ &= \mathbf{P}\,(M_t \geq -x) = 2(1 - \Phi(-x/\sqrt{t}\,)) = 2\Phi(x/\sqrt{t}\,).\end{aligned}$$

It is also obvious that $\tau_x \stackrel{d}{=} \tau_{|x|}$.

We will conclude this section by observing that, for the RV $X := e^{\mu + \sigma W_t}$ with $\mu \in \mathbf{R}$, $\sigma > 0$, for $x > 0$ one has

$$F_X(x) = \mathbf{P}\,(\mu t + \sigma W_t \leq \ln x) = \mathbf{P}\left(W_t \leq \frac{\ln x - \mu t}{\sigma}\right) = \Phi\left(\frac{\ln x - \mu t}{\sigma\sqrt{t}}\right).$$

Hence the density of X is

$$f_X(x) = \frac{dF_X(x)}{dx} = \frac{1}{\sigma x\sqrt{2\pi t}} \exp\left\{-\frac{(\ln x - \mu t)^2}{2\sigma^2 t}\right\}, \quad x > 0.$$

Of course, the last expression could also be derived using the density transformation formula (2.27). Distributions (and RVs) with such densities are called *lognormal.*[7]

11.2.4 *The Three Martingales of the BM Process*

Martingales proved to be very powerful and convenient tools of analysis. One of popular ways to use them involves applying the Optional Stopping Theorem 11.2 (which holds in continuous time as well) to suitably chosen MGs and STs. We already saw how that idea works in discrete time (p. 313), while Example 11.8 below will illustrate it in the continuous time case.

The next assertion exhibits the most useful MGs related to the BM process.

Theorem 11.6. *The following three SPs are martingales w.r.t. the natural filtration* $\{\mathcal{F}_t\}_{t\geq 0}$ *of the BM* $\{W_t\}_{t\geq 0}$:

(i) $\{W_t\}_{t\geq 0}$;

(ii) $\{Y_t := W_t^2 - t\}_{t\geq 0}$;

(iii) $\{Z_t := \exp\{uW_t - u^2 t/2\}\}_{t\geq 0}$, *where* $u \in \mathbf{R}$ *is fixed.*

Proof Assertion (i) is obvious (cf. Example 11.2). To prove (ii), observe that $\{Y_t\}_{t\geq 0}$ is integrable: $\mathbf{E}\,|Y_t| \leq \mathbf{E}\,W_t^2 + t = 2t < \infty$, and that it is clearly adapted to $\{\mathcal{F}_t\}_{t\geq 0}$. Further, for $s, t \geq 0$,

$$\begin{aligned}
\mathbf{E}\,(Y_{t+s}|\mathcal{F}_t) &= \mathbf{E}\,((W_{t+s} - W_t + W_t)^2 - (t+s)|\mathcal{F}_t) \\
&= \underbrace{\mathbf{E}\,((W_{t+s} - W_t)^2|\mathcal{F}_t)}_{=\mathbf{E}\,(W_{t+s}-W_t)^2 \text{ by CE3}} + 2\underbrace{\mathbf{E}\,((W_{t+s} - W_t)W_t|\mathcal{F}_t)}_{=W_t\mathbf{E}\,(W_{t+s}-W_t|\mathcal{F}_t) \text{ by CE2}} \\
&\quad + \underbrace{\mathbf{E}\,(W_t^2|\mathcal{F}_t)}_{=W_t^2 \text{ by CE2}} - (t+s) \\
&= s + 2W_t \times 0 + W_t^2 - (t+s) = Y_t
\end{aligned}$$

since $W_{t+s} - W_t \sim N(0, s)$ and $\mathbf{E}\,(W_{t+s} - W_t|\mathcal{F}_t) = 0$ by (i).

(iii) The SP $\{Z_t\}_{t\geq 0}$ is clearly integrable:

$$\mathbf{E}\,|Z_t| = \mathbf{E}\,Z_t = e^{-u^2 t/2}\mathbf{E}\,e^{uW_t} = 1$$

[7] In other words, an RV is said to be lognormal if its logarithm is normally distributed.

(see Example 2.4, p. 51) and adapted to $\{\mathcal{F}_t\}_{t\geq 0}$. Now note that, by CE2,

$$\frac{1}{Z_t}\mathbf{E}(Z_{t+s}|\mathcal{F}_t) = \mathbf{E}\left(\frac{Z_{t+s}}{Z_t}\,\middle|\,\mathcal{F}_t\right)$$

$$= \mathbf{E}\left(\exp\{u\underbrace{(W_{t+s}-W_t)}_{\text{independent of }\mathcal{F}_t} - \tfrac{u^2}{2}(t+s-t)\}\,|\mathcal{F}_t\right)$$

$$= \mathbf{E}\,e^{u(W_{t+s}-W_t)}e^{-u^2s/2} = 1,$$

again using the fact that $W_{t+s} - W_t \sim N(0,s)$ and Example 2.4. Thus, we showed that $\mathbf{E}(Z_{t+s}|\mathcal{F}_t) = Z_t$, which completes the proof of the theorem. □

Note the similarities and differences in the behaviour of the three MGs. "Locally" (i.e., on small time intervals), they behave very similarly to each other: they have "jerky Brownish trajectories", as all three SPs are smooth functions of the BM process, and smooth functions are "locally linear". The difference will be in the amplitude of the "local oscillations": for a smooth function f and small $s > 0$, one has $f(W_{s+t}) - f(W_t) \approx f'(W_t)(W_{t+s} - W_t)$, so for Y_t the local oscillations will tend to be large when $|W_t|$ is large and small when it is small, while for Z_t the local oscillations will tend to be very large when W_t is large and very small when W_t assumes large negative values. We will formalise this idea later, when discussing stochastic differentials.

In the long run, as $t \to \infty$, the BM process $\{W_t\}$ behaves as a simple symmetric RW: it oscillates, keeps returning to 0, the "typical values" of W_t being of the magnitude $\sqrt{t}$, so that its "excursions" away from zero tend to become longer and longer.

For the second MG, one always has $Y_t \geq -t$, with the relation turning into equality each time when the BM W_t returns to zero. So there will be arbitrary large values of t such that $Y_t = -t$, and also such that $Y_t > 0$ (and even $Y_t > t$, which follows from the law of the iterated logarithm).

For the third MG, one has

$$Z_t = \exp\{uW_t - u^2t/2\} = \exp\{t(\underbrace{uW_t/t}_{\to 0}\underbrace{-u^2/2}_{<0})\} \to 0 \quad \text{a.s. as} \quad t \to \infty$$

by the LLN. At the same time, due to the martingale property, $\mathbf{E}\,Z_t \equiv 1$ (which implies, in particular, that the family of RVs $\{Z_t\}_{t\geq 0}$ cannot be uniformly integrable, see p. 38). This means that Z_t can assume rather large values with "noticeable" probabilities.

It would be very instructive for the reader to simulate on computer a number of trajectories of $\{W_t\}_{0\leq t\leq T}$ (for a large enough T) and plot them, along with the graphs of the respective realisations of $\{Y_t\}_{0\leq t\leq T}$ and $\{Z_t\}_{0\leq t\leq T}$.

Example 11.8. We will now show how to use the exponential MG $\{Z_t\}$ to derive the Laplace transform of the first hitting time τ_x defined in (11.27). Assume w.l.o.g. that $x > 0$.

For any $u > 0$, by Theorem 11.2[8],

$$1 = \mathbf{E}\,Z_0 = \mathbf{E}\,Z_{\tau_x} = \exp\{u\underbrace{W_{\tau_x}}_{=x} - u^2\tau_x/2\} = e^{ux}\mathbf{E}\,e^{-(u^2/2)\tau_x}.$$

[8] It is easy to see that condition (11.11) holds since $W_t < x$ on the event $\{\tau_x > t\}$, so the theorem is applicable here.

Therefore, putting $s := u^2/2$ (so that $u = \sqrt{2s}$), we obtain

$$\mathbf{E}\, e^{-s\tau_x} = e^{-ux} = e^{-x\sqrt{2s}}.$$

Recalling the product property (2.70) of the Laplace transform of the sum of independent RVs, we see that, for the sum $\mathcal{T}_n := \sum_{k=1}^n \tau_1^{(k)}$ of n independent copies of τ_1, one has

$$\mathbf{E}\, e^{-s\mathcal{T}_n} = \left(\mathbf{E}\, e^{-s\tau_x}\right)^n = \left(e^{-x\sqrt{2s}}\right)^n = e^{-x\sqrt{2sn^2}} = \mathbf{E}\, e^{-sn^2\tau_1}, \quad s \geq 0,$$

meaning that $\mathcal{T}_n \stackrel{d}{=} n^2\tau_1$, which is equivalent to (11.30).

11.3 Defining the Itô Integral

We will start with the following motivating modelling argument. Suppose we have got a physical system of which the state is described by a quantity X_t such that, over a short time interval $(t, t+\Delta t)$, $\Delta t \to 0$, its change $\Delta X_t := X_{t+\Delta t} - X_t$ has the property that

$$\Delta X_t \approx a(t, X_t)\Delta t, \tag{11.31}$$

where the sloppy "$\approx$" means that ΔX_t is actually equal to the right-hand side up to an error $o(\Delta t)$ (the reader must have seen examples of such systems — including pendula, simple chemical reactions *etc.* — in first year calculus courses). Now in the limit, as $\Delta t \to 0$, one obtains that

$$\frac{\Delta X_t}{\Delta t} \to \frac{dX_t}{dt} = a(t, X_t), \tag{11.32}$$

which is just an ordinary differential equation that is sometimes more convenient to write down in the differential form as

$$dX_t = a(t, X_t)\, dt.$$

Integrating this gives an equivalent *integral equation*

$$X_t = X_0 + \int_0^t a(s, X_s)\, ds, \quad t > 0.$$

Note that we could obtain this integral equation directly from (11.31), adding up the increments of X_s over intervals of partitions of $[0, t]$ with vanishing mesh and using the "telescopic argument" on the left-hand side.

Now what if there is a random perturbation (by some form of "white noise") present in the dynamics, i.e., on the right-hand side of (11.31)? That is, one has

$$\Delta X_t \approx a(t, X_t)\Delta t + \text{random noise},$$

a possible form of the random noise being $b(t, X_t)(W_{t+\Delta t} - W_t)$, where $\{W_t\}$ is a standard BM. That form would mean that, given the current value of X_t, the random noise is Gaussian, with variance proportional to Δt, the proportionality coefficient (i.e., $b^2(t, X_t)$) depending on t and X_t (in fact, as we will see below, in this construction both coefficients a and b can dependent on the values X_s, $s \le t$, as well).

Summing up the above differences, we seem to arrive at an equation of the form

$$X_t = X_0 + \underbrace{\int_0^t a(s, X_s)\, ds}_{\text{This is OK.}} + \underbrace{\int_0^t b(s, X_s)\, dW_s}_{\text{What is that?}}, \quad t > 0.$$

As we saw in Section 11.2.2, the total variation of the BM trajectories is infinite, so the last integral cannot be defined in the sense of Lebesgue integration. In this section, we will describe the construction of the so-called *Itô integral*[9] which nevertheless makes such objects meaningful and thereby lays the foundation of stochastic calculus.

To make the presented construction of integrals of the form $\int_{t_1}^{t_2} f_t\, dW_t$ look natural, observe first that they are supposed to stand for something limiting for sums of the form

$$\sum (\text{value of } f_t) \times (\text{increment of } W_t).$$

Therefore they should have the following properties: for $0 \le t_1 < t_2$,

(i) For a non-random constat c, one has $\int_{t_1}^{t_2} cf_t\, dW_t = c \int_{t_1}^{t_2} f_t\, dW_t$.
(ii) $\int_{t_1}^{t_2} dW_t = W_{t_2} - W_{t_1}$.
(iii) For $t_1 < t_2 < t_3$, one has $\int_{t_1}^{t_3} f_t\, dW_t = \int_{t_1}^{t_2} f_t\, dW_t + \int_{t_2}^{t_3} f_t\, dW_t$.

Note that (i) implies (put there $c := 0$) that $\int_{t_1}^{t_2} 0 \cdot dW_t = 0$.

Now we will base on the above-listed desirable (if not mandatory) integral properties (i)–(iii) to construct the integral $\int_0^T f_t\, dW_t$ for an appropriate class of (random) functions $f_t = f_t(\omega)$ (extension from $\int_0^T$ to $\int_{t_1}^{t_2}$ is obvious from (iii)). We will do that in four steps.

Step 1. Take the "simplest non-trivial" f_t: for a (non-random) constant c, let

$$f_t := \begin{cases} c, & t_1 \le t < t_2, \\ 0 & \text{otherwise.} \end{cases}$$

[9]Named so after Kiyoshi Itô (also spelled as Itō, 07.09.1915–10.11.2008), a prominent Japanese mathematician who laid the foundations of stochastic calculus in his seminal 1944 paper.

Then

$$I_T(f) := \int_0^T f_t\,dW_t = \underbrace{\int_0^{t_1} 0\cdot dW_t}_{=0} + \underbrace{\int_{t_1}^{t_2} c\,dW_t}_{=c\int_{t_1}^{t_2} dW_t} + \underbrace{\int_{t_2}^{T} 0\cdot dW_t}_{=0} = c(W_{t_2} - W_{t_1}).$$

Step 2. What if, instead of $c = \text{const}$, one has an RV X in the above step function f_t? It will be crucial for the construction to assume that:

- we have an underlying filtered probability space $(\Omega, \mathcal{F}, \mathbf{F}, \mathbf{P})$,
- the random integrand $\{f_t\}_{t\ge 0}$ is adapted to $\mathbf{F}$ (see p. 304), and that
- the random integrator $\{W_t\}_{t\ge 0}$ is a BM w.r.t. the filtration $\mathbf{F}$, meaning that $\{W_t\}_{t\ge 0}$ is a BM process adapted to $\mathbf{F}$ and such that, for any $s \ge 0$, the RVs $W_t - W_s$, $t \ge s$, are independent of $\mathcal{F}_s$.

The assumption that f_t is adapted usually makes sense. Indeed, recall our interpretation of $\mathbf{F} = \{\mathcal{F}_t\}_{t\ge 0}$ as the "flow of history", so the property that f_t is $\mathcal{F}_t$-measurable means that the value of f_t is completely specified by the history of the world up to the current time t, and it is integrands of such nature that usually appear in real-life problems.

As one has $f_{t_1} = X$, the above means that X must be $\mathcal{F}_{t_1}$-measurable. That is, at time t_1 the value of X becomes known to the world and so, "from the viewpoint" of the integrator's "future" $\{W_t - W_{t_1}\}_{t\ge t_1}$ that is independent of $\mathcal{F}_{t_1}$ (and it is only those increment values that will matter for integration purposes on (t_1, T)), that value of X is basically a constant. Therefore, arguing as in Step 1, we should put

$$I_T(f) = \int_0^T f_t\,dW_t := X(W_{t_2} - W_{t_1}).$$

For our f_t and a fixed $s \ge 0$, it is obvious that $f_t\mathbf{1}_{(0,s]}(t)$, $t \ge 0$, is a step function again, and so it is easy to see that we have

$$I_s(f) := \int_0^s f_t\,dW_t = \int_0^T f_t\mathbf{1}_{(0,s]}(t)\,dW_t = X(W_{s\wedge t_2} - W_{s\wedge t_1}).$$

Observe that the SP $\{I_t(f)\}_{t\ge 0}$ is adapted to $\mathbf{F}$ and has continuous trajectories, which is obvious from the representation

$$I_t(f) = \begin{cases} 0, & t \in [0, t_1], \\ X(W_t - W_{t_1}), & t \in (t_1, t_2], \\ X(W_{t_2} - W_{t_1}), & t > t_2 \end{cases}$$

(recall that X is $\mathcal{F}_{t_1}$-measurable and hence also $\mathcal{F}_t$-measurable for all $t \ge t_1$, and that $\{W_t\}$ is adapted and has continuous trajectories).

Step 3. Next we will define the Itô integral for the so-called *simple processes* $\{f_t\}$ that are characterised by the following property: for some non-random times

$$0 = t_0 < t_1 < \cdots < t_{N-1} < t_N = T$$

and RVs $X_1, X_2, \ldots, X_N$ such that X_k is $\mathcal{F}_{t_{k-1}}$-measurable and $\mathbf{E}\, X_k^2 < \infty$, $k = 1, 2, \ldots, N$, one has

$$f_t = \sum_{k=1}^{N} X_k \mathbf{1}_{[t_{k-1}, t_k)}(t), \quad t \in [0, T].$$

Note that a simple process is adapted to $\mathbf{F}$, as the sum of adapted processes.

For a simple process $\{f_t\}$, its Itô integral is defined, in accordance with Step 2 and (iii), by

$$I_T(f) = \int_0^T f_s\, dW_s = \int_0^{t_1} + \int_{t_1}^{t_2} + \cdots + \int_{t_{N-1}}^{t_N} = \sum_{k=1}^{N} X_k (W_{t_k} - W_{t_{k-1}}).$$

As it was the case at the previous step, here we also see that the process $\{I_t(f)\}$ is adapted to $\mathbf{F}$ and has continuous trajectories, as it is the sum of processes with such properties:

$$I_f(f) = \sum_{k=1}^{N} I(f_t^{(k)}), \quad f_t^{(k)} := X_k \mathbf{1}_{[t_{k-1}, t_k)}(t). \tag{11.33}$$

Theorem 11.7. (i) *For any constants* a, b *and simple processes* f, g,

$$I_t(af + bg) = aI_t(f) + bI_t(g), \quad t \in [0, T].$$

(ii) *The SP* $\{I_t(f)\}_{t\in[0,T]}$ *is an MG:*

$$\mathbf{E}\,(I_t(f)|\mathcal{F}_s) = I_s(f), \quad 0 \le s \le t \le T,$$

and hence $\mathbf{E}\, I_t(f) = \mathbf{E}\, I_0(f) = 0$, $t \in [0, T]$.

(iii) $\mathbf{E}\, I_t^2(f) = \int_0^t \mathbf{E}\, f_s^2 ds$, $t \in [0, T]$.

The relation from (iii) is known as *Itô isometry*. Note that the square roots of both sides of that relation are L^2-norms of two related objects in two different spaces. On the right-hand side, we refer to the L^2-norm of the function $f_t(\omega)$ on the measure product space $([0, t] \times \Omega,\ \mathcal{B}([0, t]) \otimes \mathcal{F},\ \lambda \times \mathbf{P})$, λ being the Lebesgue measure. On the left, we refer to the L^2-norm of the RV $I_t(f)$, which is the "image" of that $f_t(\omega)$ under the constructed operation of Itô integration $(f \mapsto I_t(f))$ in the measure space $(\Omega, \mathcal{F}, \mathbf{P})$. The assertion tells us that they coincide, so that there is an *isometry*, i.e., a distance-preserving (in the respective L^2 spaces, and hence scalar product-preserving as well) mapping of the collection of all the f's that we consider here to the collection of their images $I_t(f)$.

Proof Assertion (i) is obvious.

(ii) As we already said, $\{I_t(f)\}$ is adapted to $\mathbf{F}$. Further, by virtue of CE1 and (11.33), we need to verify the martingale property for "single step integrands" $f_t^{(k)}$ only. So we can now assume that the SP f_t is of the form from Step 2. It will be convenient to consider three different cases, according to the location of the points s and t.

Case 1. Either [a] $t < t_1$ or [b] $s \geq t_2$:

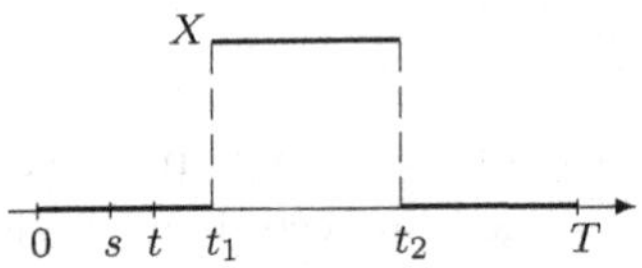

or

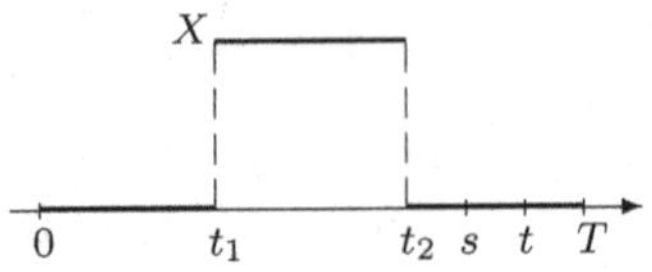

In configuration [a], obviously $I_s(f) = I_t(f) = 0$, whereas in [b] one has $I_s(f) = I_t(f) = X(W_{t_2} - W_{t_1}) = I_T(f)$. Therefore, in both situations,

$$\mathbf{E}\left(I_t(f)|\mathcal{F}_s\right) = \mathbf{E}\left(I_s(f)|\mathcal{F}_s\right) = I_s(f).$$

Case 2. Either [a] $t_1 \leq s < t < t_2$ or [b] $t_1 \leq s < t_2 \leq t$:

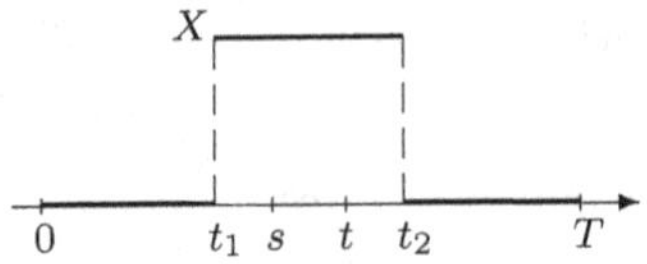

or

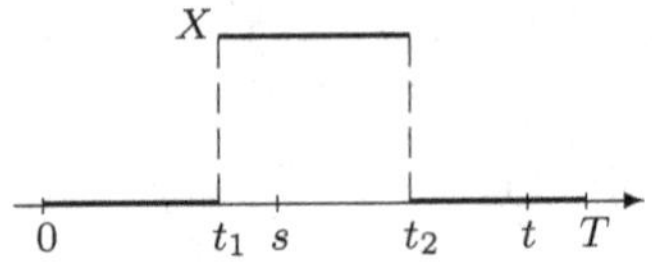

In configuration [a], one clearly has $I_u(f) = X(W_u - W_{t_1})$, $u = s, t$. So, from CE2 one obtains that

$$\begin{aligned}\mathbf{E}\left(I_t(f)|\mathcal{F}_s\right) - I_s(f) &= \mathbf{E}\left(I_t(f) - I_s(f)|\mathcal{F}_s\right) = \mathbf{E}\left(X(W_t - W_s)|\mathcal{F}_s\right)\\ &= X\mathbf{E}\left(W_t - W_s|\mathcal{F}_s\right) = X(\mathbf{E}\left(W_t|\mathcal{F}_s\right) - W_s) = 0,\end{aligned}$$

where we used CE3 and the assumption that $\{W_t\}$ is an MG w.r.t. $\mathbf{F}$. Situation [b] is considered in a similar way.

Case 3. Either [a] $s < t_1 \leq t < t_2$ or [b] $s < t_1 < t_2 \leq t$:

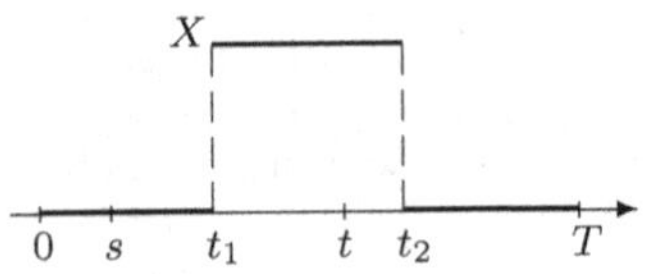

or

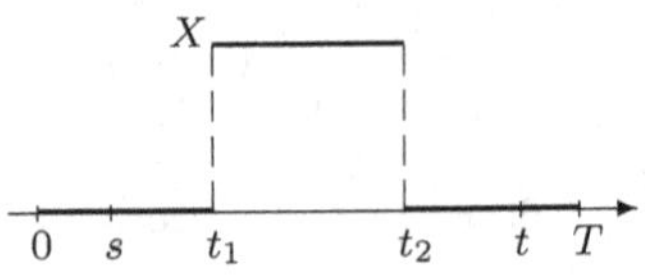

In configuration [a], using properties CE2, CE4 and the fact that X is $\mathcal{F}_{t_1}$-measurable, one has

$$\begin{aligned}\mathbf{E}\left(I_t(f)|\mathcal{F}_s\right) &= \mathbf{E}\left(X(W_t - W_{t_1})|\mathcal{F}_s\right)\\ &= \mathbf{E}\left[\mathbf{E}\left(X(W_t - W_{t_1})|\mathcal{F}_{t_1}\right)\middle|\,\mathcal{F}_s\right]\\ &= \mathbf{E}\left[X\mathbf{E}\left(W_t - W_{t_1}|\mathcal{F}_{t_1}\right)\middle|\,\mathcal{F}_s\right]\\ &= \mathbf{E}\left[X(\mathbf{E}\left(W_t|\mathcal{F}_{t_1}\right) - W_{t_1})\middle|\,\mathcal{F}_s\right] = 0,\end{aligned}$$

the last equality holding since $\{W_t\}$ is an MG, so that $\mathbf{E}\left(W_t|\mathcal{F}_{t_1}\right) - W_{t_1} = 0$. Configuration [b] is considered in a similar way.

(iii) We will only deal with the case $t = T$, as the general case is treated in the same way. Setting $\xi_k := X_k(W_{t_k} - W_{t_{k-1}})$, $k = 1, \ldots, N$, we have

$$\mathbf{E}\, I_T^2(f) = \mathbf{E}\left[\sum_{k=1}^N \xi_k\right]^2 = \sum_{k=1}^N \mathbf{E}\,\xi_k^2 + 2\sum_{1\le j<k\le N} \mathbf{E}\,\xi_j\xi_k.$$

Using properties CE4 and CE2,

$$\begin{aligned}\mathbf{E}\,\xi_k^2 &= \mathbf{E}\,\mathbf{E}\left(X_k^2(W_{t_k} - W_{t_{k-1}})^2|\mathcal{F}_{t_{k-1}}\right)\\ &= \mathbf{E}\left[X_k^2\mathbf{E}\left((W_{t_k} - W_{t_{k-1}})^2|\mathcal{F}_{t_{k-1}}\right)\right]\\ &= \mathbf{E}\left[X_k^2\mathbf{E}\left(W_{t_k} - W_{t_{k-1}}\right)^2\right] = (t_k - t_{k-1})\mathbf{E}\,X_k^2,\end{aligned}$$

where the second last equality follows from independence of $W_{t_k} - W_{t_{k-1}}$ of $\mathcal{F}_{t_{k-1}}$ and CE3.

As for the terms in the double sum, they are all zeros: since, for $j < k$, the RVs X_j, X_k and $W_{t_j} - W_{t_{j-1}}$ are $\mathcal{F}_{t_{k-1}}$-measurable, property CE4 yields

$$\begin{aligned}\mathbf{E}\,\xi_j\xi_k &= \mathbf{E}\,\mathbf{E}\left[X_j(W_{t_j} - W_{t_{j-1}})X_k(W_{t_k} - W_{t_{k-1}})\middle|\,\mathcal{F}_{t_{k-1}}\right]\\ &= \mathbf{E}\left[X_j(W_{t_j} - W_{t_{j-1}})X_k\underbrace{\mathbf{E}\left(W_{t_k} - W_{t_{k-1}}|\mathcal{F}_{t_{k-1}}\right)}_{=0 \text{ as } \{W_t\} \text{ is an MG}}\right] = 0.\end{aligned}$$

Note that we actually did not have to use CEs in the last two displayed formulae: we could simply use the fact that X_k is independent of $W_{t_k} - W_{t_{k-1}}$. The point of employing the properties of CEs was to show that independence of increments of the "integrator" process ($\{W_t\}$ in this case) is not critical for constructing stochastic integrals: the same approach would work if one uses other MGs as well.

Summarising the above computations, we conclude that

$$\mathbf{E}\, I_T^2(f) = \sum_{k=1}^N (t_k - t_{k-1})\mathbf{E}\,X_k^2 = \int_0^T \mathbf{E}\, f_t^2 dt,$$

since $\mathbf{E}\, f_t^2$ is a step function equal to $\mathbf{E}\,X_k^2$ on $[t_{k-1}, t_k)$, $k = 1, \ldots, N$.

□

Step 4. Finally, one can extend the definition of the Itô integral to the class of adapted integrands f_s such that

$$\int_0^T \mathbf{E}\, f_s^2\, ds < \infty, \tag{11.34}$$

using approximations of such integrands by simple processes (in a sense, similarly to the approach used to construct the Lebesgue integral). It turns out that, for any such integrand f_s there exists a sequence $\{f_s^{(n)}\}_{n\geq 1}$ of simple processes such that

$$\int_0^T \mathbf{E}\,(f_s - f_s^{(n)})^2\, ds \to 0 \quad \text{as} \quad n \to \infty. \tag{11.35}$$

It will then follow from Itô isometry (Theorem 11.7(iii)) that the sequence $\{I_T(f^{(n)})\}_{n\geq 1}$ will converge in L^2 to a certain limit $I_T(f)$, which actually will be the same for any sequence $\{f^{(n)}\}_{n\geq 1}$ having property (11.35). Moreover, the same will also be true for the sequences $\{I_t(f^{(n)})\}_{n\geq 1}$, $t \in [0,T]$, of which the respective limits we will denote by $I_t(f)$.

It is these limiting RVs that are called the Itô integrals of our general integrands f:

$$\int_0^t f_s\, dW_s := I_t(f), \quad t \in [0,T].$$

In view of the above-mentioned fact that the limits are independent of the choice of approximating sequences, the definition is consistent.

It turns out that thus defined Itô integrals for adapted integrands f, g satisfying (11.34) inherit all the nice properties of such integrals of simple processes:

II1. *Linearity*: for constant a, b,

$$I_t(af + bg) = aI_t(f) + bI_t(g), \quad t \in [0,T].$$

II2. $\{I_t(f)\}_{t\in[0,T]}$ is an *adapted process with continuous trajectories.*[10]

II3. The SP $\{I_t(f)\}_{t\in[0,T]}$ is an *MG* w.r.t. $\mathbf{F}$:

$$\mathbf{E}\,(I_t(f)|\mathcal{F}_s) = I_s(f), \quad 0 \leq s \leq t \leq T,$$

and hence $\mathbf{E}\, I_t(f) = \mathbf{E}\, I_0(f) = 0$.

[10]To be more precise, this property must be stated as "there exists an equivalent modification of $\{I_t(f)\}_{t\in[0,T]}$ that is adapted and has continuous trajectories." Indeed, the RVs $I_t(f)$ were defined as L^2 limits only, so one cannot claim any path properties for them. In this exposition, though, we agreed to omit such technicalities.

II4. The *Itô isometry*:

$$\mathrm{Var}\,(I_t(f)) = \mathbf{E}\, I_t^2(f) = \int_0^t \mathbf{E}\, f_s^2\, ds, \quad t \in [0,T].$$

Moreover, for any f, g satisfying condition (11.34),

$$\mathrm{Cov}\,(I_t(f), I_t(g)) = \mathbf{E}\, I_t(f) I_t(g) = \int_0^t \mathbf{E}\, f_s g_s\, ds, \quad t \in [0,T]$$

(see Problem 22 on p. 347).

Remark 11.1. In the important special case where the integrand f_t is non-random, one has

$$I_t(f) = \int_0^t f_s dW_s \sim N\Big(0, \int_0^t f_s^2 ds\Big).$$

Moreover, the process $\{I_t(f)\}_{t\in[0,T]}$ will have independent increments, and so it will be nothing else but a time-changed Wiener process: for some non-decreasing function $a(t) \geq 0$ and BM $\{\widetilde{W}_t\}$, one has $I_t(f) = \widetilde{W}_{a(t)}$.

Indeed, a non-random f_t can be approximated by a sequence of non-random step functions $f_t^{(n)} = c_k^{(n)}$, $t \in [t_{k-1}^{(n)}, t_k^{(n)})$, $k = 1, \ldots, n$. For each of them,

$$Z_n := I_t(f^{(n)}) = \sum_{k=1}^{n} c_k^{(n)} (W_{t_k^{(n)}} - W_{t_{k-1}^{(n)}})$$

is obviously Gaussian as a sum of independent normal RVs, with zero mean and variance $\sigma_n^2 := \sum_{k=1}^n (c_k^{(n)})^2 (t_k^{(n)} - t_{k-1}^{(n)})$. Therefore the L^2 limit $I_t(f)$ can only be Gaussian as well. To see that, recall that if Z_n converge, as $n \to \infty$, to $I_t(f)$ in L^2, then they also converge in distribution (see p. 43), and so their ChFs converge point-wise to the ChF of $I_t(f)$. For $Z_n \sim N(0, \sigma_n^2)$, one has

$$\varphi_{Z_n}(u) = e^{-\sigma_n^2 u^2/2}, \quad u \in \mathbf{R}.$$

Clearly, this expression converges to a limit iff, for some $\sigma \geq 0$, one has $\sigma_n^2 \to \sigma^2$ as $n \to \infty$, and then $\varphi_{I_t(f)}(u) = e^{-\sigma^2 u^2/2}$, $u \in \mathbf{R}$, which means that $I_t(f) \sim N(0, \sigma^2)$.

The values of the parameters of the distribution of $I_t(f)$ follow from II3 and II4.

Finally, that the increments of the SP $\{I_t(f^{(n)})\}_{t\in[0,T]}$ in the case of non-random simple process $f_t^{(n)}$ are independent and normally distributed is obvious. Likewise, it is also clear that the limiting process should then have independent Gaussian increments as well. Therefore, both processes $\{I_t(f)\}_{t\in[0,T]}$ and $\{W_{a(t)}\}_{t\in[0,T]}$ with $a(t) := \int_0^t f_s^2\, ds$ are Gaussian, have zero mean functions and the same covariance functions (the latter follows from II4 and (11.19)), which means that the SPs have the same distribution.

Example 11.9. *Find the distribution of* $X := \int_0^t s\, dW_s$. As the integrand is non-random, X is normally distributed with zero mean and variance

$$\mathbf{E}\, X^2 = \int_0^t s^2\, ds = \frac{t^3}{3}.$$

That is, $X \sim N(0, t^3/3)$.

Example 11.10. *Compute* $\int_0^t W_s\, dW_s$. We will do this be constructing a sequence of approximating simple processes $f^{(n)}$ and then computing the limit of $I_t(f^{(n)})$.

Set $f_s^{(n)} := W_{t(k-1)/n}$ for $t(k-1)/n \le s < tk/n$, $k = 1, 2, \ldots, n$. Then, by the definition of Itô integral for simple processes,

$$\int_0^t f_t^{(n)}\, dW_s = \sum_{k=1}^n W_{t(k-1)/n}(W_{tk/n} - W_{t(k-1)/n})$$
$$= \frac{W_t^2}{2} - \frac{1}{2}\underbrace{\sum_{k=1}^n (W_{tk/n} - W_{t(k-1)/n})^2}_{\to t,\ \text{cf. (11.24)}} \to \frac{W_t^2}{2} - \frac{t}{2}.$$

The second equality holds since, provided that $a_0 = 0$, one has

$$\sum_{k=1}^n a_{k-1}(a_k - a_{k-1}) = -\left(\frac{1}{2}a_0^2 - a_0a_1 + \frac{1}{2}a_1^2\right) - \left(\frac{1}{2}a_1^2 - a_1a_2 + \frac{1}{2}a_2^2\right) - \cdots$$
$$- \left(\frac{1}{2}a_{n-1}^2 - a_{n-1}a_n + \frac{1}{2}a_n^2\right) + \frac{1}{2}a_n^2 = \frac{1}{2}a_n^2 - \frac{1}{2}\sum_{k=1}^n (a_k - a_{k-1})^2.$$

Thus the desired Itô integral is nothing else but the second MG of the BM process (Theorem 11.6(ii)).

Now suppose for a moment that the usual rules of differentiation and integration were applicable in Itô calculus and, in particular, that one had $d(W_s)^2 = 2W_s dW_s$. That would imply that

$$\int_0^t W_s\, dW_s = \int_0^t \frac{1}{2}\, d(W_s)^2 = \frac{W_t^2}{2},$$

which is clearly wrong (as the resulting process is not even an MG!). So one concludes that, say, the usual chain rule fails in stochastic calculus. There is, however, a modification of that formula that works for Itô integrals!

11.4 The Itô Formula

One says that $\{X_t\}_{t\in[0,T]}$ is an *Itô process* (on a filtered probability space $(\Omega, \mathcal{F}, \mathbf{F}, \mathbf{P})$ with a BM $\{W_t\}_{t\in[0,T]}$ on it) if

$$X_t = X_0 + \int_0^t a_s\, ds + \int_0^t b_s\, dW_s, \quad t \in [0, T], \tag{11.36}$$

where

- X_0 is $\mathcal{F}_0$-measurable,

• $\{a_t = a_t(\omega)\}_{t\in[0,T]}$ is an **F**-adapted process with measurable trajectories such that $\int_0^T |a_s|\,ds < \infty$,

• $\{b_t \equiv b_t(\omega)\}_{t\in[0,T]}$ is an **F**-adapted process such that $\int_0^T \mathbf{E}\, b_s^2\, ds < \infty$ (so that the Itô integral of b_t is defined).

One says that Itô process (11.36) has on $[0,T]$ *stochastic differential*

$$dX_t = a_t dt + b_t dW_t, \tag{11.37}$$

where a_t is referred to as the *drift coefficient*, and b_t as the *diffusion coeffieicent* of the process $\{X_t\}$.

Thus, whenever you see a differential formula like (11.37), you should keep in mind that what it actually means is an integral relation like (11.36).

Theorem 11.8. *Assume that $\{X_t\}_{t\in[0,T]}$ is an Itô process with stochastic differential* (11.37), *$f(x)$ is a twice continuously differentiable function. Then $Y_t := f(X_t)$, $t \in [0,T]$, is also an Itô process, with stochastic differential*

$$dY_t \equiv df(X_t) = \left[f'(X_t)a_t + \frac{1}{2}f''(X_t)b_t^2\right]dt + f'(X_t)b_t dW_t. \tag{11.38}$$

The formula is actually quite easy to understand and memorise. Indeed, introduce the following "multiplication table" for differentials:

$$\begin{aligned} dt \cdot dt &= (dt)^2 = 0, \\ dt \cdot dW_t &= 0, \\ dW_t \cdot dW_t &= (dW_t)^2 = dt. \end{aligned} \tag{11.39}$$

The meaning of the above is that, in approximating integral sums, the eventual contribution of the terms corresponding to the first two lines will be negligibly small compared to the contribution from terms corresponding to dt and dW_t (we will discuss that shortly).

Using the table, the Itô formula (11.38) can be re-written as

$$df(X_t) = f'(X_t)dX_t + \frac{1}{2}f''(X_t)(dX_t)^2. \tag{11.40}$$

Indeed, given the process has stochastic differential (11.37), one has from multiplication table (11.39) that

$$(dX_t)^2 = a_t^2\underbrace{(dt)^2}_{=0} + 2a_tb_t\underbrace{dt\cdot dW_t}_{=0} + b_t^2\underbrace{(dW_t)^2}_{=dt} = b_t^2 dt,$$

so that (11.40) turns into (11.38).

Relation (11.40) can be interpreted as just Taylor's formula for f with two terms and differentials instead of differences. While in usual calculus, when one uses the Taylor expansion to represent the function's increment corresponding to a finite increment of the variable and then passes to the limit (whereby finite differences turn into infinitesimal quantities), only the first term of the expansion will eventually matter, in stochastic calculus one must retain the first two terms.

To prove (11.38), one needs to show that a relation of the form (11.36) holds for $\{Y_t\}$, with the drift and diffusion coefficients specified by (11.38). To do that, one can use the telescoping argument to write

$$Y_t - Y_0 = f(X_t) - f(X_0) = \sum_{k=1}^{n} (f(X_{t_k}) - f(X_{t_{k-1}})),$$

for a partition $0 = t_0 < t_1 < \cdots < t_n = t$ of $[0, t]$, apply the Taylor formula with two terms and reminder in Peano's form to the function increments in the sum, and then analyse what happens in the limit, when the partition mesh tends to zero. Using Δ to denote the differences of the respective quantities, the main outcome of that analysis will, roughly speaking, be that $\sum_k (\Delta t_k)^2 \to 0$ (which corresponds to the "multiplication table rule" $(dt)^2 = 0$), $\sum_k \Delta t_k \cdot \Delta W_{t_k} \to 0$ (the shorthand for that being $dt \cdot dW_t = 0$) and that $\sum_k (\Delta W_{t_k})^2 \to t$ (which corresponds to our $(dW_t)^2 = dt$).

Example 11.11. For $f(x) = \frac{1}{2}x^2$, $X_t = W_t$, one has $f'(x) = x$, $f''(x) = 1$, so that, according to (11.40), for $Y_t := f(X_t)$ one has

$$dY_t = d\left(\frac{1}{2}W_t^2\right) = f'(W_t)dW_t + \frac{1}{2}f''(W_t)(dW_t)^2 = dW_t + \frac{1}{2}\,dt,$$

which is in perfect agreement with the result of Example 11.10, of course.

Example 11.12. For $f(x) = e^x$ and $X_t = W_t$, one has $f'(x) = f''(x) = e^x$, so that, according to (11.40), for $Y_t := f(X_t)$ one has

$$dY_t = de^{W_t} = e^{W_t}dW_t + \frac{1}{2}\,e^{W_t}(dW_t)^2 = Y_t dW_t + \frac{1}{2}Y_t dt,$$

which actually looks like some kind of differential equation! We will discuss such equations in the next section.

The Itô formula (11.38) can be extended to the case of several variables, including situations where f depends on time as well. First we will state a version involving one univariate Itô process *and* the time variable. It can be proved in essentially the same way as the simpler Theorem 11.8. We will use the compact notation $\partial_t := \frac{\partial}{\partial t}$, $\partial_{xx} := \frac{\partial^2}{\partial x^2}$ *etc.* for partial derivatives.

Theorem 11.9. *Let $f(t, x)$ have continuous partial derivative $\partial_t f$ and be twice continuously differentiable in x, and let $\{X_t\}$ be an Itô process with stochastic differential* (11.37). *Then $Y_t := f(t, X_t)$ is also an Itô process, with stochastic differential*

$$dY_t = \partial_t f(t, X_t)dt + \partial_x f(t, X_t)dX_t + \frac{1}{2}\,\partial_{xx} f(t, X_t)(dX_t)^2. \qquad (11.41)$$

A simple calculation using (11.39) shows that the right-hand side of (11.41) equals

$$\Big[\partial_t f(t,X_t) + a_t\partial_x f(t,X_t) + \frac{1}{2}\,b_t^2\partial_{xx} f(t,X_t)\Big]\,dt + b_t\partial_x f(t,X_t)dW_t.$$

Example 11.13. For the geometric BM process (11.17), where we put $Z_0 := 1$ for simplicity, one has $Z_t = f(t,W_t)$, with $f(t,x) = e^{\mu t+\sigma x}$. Since $\partial_t f = \mu f$, $\partial_x f = \sigma f$, $\partial_{xx} f = \sigma^2 f$, one obtains from (11.41) that

$$\begin{aligned} dZ_t &= \mu f(t,W_t)dt + \sigma f(t,W_t)dW_t + \frac{\sigma^2}{2} f(t,W_t)(dW_t)^2 \\ &= \left(\mu + \frac{\sigma^2}{2}\right) Z_t dt + \sigma Z_t dW_t. \end{aligned}$$

Treating the case where f is a function of several Itô processes on a common filtered probability space, driven by different (possibly correlated) BM processes on it, is somewhat beyond the scope of the present text. The interested reader is referred to the books listed at the end of this chapter. We will just consider the following simple situation, that stresses the difference between the "usual" and Itô calculus.

Theorem 11.10. (The product rule of Itô calculus) *Let $\{X_t\}$ and $\{Y_t\}$ be two Itô processes on a common filtered probability space, satisfying*

$$dX_t = a_t dt + b_t dW_t, \qquad dY_t = \widetilde{a}_t dt + \widetilde{b}_t dW_t,$$

with a common BM $\{W_t\}$. Then their product $Z_t := X_tY_t$ is also an Itô process of which the stochastic differential equals

$$dZ_t \equiv d(X_tY_t) = Y_t dX_t + X_t dY_t + dX_t \cdot dY_t. \tag{11.42}$$

As the reader can easily verify using (11.39), the right-hand side here is equal to

$$(Y_t a_t + X_t \widetilde{a}_t + b_t \widetilde{b}_t)dt + (Y_t b_t + X_t \widetilde{b}_t)dW_t.$$

The product formula (11.42) can be proved in the same way as the "single-process" versions of the Itô formula. Informally, one can write the infinitesimal increment of Z_t as

$$d(X_tY_t) = (X_t + dX_t)(Y_t + dY_t) - X_tY_t,$$

and then just expand it.

Remark 11.2. If one of the processes $\{X_t\}$ and $\{Y_t\}$ is *differentiable* (and so contains no Itô integral term), we obtain the ordinary product rule of calculus, as in that case $dX_t{\cdot}dY_t = 0$ (since one of the stochastic differentials does not include the dW_t-term!). Therefore, in such a case one can use integration by parts when dealing with Itô integrals.

Example 11.14. For $Z_t = W_t e^{W_t}$, letting $X_t := W_t$, $Y_t := e^{W_t}$ and using the result of Example 11.12, we obtain:

$$\begin{aligned} dZ_t &= e^{W_t} dW_t + W_t de^{W_t} + dW_t \cdot de^{W_t} \\ &= e^{W_t} dW_t + W_t\Big(e^{W_t} dW_t + \frac{1}{2} e^{W_t} dt\Big) + dW_t \cdot \Big(e^{W_t} dW_t + \frac{1}{2} e^{W_t} dt\Big) \\ &= \Big(1 + \frac{1}{2} W_t\Big) e^{W_t} dt + (1 + W_t) e^{W_t} dW_t. \end{aligned}$$

Remark 11.3. Observe that an Itô process

$$X_t = X_0 + \int_0^t a_s\, ds + \int_0^t b_s\, dW_s, \quad t \in [0, T],$$

is an MG iff $a_t = 0$, $t \in [0, T]$. That the last condition is sufficient for the martingale property is obvious from II3.

Proving that it is also necessary is more difficult. We will only do that under the simplifying assumption that $\{a_t\}_{t \in [0,T]}$ is a bounded SP with continuous trajectories.

Note that if $\{X_t\}$ is an MG then, as $\{I_t(b)\}$ is also an MG, the SP $\left\{Y_t := \int_0^t a_s\, ds\right\}$ must be an MG as well. Therefore, for any $t \in [0, T]$,

$$0 \equiv Y_0 = \mathbf{E}\, Y_0 = \mathbf{E}\, Y_t = \mathbf{E} \int_0^t a_s\, ds = \int_0^t \mathbf{E}\, a_s\, ds. \tag{11.43}$$

Since a_t has continuous trajectories and is bounded, it follows from the dominated convergence theorem (p. 38) that $\mathbf{E}\, a_t$ is a continuous function. Differentiating (11.43) yields then that $\mathbf{E}\, a_t \equiv 0$. Therefore if, for a given $t \in (0, T]$, one has $\mathbf{P}\,(a_t = 0) < 1$, then there exists a $\delta > 0$ such that $p := \mathbf{P}\,(A) > 0$ for the event $A := \{a_t > \delta\}$ (how otherwise will one get $\mathbf{E}\, a_t = 0$ for such a t?).

The martingale property of $\{Y_t\}$ means that, for $t, t + h \in [0, T]$, $h > 0$,

$$0 = \mathbf{E}\,(Y_{t+h} | \mathcal{F}_t) - Y_t = \mathbf{E}\,(Y_{t+h} - Y_t | \mathcal{F}_t) = \mathbf{E}\left(\int_t^{t+h} a_s\, ds \,\middle|\, \mathcal{F}_t\right).$$

Since $A \in \mathcal{F}_t$ (as $\{a_t\}$ is adapted), the above implies (cf. p. 56) that

$$0 = \mathbf{E}\left(\int_t^{t+h} a_s\, ds; A\right) = \mathbf{E} \int_t^{t+h} a_s \mathbf{1}_A\, ds = \int_t^{t+h} \mathbf{E}\,(a_s \mathbf{1}_A)\, ds,$$

where we could again use Fubini's theorem because a_s is bounded. But the last integral cannot be zero since $\mathbf{E}\,(a_s \mathbf{1}_A) \to \mathbf{E}\,(a_t \mathbf{1}_A) > \delta p > 0$ as $s \searrow t$ (again by the dominated convergence theorem), and so $\mathbf{E}\,(a_s \mathbf{1}_A) > 0$ for $s \in [t, t + h]$ once $h > 0$ is small enough! This contradiction proves that our initial assumption that $\mathbf{P}\,(a_t = 0) < 1$ must be wrong, i.e., $a_t = 0$ for all $t \in [0, T]$ indeed.

11.5 Stochastic Differential Equations

By a *stochastic differential equation* (SDE) one means an equation involving stochastic differentials. We will only deal here with SDEs of the form

$$dX_t = a(t, X_t)dt + b(t, X_t)dW_t, \quad t \in [0, T], \quad X_0 = x_0, \tag{11.44}$$

where $a(t,x)$ and $b(t,x)$, $t \geq 0$, $x \in \mathbf{R}$, are non-random functions, x_0 is a given non-random initial condition, and, as usual, $\{W_t\}$ is a BM given on a filtered probability space $(\Omega, \mathcal{F}, \mathbf{F}, \mathbf{P})$.

An Itô process $\{X_t\}_{t\in[0,T]}$ on the same space $(\Omega, \mathcal{F}, \mathbf{F}, \mathbf{P})$ is said to be a solution[11] to SDE (11.44) if

$$X_t = x_0 + \int_0^t a(s, X_s)ds + \int_0^t b(s, X_s)dW_s, \quad t \in [0,T]. \tag{11.45}$$

Equation (11.44) will have a unique solution provided that the functions a and b are regular enough, e.g., if they are measurable and there exists a $C < \infty$ such that[12]

$$\left.\begin{array}{l} |a(t,x)| + |b(t,x)| \leq C(1+|x|), \\ |a(t,x) - a(t,y)| + |b(t,x) - b(t,y)| \leq C|x-y| \end{array}\right\} \quad \text{for } t \in [0,T],\ x, y, \in \mathbf{R}.$$

Remark 11.4. It is important to point out that SDEs rarely admit solutions of an analytic closed form (one exceptional class being linear SDEs, of which a special case is considered in Example 11.15 below). The main role of SDEs is to provide a way to specify a large and important class of SPs.

Example 11.15. *The Ornstein–Uhlenbeck process.*[13] We will start with a physical motivation for introducing the process (which was originally suggested to describe the velocity of a massive Brownian particle in the presence of friction).

Suppose that a particle of unit mass is moving through a fluid, in the presence of the frictional drag proportional to velocity (with proportionality coefficient $\alpha > 0$) and random noise due to collisions with (much smaller) molecules of the fluid. Denoting by X_t the particle's velocity at time t (as we will be dealing with a univariate process, think about X_t as the projection of the particle's velocity on a coordinate axis) and modelling the collision forces by a BM process' increments, one obtains from Newton's second law of motion that

$$dX_t = -\alpha X_t dt + \sigma dW_t. \tag{11.46}$$

[11]Such solutions are called *strong*. One says that (11.44) has a *weak* solution if there *exists* a filtered probability space, with a BM process $\{\widetilde{W}_t\}$ and an Itô process $\{\widetilde{X}_t\}$ on it, such that the latter has stochastic differential of the form (11.44) there, with both W_t and X_t replaced in it with $\widetilde{W}_t$ and $\widetilde{X}_t$, respectively.

[12]See, e.g., Section 5.2 in Øksendal (2003).

[13]Named after Dutch physicists Leonard Ornstein (12.11.1880–20.05.1941) and George Eugene Uhlenbeck (06.12.1900–31.10.1988), whose paper studying a free particle in Brownian motion, moving in a rarefied gas and affected by a friction force proportional to the pressure, was published in 1930.

One can solve that equation using the variation-of-constants method. First consider the homogeneous equation

$$dX_t = -\alpha X_t dt,$$

which clearly admits solution of the form $X_t = Ce^{-\alpha t}$, $C =$ const. Now assume that actually $C = C_t$ is also a function of t (more precisely, that it is an Itô process!), and look for solutions of (11.46) of the form $X_t = C_t e^{-\alpha t}$. By the product rule (11.42), as $e^{-\alpha t}$ is non-random,

$$dX_t = d(C_t e^{-\alpha t}) = e^{-\alpha t} dC_t - \alpha C_t e^{-\alpha t} dt = e^{-\alpha t} dC_t - \alpha X_t dt,$$

so that substituting this into (11.46) leads to $dC_t = \sigma e^{\alpha t} dW_t$. Therefore,

$$X_t = C_t e^{-\alpha t} = \left[C_0 + \int_0^t \sigma e^{\alpha s} dW_s \right] e^{-\alpha t} = e^{-\alpha t} x_0 + \sigma e^{-\alpha t} \int_0^t e^{\alpha s} dW_s,$$

where we used the initial condition $X_0 = x_0$ to determine the value of C_0.

Since $\int_0^t e^{2\alpha s} ds = (e^{2\alpha t} - 1)/(2\alpha)$, we see from Remark 11.1 that

$$X_t \sim N\big(e^{-\alpha t} x_0, \sigma_{\alpha,t}^2\big), \quad \sigma_{\alpha,t}^2 := \frac{\sigma^2}{2\alpha}(1 - e^{-2\alpha t}). \tag{11.47}$$

Thus, the distribution of X_t converges to $N(0, \sigma^2/(2\alpha))$ as $t \to \infty$, which is actually a display of the *ergodicity* of the process: as time passes, the SP "forgets" where it started, its distribution approaching the stationary one.

It is not hard to show that solutions to SDEs of the form (11.44) are in fact Markov processes (we will discuss them in more detail in the next chapter). Our argument deriving the distribution of X_t given X_0 actually yields the transition probabilities of the process as well: for $0 \le s < t$ and $x, y \in \mathbf{R}$,

$$\mathbf{P}\,(X_t \in dy | X_s = x) = \frac{1}{\sqrt{2\pi}\,\sigma_{\alpha,t}} \exp\left\{ -\frac{\big(y - xe^{-\alpha(t-s)}\big)^2}{2\sigma_{\alpha,t}^2} \right\} dy.$$

Therefore, it follows from the product formula (11.22) (which holds for general Markov processes possessing transition densities) that $\{X_t\}$ is a Gaussian process (i.e., all its FDDs are normal).

Example 11.16. *The Vasicek interest rate model.*[14] The "mean-reverting" property of the Ornstein-Uhlenbeck process (i.e., the tendency of the SP's trajectory to keep returning to its "historic average value") made it a candidate for a mathematical model of the interest rate dynamics. In that simple model, the spot interest rate r_t is assumed to satisfy the SDE

$$dr_t = a(b - r_t)dt + \sigma dW_t, \quad t > 0,$$

[14]The first model to capture "mean reversion" of interest rates. Suggested by O. Vasicek in 1977.

where a, b, σ and r_0 are all positive constants.

Putting $X := r_t - b$, one has $dX_t = dr_t$ from Itô's formula, so that

$$dX_t = -aX_t dt + \sigma dW_t, \quad t > 0,$$

meaning that $\{r_t - b\}$ is an Ornstein–Uhlenbeck process, with (cf. (11.47))

$$r_t \sim N\big(b + e^{-at}(r_0 - b), \sigma_{a,t}^2\big).$$

The model has an obvious deficiency: with a positive probability, the interest rate r_t can assume negative values, which is undesirable. This is fixed in the model presented in the next example.

Example 11.17. *The Cox–Ingersoll–Ross interest rate model.*[15] Now one assumes that

$$dr_t = a(b - r_t)dt + \sigma\sqrt{r_t}dW_t, \quad t > 0,$$

where a, b, σ and r_0 are again all positive constants. The effect of having the factor $\sqrt{r_t}$ in the diffusion coefficient is that it "freezes" the random oscillations as r_t approaches zero, and so the positive drift term becomes dominating. Hence the model will never produce negative interest rate values. Moreover, r_t will never turn into zero provided that $2ab \geq \sigma^2$.

11.6 Recommended Literature

GIKHMAN, I.I. AND SKOROKHOD, A.V. *Introduction to the Theory of Random Processes.* Dover, 1996. [*An excellent classical introduction to the theory of random processes.*]

KARATZAS, I. AND SHREVE, S. *Brownian Motion and Stochastic Calculus.* 2nd end. Springer, New York, 1991. [*A rather advanced text. First read the other books recommended in this section!*]

MIKOSCH, T. *Elementary Stochastic Calculus with Finance in View.* World Scientific, Singapore, 1999. [*A rather gentle introduction into stochastic calculus.*]

MÖRTERS, P. AND PERES, Y. *Brownian Motion.* Cambridge University Press, Cambridge, 2010. [*A compendium of key mathematical results on the Brownian motion process.*]

ØKSENDAL, B.K. *Stochastic Differential Equations: An Introduction with Applications.* 6th edn. Springer, Berlin, 2003. [*A popular introduction to stochastic calculus and its applications.*]

[15]Suggested by J. C. Cox, J. E. Ingersoll and S. A. Ross in 1985.

ROGERS, L.C.G. AND WILLIAMS, D. *Diffusions, Markov Processes, and Martingales: Volume 1, Foundations.* 2nd edn. Cambridge University Press, Cambridge, 2000. [*Together with its companion second volume, this book provides thorough foundations of important parts of the theory of random processes touched upon in the present chapter.*]

ROGERS, L.C.G. AND WILLIAMS, D. *Diffusions, Markov Processes and Martingales: Volume 2, Itô Calculus.* 2nd edn. Cambridge University Press, Cambridge, 2000.

11.7 Problems

1. Let $X_1, X_2, \ldots$ be i.i.d. RVs with $\mathbf{E}\, X_1 = \mu$, $\mathrm{Var}\,(X_1) = \sigma^2 < \infty$. Put $S_0 := 0$, $S_n := S_{n-1} + X_n$, $n \geq 1$. Compute

   (i) $\mathbf{E}\,(S_{n+m}|\, S_n)$, $m, n = 0, 1, 2, \ldots$;

   (ii) $\mathbf{E}\,(X_1|\, S_n)$, $n \geq 1$;

   (iii) $\mathbf{E}\,(S_{n+m}^2|\, S_n)$, $m, n = 0, 1, 2, \ldots$;

   (iv) $\mathbf{E}\,(S_m|\, S_n)$, $m = 0, 1, \ldots, n$.

   Hints. (ii) $\mathbf{E}\,(X_1|\, S_n) = \mathbf{E}\,(X_2|\, S_n)$ $(n \geq 2)$ *etc.* by symmetry. (iv) You may wish to use the result of one of the parts (i)–(iii) above.

2. Let $\{N_t\}_{t\geq 0}$ be a Poisson process with rate $\lambda > 0$, $\mathcal{F}_t = \sigma\{N_s,\, 0 \leq s \leq t\}$ the "history" of the process up to the time t. Using the properties of the Poisson process and conditional expectations, find

   (i) $\mathbf{E}\,(N_{t+s}|\, \mathcal{F}_t)$, $s, t \geq 0$;

   (ii) $\mathbf{E}\,(N_{t+s}^2|\, \mathcal{F}_t)$, $s, t \geq 0$;

   (iii) $\mathbf{E}\,(N_s|\, \mathcal{F}_t)$ and $\mathbf{E}\,(N_s^2|\, \mathcal{F}_t)$, $0 \leq s \leq t$;

   (iv) $\mathbf{E}\,(N_s|\, N_t)$ and $\mathbf{E}\,(N_s^2|N_t)$, $0 \leq s \leq t$.

   Hint. It is not much different from the previous problem, is it?

3. Let $\{X_t\}_{t=0,1,\ldots,T}$ be a positive SP adapted to a filtration $\mathbf{F} = \{\mathcal{F}_t\}$. In each of the following cases, say if the RV τ is an ST w.r.t. $\mathbf{F}$ (if the condition in the definition of the random time τ in (iii)–(iv) is never met for $t \leq T$, we just put $\tau := T$ to avoid any inconvenience). Explain (e.g., expressing events $\{\tau \leq t\}$ in terms of the RVs X_k, $k = 1, 2, \ldots$).

   (i) $\tau := m = \text{const}$;

   (ii) $\tau := \tau_1 \wedge \tau_2$, where τ_j are STs, $j = 1, 2$;

   (iii) $\tau := \min\{t \geq 0 : X_{t+1}/X_t > 1\}$;

   (iv) $\tau := \min\left\{t \geq 0 :\ \sum_{k=0}^t X_k > X_t^2\right\}$;

(v) $\tau := \max\{t \le T : X_t > 10\}$.

4. Let Y be an integrable RV (i.e., $\mathbf{E}|Y| < \infty$) on a filtered probability space $(\Omega, \mathcal{F}, \mathbf{F} = \{\mathcal{F}_t\}_{t\ge 0}, \mathbf{P})$. Show that $X_t := \mathbf{E}(Y|\mathcal{F}_t)$ is an MG.[16]

5. Let $\{X_t\}_{t\ge 0}$ be a square-integrable (i.e., $\mathbf{E}X_t^2 < \infty$) MG. Show that the process has *orthogonal increments* in the sense that, for any $0 \le t_1 \le t_2 \le t_3 \le t_4$, one has $\mathbf{E}(X_{t_2} - X_{t_1})(X_{t_4} - X_{t_3}) = 0$.

6. Let $S_0 := 0$, $S_n := Y_1 + \cdots + Y_n$, $n \ge 1$, Y_j being i.i.d. RVs with $\mathbf{E}Y_j = 0$, $\mathrm{Var}(Y_j) = \sigma^2 < \infty$. Show that $X_n := S_n^2 - n\sigma^2$, $n \ge 0$, is an MG (i) with respect to the filtration $\mathcal{F}_n = \sigma(Y_1, \ldots, Y_n)$; (ii) with respect to the natural filtration $\mathcal{F}_n' = \sigma(X_1, \ldots, X_n)$.

   Hint. (ii) Use the result of (i) and the fact that $\mathcal{F}_n' \subset \mathcal{F}_n$ (why does the last relation hold?).

7. Denote by $\{N_t\}_{t\ge 0}$ a Poisson process with rate $\lambda > 0$. Show that all three processes (i) $N_t - \lambda t$; (ii) $(N_t - \lambda t)^2 - \lambda t$; (iii) $\exp\{uN_t - \lambda t(e^u - 1)\}$ (u is a fixed real number) are MGs w.r.t. the filtration $\mathcal{F}_n = \sigma(N_s, s \le t)$.

8. Let $S_0 := 0$, $S_n := Y_1 + \cdots + Y_n$, $n \ge 1$, Y_j being i.i.d. RVs with $\mathbf{P}(Y_1 = 1) = 1 - \mathbf{P}(Y_1 = -1) = 1/2$. Denote by $\tau := \min\{n \ge 0 : S_n = a \text{ or } S_n = b\}$ the first time the RW S_n hits one of the (integer) barriers $a < 0 < b$. Use Theorem 11.2, without verifying its conditions in detail, to:

   (i) find the distribution of S_τ;[17]

   (ii) compute $\mathbf{E}\tau$.

   Hints. (i) Use the martingale $X_n := S_n$. (ii) Use the result of (i) and the MG from Problem 6.

9. Suppose you are playing a "fair game" betting \$1 at each play (in which you win/lose w.p. $\frac{1}{2}$, independently of the past). Then $\{X_n :=$ your fortune after n plays$\}_{n\ge 0}$ is an MG w.r.t. its "history" $\mathbf{F} = \{\mathcal{F}_0, \mathcal{F}_1, \mathcal{F}_2, \ldots\}$.

   Now suppose that, for each play $n = 1, 2, \ldots$, your stake can be an *arbitrary bounded amount* Y_n, but you have to decide how much to stake *before* that play, i.e., basing on the history up to play $n-1$ (inclusive). Mathematically, this means that, for any n, $|Y_n| \le C_n = \text{const} < \infty$, and $\{Y_n\}_{n\ge 1}$ is a predictable process (cf. p. 307).

   In that case, you win the amount $Y_n(X_n - X_{n-1})$ on play n (as it was $X_n - X_{n-1}$ when staking \$1 each time), and hence your total net gain after n plays is

$$Z_n := \sum_{k=1}^{n} Y_k(X_k - X_{k-1}), \quad n = 1, 2, \ldots; \quad Z_0 = 0.$$

---

[16]This MG is referred to as a *Lévy martingale.* One can show that $X_t \to \mathbf{E}(Y|\mathcal{F}_\infty)$ a.s. as $t \to \infty$, where $\mathcal{F}_\infty := \sigma(\bigcup_{n\ge 0} \mathcal{F}_t)$ (generally speaking, $\mathcal{F}_\infty$ is NOT the same as $\bigcup_{n\ge 0} \mathcal{F}_t$!). In fact, *any uniformly integrable MG is the Lévy MG* for some Y.

[17]Note that we already solved that (gambler's ruin) problem using a different approach in Example 3.21 (see the second case in (3.44)).

The process $\{X_n\}_{n\geq 0}$ is called the *martingale transform*[18] of X_n by Y_n.

(i) Show that $\{Z_n\}_{n\geq 0}$ is an MG w.r.t. $\mathbf{F}$.

(ii) The betting strategy of doubling when losing is called "martingale". You begin with a unit stake, win each play w.p. $\frac{1}{2}$ (regardless of the results of all the previous plays), and double your stake for the next play when losing. After each win, you reset the stake size to one. Represent your net gain process $\{Z_n\}$ when using this strategy as a martingale transform: specify the processes $\{X_n\}$ and $\{Y_n\}$.

(iii) Under the assumptions of part (ii), assume that you stop at the time τ of the first win. Find the distribution and expectation of the ST τ and verify if the statement of Theorem 11.2 holds for the process $\{Z_n\}$ and ST τ.

(iv) Compute the expectation $\mathbf{E}(Z_n; \tau > n)$ and hence verify if the general condition (11.11) sufficient for (11.9) is satisfied in this case.

10. Let $\{S_n\}_{n\geq 0}$ be an RW defined as follows: starting at $S_0 = 0$, the walking particle at each transition goes up 1 w.p. $p = \frac{6}{7}$ and down 2 w.p. $1 - p = \frac{1}{7}$.

(i) Show that $X_n := 2^{-S_n}$, $n = 0, 1, 2, \ldots$, is an MG.

(ii) Introduce the ST $\tau := \min\{n \geq 0 : X_n \leq 0.1\}$. Use Theorem 11.2 (without verifying its conditions) and the MG $\{Z_n := S_n - \mathbf{E}\, S_n\}$ to compute $\mathbf{E}\,\tau$.

Hint. (ii) First express the ST τ in terms of the random walk S_n. What are the possible values of S_τ? Note that if $S_n > S_{n-1}$, then $S_n = S_{n-1} + 1$.

11. Let $\{S_n\}_{n\geq 0}$ be a simple RW: starting at some initial point S_0, the walking particle at each transition goes up 1 w.p. $p \in (0,1)$ and down 1 w.p. $q = 1-p$. Assume that $p \neq \frac{1}{2}$.

(i) Show that $\{X_n := (q/p)^{S_n}\}$ is an MG.

(ii) Suppose the walk starts at $S_0 = 0$ and stops at the time $\tau := \min\{n \geq 0 : S_n = a \text{ or } S_n = b\}$, where $a < 0 < b$ are integers. Use Theorem 11.2 and the MG from part (i) to find the distribution of the RV S_τ, and then Theorem 11.2 and the MG $\{Z_n := S_n - n(p-q)\}$ to compute $\mathbf{E}\,\tau$.

12. Show that the set $\{U_1, U_2, \ldots\}$, where U_j are i.i.d. $U(0,1)$-RVs, is everywhere dense in [0,1] with probability 1.

Hint. You may wish to use the Glivenko-Cantelli theorem (2.87).

13. Find the distribution of $X := 2W_{t_1} - W_{t_2}$, $0 < t_1 < t_2$.

14. Find the distribution of $X := W_0 + W_2 - W_3 + 2W_4$.

15. Derive the BM's FDD density $f_{t_1,\ldots,t_k}(x_1, \ldots, x_n)$ (see (11.22)) using (2.33) and the observation that $(W_{t_1}, W_{t_2}, \ldots, W_{t_n})$ is the result of a simple linear transformation of the vector $(W_{t_1} - W_{t_0}, W_{t_2} - W_{t_1}, \ldots, W_{t_n} - W_{t_{n-1}})$ with

[18] Martingale transforms are discrete analogues of stochastic integrals and play an important role in the mathematical theory of finance in discrete time.

independent components, so that the latter vector's density is just the product of the densities of the increments $W_{t_k} - W_{t_{k-1}}$, $k = 1, 2, \ldots, n$.

16. Note that the transformation of the standard normal vector $\boldsymbol{Z} \in \mathbf{R}^n$ into the vector of the values of the standard BM from the simulation algorithm on p. 320 can be written in the matrix form as $(W_{t_1}, \ldots, W_{t_n}) = \boldsymbol{Z}A$, $A \in \mathbf{R}^{n \times n}$. Specify the matrix A. How will the above matrix representation change if you were to directly simulate the vector $(X_{t_1}, \ldots, X_{t_n})$, where $\{X_t\}$ is the arithmetic BM (11.16)?

17. Compute the joint densities of (i) $(2W_3, W_5)$ and (ii) $(W_2, 2W_3, W_5)$.

18. Use Theorem 11.4 to show that $\{\widetilde{W}_t := tW_{1/t}\}_{t \geq 0}$ is a standard BM process proved that $\{W_t\}_{t \geq 0}$ is such.

19. Let $\tau := \min\{t > 0 : W_t = \pm\sqrt{a + bt}\,\}$ be the first time the BM crosses one of the two parabolic boundaries $\pm\sqrt{a + bt}$, $t \geq 0$, where $a > 0$ and $b \in (0, 1)$ are some constants. Use Theorems 11.6 and 11.2 to compute $\mathbf{E}\,\tau$.

20. Denote by $\tau := \min\{t > 0 : W_t \leq 2t - 4\}$ the first time the BM process crosses the boundary $v_t := 2t - 4$, $t \geq 0$. Using the three martingales of the Brownian motion (Theorem 11.6) and Theorem 11.2 (do not verify the conditions of the theorem), compute for the stopping time τ its:

    (i) mean value $\mathbf{E}\,\tau$;

    (ii) variance $\mathrm{Var}\,(\tau)$;

    (iii) Laplace transform $l_\tau(s) = \mathbf{E}\, e^{-s\tau}$, $s \geq 0$.

    (iv) Compute also $\mathbf{E}\, W_\tau$ and $\mathbf{E}\, W_\tau^2$.

21. Denote by $\tau := \min\{t > 0 : W_t = a \text{ or } W_t = b\}$ the first time the standard BM process takes one of the values a or b $(a < 0 < b)$. Using the three martingales of the BM and Theorem 11.2 (do not verify the conditions of the theorem),

    (i) find the distribution of W_τ;

    (ii) compute the mean value $\mathbf{E}\,\tau$;

    (iii) compute the Laplace transform $l_\tau(s) = \mathbf{E}\, e^{-s\tau}$, $s \geq 0$, in the case when $a = -1$, $b = 1$;

    (iv) use the result of part (iii) to compute the mean $\mathbf{E}\,\tau$ when $a = -1$, $b = 1$. Compare the result with that for question (ii).

    (v) Use the result of part (iii) to compute the variance $\mathrm{Var}\,(\tau)$ when $a = -1$, $b = 1$.

22. Let f_t and g_t be simple processes on $[0, T]$ given on a common filtered probability space, with a BM $\{W_t\}_{t \geq 0}$ given on it. Show by a direct calculation that $\mathbf{E}\, I_t(f) I_t(g) = \int_0^t \mathbf{E}\, f_s g_s\, ds$, $t \in [0, T]$.

23. Let g_t be a non-random function on $[0,T]$ satisfying $\int_0^T g_t^2 dt < \infty$. Use Itô's formula to show that the following process is an MG:

$$Y_t = \exp\left\{\int_0^t g_s dW_s - \frac{1}{2}\int_0^t g_s^2 ds\right\}, \quad t \in [0,T].$$

24. Compute the stochastic differential $d\cos(W_t)$.

25. Put $X_t := t + W_t$, $t \geq 0$.

    (i) Apply Itô's formula to compute the stochastic differential de^{-2X_t}.

    (ii) Is the process $Y_t := e^{-2X_t}$, $t \geq 0$, a martingale? Explain.

26. The price S_t of a risky asset evolves according to the SDE

$$dS_t = 0.2S_t dt + S_t dW_t, \quad t \geq 0, \quad S_0 = 5$$

    (this is a special case of the so-called *Black–Scholes framework* to be discussed in Chapter 13).

    (i) It is suspected that the SDE has a solution of the form $S_t = ce^{at+bW_t}$, where a, b and c are some constants. Use Itô's formula to verify this suspicion and find the values of the constants a, b and c in the solution.

    (ii) Show that the process $X_t = 1/S_t$ satisfies the SDE

$$dX_t = 0.8X_t dt - X_t dW_t, \quad t \geq 0, \quad X_0 = 0.2.$$

27. The "stochastic volatility" Heston model assumes that the "variance process" $\{V_t\}_{t\geq 0}$ follows the SDE

$$dV_t = (1 - V_t)dt + 2\sqrt{V_t}\, dW_t, \quad t \geq 0, \quad V_0 = 1.$$

    (i) Derive an SDE for the "volatility process" $Z_t = \sqrt{V_t}$, $t \geq 0$, and find the initial condition for the SDE (i.e., the value Z_0).

    (ii) Show that the SP $Z_t := e^{-t/2}\left(1 + \int_0^t e^{s/2} dW_s\right)$ satisfies the SDE and the initial condition you derived in part (i).

Chapter 12

Diffusion Processes

12.1 Definitions

One way of introducing *diffusion processes* (or simply *diffusions*[1]) is to define them as solutions to SDEs of the form (11.44), i.e., processes admitting representation (11.45). Note the key difference between that representation and the one for Itô processes (11.36): in the latter, the drift and diffusion coefficients can depend on the past, whereas in the former the coefficients are functions of time and the current value of the process only. The main consequence of that is that diffusions are Markov processes, whereas the Itô processes in the general case are not. The Markov property enables one to use the powerful mathematical machinery[2], as we already saw in Chapters 3 and 6, when dealing with Markov chains and continuous time jump Markov processes. Unlike the latter, diffusion processes have continuous trajectories, and so the machinery will now mostly rely on differential equations in both time and space variables. Before we proceed to discussing them, we will present the classical approach to introducing diffusion processes, as it provides additional insight into the nature of the processes and highlights their applicability range.

Recall relation (6.11) (p. 174) that describes the probabilities of a Markov jump process $\{X_t\}_{t\geq 0}$ transitions, from a given state at time t to its possible states at time $t+h$, $h \searrow 0$. Fix a state y of the process and note that $\mathbf{P}\,(X_t = y) = \mathbf{E}\, f(X_t)$, where $f(z) = \delta_{zy}$ is Kronecker's delta.

[1]As the very name suggests, these are processes that model the physical phenomena of diffusion, a transport process which is a result of the random movement of the diffusing particles. See also p. 65.

[2]Basically, the so-called semi-group theory.

Hence one can re-write (6.11) as

$$\mathbf{E}\left(f(X_{t+h}) - f(X_t)|X_t = x\right) = a_{xy}h + o(h). \tag{12.1}$$

It turns out that diffusion processes also admit characterisation in terms of relations of the form (12.1), but with a different choice of functions f.

Keeping in mind the goal of modelling by our SP $\{X_t\}$ the behaviour of a diffusing physical particle, we will first of all try to exclude the possibility of jumps assuming that $\{X_t\}$ is a Markov process with transition probabilities such that

$$\forall \delta > 0, \quad \mathbf{P}\left(|X_{t+h} - X_t| > \delta|X_t = x\right) = o(h) \quad \text{as } h \searrow 0. \tag{12.2}$$

Without that assumption, the process $\{X_t\}$ can have jumps (cf. (6.11))!

Next, we want our model to have a finite "average drift". Similarly to (12.1) (but now with $f(x) \equiv x$), we will require to that end that[3]

$$\mathbf{E}\left(X_{t+h} - X_t|X_t = x\right) = \mu h + o(h) \quad \text{as } h \searrow 0 \tag{12.3}$$

for some (nice enough) drift function $\mu = \mu(t, x)$.

Finally, we will require in addition[4] that (again, similarly to (12.1), but now with $f(x) \equiv x^2$), for some $\nu = \nu(t, x)$,

$$\mathbf{E}\left(X_{t+h}^2 - X_t^2|X_t = x\right) = \nu h + o(h) \quad \text{as } h \searrow 0.$$

Given that (12.3) holds, the above relation is equivalent to

$$\mathbf{E}\left((X_{t+h} - X_t)^2|X_t = x\right) = \sigma^2 h + o(h) \quad \text{as } h \searrow 0, \tag{12.4}$$

where $\sigma^2 = \sigma^2(t, x) = \nu(t, x) - 2\mu(t, x)x$. Note that, for a process $\{X_t\}$ with smooth trajectories, one would instead have

$$\mathbf{E}\left((X_{t+h} - X_t)^2|X_t = x\right) = \mu h^2 + o(h^2) = o(h)$$

instead of (12.4) (we would actually have (12.4) with $\sigma^2 = 0$).

It is not difficult to verify that conditions (12.2)–(12.4) will be satisfied for an SP $\{X_t\}$ solving the SDE

$$dX_t = \mu(t, X_t)dt + \sigma(t, X_t)dW_t. \tag{12.5}$$

Indeed, just recall that the above equation basically means that, for small $h > 0$,

$$X_{t+h} - X_t \approx \mu(t, X_t)h + \sigma(t, X_t)(W_{t+h} - W_t), \tag{12.6}$$

[3]The left-hand side of (12.3) should actually be $\mathbf{E}\left(X_{t+h} - X_t; |X_{t+h} - X_t| < \delta|X_t = x\right)$ for some $\delta > 0$, to avoid integrability issues. However, we sometimes allow ourselves to be somewhat sloppy.

[4]One can actually show that, under broad regularity conditions on transition probabilities, condition (12.2) will imply both (12.3) and (12.4) (see, e.g., Vol. 2, Section X.4 of W. Feller's textbook listed on p. 68), but this is far beyond the scope of this text.

and so, provided that μ and σ are continuous functions of their arguments (which is the standard assumption), relations (12.2)–(12.4) follow.

And the other way around, once (12.2)–(12.4) are satisfied for $\{X_t\}$, one can show that the process satisfies the above SDE. One way to demonstrate that is to use (12.2)–(12.4) to derive differential equations to be discussed in the next section, and then to show that they imply (12.5). In fact, one can think about that as a "localised version" of the CLT stating that the conditional (given the present value) distributions of the increments of the SP over short time intervals are "almost normal", as can be seen from (12.6): given $X_t = x$, the process increment's distribution is close to $N(\mu(t,x)h, \sigma^2(t,x)h)$.

12.2 Kolmogorov Differential Equations and Generators

We already met Kolmogorov differential equations in Chapter 6, where we derived them for continuous jump Markov process. In particular, for time-inhomogeneous processes with transition probability matrices $P(s,t)$, we obtained equations (6.15) and (6.18):

$$\partial_t P(s,t) = P(s,t)A(t), \qquad \partial_s P(s,t) = -A(s)P(s,t), \tag{12.7}$$

the former known as the *forward Kolmogorov equation* (FKE), the latter as the *backward Kolmogorov equation* (BKE). Here "forward/backward" refers to the time variable w.r.t. which we differentiate (t and s, respectively), and $A(t)$ is the time t matrix of transition rates of the process (see (6.16)), also known as its (infinitesimal) generator.

The notion of generator is defined for general Markov processes as well, via relations similar to (12.1) that describe the asymptotic behaviour of $\mathbf{E}(f(X_{t+h}) - f(X_t)|X_t = x)$ as $h \searrow 0$, for different "test functions" f. Generators are important tools for analysing Markov processes and computing their characteristics. For diffusions, however, generators are *differential operators* rather than matrices.

BKE. For a function $\psi : \mathbf{R} \to \mathbf{R}$ and $t \in (0,T]$, $x \in \mathbf{R}$, introduce

$$v(s,x) := \mathbf{E}_{s,x}\psi(X_t) := \mathbf{E}(\psi(X_t)|X_s = x), \quad s \in [0,t], \ x \in \mathbf{R}$$

(we assume that ψ is nice enough, so that the right-hand side here exists and is finite). First we note that, for $h \in [0, t-s]$, one has

$$\begin{aligned} v(s,x) &= \mathbf{E}_{s,x}\mathbf{E}(\psi(X_t)|\mathcal{F}_{s+h}) && \text{[by CE4]} \\ &= \mathbf{E}_{s,x}\mathbf{E}(\psi(X_t)|X_{s+h}) && \text{[by Markov property]} \\ &= \mathbf{E}_{s,x}v(s+h, X_{s+h}). && \text{[by the definition of } v\text{]} \end{aligned} \tag{12.8}$$

Next we apply Itô's formula from Theorem 11.9[5] to the SP $Y_r := v(r, X_r)$:

$$Y_{s+h} - Y_s = \int_s^{s+h} \partial_r v(r, X_r)\, dr + \int_s^{s+h} \partial_x v(r, X_r)\, dX_r + \frac{1}{2}\int_s^{s+h} \partial_{xx} v(r, X_r)\,(dX_r)^2$$

$$= \int_s^{s+h} \big[\partial_r v(r, X_r) + \partial_x v(r, X_r)\mu(r, X_r) + \frac{1}{2}\,\partial_{xx} v(r, X_r)\sigma^2(r, X_r)\big]\, dr + \int_s^{s+h} \partial_x v(r, X_r)\, dW_r.$$

Taking the conditional expectations $\mathbf{E}_{s,x}$ of both sides of that equality, using the above observation (12.8) that $\mathbf{E}_{s,x} Y_{s+h} = v(s, x) \equiv \mathbf{E}_{s,x} Y_s$ and property II3 (implying that Itô integrals have zero means), and setting

$$f(r) := \partial_r v(r, X_r) + \mu(r, X_r)\partial_x v(r, X_r) + \frac{1}{2}\,\sigma^2(r, X_r)\partial_{xx} v(r, X_r),$$

we conclude that

$$0 = \mathbf{E}_{s,x}\int_s^{s+h} f(r)\, dr = \int_s^{s+h} \mathbf{E}_{s,x} f(r)\, dr$$

by Fubini's theorem. So, by the mean value theorem[6], $0 = \mathbf{E}_{s,x} f(s^*)$ for some $s^* \in [s, s+h]$. Now letting $h \searrow 0$, we see that $0 = \mathbf{E}_{s,x} f(s)$. Since $\mathbf{E}_{s,x}\partial_s v(s, X_s) = \partial_s v(s, x)$ *etc.*, we arrive at the BKE for diffusions:

$$\partial_s v(s, x) = -\mu(s, x)\partial_x v(s, x) - \frac{1}{2}\,\sigma^2(s, x)\partial_{xx} v(s, x), \tag{12.9}$$

which our function $v(s, x) = \mathbf{E}_{s,x}\psi(X_t)$ satisfies for $s < t$, together with the boundary condition $v(t, z) = \psi(z)$, $z \in \mathbf{R}$.

Remark 12.1. For any solution v to the PDE (12.9) and diffusion $\{X_t\}$ driven by (12.5), introduce the process $Y_t := v(t, X_t)$. Then, by Itô's formula (11.41), the new process will have the stochastic differential

$$dY_t = \partial_t v(t, X_t)dt + \partial_x v(t, X_t)dX_t + \frac{1}{2}\,\partial_{xx} v(t, X_t)(dX_t)^2$$

$$= \underbrace{\Big[\partial_t v(t, X_t) + \mu(t, X_t)\partial_x v(t, X_t) + \frac{1}{2}\,\sigma^2(t, X_t)\partial_{xx} v(t, X_t)\Big]}_{=0 \text{ by}(12.9)} dt$$

$$+ \sigma(t, X_t)\partial_x v(t, X_t)dW_t = \sigma(t, X_t)\partial_x v(t, X_t)dW_t, \tag{12.10}$$

which means, owing to II3, that $\{Y_t\}$ is an MG (provided that the respective integrability conditions are satisfied).

[5]To do so, one formally needs to verify the condition of the theorem. This is one of those technical details that we prefer to omit in this exposition.

[6]Formally, need the integrand to be continuous for that. It is, under broad assumptions.

Equation (12.9) belongs to the class of the so-called *parabolic PDEs*. This is a well-studied class of PDEs, and there exists extensive literature devoted to the theory and applications of the equations. One may wish to start with A. Friedman's book entitled *Partial Differential Equations of Parabolic Type*, Dover, 2008 (which is a republication of the text originally published in 1964 by Prentice-Hall, Englewood Cliffs, NJ).

Introducing the differential operator

$$A(s) := \mu(s,x)\partial_x + \frac{1}{2}\sigma^2(s,x)\partial_{xx} \tag{12.11}$$

(which is called the *generator* of the diffusion process given by (12.5)), we see that our newly derived equation (12.9) (for v) has the same form as the second equation in (12.7) (for P), which is no wonder though as both are special cases of the general BKE for Markov processes.

Remark 12.2. Equation (12.9) also holds for the function

$$v(s,x) := \mathbf{E}_{s,x}\psi(\tau, X_\tau), \quad (s,x) \in D,$$

where $\tau := \inf\{t \geq s : (t, X_t) \in \partial D\}$ is the first time after (the "present time") s when the trajectory of the process $\{X_t\}_{t\geq 0}$ hits the boundary ∂D of a (nice enough) set $D \ni (0, X_0)$ in the plane of (t,x)-values, $\psi(t,y)$, $(t,y) \in \partial D$, being a function given on that boundary. The situation we dealt with above is just a special case of that general setup, where we had $D = \{(s,x) \in \mathbf{R}^2 : 0 < s < t\}$ (so that $\tau \equiv t$).

Example 12.1. *Boundary crossing probabilities.* Let $\{X_t\}_{t\geq 0}$ be a diffusion process solving the SDE (12.5) and, for some $z_- < z_+$ and $T \in (0,\infty)$,

$$D := [0,T] \times [z_-, z_+]$$

be the rectangular area corresponding to time values between 0 and T and space values between the levels z_- and z_+. Set the boundary value function ψ as follows:

$$\begin{aligned} \psi(t,z_-) &:= 0, \quad t \in [0,T], \\ \psi(t,z_+) &:= 1, \quad t \in [0,T], \\ \psi(T,x) &:= 0, \quad x \in (z_-, z_+), \end{aligned}$$

and let $\tau_\pm := \inf\{t > 0 : X_t = z_\pm\}$ be the first times when the process $\{X_t\}$ hits the respective barriers $z_\pm$. Then, for $x_0 \in (z_-, z_+)$,

$$v(0,x_0) := \mathbf{E}_{0,x_0}\psi(\tau, X_\tau) = \mathbf{P}\left(\tau_+ < \tau_- \wedge T \,|\, X_0 = x_0\right).$$

Furthermore, setting ψ as

$$\begin{aligned} \psi(t,z_-) &:= 0, \quad t \in [0,T], \\ \psi(t,z_+) &:= 0, \quad t \in [0,T], \\ \psi(T,x) &:= 1, \quad x \in (z_-, z_+), \end{aligned}$$

will lead to the probability of not touching the barriers at $z_{\pm}$ on the time interval $[0,T]$: for $x_0 \in (z_-, z_+)$,

$$v(0,x_0) := \mathbf{E}_{0,x_0}\psi(\tau, X_\tau) = \mathbf{P}\left(\tau_- \wedge \tau_+ > T \mid X_0 = x_0\right).$$

As we will see in Section 13.8, such functions can be used in pricing the so-called "barrier options".

Example 12.2. *Transition densities.* For fixed $t > 0$, $y \in \mathbf{R}$, and $h > 0$, consider the boundary value function $\psi(z) = \frac{1}{2h}\mathbf{1}_{(y-h,y+h)}(z)$, $z \in \mathbf{R}$. Then the function

$$\begin{aligned} v(s,x) &= \mathbf{E}_{s,x}\psi(X_t) = \frac{1}{2h}\mathbf{E}_{s,x}\mathbf{1}_{(y-h,y+h)}(X_t) \\ &= \frac{1}{2h}\mathbf{P}\left(X_t \in (y-h, y+h) | X_s = x\right) \end{aligned}$$

satisfies (12.9) with the boundary condition $v(t,x) = \psi(x)$, $x \in \mathbf{R}$. Passing now to the limit as $h \searrow 0$, one can show that the limiting for $v(s,x)$ function will still satisfy equation (12.9) (now with the boundary condition of the form $v(t,x) = \delta_y(x)$, the *Dirac*[7] *delta function*), while its meaning is given by

$$p(s,x;t,y) := \lim_{h\searrow 0} \frac{1}{2h}\mathbf{P}\left(X_t \in (y-h,y+h)\right) = \frac{1}{dy}\mathbf{P}\left(X_t \in dy | X_s = x\right),$$

which is the transition density of the process (one can show that it exists). Thus, the transition density $p(s,x;t,y)$ of a diffusion process satisfies the respective BKE as a function of (s,x), when (t,y) is kept fixed.

In the case of the BM process, one has (see (11.21)) that, for $0 \le s < t$,

$$v(s,x) := p(s,x;t,y) = \frac{1}{\sqrt{2\pi(t-s)}}\exp\left\{-\frac{(y-x)^2}{2(t-s)}\right\} =: f(t-s, y-x),$$

where $f(r,z)$, $z \in \mathbf{R}$, is the density of $N(0,r)$, $r > 0$. By the chain rule,

$$\partial_s v(s,x) = -\partial_r f(t-s, y-x), \quad \partial_{xx} v(s,x) = \partial_{zz} f(t-s, y-x),$$

and since $\mu(s,x) \equiv 0$, $\sigma(s,x) \equiv 1$ for the BM process, we see that the BKE for the transition density in this case turns into the *heat equation* (to be discussed below) for $f = f(r,z)$:

$$\partial_r f = \frac{1}{2}\partial_{zz} f, \quad (r,z) \in (0,\infty) \times \mathbf{R}. \tag{12.12}$$

[7] After Paul Adrien Maurice Dirac (08.08.1902–20.10.1984), an outstanding English theoretical physicist who shared the 1933 Nobel Prize for physics with Erwin Schrödinger "for the discovery of new productive forms of atomic theory". The delta function on the real line has the property that it is zero everywhere except at zero, but its integral over the entire real line equals one. One can easily show that no ordinary function can have the postulated property, but the concept itself makes perfect sense as a so-called *generalised function*. Delta function was introduced by P. Dirac as a "convenient notation" in 1930; he called it so by analogy with the discrete Kronecker delta.

Remark 12.3. Suppose we need to solve a boundary value problem for a PDE of the form (12.9), $s \in [0, T]$, with the boundary condition $v(T, x) = \psi(x)$ for a given function ψ. Once could do (and does!) that numerically, using the Monte Carlo method to estimate the expectations $\mathbf{E}_{s,x}\psi(X_T)$. The "brute force approach" would involve simulating (which is in fact done in discrete time) a large number n of trajectories $\{X_t^{(k)}\}_{t\in[s,T]}$, $k = 1, \ldots, n$, of the process, all starting at the same point $X_s^{(k)} = x$ at time s, and then taking the mean value $n^{-1}\sum_{k\le n}\psi(X_T^{(k)})$ as an estimator for $v(s, x)$. However, there are much smarter Monte Carlo alternatives to that.

FKE.[8] The equation (for unknown function $u = u(t, y)$) has the following form:

$$\partial_t u(t, y) = -\partial_y(\mu(t, y)u(t, y)) + \frac{1}{2}\,\partial_{yy}(\sigma^2(t, y)u(t, y)). \tag{12.13}$$

Setting

$$A^*(t)(\,\cdot\,) := -\partial_y(\mu(t, y)\,\cdot\,) + \frac{1}{2}\,\partial_{yy}(\sigma^2(t, y)\,\cdot\,) \tag{12.14}$$

(note that $A^*(t)$ is actually the so-called *adjoint operator* of the operator $A(s)$ from (12.11)), equation (12.13) (for u) takes the form

$$\partial_t u(t, y) = -A^*(t)u(t, y). \tag{12.15}$$

This is actually an analog of the first equation in (12.7) (for P), the change of order of factors in that first equation (compared to the second one in (12.7)) corresponding to switching from $A(s)$ to its adjoint $A^*(t)$.

One can show that, under suitable assumptions on the smoothness of the drift and diffusion coefficients, the function $u(t, y) := p(s, x; t, y)$ obtained from the transition density of our diffusion process by fixing its "backward variables" (s, x), will satisfy the FKE (12.13).

Indeed, consider a function $f : \mathbf{R} \to \mathbf{R}$ such that

$$\left.\begin{array}{l} f \text{ is twice continuously differentiable and,} \\ \text{for some finite } a < b, \text{ one has } f(y) = 0,\ y \notin [a, b] \end{array}\right\}. \tag{12.16}$$

[8] Also known (mostly to physicists) as the *Fokker–Plank equation*, named after Adriaan Daniël Fokker (17.08.1887–24.09.1972), a Dutch physicist and musician (and a cousin of the famous aircraft designer A. H. G. Fokker), and Max Karl Ernst Ludwig Planck (23.04.1858–04.10.1947), an outstanding German theoretical physicist, the originator of quantum theory (Nobel Prize in Physics, 1918), who was famously advised by his physics professor at the university not to go into physics as "in this field, almost everything is already discovered, and all that remains is to fill a few holes". The equation first appeared in A. D. Fokker's 1913 PhD thesis that dealt with Brownian motions of electrons in a radiation field. The general theory of the FKEs was systematically developed in a famous paper by A. N. Kolmogorov published in 1931.

Applying Itô's formula (11.38) to the process $Y_t := f(X_t)$, we obtain that $dY_t = df(X_t) = \big(\mu(t, X_t)f'(X_t) + \frac{1}{2}\sigma^2(t, X_t)f''(X_t)\big)dt + \sigma(t, X_t)f'(X_t)dW_t$, so that, for $t > s$,

$$f(X_t) - f(X_s) = \int_s^t \mu(r, X_r)f'(X_r)\,dr + \frac{1}{2}\int_s^t \sigma^2(r, X_r)f''(X_r)\,dr + \int_s^t \sigma(r, X_r)f'(X_r)\,dW_r.$$

Taking conditional expectations $\mathbf{E}_{s,x}$ of the both sides, and using II3 and Fubini's theorem yields

$$\begin{aligned}\mathbf{E}_{s,x}f(X_t) - \mathbf{E}_{s,x}f(X_s) &\equiv \int f(y)u(t,y)dy - f(x)\\ &= \mathbf{E}_{s,x}\int_s^t \mu(r, X_r)f'(X_r)\,dr + \frac{1}{2}\mathbf{E}_{s,x}\int_s^t \sigma^2(r, X_r)f''(X_r)\,dr\\ &= \int_s^t \left[\int \mu(r,y)f'(y)u(r,y)dy\right]dr + \frac{1}{2}\int_s^t\left[\int \sigma^2(r,y)f''(y)u(r,y)dy\right]dr.\end{aligned} \tag{12.17}$$

Integrating by parts, one obtains

$$\int \mu(r,y)f'(y)u(r,y)dy = \underbrace{\Big[\mu(r,y)u(r,y)f(y)\Big]_{-\infty}^{\infty}}_{=0 \text{ by } (12.16)} - \int f(y)\partial_y(\mu(r,y)u(r,y))\,dy.$$

Similarly, integrating by parts and using (12.16) twice, one gets

$$\int \sigma^2(r,y)f''(y)u(r,y)dy = \int f(y)\partial_{yy}(\sigma^2(r,y)u(r,y))\,dy.$$

Substituting the last two relations back into (12.17) and again using Fubini's theorem yields

$$\int f(y)\left\{u(t,y) + \int_s^t \left[\partial_y(\mu(r,y)u(r,y)) - \frac{1}{2}\partial_{yy}(\sigma^2(r,y)u(r,y))\right]dr\right\}dy = f(x).$$

Differentiating the last equality w.r.t. t, one obtains that

$$\int f(y)\left[\partial_t u(t,y) + \partial_y(\mu(r,y)u(r,y)) - \frac{1}{2}\partial_{yy}(\sigma^2(r,y)u(r,y))\right]dy = 0.$$

As this holds for *any* function f satisfying (12.16), one must have $[\cdots] \equiv 0$, completing the derivation of (12.13).

Assuming that the expression in $[\cdots]$ is continuous in y (one can show that it will be continuous once the drift and diffusion coefficients are smooth enough), the last claim is a special case of the following simple useful result.

Lemma 12.1. *Assume that $\varphi : \mathbf{R} \to \mathbf{R}$ is continuous and, for any infinitely many times continuously differentiable function f satisfying the second condition from* (12.16), *one has $\int f(y)\varphi(y)\,dy = 0$. Then $\varphi(y) \equiv 0$.*

Proof We will prove the lemma by contradiction. Indeed, let there exist a y_0 such that $\varphi(y_0) \neq 0$; w.l.o.g, we can assume that $c := \varphi(y_0) > 0$. Then, by continuity, there exists a $\delta > 0$ such that $\varphi(y) > c/2$ for $y \in (y_0 - \delta, y_0 + \delta)$. One can easily show that the function

$$f(y) := \begin{cases} 0, & y \notin (y_0 - \delta, y_0 + \delta), \\ \exp\{-(y - y_0)^2/(\delta^2 - (y - y_0)^2)\}, & y \in (y_0 - \delta, y_0 + \delta), \end{cases}$$

will satisfy the conditions on f stated in the lemma. As $f(y) > 0$ for all $y \in (y_0 - \delta, y_0 + \delta)$, one has

$$\int f(y)\varphi(y)\,dy \geq \frac{c}{2}\int f(y)\,dy > 0,$$

which contradicts the lemma conditions and so implies that our assumption that $\varphi(y) \not\equiv 0$ was wrong. □

In the special case of the BM process, the function $u(t, y) := p(s, x; t, y)$ (the transition density of the BM, with fixed "backward variables") satisfies the FKE

$$\partial_t u(t, y) - \frac{1}{2}\partial_{yy}\, u(t, y) = 0, \tag{12.18}$$

which has the same form as (12.12), i.e., is the heat equation.[9] This is a mathematical model for the distribution of heat in homogeneous physical matter over time (and for the diffusion phenomena as well). To illustrate the relationship between that problem and the BM process, suppose we have an infinitely long (and very thin) homogeneous rod, y being the "longitudinal coordinate" of a location on it, and denote by $u(t, y)$ the temperature of the rod at time t at location y. Assuming that the rod is thermally insulated (so that no heat dissipates from it into space, i.e., $\int u(t, y)\,dy = \text{const}$), Fourier's law of heat transfer and energy conservation imply that the temperature function $u(t, y)$ will satisfy (12.18). The solution to (12.18) with the initial condition $u(0, y) = \delta(x - y)$ (the Dirac delta function) will, roughly speaking, correspond to a situation where, at time $t = 0$, one applies a very hot blow torch to a tiny spot of the (very cold at that time) rod at location x. On the other hand, in the diffusion processes context, this solution will be the density of $X_t := x + W_t$, the BM process starting at $X_0 = x$ (the Dirac delta function can be interpreted as the "density" of the distribution degenerate at 0). This can hardly be surprising, as the two physical phenomena are so closely related.

Figure 12.1 shows the 3D and contour plots of the above $u(t, y)$, when $x = 0$. The vertical cross-sections of the surface on the left pane corresponding to fixed t-values are nothing but the densities of $N(0, t)$, giving also the heat distributions at those times. If you saw pictures of comets (close to their nuclei) or aerials photos of strong local fires, in presence of wind, the right pane may look familiar to you.

[9] We are dealing here with the univariate case only; the multivariate version of the heat equation (that will describe multivariate BM processes) will have ∂_{yy} replaced in it with the *Laplace differential operator*.

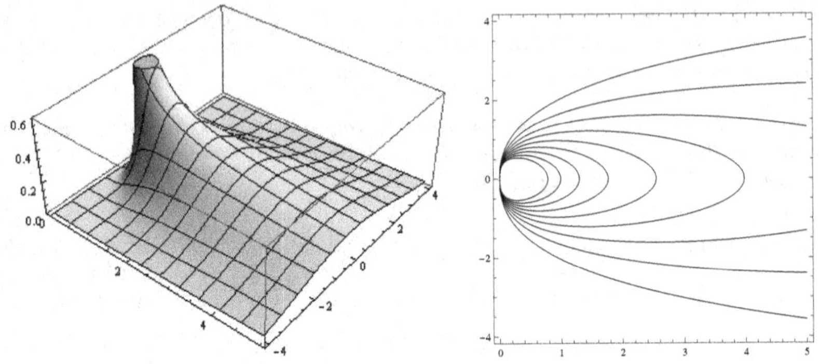

Fig. 12.1 The 3D and contour plots of the transition density $p(0,0;t,y)$ for the BM process for $t \in (0,4)$, $y \in (-4,4)$. Along with the probability density, the plots depict the dynamics of heat dissipation in a thin rod, the initial condition being an "overheated point" at zero.

12.3 Stationary Distributions

In this section we will only deal with time-homogeneous diffusions, i.e., with diffusion processes whose dynamics do not change with time and are governed by SDEs of the form

$$dX_t = \mu(X_t)dt + \sigma(X_t)dW_t, \tag{12.19}$$

with coefficients that depend on the current state of the SP, but not on the time. Such a diffusion is an continuous homogeneous Markov process and, as it was the case for Markov chains and jump Markov processes, one of the important questions concerning the SP is whether it is ergodic and so has a limiting (as $t \to \infty$) distribution independent of the initial state. Such a distribution is called *stationary* for the process, and it is not hard to show that if X_0 follows that distribution then the process $\{X_t\}_{t\geq 0}$ will be strictly stationary (cf. Section 3.4).

Conditions for ergodicity of diffusions, as for all Markov processes, can be stated in terms of renewal phenomena: roughly speaking, there ought to be a recurrent state with a finite mean recurrence time. For diffusions such conditions can be stated in terms of divergence/convergence of certain integrals involving the drift and diffusion coefficients.[10] Here we will only discuss how to compute the stationary distribution of a diffusion process (when it exists, of course), at our usual (not too high) level of rigour.

[10]For more detail, the interested reader is referred to Pinsky (2003) (see Section 11.6).

Recall our derivation of equation (6.10) for the stationary distribution π of a jump Markov process and its form: $0 = \pi A$. The right-hand side here is akin to what appears on the right-hand side of the FKE for such SPs (the first equation in (12.7)), and the reason for that is transparent: for ergodic processes, as $t \to \infty$, the transition probabilities tend to the stationary distributions and so, roughly speaking, their dependence on the "forward time" t vanishes, turning the left-hand side of that equation into zero. Likewise for diffusions, and the equation for the stationary density $\pi(y)$ in this case will, as one could expect from the above analogy between the first equation in (12.7) and (12.15), be of the form $0 = A^*\pi$.

More formally, arguing as on p. 355, for a function f satisfying (12.16) and $t > 0$, one obtains

$$f(X_{t+1}) - f(X_t) = \int_t^{t+1} \mu(X_r)f'(X_r)\,dr + \frac{1}{2}\int_t^{t+1} \sigma^2(X_r)f''(X_r)\,dr + \int_t^{t+1} \sigma(X_r)f'(X_r)\,dW_r.$$

Taking conditional expectations $\mathbf{E}_{0,x}$ of the both sides, and using II3 and Fubini's theorem yields

$$\mathbf{E}_{0,x}f(X_{t+1}) - \mathbf{E}_{0,x}f(X_t) = \int_t^{t+1} \mathbf{E}_{0,x}\left[\mu(X_r)f'(X_r)\,dr + \frac{1}{2}\sigma^2(X_r)f''(X_r)\right]dr. \quad (12.20)$$

Now if the process $\{X_t\}$ is ergodic, so that X_t has a limiting distribution as $t \to \infty$ which is independent of X_0, then it follows from (2.40) that

$$\lim_{t\to\infty} \mathbf{E}_{0,x}f(X_{t+1}) = \lim_{t\to\infty} \mathbf{E}_{0,x}f(X_t) =: \mathbf{E}\,f(X_\infty),$$

denoting by X_∞ an RV following that limiting distribution. Likewise, assuming that $\mu(y)f'(y) + \frac{1}{2}\sigma^2(y)f''(y)$ is a bounded continuous function, for the integrand in (12.20) one has

$$\lim_{t\to\infty} \mathbf{E}_{0,x}[\cdots] = \mathbf{E}\left[\mu(X_\infty)f'(X_\infty)\,dr + \frac{1}{2}\sigma^2(X_\infty)f''(X_\infty)\right].$$

The last two displayed formulae and the dominated convergence theorem mean that letting $t \to \infty$ in (12.20) yields

$$\mathbf{E}\left[\mu(X_\infty)f'(X_\infty) + \frac{1}{2}\sigma^2(X_\infty)f''(X_\infty)\right] = 0. \quad (12.21)$$

Now assume that the (limiting) distribution of X_∞ has density $\pi(y)$. Re-writing (12.21) as

$$\int \mu(y)f'(y)\pi(y)\,dy + \frac{1}{2}\int \sigma^2(y)f''(y)\pi(y)\,dy = 0$$

and integrating by parts, as in our derivation of the FKE, we obtain

$$\int f(y)\Big[(\mu(y)\pi(y))' - \frac{1}{2}(\sigma^2(y)\pi(y))''\Big]dy = 0,$$

where we switched from the partial derivatives notation ∂_y, ∂_{yy} to the usual $'$, $''$ since we are now dealing with univariate functions. As this holds for *any* function f satisfying (12.16), by Lemma 12.1 one must have $[\cdots] \equiv 0$.

Thus we showed that, indeed, the stationary density of our process must satisfy equation of the form $0 = A^*\pi$ (note that, as the process is homogeneous, the expression for $A^*(t)$ in (12.14) does not depend on t), i.e., equation

$$(\mu(y)\pi(y))' - \frac{1}{2}(\sigma^2(y)\pi(y))'' = 0. \tag{12.22}$$

Formally, one would also need to specify the state space of the diffusion process which will be an interval with end points $x_- < x_+$ (it can be open, semi-open, closed, one can have $x_- = -\infty$ *etc.*; we will not discuss here neither diffusion processes' behaviour close the boundary of their state space nor any other delicate issues of the theory), and specify that (12.22) holds for $y \in (x_-, x_+)$.

This is a second order linear ordinary differential equation (ODE) for π, which can be easily solved, with the formula for the general solution involving two constants.

Indeed, setting $p(y) := \sigma^2(y)\pi(y)$ and integrating equation (12.22), one obtains

$$p'(y) - \frac{2\mu(y)}{\sigma^2(y)}\,p(y) = C_1, \quad C_1 = \text{const.}$$

Using the standard methods for integration (involving integrating factors or variation of constants), we obtain from here that, for any $y_0 \in (x_-, x_+)$,

$$(g_{y_0}(y)p(y))' = C_1 g_{y_0}(y), \quad \text{where} \quad g_{y_0}(y) := \exp\Big\{-\int_{y_0}^{y}\frac{2\mu(z)}{\sigma^2(z)}\,dz\Big\}.$$

Integrating now the last differential equation yields

$$\pi(y) = \frac{p(y)}{\sigma^2(y)} = \frac{C_1 S_{y_0}(y) + C_2}{\sigma^2(y)g_{y_0}(y)}, \quad C_j = \text{const}, \quad j = 1, 2, \tag{12.23}$$

where

$$S_{y_0}(y) = \int_{y_0}^{y} g_{y_0}(u)\,du \equiv \int_{y_0}^{y}\exp\Big\{-\int_{y_0}^{u}\frac{2\mu(z)}{\sigma^2(z)}\,dz\Big\}du \tag{12.24}$$

is the so-called *scale function* of the diffusion $\{X_t\}$.

Remark 12.4. Clearly, $S'_{y_0}(y) = g_{y_0}(y)$, $S''_{y_0}(y) = g'_{y_0}(y) = -\frac{2\mu(y)}{\sigma^2(y)}g_{y_0}(y)$, and hence the (time-independent) function $v(s,x) := S_{y_0}(x)$ solves the BKE which, in the case of time-independent functions $v(s,x) =: V(x)$, takes the form

$$0 = \mu(x)V'(x) + \frac{1}{2}\sigma^2(x)V''(x). \tag{12.25}$$

Therefore, by Remark 12.1, the (continuous) SP $Y_t := S_{y_0}(X_t)$ is an MG. This remark will come handy later on.

In most applications of ODEs, integration constants are determined from initial and/or boundary conditions. In this case, however, one uses instead the two conditions ensuring that π is a probability density:

$$\pi(y) \geq 0, \qquad \int \pi(y)\,dy = 1. \tag{12.26}$$

If it is possible to choose integration constants C_j to meet these conditions, then a stationary density exists and is given by the general formula (12.23) with the chosen constants' values; if not, no stationary density exists.

Example 12.3. In the BM process case, one has $\mu(y) \equiv 0$, $\sigma(y) \equiv 1$, and so equation (12.22) turns into $\pi'(y) - \frac{1}{2}\pi''(y) = 0$. This is a second order homogeneous linear ODE with constant coefficients, and it can be easily shown that its general solution is of the form $\pi(y) = C_1e^{2y} + C_2$, $y \in \mathbf{R}$ (we could also use (12.23) to derive that formula). It is clear that no choice of C_j can ensure (12.26), which means that the BM cannot have stationary density (which was obvious anyway, as $W_t \sim N(0,t)$).

Example 11.15 (continued). For the Ornstein–Uhlenbeck process, one has $\mu(y) = -\alpha y$, $\sigma^2(y) = \sigma^2 = \text{const}$, yielding $g_0(y) = e^{\beta y^2}$, $\beta := \alpha/\sigma^2$, when one chooses $y_0 := 0$. Therefore, in this case, the general solution (12.23) is

$$\pi(y) = c_1e^{-\beta y^2}\int_0^y e^{\beta z^2}dz + c_2e^{-\beta y^2}, \quad y \in \mathbf{R},$$

where we set $c_j := C_j\sigma^{-2}$ (still constants!). Clearly, we have here $\pi(y) < 0$ for all large enough negative y (and so (12.26) fails) unless we put $C_1 := 0$. Thus $\pi(y) = c_2e^{-\beta y^2}$, $y \in \mathbf{R}$, i.e., $\pi(y)$ is the density of $N(0, \sigma^2/(2\alpha))$, confirming the result of our earlier direct calculation of the limiting density on p. 342.

12.4 The Method of Differential Equations

In this section we will show how to use the BKE we derived in Section 12.2 to compute important characteristics of homogeneous diffusions, i.e., SPs $\{X_t\}$ governed by the SDE (12.19). We will be dealing with the situation referred to in Remark 12.2, assuming that the region D is the strip between two levels: $D := \{(t,x) : t > 0, x \in (a,b),\}$ for some $a < b$, so that

$$\tau := \inf\{t \geq s : X_t = a \text{ or } X_t = b\}.$$

We will consider three different choices of the boundary value function ψ on ∂D that lead to functions $v(s,x) := \mathbf{E}_{s,x}\psi(\tau, X_\tau)$ with important probabilistic interpretations.

Case 1: $\psi(t,a) \equiv 0$, $\psi(t,b) \equiv 1$, $t > 0$.

In this case,

$$v(s,x) = \mathbf{E}_{s,x}\mathbf{1}_{\{X_\tau = b\}} = \mathbf{P}_{s,x}(X_\tau = b) \equiv \mathbf{P}_{0,x}(X_\tau = b) =: V(x)$$

is the probability for the process to hit level b prior to hitting a, starting from the point x (as the process is time-homogeneous, it does not matter at what time it starts from x, so the probability is a function of x only). Therefore the function $V(x)$ will satisfy equation (12.25) (the "time-independent version of the BKE") with the obvious boundary conditions $V(a) = \mathbf{P}_{0,a}(X_\tau = b) = 0$, $V(b) = \mathbf{P}_{0,b}(X_\tau = b) = 1$. Like (12.22), that equation admits a closed-form solution, its general solution being of the form

$$V(x) = C_1 S_{x_0}(x) + C_2,$$

where $S_{x_0}(x)$ is the scale function (12.24) (cf. Remark 12.4), $x_0 \in [a,b]$ being an arbitrary fixed point.

Using the boundary conditions to find the constants C_j, we obtain

$$V(x) = \frac{S_{x_0}(x) - S_{x_0}(a)}{S_{x_0}(b) - S_{x_0}(a)}, \quad x \in (a,b). \tag{12.27}$$

In particular, choosing $x_0 := a$ yields

$$V(x) = S_a(x)/S_a(b), \quad x \in (a,b). \tag{12.28}$$

Remark 12.5. That this probability has such a simple form in terms of the scale function is no surprise in view of Remark 12.4. Indeed, since $\{S_a(X_t)\}_{t\geq 0}$ is a continuous MG, by the Optional Stopping Theorem 11.2

(condition (11.11) can easily be verified for our first hitting time τ), one has, for $x \in (a, b)$,

$$S_a(x) = \underbrace{\mathbf{E}_x S_a(X_0) = \mathbf{E}_x S_a(X_\tau)}_{\text{by Theorem 11.2}} = S_a(a)(1 - V(x)) + S_a(b)V(b).$$

Since clearly $S_a(a) = 0$, the representation (12.28) follows!

Example 12.4. For the BM process $\{X_t = x + W_t\}$ starting at $X_0 = x$, one has $\mu(x) \equiv 0$, so that $S_a(x) = x - a$ and hence

$$V(x) = \frac{x - a}{b - a}, \quad x \in (a, b). \tag{12.29}$$

Note that this is a "continuous version" of the result (3.44) in the symmetric case ($p = \frac{1}{2}$) of the classical gambler's ruin problem (we used difference equations, rather than the differential ones, to derive it in Example 3.21).

Note also that we could immediately obtain the same answer by directly applying the Optional Stopping Theorem 11.2 to our process (which is an MG by Theorem 11.6(i)): $x = \mathbf{E} X_0 = \mathbf{E} X_\tau = bV(x) + a(1 - V(x))$, which yields (12.29), cf. Problem 21(i) on p. 347. This, of course, is merely a special case of the martingale approach explained in Remark 12.5. We used the same approach to solve the gambler's ruin problem in Problem 8 on p. 345.

Example 12.5. It is clear that one will obtain the same answer (12.29) for *any diffusion with zero drift*, i.e., when $\mu(x) \equiv 0$ (both the differential equations method and the martingale method can be used).

Example 12.6. Consider an arithmetic BM process $X_t := x + \mu t + \sigma W_t$ starting at $X_0 = x$ (here both μ and σ are constants). The case $\mu = 0$ was dealt with earlier, while when $\mu \neq 0$, solution (12.28) becomes

$$V(x) = \frac{e^{-\beta x} - e^{-\beta a}}{e^{-\beta b} - e^{-\beta a}}, \quad x \in (a, b), \quad \beta := \frac{2\mu}{\sigma^2}.$$

One can use this result to derive the probability of *ever reaching* level $b > 0$ by an arithmetic BM starting at zero. Indeed, that probability can be obtained from (12.29) (when $\mu = 0$) and the above formula (when $\mu \neq 0$) by letting there $x := 0$ and $a \to -\infty$ (why?), which yields

$$\mathbf{P}\left(M := \max_{t \geq 0}(\mu t + \sigma W_t) > b\right) = \begin{cases} 1 & , \ \mu \geq 0, \\ e^{2\mu b/\sigma^2} & , \ \mu < 0. \end{cases}$$

Thus, in the presence of negative drift, the global maximum of the arithmetic BM process is exponentially distributed! One could actually infer that fact from the memoryless property of the distribution of M, which is characteristic of the exponential distribution (see Problem 1 on p. 167). The property can be derived from the observation that, for any fixed $x, y > 0$, on the event $\{M > x + y\}$ one has $\tau_x := \inf\{t > 0 : X_t = x\} < \infty$ (indeed, as the trajectories of X_t are continuous, to reach level $x + y$ they must hit x at some time!). Therefore, by the TPF,

$$\begin{aligned}
\mathbf{P}(M > x+y) &= \int_0^\infty \mathbf{P}(M > x+y \mid \tau_x = t)\mathbf{P}(\tau_x \in dt) \\
&= \int_0^\infty \mathbf{P}\big(\underbrace{W_t}_{=x} + \underbrace{\max_{s\ge t}(\mu s + \sigma(W_s - W_t))}_{\stackrel{d}{=}M,\ \text{independent of } \tau_x} > x + y \mid \tau_x = t\big)\mathbf{P}(\tau_x \in dt) \\
&= \int_0^\infty \mathbf{P}(M > y)\mathbf{P}(\tau_x \in dt) = \mathbf{P}(M > y)\int_0^\infty \mathbf{P}(\tau_x \in dt) \\
&= \mathbf{P}(M > y)\mathbf{P}(M > x),
\end{aligned}$$

which means that (5.4) holds for the RV M, and so M is exponentially distributed.

Example 12.7. For the Ornstein–Uhlenbeck process (11.46),

$$\int_a^y \frac{2\mu(z)}{\sigma^2(z)}\,dz = -\frac{2\alpha}{\sigma^2}\int_a^y z\,dz = \beta(a^2 - y^2), \quad \beta := \frac{\alpha}{\sigma^2} > 0,$$

so that $S_a(x) = e^{-\beta a^2}\int_a^x e^{\beta y^2}dy$. Therefore, by (12.28),

$$V(x) = \frac{\int_a^x e^{\beta y^2}dy}{\int_a^b e^{\beta y^2}dy}, \quad x \in (a,b).$$

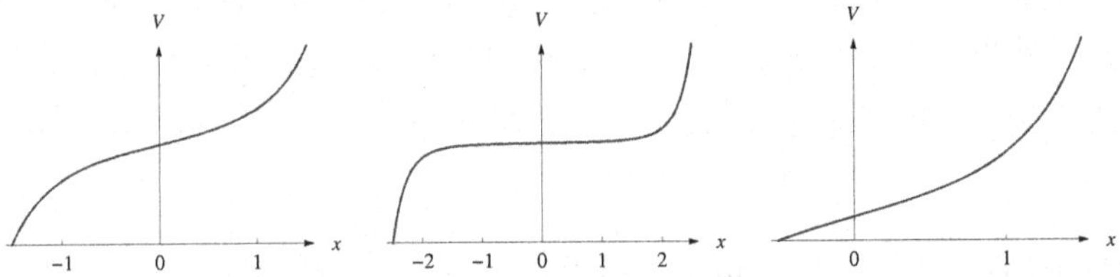

Fig. 12.2 The plots of $V(x)$ = the probability of hitting level $b > a$ prior to hitting a by the Ornstein-Uhlenbeck process (11.46) with $\alpha = \sigma = 1$, starting at $x \in (a,b)$, for three different intervals $(a,b) = (-1.5, 1.5)$, $(-2, 2)$ and $(-0.5, 1.5)$ (from left to right).

It would be quite instructive to plot the function $V(x)$ for a number of different intervals (a, b), which we did in Fig. 12.2 (choosing $\alpha = \sigma = 1$, so that $\beta = 1$ as well). In the left pane, $(a,b) = (-1.5, 1.5)$, and the plot of $V(x)$ is not very different from the straight line (12.29) that would correspond to the BM process. This is because the effect of the drift coefficient $-\alpha X_t$ in that region is relatively small, and the values of $V(x)$ only depend on the dynamics of the diffusing inside (a, b). In the middle, the interval is still symmetric, but somewhat wider: $(a,b) = (-2.5, 2.5)$, and the plot is already very different: in the middle part (roughly between ± 1.5) it is flat, i.e., starting from those x-values, the probability of first reaching b is practically one and the same as for $x = 0$, which is equal to $\frac{1}{2}$. The latter value is due to the symmetry of the SP, while the flatness is explained by the observation that, for an interval $(-b, b)$ of such width, the effect of the drift term will make it very likely that the process hits the point 0 prior to reaching any of the boundaries $\pm b$, and once the process hits zero, its changes of first reaching b are $\frac{1}{2}$. Close to the boundary points $\pm b$, the plot is quite steep, indicating that even a small change in the location of the initial point will make difference. Again, this can be explained by the effect of the drift coefficient: it is quite large in that region, so if the process does not hit the nearby boundary point very soon, it will drift away to hit zero first. Finally, $(a, b) = (-0.5, 1.5)$ for the right pane, resulting an asymmetric plot. The reader may wish to interpret the plot on his/her own, using the above observations concerning the two symmetric cases.

Case 2: $\psi(t, a) \equiv \psi(t, b) \equiv t,\ t > 0.$

In this case,

$$v(s, x) = \mathbf{E}_{s,x}\psi(\tau, X_\tau) = \mathbf{E}_{s,x}\tau = s + \mathbf{E}_{0,x}\tau =: s + U(x).$$

Therefore, $\partial_s v(s, x) = 1$, $\partial_x v(s, x) = U'(x)$, $\partial_{xx} v(s, x) = U''(x)$, and hence the BKE becomes the ODE for the new function U:

$$1 = -\mu(x)U'(x) - \frac{1}{2}\sigma^2(x)U''(x), \tag{12.30}$$

with the obvious boundary conditions $U(a) = U(b) = 0$ (as $\tau = 0$ when the process starts at a boundary point). The general solution to the boundary problem is somewhat messy, so we will only consider a couple of examples.

Example 12.4 (continued). For the BM process, the equation turns into $1 = -\frac{1}{2}U''(x)$, which has the general solution $U(x) = -x^2 + C_1 x + C_2$. Using the boundary conditions to determine the constants C_j, we find that

$$U(x) = (b - x)(x - a), \quad x \in (a, b).$$

Again, the same answer can easily be obtained using the Optional Stopping Theorem, but this time we have to apply it to the second MG $\{Y_t := W_t^2 - t \equiv (X_t - x)^2 - t\}_{t \ge 0}$ of the BM (see Theorem 11.6(ii)):

$$\begin{aligned} 0 &= \mathbf{E}_x((X_0 - x)^2 - 0) = \mathbf{E}_x((X_\tau - x)^2 - \tau) \\ &= (a - x)^2\mathbf{P}_x(X_\tau = a) + (b - x)^2\mathbf{P}_x(X_\tau = b) - \mathbf{E}_x\tau \\ &= (a - x)^2(1 - V(x)) + (b - x)^2 V(x) - U(x), \end{aligned} \tag{12.31}$$

and then substitute into the last expression the value of $V(x)$ from (12.29).

Example 12.8. Consider the "square root process" with stochastic differential (12.19) with $\mu(x) \equiv \mu = \text{const}$ and $\sigma(x) = \sqrt{2\mu x}$, $x > 0$ (as we will see in Section 12.5.1 below, such processes approximate the dynamics of *branching processes* starting with a large number of individuals/particles). Our equation (12.30) for $U(x)$ corresponding to some $b > a > 0$ becomes

$$1 = -\mu U'(x) - \frac{1}{2} \times 2\mu x U''(x) = -\mu(U'(x) + xU''(x)) = -\mu(xU'(x))'.$$

Integrating the above equation, we obtain $U'(x) = -1/\mu + C_1/x$, so that

$$U(x) = -\frac{x}{\mu} + C_1 \ln x + C_2.$$

Now the boundary conditions $U(a) = U(b) = 0$ yield that

$$U(x) = -\frac{x-a}{\mu} + \frac{b-a}{\mu}\frac{\ln(x/a)}{\ln(b/a)}, \quad x \in (a, b).$$

Case 3: $\psi(t, a) \equiv \psi(t, b) \equiv e^{-\lambda t}$, $t > 0$, for a fixed parameter $\lambda > 0$.
In this case,

$$v(s, x) = \mathbf{E}_{s,x}\psi(\tau, X_\tau) = \mathbf{E}_{s,x}e^{-\lambda\tau} = e^{-\lambda s}\mathbf{E}_{0,x}e^{-\lambda\tau} =: e^{-\lambda s}W(x).$$

Computing the partial derivatives of that $v(s, x)$ yields the ODE

$$\frac{1}{2}\sigma^2(x)W''(x) + \mu(x)W'(x) - \lambda W(x) = 0$$

for $W(x)$, with obvious boundary conditions $W(a) = W(b) = 1$.

Example 12.4 (continued). For the BM process, the above equation turns into $W''(x) - 2\lambda W(x) = 0$, which has the general solution

$$W(x) = C_1 e^{\sqrt{2\lambda}x} + C_2 e^{-\sqrt{2\lambda}x},$$

and one uses the boundary conditions to find the C_j's.

Letting $a \to -\infty$, we will end up with the single boundary problem: now $\tau = \inf\{t > 0 : x + W_t = b\}$, and $W(x) = \mathbf{E}_{0,x}e^{-\lambda\tau}$, with the single boundary condition $W(b) = 1$. That condition, together with the obvious observation that one must have $C_2 = 0$ (for otherwise one would have $|W(x)| \to \infty$ as $x \to \infty$), yields that $W(x) = e^{-\sqrt{2\lambda}(b-x)}$, $x \le b$. In particular, for the BM process starting at zero, we obtain $W(0) = e^{-\sqrt{2\lambda}b}$, which is clearly the same expression for the LT $\mathbf{E}\, e^{-\lambda\tau_b}$ as we derived in Example 11.8 for $\tau_b := \inf\{t > 0 : W_t = b\}$ using the third MG of the BM.

Remark 12.6. Cases 1–3 are actually special cases of what is known as the

Feynman[11]*–Kac*[12] *formula* which can be stated as follows: in the general situation from Remark 12.2 (p. 353), for a diffusion process $\{X_t\}$ solving the SDE (12.5) and nice enough functions $f(s,x)$ and $g(s,x)$, $(s,x) \in D$, the function

$$v(s,x) := \mathbf{E}_{s,x}\left[\int_s^\tau e^{-\int_s^u f(r,X_r)dr} g(u,X_u)\,du + e^{-\int_s^\tau f(r,X_r)dr}\psi(\tau,X_\tau)\right]$$

is the unique solution to the PDE

$$\partial_s v(s,x) + \mu(s,x)\partial_x v(s,x) + \frac{1}{2}\sigma^2(s,x)\partial_{xx}v(s,x) = f(s,x)v(s,x) - g(s,x)$$

in the region $(s,x) \in D$, with the boundary condition $v(s,x) = \psi(s,x)$, $(s,x) \in \partial D$. Like the BKE, it can be derived using Itô's formula.

As a simple exercise, the reader is suggested to figure out what choices of functions f and g will yield our special Cases 1–3 discussed in this section.

12.5 Some Applications

12.5.1 *Branching Processes*

We will only consider the simplest branching process model, the so-called Galton[13]–Watson[14] process.[15] Branching processes find their applications in such areas as biology, chemistry, nuclear physics *etc.* Recall that we

[11]After Richard Phillips Feynman (11.05.1918-15.02.1988), an outstanding American theoretical physicist who, in particular, shared the 1965 Nobel Prize for physics with J. Schwinger and S.-I. Tomonaga for the development of quantum electrodynamics. You may wish to read *"Surely You're Joking, Mr. Feynman!": Adventures of a Curious Character* by R. Feynman (and R. Leighton), to learn more about Feynman's character, and the famous textbook *The Feynman Lectures on Physics* by R. Feynman (and R. B. Leighton and M. Sands), to learn more about physics.

[12]After Mark (Marek) Kac (03[16].08.1914–26.10.1984), a prominent Polish–American mathematician, whose main contributions were in probability theory (and probabilistic number theory as well).

[13]After Sir Francis Galton (16.02.1822–17.01.1911), a famous English scientist and inventor with very broad interests (in particular, he was a pioneer in eugenics and devised a method for classifying fingerprints).

[14]After Rev. Henry William Watson (25.02.1827–11.01.1903), an English mathematician who, in particular, co-authored with F. Galton the 1875 paper *On the Probability of the Extinction of Families*, one of the first in the studies of the branching phenomena.

[15]The process is sometimes also called the Bienaymé-Galton-Watson process, after Irénée-Jules Bienaymé (28.08.1796–19.10.1878), a French statistician who considered the name extinction problem earlier and knew its solution already by 1845. The term "branching process" itself was coined by A.N. Kolmogorov at his seminar at the Moscow State University in 1946-1947.

already encountered a simple continuous time branching process model in Example 6.5.

Consider a population of individuals (often referred to as particles), that evolves in discrete time. Denote by Z_k the size of the population at time $k = 0, 1, 2, \dots$ Each time step, there is a "change of generations": each of the existing individuals produces a random number of children and disappears. Denote by $\xi_{k,j}$ the number of children of the jth individual in the kth generation, so that $j = 1, 2, \dots, Z_k$ (provided that $Z_k > 0$), and assume that the $\xi_{k,j}$'s are i.i.d. RVs. If $Z_k = 0$, then the population has become extinct by the time k, and it remains extinct forever (although there are versions of the model where immigration is possible). The $(k+1)$st generation consists of all the children of the kth generation individuals.

Thus, we have the following dynamics for the population size:

$$Z_{k+1} = \sum_{j=1}^{Z_k} \xi_{k,j}, \quad k = 0, 1, 2, \dots,$$

with an initial value $Z_0 > 0$ (we follow the usual convention that $\sum_{j=1}^{0} = 0$). We are interested in the behaviour of the distribution of Z_n as $n \to \infty$ and, moreover, in that of the whole trajectory $\{Z_0, Z_1, \dots, Z_n\}$ as $n \to \infty$.

The most interesting case is actually when the process is "near-critical", i.e., when the mean offspring number $\mathbf{E}\,\xi_{k,j}$ is close to one (if it is noticeably less than one, the process vanishes exponentially fast; if it is noticeably greater than one, the process "explodes" at an exponential rate). A convenient way to formalise the concept is to assume that, for the trajectory of length n (i.e., for $\{Z_0, Z_1, \dots, Z_n\}$), the distribution of $\xi_{k,j} = \xi_{k,j}^{(n)}$ actually depends on n in such a way that $\mathbf{E}\,\xi_{k,j} = 1 + \frac{\alpha}{n}$ for a fixed $\alpha \in \mathbf{R}$ (for brevity, the superscript $^{(n)}$ will be omitted, and we will use just ξ to denote an RV that has the offspring distribuion). Such an assumption ensures that the population will be "almost stable" (at least, on short time intervals; when this assumption is not met, one will have to use other approximations to the distributions of interest). Indeed, note that, by the TPF,

$$\mathbf{E}\,Z_{k+1} = \mathbf{E}\left(\sum_{j=1}^{Z_k} \xi_{k,j}\right) = \sum_{m=0}^{\infty} \underbrace{\mathbf{E}\left(\sum_{j=1}^{Z_k} \xi_{k,j} \middle| Z_k = m\right)}_{=\mathbf{E}\sum_{j=1}^{m} \xi_{k,j} = m\mathbf{E}\,\xi} \mathbf{P}\,(Z_k = m)$$

$$= \mathbf{E}\,\xi \sum_{j=1}^{m} m\mathbf{P}\,(Z_k = m) = \mathbf{E}\,\xi\,\mathbf{E}\,Z_k = \cdots = (\mathbf{E}\,\xi)^{k+1}\mathbf{E}\,Z_0,$$

so that for generation numbers k comparable to n, i.e., such that $k/n = t + o(1)$ as $n \to \infty$, one will have

$$\mathbf{E}\, Z_k = \left(1 + \frac{\alpha}{n}\right)^k \mathbf{E}\, Z_0 = \left(1 + \frac{\alpha}{n}\right)^{tn+o(n)} \mathbf{E}\, Z_0 = e^{\alpha t + o(1)} \mathbf{E}\, Z_0.$$

To obtain a diffusion process approximation to the dynamics of our branching process, we will assume in addition that $\mathrm{Var}\,(\xi) \to s^2 < \infty$ (in fact, one also needs that $\xi^2 = (\xi^{(n)})^2$ are uniformly integrable), that the branching process starts with a "large number" of individuals, meaning that $Z_0/n \to x_0 > 0$ as $n \to \infty$, and that the time "runs fast", meaning that generations change each $1/n$ time units. Thus, we will now consider the scaled process

$$X_t^{(n)} := \frac{Z_k}{n}, \quad t \in \left[\frac{k}{n}, \frac{k+1}{n}\right), \quad k = 0, 1, \ldots, n,$$

which can alternatively we written as $X_t^{(n)} := n^{-1} Z_{\lfloor nt \rfloor}$, where $\lfloor s \rfloor$ stands for the integer part of s.

It turns out that, as $n \to \infty$, the distribution of $\{X_t^{(n)}\}_{t \in [0,1]}$ will weakly converge to that of a certain diffusion process (cf. this with the functional CLT mentioned in Section 2.10). We will not prove that fact, but will show instead that (12.3) and (12.4) are "almost satisfied" for $\{X_t^{(n)}\}$ for large n, thus making the claim plausible. We will not discuss (12.2) here, though, as it is somewhat more technical.

Take $h := n^{-1}$ and, assuming that nx and $k := nt$ are integer, consider

$$\begin{aligned}\mathbf{E}\,(X_{t+h}^{(n)} - X_t^{(n)} | X_t^{(n)} = x) &= \mathbf{E}\,(n^{-1} Z_{k+1} - x \,| Z_k = nx) \\ &= \mathbf{E}\left(\frac{1}{n}\sum_{j=1}^{nx} \xi_{k,j} - x\right) = \frac{nx}{n}\left(1 + \frac{\alpha}{n}\right) - x = \alpha x h = \mu(x) h,\end{aligned}$$

if we put $\mu(x) := \alpha x$, $x \geq 0$.

Further, we similarly have

$$\begin{aligned}\mathbf{E}\,((X_{t+h}^{(n)} - X_t^{(n)})^2 | X_t^{(n)} = x) &= \mathbf{E}\left(\frac{1}{n}\sum_{j=1}^{nx} \xi_{k,j} - x\right)^2 \\ &= \frac{1}{n^2}\mathrm{Var}\left(\sum_{j=1}^{nx} \xi_{k,j}\right) + \Bigg[\underbrace{\mathbf{E}\left(\frac{1}{n}\sum_{j=1}^{nx} \xi_{k,j} - x\right)}_{=\alpha x h}\Bigg]^2 \\ &= \frac{1}{n^2} \times nx \mathrm{Var}\,(\xi) + (\alpha x h)^2 = s^2 x h + o(h) = \sigma^2(x) h + o(h),\end{aligned}$$

if we put $\sigma(x) := s\sqrt{x}$, $x \ge 0$.

Comparing these relations (that hold for our special choice of h) to (12.3) and (12.4), it is reasonable to expect that the process $\{X_t^{(n)}\}_{t\in[0,1]}$ will be close to a diffusion with $\mu(x) = \alpha x$, $\sigma(x) = s\sqrt{x}$. And indeed, under the above conditions, one can prove convergence in distribution of $\{X_t^{(n)}\}_{t\in[0,1]}$ to the diffusion process $\{X_t\}_{t\in[0,1]}$ solving the SDE

$$dX_t = \alpha X_t dt + s\sqrt{X_t}dW_t, \quad t \in [0,1]; \quad X_0 = x_0. \tag{12.32}$$

We already dealt with a special case of such a process in Example 12.8. This is one of a few diffusion processes for which transition probabilities are available in a closed analytic form.

Note that the process $\{X_t\}$ can actually *reach* the boundary $x = 0$, which is an absorbing state for the diffusion (once it hits the point, it stays there forever). This would not be so if we replaced the diffusion coefficient $s\sqrt{X_t}$ in (12.32) with sX_t, as in that case we would obtain a geometric BM process, cf. Example 11.13, which is always positive. This is so because, in the latter case, the amplitude of the "local" random oscillations of the process, which is specified by the diffusion coefficient, vanishes much faster as X_t approaches zero than in the former one. As the drift coefficient also vanishes, the trajectory of the SP kind of "flattens out" in vicinity of zero. Yet in the case of the "square root process" the small random oscillations produced by the dW_t-term prove to be strong enough for the process to eventually hit the boundary point $x = 0$, whereas in the geometric BM they are too small for that!

The event $\{X_t = 0\}$ means that the population modelled by the process has become extinct by the time t. One of the most important problems for such processes is to find the eventual extinction probability

$$V(x) := \mathbf{P}_{0,x}(X_t = 0 \text{ for some } t < \infty)$$

given the initial state of the process was x. This is the boundary crossing problem we dealt with in Section 12.4, Case 1, with boundaries $b = 0$, $a = +\infty$ (we had there $a < b$, but this is clearly irrelevant). Since here $2\mu(x)/\sigma^2(x) = 2\alpha s^{-2} =: \beta$, we obtain from (12.24) that the scale function for our process has the form

$$S_{x_0}(x) = \int_{x_0}^{x} e^{-\beta(u-x_0)}du = \beta^{-1}(1 - e^{-\beta(x-x_0)}), \quad \beta \ne 0,$$

and $S_{x_0}(x) = x - x_0$ when $\beta = 0$. Therefore, according to (12.27), one has, for $x > 0$,

$$V(x) = \left[\frac{(1-e^{-\beta(x-x_0)}) - (1-e^{-\beta(a-x_0)})}{(1-e^{-\beta(b-x_0)}) - (1-e^{-\beta(a-x_0)})}\right]_{a=\infty, b=0} = \begin{cases} 1, & \beta < 0, \\ e^{-\beta x}, & \beta > 0, \end{cases}$$

and $V(x) = 1$ when $\beta = 0$. Thus, if $\alpha \le 0$ then extinction is certain, while when $\alpha > 0$ (so that the mean offspring number is greater than one), the extinction probability is given by the exponential function $e^{-2\alpha s^{-2}x}$, $x > 0$.

That the extinction probability should decay as an exponential function actually follows from the branching nature of the process: note that we can think of Z_k as the sum of the kth generation sizes of Z_0 independent copies of our branching process starting with one particle each. As the extinction of the whole population is clearly equivalent to extinction of all the sub-populations formed by the descendants of the initial particles, it follows from independence that $\mathbf{P}(Z_k = 0|Z_0 = z) = (\mathbf{P}(Z_k = 0|Z_0 = 1))^z$. Likewise, the limiting diffusion process has the property that if $X_t(x_j)$, $j = 1, 2$, are independent diffusions with stochastic differentials as in (12.32) and initial values x_j, then their sum $X_t(x_1) + X_t(x_2)$ will also have stochastic differential as in (12.32) and the initial value $x_1 + x_2$. As $X_t(x_1) + X_t(x_2)$ turns into zero iff each of the summands is zero, by independence one has $\mathbf{P}(X_t(x_1) + X_t(x_2) = 0) = \mathbf{P}(X_t(x_1) = 0)\mathbf{P}(X_t(x_2) = 0)$. Letting $t \to \infty$, we obtain $V(x_1 + x_2) = V(x_1)V(x_2)$, which implies that $V(x)$ must be an exponential function (cf. Problem 1 on p. 167).

12.5.2 *The Wright-Fisher Model*

The Wright[16]-Fisher[17] model was suggested to describe gene frequency fluctuations in populations. We will consider the version of the model for a haploid population, i.e., when each individual has one copy of the gene that has two alleles, A and a. The model assumes that all generations are of the same size n, generations do not overlap, and each copy of the gene found in the new generation is drawn independently from all n copies of the gene in the previous one, according to a random mechanism which is affected by *mutation* and *selection* as follows. Firstly, in each individual, a gene of type A (if it is present) mutates into a w.p. $\alpha \geq 0$, while a gene of type a (if it is present) mutates into A w.p. $\beta \geq 0$. Secondly, gene A is superior to a in the sense that the chances of a to be transmitted to the new generation are not higher than those for A: denoting by R the event of successful reproduction, by P_A the event that the parent individual had gene A, and by P_a that it had gene a, we stipulate that, for a fixed $s \geq 0$,

$$\mathbf{P}(R|P_A) = (1+s)\mathbf{P}(R|P_a) =: (1+s)q.$$

Of course, the above model is just a rather crude approximation to real-life situations, but it is still of interest when analysing advance and fixation of genes in populations.[18]

Now suppose that originally there were i individuals with gene A in the current generation (so that $n - i$ individuals had gene a). Due to mutation, by the time of reproduction the probability of a randomly chosen individual

[16]After Sewall Green Wright (21.12.1889–03.03.1988), a prominent American geneticist, one of the founders of theoretical population genetics.
[17]After Sir Ronald Aylmer Fisher (17.02.1890–29.07.1962), an outstanding English scientist, who contributed to many areas including statistics (design of experiments, ANOVA, the method of maximum likelihood, Fisher information *etc.*) and evolutionary biology, one of the founders of theoretical population genetics.
[18]A more adequate model would involve two variables, the gene frequency and population size. The first model of such kind was suggested by W. Feller in 1951.

(i.e., a "candidate parent") to have gene A will, by the TPF, be

$$\mathbf{P}(P_A) = (1-\alpha) \times \frac{i}{n} + \beta \times \frac{n-i}{n},$$

while the probability that it has gene a will be

$$\mathbf{P}(P_a) = \alpha \times \frac{i}{n} + (1-\beta) \times \frac{n-i}{n}.$$

Taking into account selectivity, the probability that an individual from the new generation will have gene A will, by the Bayes formula, be

$$\begin{aligned}\mathbf{P}(A|R) &= \frac{\mathbf{P}(R|A)\mathbf{P}(P_A)}{\mathbf{P}(R|A)\mathbf{P}(P_A) + \mathbf{P}(R|a)\mathbf{P}(P_a)} \\ &= \frac{(1+s)q\big[(1-\alpha)\frac{i}{n} + \beta\frac{n-i}{n}\big]}{(1+s)q\big[(1-\alpha)\frac{i}{n} + \beta\frac{n-i}{n}\big] + q\big[\alpha\frac{i}{n} + (1-\beta)\frac{n-i}{n}\big]} \\ &= \frac{(1+s)\big[(1-\alpha)\frac{i}{n} + \beta\frac{n-i}{n}\big]}{1 + s\big[(1-\alpha)\frac{i}{n} + \beta\frac{n-i}{n}\big]} =: p_i.\end{aligned}$$

The above argument provides justification for the Wright-Fisher model to postulate that, given there were i individuals with gene A (and $n-i$ ones with gene a) in the kth generation, the $(k+1)$st generation consists of n individuals each having gene A w.p. p_i, independently of others (the remaining ones having gene a). Thus, the dynamics of the numbers Z_k of individuals with gene A in generations $k = 0, 1, 2, \ldots$ are described by a Markov chain with the state space $\{0, 1, 2, \ldots, n\}$ and transition probabilities

$$\mathbf{P}(Z_{k+1} = m | Z_k = i) = \binom{n}{m} p_i^m (1-p_i)^{n-m}, \quad i, m = 0, 1, \ldots, n.$$

Note that $X_t^{(n)} := n^{-1} Z_{\lfloor nt \rfloor} \in [0,1]$ is the relative frequency of gene A in the $\lfloor nt \rfloor$th generation. Assuming that both selectivity and mutation are subtle, which is formalised by stipulating that

$$\alpha = \frac{\gamma_1}{n}, \quad \beta = \frac{\gamma_2}{n}, \quad s = \frac{\rho}{n} \quad \text{for some constant } \gamma_j, \rho \geq 0,\ j = 1, 2,$$

one can demonstrate that, as $n \to \infty$, the distribution of the scaled process $\{X_t^{(n)}\}_{t \in [0,1]}$ converges to the distribution of a diffusion process. As in the previous section, we will not prove that but, rather, show that (12.3) and (12.4) are satisfied for our process $\{X_t^{(n)}\}$ for $h := n^{-1}$.

Given $X_t^{(n)} = x$, $x = i/n$, the RV $nX_{t+h}^{(n)}$ has the binomial distribution B_{n,p_i}, with mean np_i and variance $np_i(1-p_i)$. Therefore,

$$\begin{aligned}
\mathbf{E}\,(X_{t+h}^{(n)} - X_t^{(n)} | X_t^{(n)} = x) &= n^{-1} \times np_i - x = p_i - x \\
&= \frac{\left(1+\frac{\rho}{n}\right)\left[(1-\frac{\gamma_1}{n})x + \frac{\gamma_2}{n}(1-x)\right]}{1+\frac{\rho}{n}\left[(1-\frac{\gamma_1}{n})x + \frac{\gamma_2}{n}(1-x)\right]} - x \\
&= \frac{\left(1+\frac{\rho}{n}\right)\left[x - \frac{1}{n}(\gamma_1 x - \gamma_2(1-x))\right]}{1+\frac{\rho}{n}\left[x - \frac{1}{n}(\gamma_1 x - \gamma_2(1-x))\right]} - x \\
&= (\rho x(1-x) - \gamma_1 x + \gamma_2(1-x))h + o(h) = \mu(x)h + o(h)
\end{aligned}$$

if we put $\mu(x) := \rho x(1-x) - \gamma_1 x + \gamma_2(1-x)$, $x \in [0,1]$. Note that we showed that $p_i = x + \mu(x)h + o(h)$.

Using the same observation concerning the distribution of $nX_{t+h}^{(n)}$ and the above result for the means, one has

$$\begin{aligned}
\mathbf{E}\left((X_{t+h}^{(n)} - X_t^{(n)})^2 \,\middle|\, X_t^{(n)} = x\right) &= n^{-2} \times np_i(1-p_i) + \left(n^{-1} \times np_i - x\right)^2 \\
&= x(1-x)h + o(h) + \left(\mu(x)h + o(h)\right)^2 = \sigma^2(x)h + o(h),
\end{aligned}$$

if we put $\sigma(x) := \sqrt{x(1-x)}$, $x \in [0,1]$.

Comparing these relations with (12.3) and (12.4), one can expect (and this is the case indeed) that the limiting process $\{X_t\}$ will satisfy the SDE

$$dX_t = (\rho X_t(1-X_t) - \gamma_1 X_t + \gamma_2(1-X_t))\,dt + \sqrt{X_t(1-X_t)}dW_t. \tag{12.33}$$

Consider the simple special case of one-way mutation: $\gamma_1 =: \gamma > 0$, $\gamma_2 = 0$ (so that A can mutate into a, but not the other way around) in the absence of selection ($\rho = 0$). Then equation (12.33) simplifies to

$$dX_t = -\gamma X_t dt + \sqrt{X_t(1-X_t)}dW_t.$$

From our discussion of the boundary behaviour of the diffusion approximating branching processes (p. 370; note that the behaviour of the diffusion coefficient, as the process approaches any of the boundaries 0 or 1, is the same as for that process: it is given by the square root of the distance to the boundary), one can expect "fixation" of gene a (i.e., disappearance of A from the populations) w.p. 1, and this is the case indeed. Note, however, that prior to the fixation of a, it may still happen that all the individuals in the population will have gene A (i.e., $X_t = 1$). What is the probability of that event, given that $X_0 = x \in (0,1)$?

This is again the boundary crossing problem we dealt with in Section 12.4, Case 1, now with boundaries $b = 1$, $a = 0$. Since here $2\mu(x)/\sigma^2(x) = -2\gamma/(1-x)$, we obtain from (12.24) that the scale function for our process has the form

$$\begin{aligned}
S_0(x) &= \int_0^x \exp\left\{2\gamma \int_0^u \frac{dz}{1-z}\right\}du = \int_0^x \exp\{-2\gamma \ln(1-u)\}du \\
&= \frac{1}{1-2\gamma}\left[(1-u)^{1-2\gamma}\right]_0^x = \frac{(1-x)^{1-2\gamma} - 1}{1-2\gamma}, \quad x \in [0,1],
\end{aligned}$$

assuming that $\gamma \neq \frac{1}{2}$ (the reader may wish to consider the case $\gamma = \frac{1}{2}$ on her/his own). Therefore, according to (12.28), one has, for $b \in (0,1]$ and $x \in [0,b]$,

$$V(x) = \frac{S_0(x)}{S_0(b)} = \frac{(1-x)^{1-2\gamma}-1}{(1-b)^{1-2\gamma}-1}.$$

To find the desired probability for our diffusion to ever hit the boundary $b = 1$, it remains to compute $\lim_{b \nearrow 1} V(x)$. The answer will depend on the value of γ.

If $\gamma > \frac{1}{2}$, then $1 - 2\gamma < 0$ and so clearly $\lim_{b \nearrow 1} V(x) = 0$. Thus, if the rate of the mutation $A \mapsto a$ is high enough, the population will never have all its individuals with gene A.

However, if $\gamma < \frac{1}{2}$, then $1 - 2\gamma > 0$ and hence $\lim_{b \nearrow 1} V(x) = 1 - (1-x)^{1-2\gamma}$. With that probability, our diffusion process starting at point x will hit the upper boundary (all genes are A) prior to eventual fixation at zero (only genes a are present in the population).

12.5.3 *The Brownian Bridge Process*

Along with the BM process, the Browning bridge (BB) process, also known as the tied-down BM process or conditional BM process, is one of the most popular "building blocks" in stochastic modelling.

The standard BB process $\{W_t^0\}_{t\in[0,1]}$ can be defined as the *standard BM process conditioned to be at zero at time* $t = 1$, which can symbolically be written as

$$\{W_t^0\}_{t\in[0,1]} \stackrel{d}{=} \big(\{W_t\}_{t\in[0,1]} \,|\, W_0 = 0\big).$$

Of course, the conditioning event is of zero probability, but, as we did in (2.79), in such a case the distribution can be found as the limit of conditional distributions given the events $\{|W_0| < \varepsilon\}$ (of positive probability) as $\varepsilon \searrow 0$.

Will the BB process still be a diffusion? To get an insight into its behaviour, for a fixed $t \in (0,1)$, take a small $h > 0$ such that $t + h < 1$, and consider the increment of the BB process on the time interval $(t, t+h)$. Note that the conditional distribution of W_{t+h}^0 given $W_t^0 = x$ will then be the same as the conditional distribution of W_{t+h} given $W_t = x$ and $W_1 = 0$, which has density that can be easily computed. Indeed, denoting by $f_{t_1,\ldots,t_k}(x_1,\ldots,x_n)$ the density of $(W_{t_1},\ldots,W_{t_n})$, we obtain from (11.22) and (2.80) that the desired conditional density at point y will be given by

$$\begin{aligned}\frac{f_{t,t+h,1}(x,y,0)}{f_{t,1}(x,0)} &= \frac{p_t(0,x)p_h(x,y)p_{1-t-h}(y,0)}{p_t(0,x)p_{1-t}(x,0)}\\ &= C(t,h,x)\exp\left\{-\frac{(y-x)^2}{2h} - \frac{y^2}{2(1-t-h)}\right\},\end{aligned}$$

where $C(t,h,x)$ is a quantity depending on its arguments only and, setting $u := 1 - h/(1-t) = 1 + o(1)$ as $h \to 0$, the argument of the exponent is equal to

$$-\frac{y^2}{2hu} + \frac{yx}{h} - \frac{x^2}{2h} = -\frac{(y-xu)^2}{2hu} + \frac{x^2(u-1)}{2h}.$$

So the density of $(W^0_{t+h} \mid W^0_t = x)$ at y is equal to $C_1(t,h,x)e^{-(y-xu)^2/(2hu)}$, and hence the conditional distribution of $W^0_{t+h} - W^0_t$ given that $W^0_t = x$ is

$$N(xu - x, hu) = N\biggl(-\frac{xh}{1-t}, h + o(h)\biggr).$$

Comparing this with (12.6) leads us to the second (equivalent) definition of the BB process as the *solution to the SDE*

$$dX_t = -\frac{X_t}{1-t}\,dt + dW_t, \quad X_0 = 0. \tag{12.34}$$

Note the form of the drift coefficient: as its denominator vanishes as t approaches one, it represents an ever growing force "pushing" the process towards zero. So it can be considered as a kind of control of which the sole purpose is to eventually "suppress" the randomness emanating from the dW_t term to ensure that $X_1 = 0$.

Clearly, the BB is a Gaussian process. Indeed, the distribution of $(W^0_{t_1}, \ldots, W^0_{t_n})$, $0 \le t_1 < \cdots < t_n < 1$, can be obtained by conditioning the Gaussian distribution of $(W_{t_1}, \ldots, W_{t_n})$ on $W_1 = 0$, and such a conditional distribution is always Gaussian as well, as one can derive from (2.80) and (2.61). As any Gaussian process $\{X_t\}$ is uniquely determined by its mean function $m_X(t) = \mathbf{E}\,X_t$ and covariance function $\gamma_X(s,t) := \mathrm{Cov}\,(X_s, X_t)$ (cf. p. 318), one can also define the BB process by specifying these functions. We just have to compute them first.

That could be done directly, by computing conditional expectations basing on the original definition of the process. It would perhaps be more instructive to do that by taking expectations of the both sides of the SDE for the process written in the integral form, see (11.45). In the case of the SDE (12.34), that form is

$$X_t = X_0 - \int_0^t \frac{X_s}{1-s}\,ds + \int_0^t dW_s = -\int_0^t \frac{X_s}{1-s}\,ds + W_t, \quad t \in [0,1].$$

Taking expectations and using Fubini's theorem yields

$$m_X(t) = -\int_0^t \frac{m_X(s)}{1-s}\,ds.$$

So m_X is differentiable and satisfies $m'_X(t) = -m_X(t)/(1-t)$, or

$$(\ln m_X(t))' = -\frac{1}{1-t} = (\ln(1-t))'.$$

Therefore, $m_X(t) = C(1-t)$. As $m_X(0) = 0$, we get $m_X(t) = 0$, $t \in [0,1]$.

To compute the variance of the process, apply Itô's formula (11.40) to the process $Y_t := X_t^2 = f(X_t)$ with $f(x) = x^2$ (so that $f'(x) = 2x$, $f''(x) = 2$) to get

$$d(X_t^2) = f'(X_t)dX_t + \frac{1}{2}f''(X_t)(dX_t)^2 = \left(1 - \frac{2X_t^2}{1-t}\right)dt + 2X_t dW_t,$$

so that

$$X_t^2 = X_0^2 + \int_0^t \left(1 - \frac{2X_s^2}{1-s}\right)ds + 2\int_0^t X_s dW_s.$$

Taking expectation on both sides and using Fubini's theorem and **II3** yields

$$\mathbf{E}\,X_t^2 = \int_0^t \left(1 - \frac{2\mathbf{E}\,X_s^2}{1-s}\right)ds.$$

Differentiating this relation, we obtain that

$$(\mathrm{Var}\,(X_t))' = 1 - \frac{2}{1-t}\,\mathrm{Var}\,(X_t), \quad t \in (0,1); \quad \mathrm{Var}\,(X_0) = 0.$$

One can easily solve this first order linear ODE to get

$$\mathrm{Var}\,(X_t) = t(1-t), \quad t \in (0,1).$$

To compute the covariances, one can take the expectations of the both sides of the obvious relation $X_sX_t = X_s\left(X_s + \int_s^t dX_v\right)$, $0 \le s < t \le 1$, to obtain

$$\gamma_X(s,t) = \mathbf{E}\,X_sX_t = \mathrm{Var}\,(X_s) - \int_s^t \frac{\mathbf{E}\,X_sX_v}{1-v}\,dv + \underbrace{\mathbf{E}\int_s^t X_s dW_v}_{=0 \text{ by II3}}$$

$$\equiv \mathrm{Var}\,(X_s) - \int_s^t \frac{\gamma_X(s,v)}{1-v}\,dv. \tag{12.35}$$

Keeping s fixed, differentiate (12.35) w.r.t. to t to get

$$\partial_t\gamma_X(s,t) = -\frac{\gamma_X(s,t)}{1-t}, \quad t \in (s,1); \quad \gamma_X(s,s) = \mathrm{Var}\,(X_s) = s(1-s).$$

Therefore the covariance function of the BB process $X_t = W_t^0$ equals

$$\gamma_X(s,t) = s(1-t), \quad 0 \le s \le t \le 1. \tag{12.36}$$

Thus we arrive at the following result parallel to Theorem 11.4 and giving one more equivalent definition of the BB process.[19]

Theorem 12.1. *A Gaussian SP* $\{X_t\}_{t\in[0,1]}$ *is a BB process iff* $m_X(t) \equiv 0$ *and* $\gamma_X(s,t) = s \wedge t - st$, $s,t \in [0,1]$.

[19]See our footnote on p. 318 regarding continuous modifications.

However, it is easy to verify that the process

$$X_t := W_t - tW_1, \quad t \in [0,1], \tag{12.37}$$

has the same mean and covariance functions. Thus, representation (12.37) gives us one more (equivalent) definition of the BB process.

One can give yet one more (equivalent) definition of the BB process, also based on Theorem 12.1. Namely, the BB process can be defined by

$$X_t := (1-t)W_{t/(1-t)}, \quad t \in [0,1] \tag{12.38}$$

(by continuity, the value of the right-hand side of (12.38) at $t = 1$ is zero due to the LLN).

Indeed, clearly $\mathbf{E}\, X_t = 0$, while by Theorem 11.4, for $0 \le s \le t \le 1$,

$$\mathbf{E}\, X_s X_t = (1-s)(1-t)\underbrace{\mathbf{E}\, W_{s/(1-s)} W_{t/(1-t)}}_{\frac{s}{1-s}\wedge\frac{t}{1-t}=\frac{s}{1-s}} = s(1-t).$$

An alternative way to prove that (12.38) is a BB process is to show that its stochastic differential has the form (12.34) (with a different BM in it!).

Remark 12.7. The process we discussed in this section is the *standard BB*. More generally, one also deals with BBs that "bridge" arbitrary given points (t_j, x_j), $j = 1, 2$, with $0 \le t_1 < t_2$ in the following sense: these are processes $\{X_t\}_{t\in[t_1,t_2]}$ whose distributions coincide with the conditional distributions of the segments $\{W_t\}_{t\in[t_1,t_2]}$ of the BM trajectory given $W_{t_j} = x_j$, $j = 1, 2$:

$$\{X_t\}_{t\in[t_1,t_2]} \stackrel{d}{=} \big(\{W_t\}_{t\in[t_1,t_2]} \,|\, W_{t_1} = x_1,\, W_{t_2} = x_2\big).$$

Using conditional densities or covariance functions, it is not hard to show that such a bridge can be obtained as a time/space transformation of the standard BB of the form

$$\frac{t_2 - t}{t_2 - t_1}\, x_1 + \sqrt{t_2 - t_1}\, W^0_{(t-t_1)/(t_2-t_1)} + \frac{t - t_1}{t_2 - t_1}\, x_2, \quad t \in [t_1, t_2].$$

Note that, in the absence of the middle term, the above will be just a parametric representation of the straight line segment connection the points (t_j, x_j), $j = 1, 2$. The middle term superimposes on it a version of the BB process that is scaled in such a way that it "straddles" now the interval $[t_1, t_2]$ (turning into zero at its end points) and has the same diffusion coefficient as the BM (this what the square root factor is for).

A remarkable property of the BB process is that there is a simple formula for the probability of its hitting a linear boundary. For arbitrary $b_j \ge 0$,

$j = 1, 2$, the probability that W_t^0 will cross the linear boundary with end points $(0, b_1)$ and $(1, b_2)$ is, using (12.38) and the time change $s := \frac{t}{1-t}$,

$$\begin{aligned}
&\mathbf{P}\left(\sup_{t\in[0,1]} (W_t^0 - b_1 - (b_2 - b_1)t) > 0\right) \\
&\quad = \mathbf{P}\left(\sup_{t\in[0,1]} ((1-t)W_{t/(1-t)} - b_1 - (b_2 - b_1)t) > 0\right) \\
&\quad = \mathbf{P}\left[\sup_{s\ge 0}\left(\frac{1}{1+s}W_s - b_1 - \frac{(b_2 - b_1)s}{1+s}\right) > 0\right] \\
&\quad = \mathbf{P}\left(\sup_{s\ge 0}(W_s - b_2 s) > b_1\right) = e^{-2b_1 b_2} \qquad (12.39)
\end{aligned}$$

from Example 12.6.

Example 12.9. *Empirical distribution functions.* We already introduced the concept of EDF in Example 2.6, see (2.85). Equivalently, the EDF for a sample $Y_1, \ldots, Y_n \sim F$ can be written as

$$F_n^*(t) := \frac{1}{n}\sum_{j=1}^{n} \mathbf{1}_{\{Y_j \le t\}}, \quad t \in \mathbf{R}.$$

From the statistical viewpoint, one can think about the EDF F_n^* as an estimator for the (unknown) theoretical DF F.

It is important to realise that most popular *statistics* (i.e., functions $S = S(Y_1, \ldots, Y_n)$ of the sample) admit simple representations in terms of the EDF. For example, the sample mean is just the mean of the EDF:

$$\overline{Y} := \frac{1}{n}\sum_{j=1}^{n} Y_j = \int t\, dF_n^*(t),$$

the sample variance is the variance of the EDF:

$$s^2 := \frac{1}{n}\sum_{j=1}^{n}(Y_j - \overline{Y})^2 = \overline{Y^2} - \left(\overline{Y}\right)^2 = \int t^2 dF_n^*(t) - \left(\int t\, dF_n^*(t)\right)^2,$$

sample quantiles are basically the quantiles of the EDF, *etc.*

The idea of the powerful "substitution (or plug-in) method"[20] is that if you are interested in a parameter θ of the DF F that can be expressed as a functional of the DF: $\theta = G(F)$ for some nice enough G, then the statistic $\theta_n^* := G(F_n^*)$ will be a good estimator for θ. More formally, provided that $Y_1, \ldots, Y_n$ are i.i.d. (and under weaker dependence assumptions as well),

[20]The systematic studies of such estimators go back to R. von Mises (1947).

as the sample size n goes to infinity, the EDFs converge uniformly on $\mathbf{R}$ w.p. 1 to the theoretical DF by the Glivenko-Cantelli theorem (see p. 59). So if the functional G is continuous at the "point" F in the uniform metric in the sense that $G(F_n) \to G(F)$ if $\sup_{t\in\mathbf{R}} |F_n(t) - F(t)| \to 0$ as $n \to \infty$, then $G(F_n^*) \to G(F) = \theta$ a.s., meaning that the estimator θ_n^* is *strongly consistent.* Moreover, this general approach also allows one to infer the asymptotic normality of a large class of estimators.

Indeed, recall that if Q is the quantile function (10.4) of the DF F and $U \sim U(0,1)$, then $Q(U) \sim F$ by Theorem 10.1. Therefore, w.l.o.g. one can assume that $Y_i = Q(U_i)$, where $U_1, \dots, U_n$ is an i.i.d. $U(0,1)$-sample. Then, using (10.5), one has

$$F_n^*(t) = \frac{1}{n}\sum_{i=1}^n \mathbf{1}_{\{Q(U_i)\le t\}} = \frac{1}{n}\sum_{i=1}^n \mathbf{1}_{\{U_i \le F(t)\}} =: U_n^*(F(t)), \qquad (12.40)$$

which is the value of the EDF U_n^* of the uniform sample $U_1, \dots, U_n$ at the point $F(t)$. Thus, one only needs to derive the asymptotic behaviour of U_n^* as $n \to \infty$ to get that for F_n^*, as the latter function is obtained from the former one by change of variables.

We already noted on p. 61 that, due to the de Moivre–Laplace theorem and (2.86),

$$\sqrt{n}(U_n^*(t) - t) \equiv \frac{1}{\sqrt{n}}\sum_{i=1}^n (\mathbf{1}_{\{U_i\le t\}} - t) \overset{\text{distr}}{\longrightarrow} N(0, t(1-t)) \quad \text{as } n \to \infty,$$

which immediately implies, in view of the preceding representation for F_n^*, that the distribution of (2.89) converges to $N(0, F(t)(1 - F(t)))$. The assertion can be easily extended (using, e.g., the machinery of ChFs) to the multivariate case, by showing that, for any $0 < t_1 < \cdots < t_k < 1$, the distribution of the vector

$$\sqrt{n}(U_n^*(t_1) - t_1, \dots, U_n^*(t_k) - t_k) \equiv \frac{1}{\sqrt{n}}\sum_{i=1}^n (\mathbf{1}_{\{U_i\le t_1\}} - t_1, \dots, \mathbf{1}_{\{U_i\le t_k\}} - t_k)$$

will converge as $n \to \infty$ to that of a Gaussian random vector $(X_{t_1}, \dots, X_{t_k})$ such that, for $j, m = 1, \dots, k$,

$$\mathbf{E}\, X_{t_j} = \mathbf{E}\,(\mathbf{1}_{\{U_1\le t_j\}} - t_j) = \mathbf{P}\,(U_1 \le t_j) - t_j = 0$$

and

$$\begin{aligned}\mathbf{E}\, X_{t_j} X_{t_m} &= \mathbf{E}\,(\mathbf{1}_{\{U_1\le t_j\}} - t_j)(\mathbf{1}_{\{U_1\le t_m\}} - t_m) \\ &= \mathbf{E}\, \underbrace{\mathbf{1}_{\{U_1\le t_j\}}\mathbf{1}_{\{U_1\le t_m\}}}_{=\mathbf{1}_{\{U_1\le t_j\wedge t_m\}}} - t_j t_m = t_j \wedge t_m - t_j t_m.\end{aligned}$$

That is, the limiting vector $(X_{t_1}, \ldots, X_{t_k})$ has the same distribution as $(W^0_{t_1}, \ldots, W^0_{t_k})$. So, it should come as no surprise that the empirical process $\{E^{(n)}_t\}_{t\in[0,1]}$ defined by (2.89) converges in distribution to the time-changed BB process $\{W^0_{F(t)}\}_{t\in[0,1]}$, which results in the "large sample" EDF distributional approximation (2.92). Now if the functional G in the representation $\theta = G(F)$ is differentiable, with derivative G' (our G is a functional, i.e., a function whose argument is a function as well, so G' is actually a linear operator), then (2.92) will imply that

$$\begin{aligned}\theta^* &= G(F^*_n) \approx G\big(F + n^{-1/2}W^0_{F(\cdot)}\big)\\ &\approx G(F) + n^{-1/2}G'(F)W^0_{F(\cdot)} = \theta + n^{-1/2}G'(F)W^0_{F(\cdot)}.\end{aligned}$$

This means that the distribution of $\sqrt{n}(\theta^* - \theta)$ tends to a normal one as $n \to \infty$, which is exactly the highly desirable asymptotic normality property of the estimator θ^*.

To illustrate the above general approach, consider the following *Smirnov statistic*:[21]

$$D^+_n := \sup_{x\in\mathbf{R}} (F^*_n(x) - F(x)).$$

We see from (12.40) that if F is a continuous DF, then D^+_n has the same distribution as

$$\sup_{x\in\mathbf{R}} [U^*_n(F(x)) - F(x)] = \sup_{t\in[0,1]} (U^*_n(t) - t).$$

But the distribution of the latter RV clearly does not depend on F (provided that it is actually true that $Y_i \sim F$, of course), which is very handy if one uses the statistic for goodness-of-fit testing, i.e., testing the null hypothesis H_0 stating that the elements of the original sample $Y_1, \ldots, Y_n$ really follow the DF F.

Indeed, to construct a test with a critical region of the form $D^+_n > d_\alpha$ (i.e., the test rejects H_0 when the inequality is satisfied) with significance level α (meaning that the error of the first kind — rejecting H_0 when it is true — has probability α), we just have to find d_α such that

$$\mathbf{P}\,(D^+_n > d_\alpha) = \alpha.$$

As the distribution of D^+_n does not depend on F (given that the DF F is continuous!), we just have to compute the value $d_\alpha = d_\alpha(n)$ once and for all: it can then be used for size n samples for any F. For large n values, one can compute an approximate value of d_α using the convergence of empirical processes that we discussed in Example 12.9. Since, for a fixed $b > 0$, one has, as $n \to \infty$,

$$\mathbf{P}\,(\sqrt{n}D^+_n > b) = \mathbf{P}\left(\sup_{t\in[0,1]} E^{(n)}_t > b\right) \to \mathbf{P}\left(\sup_{t\in[0,1]} W^0_t > b\right) = e^{-2b^2}$$

from (12.39) with $b_1 = b_2 = b$, it follows that $d_\alpha(n) \approx \sqrt{|\ln\alpha|/(2n)}$ for large n values. For small and moderate n, the values of $d_\alpha(n)$ are available from tables and statistical software.

[21] After Nikolai Vasilyevich Smirnov (04.10.1900–02.06.1966), a prominent Russian probabilist and statistician.

More popular is the *Kolmogorov-Smirnov test*, which is a symmetric version of the above test. It is based on the "two-sided statistic" $D_n^+ := \sup_{x\in\mathbf{R}} |F_n^*(x) - F(x)|$. By the same token, for any fixed $b > 0$, one has

$$\mathbf{P}\left(\sqrt{n}D_n > b\right) \to \mathbf{P}\left(\sup_{t\in[0,1]} |W_t^0| > b\right) \quad \text{as } n \to \infty.$$

The distribution on the right-hand side is referred to as the *Kolmogorov distribution* and is available as the sum of an infinite series.

12.6 Recommended Literature

Please refer to the list in Section 11.6. In addition, we would like to mention here the following books:

KARLIN, S. AND TAYLOR, H. *A Second Course in Stochastic Processes.* Academic Press, New York, 1981. [*Among other nice features, the book has an insightful introductory chapter on diffusion processes (presenting, in particular, the method of differential equations).*]

PINKSY, R.G. *Positive Harmonic Functions and Diffusion.* Cambridge University Press, Cambridge, 2008. [*A self-contained account of the construction and basic properties of diffusion processes.*]

12.7 Problems

1. Let $X_t := x_0 + t + 2W_t$, $t \geq 0$, be an arithmetic BM process with a unit drift, starting at the point $X_0 = x_0$.

   (i) Using independence of the increments of the BM process, find the transition density $p(s, x; t, y) := \frac{d}{dy}\mathbf{P}(X_t \leq y |\, X_s = x)$ of the process $\{X_t\}$ $(0 \leq s < t)$.

   (ii) Write down the forward and backward Kolmogorov equations for the process $\{X_t\}$.

   (iii) Verify (by the direct computation of the respective derivatives) that the function $u(t, y) := p(0, x_0; t, y)$, where $p(\cdots)$ is the density you found in part (i), satisfies the forward equation you wrote down in part (ii). What could you say about the initial condition that solution satisfies? [In other words: what happens to $u(t, y)$ as $t \to 0$?]

2. Let $\{Y_t\}$ be an Ornstein–Uhlenbeck process satisfying

$$dY_t = -\alpha Y_t dt + \sigma dW_t, \qquad \alpha, \sigma > 0, \quad Y_0 = x. \tag{12.41}$$

   (i) Write down the forward and backward Kolmogorov equations for $\{Y_t\}$.

(ii) Using the product rule of the Itô calculus, show that

$$Y_t := e^{-\alpha t} Z_{\sigma^2(e^{2\alpha t}-1)/(2\alpha)}, \qquad dZ_t = dW_t, \quad Z_0 = x,$$

is an Ornstein–Uhlenbeck process satisfying equation (12.41) (perhaps with a different Brownian motion process $\widetilde{W}_t$).

(iii) Use part (ii) to derive the density of Y_t, $t > 0$. Find the limit of the density as $t \to \infty$ and comment on the significance of the existence of the limit from the viewpoint of the long-run behaviour of the Ornstein–Uhlenbeck process.

Hint: (ii) You may wish to use Remark 11.1.

3. Let $\{X_t\}_{t\geq 0}$ be a diffusion process taking values in $(0, \infty)$, described by the SDE

$$dX_t = (1 - X_t)\, dt + \sqrt{2X_t}\, dW_t.$$

(i) Write down the forward and backward Kolmogorov equations for the SP.

(ii) Use the method of differential equations to derive the stationary density of the process.

4. A diffusion process $\{X_t\}_{t\geq 0}$ is given by the following SDE:

$$dX_t = \frac{1}{2} X_t dt + dW_t.$$

(i) Write down the backward and forward Kolmogorov equations for this SP.

(ii) Use the backward Kolmogorov equation to derive a differential equation for the function $V(x) = \mathbf{P}\left(X_\tau = b \middle| X_0 = x\right)$, where $\tau := \min\{t \geq 0 : X_t = a \text{ or } X_t = b\}$, $a < x < b$, is the first time the SP hits one of the barriers a or b. Specify boundary conditions for the equation and solve it.

(iii) Sketch the plots of $V(x)$, $x \in [a, b]$, when [a] $(a, b) = (-2, 2)$; [b] $(a, b) = (-1, 1)$; [c] $(a, b) = (0, 2)$; [d] $(a, b) = (2, 4)$. Comment briefly on the differences in shape.

(iv) Find $\mathbf{P}\left(\max_{t\geq 0} X_t > b \middle| X_0 = x\right)$, $x, b \in \mathbf{R}$.

Hints: (ii)–(iv) You may find it convenient to use the standard normal DF in presenting your results.

5. A diffusion process $\{X_t\}_{t\geq 0}$ is given by the SDE

$$dX_t = -\alpha X_t\, dt - \sqrt{1 - X_t^2}\, dW_t, \quad X_0 = x \in (-1, 1),$$

where α is a fixed real number.

(i) Write down the backward and forward Kolmogorov equations for $\{X_t\}$.

(ii) One can show that the process $\{X_t\}$ is *ergodic* when $\alpha > 0$. Derive the stationary density $\pi(y)$ of the process in this case (it suffices to give the answer in the form $\pi(y) = Cf(y)$, where $f(y)$ is given explicitly, and C is a constant, no need to compute the latter!). Sketch the plots of $\pi(y)$ when [a] $\alpha = 1/2$, [b] $\alpha = 1$, [c] $\alpha = 2$ and [d] $\alpha = 3$. Comment on the character of the dependence of the density shape on α. What do you think could explain the observed effect?

(iii) Let $-1 < a < b < 1$. For the values [a] $\alpha = 1/2$ and [b] $\alpha = 1$ compute the probability that, starting at a given point $X_0 = x \in (a, b)$, the process $\{X_t\}$ will hit a barrier at the level b before hitting one at the level a.

(iv) In the case $\alpha = 0$, compute the expected time till the process $\{X_t\}$ starting at a given point $X_0 = x \in (-1, 1)$ hits one of the points ± 1.

6. Derive ODEs for the first two moment functions $m_X(t) := \mathbf{E}\, X_t$ and $m_{X^2}(t) := \mathbf{E}\, X_t^2$ for the continuous branching process solving SDE (12.32). Solve these equations to compute the moment functions.

   Hint: Cf. p. 375.

7. The diffusion process $\{X_t\}_{t \geq 0}$ given by the SDE

$$dX_t = X_t(1 - X_t)dt + \sqrt{X_t(1 - X_t)}\, dW_t, \quad X_t \in (0, 1),$$

   corresponds to the Wright-Fisher gene frequency model involving selection only (this is a special case of (12.33), with $\rho = 1$ and $\gamma_1 = \gamma_2 = 0$).

   (i) Write down the backward and forward Kolmogorov equations for $\{X_t\}$.

   (ii) Assuming that the initial frequency was $X_0 = x \in (0, 1)$, derive the probability of the eventual fixation of the gene frequency at one (meaning that $X_t = 1$ for all $t \geq \tau$ for some time τ). That event occurs if the process X_t hits the boundary point 1 before hitting point 0.

8. A diffusion process $\{X_t\}_{t \geq 0}$ given by the SDE

$$dX_t = (-\gamma_1 X_t + \gamma_2(1 - X_t))dt + \sqrt{X_t(1 - X_t)}\, dW_t, \quad X_t \in (0, 1),$$

   with $\gamma_j \geq 0$ arises in the Wright-Fisher model for the fluctuation of gene frequency without selection differences (this is a special case of (12.33), with $\rho = 0$). One can show that the process has an absolutely continuous stationary distribution.

   (i) Write down the forward and backward Kolmogorov equations for $\{X_t\}$.

   (ii) Consider the special case when $\gamma_1 = \gamma_2 = 1$. Write down an ODE for the stationary density and solve the equation. Plot the stationary density.

9. Verify that the process $\{X_t\}$ specified by (12.37) has the mean and covariance functions from the statement of Theorem 12.1.

Chapter 13

Elements of Mathematical Finance

13.1 Introductory Remarks

The purpose of this chapter is to give a gentle introduction into the area of mathematical finance that is concerned with computing the so-called fair (or risk-free, or arbitrage free) prices of financial derivatives. One can say that the development of that area began with the 1973 seminal papers by F.S. Black, M.S. Scholes and R.C. Merton, and that it received a substantial boost in the late 1970s when a link[1] between arbitrage free pricing and martingales was discovered. Since then the field has became one of the largest and most important areas of applied stochastic modelling, and this is the main reason for including the present chapter in this book.

We will not give here any substantial historical background and/or description of the technical detail of the derivatives' trading and pricing (to learn about that, the interested reader is referred to the literature listed in Section 13.9 and further references therein), neither will we dwell on potential and real dangers that are related to the use and abuse of financial derivatives and the role thereof in financial scandals (the interested reader may wish to find that information elsewhere). What we will present are the basic ideas of arbitrage free pricing and implementation thereof in the so-called binomial and Black–Scholes markets.

Let us start with a few basic concepts. Derivatives (a.k.a. "contingent claims") are financial instruments that are called so because they have no intrinsic value, but derive their value from something else, which can be a stock, index, commodity or interest rate, called the "underlying". A derivative is basically a contract between two parties that specifies conditions

[1]That relationship was first noted in: Harrison, J.M. and Kreps, D.M. (1979) Martingales and arbitrage in multiperiod securities markets. *J. Econ. Theory.* 20, 381–408.

under which payments are to be made between the parties. The conditions include the dates, resulting values and definitions of the underlying variables, the parties' contractual obligations, and the notional amount. Derivative contracts can be traded at specialised derivatives exchanges (or other exchanges) or can be privately traded "over-the-counter" (OTC).[2]

There exists a variety of derivative contracts, including the so-called forwards, futures, options, swaps *etc.* We will mostly deal with *options* in this chapter. An option is a contract which gives the buyer (the owner) the right, but not the obligation, to buy (in the case of *call options*) or sell (in the case of *put options*) the underlying asset at a specified "strike price" (a.k.a. the "exercise price") on or before a specified date (the expiry of the option). The seller of the option has the corresponding obligation to fulfill the transaction (i.e., sell in the case of a call option, or buy in the case of a put option) if the option owner decides to "exercise" the option. In that case, the respective financial transaction is to be carried out immediately, and then the option contract is terminated.

Strictly speaking, the above defines so-called "vanilla options", named so because they have no special features (as vanilla was the default ice cream flavour in some parts of the world). There exists, however, a variety of more complex options that are called "exotic". Examples of exotic options are the so-called Asian options, where the payoff is determined by the average underlying price over a specified time period, and barrier options for which the payoff depends on whether or not the price process crossed a specified barrier during a specified time period. We will discuss the problem of pricing the latter variety of options in Section 13.8.

Why would one want to purchase an option? There may be several reasons for that, as we already pointed out in Example 4.2. The most noble one is *hedging*,[3] i.e., investing with the aim to reduce the risk of adverse price movements in an asset. For instance, airlines buy options on jet fuel (as fuel prices can be quite volatile and so constitute a serious risk to airlines[4]), Australian university libraries buy options on foreign currency

[2]To give an idea of the traded derivatives volumes, as of June 2011, the notional amounts of derivatives outstanding were estimated to be around US$ 700 trillion (how many zeros will one need to write that number in the decimal notation?) for OTC derivatives and US$ 83 trillion for derivatives traded on exchanges (*The Economist*, 07.04.2012). Is that much? As of January 2012, the total market capitalisation of the companies listed at the NASDAQ Stock Exchange market (which is the second-largest stock exchange in the world) was US$ 4.45 trillion.

[3]In this context, to *hedge* means "to cover oneself against loss (on a bet *etc.*) by betting, *etc.*, on the other side" (*Shorter Oxford English Dictionary*).

[4]Thus, from 1998 to 2008, by buying options on jet fuel, *Southwest Airlines* "saved US$ 3.5 billion over what it would have spent if it had paid the industry's average price for jet fuel. That's equal to about 83% of the company's profits over the last $9\frac{1}{2}$ years" (*USA Today*, 24.07.2008).

exchange rates (as subscription prices of most periodicals are set in US$ or euro rather than in AU$) *etc.*

Another major reason for buying options is *speculation*, when one invests opportunistically in the hope to make a profit, the underlying being irrelevant to the investor (note that speculation is actually needed to make hedging possible: one cannot lay off risks unless there is someone willing to take them on!).

Example 13.1. On 04.12.2013, the closing *Apple Inc.* share price was US$ 565 (the stock was traded at NASDAQ). At the same time, the price of a call option on that stock with strike $K =$ US$ 600 and expiry date 14.03.2014 was US$ 20. Suppose a speculator buys such a call on 04.12.2013. What will happen at the expiry date? If the then stock price S will be less than K, the speculator will do nothing as she could buy the same stock for less at the market: she will let the option expire and lose the US$ 20 she paid for it (note that the loss is relatively small). However, if $S > K$, the call holder will exercise it, getting a share of the stock for the contractual US$ 600. If she then immediately resells the share at the market price, the amount she will end up receiving on expiry will be $S - K$ (ignoring transactions costs), which can (at least, in theory) be arbitrary large.

Some people argue that the first recorded instance of option purchase was the episode described in Aristotle's[5] *Politics* (Book I, Chapter IX) concerning Thales the Milesian[6] making the point that philosophy could be useful, contrary to what the general public thought: "he, perceiving by his skill in astrology that there would be great plenty of olives that year, while it was yet winter, having got a little money, he gave earnest for all the oil works that were in Miletus and Chios, which he hired at a low price, there being no one to bid against him; but when the season came for making oil, many persons wanting them, he all at once let them upon what terms he pleased; and raising a large sum of money by that means, convinced them that it was easy for philosophers to be rich if they chose it, but that that was not what they aimed at."

As any rights, options (and other derivatives as well) come at a price. One can say that, for derivatives traded at exchanges, the price is determined by the market. But it may not be so for OTC options, of which the prices are negotiated. However, how would one know what would be the "fair price" for a derivative? This is actually one of the main questions in

[5]Aristotle (384 BC–322 BC), an ancient Greek philosopher and scientist, one of the greatest intellectual figures of Western history (who actually tutored Alexander the Great for several years). For an accessible introduction to Aristotle, see Ackrill, J.L. *Aristotle the Philosopher*, Oxford University Press, Oxford, 1981.

[6]Thales of Miletus (c. 624 BC–c. 546 BC), an ancient Greek philosopher whom Aristotle regarded as the first philosopher in the Greek tradition.

mathematical finance, and in this chapter we will try to explain the basic ideas behind pricing such financial products.

One may wish to consider purchasing an option as a game of chance (which it actually is) and, as it is usually done in gambling, compute the fair price of the option as the expectation of its payoff. That, however, will not be quite helpful for hedging purposes, as such an approach does not allow one to really control risks.

The current mainstream approach to derivatives pricing used in mathematical finance is based on a different idea, defining the fair price of an option as the price that excludes *arbitrage possibilities*, i.e., possibilities of making a risk-free profit. It turns out that, under broad assumptions on the financial market, one can use underlying securities to construct a dynamic portfolio that will exactly replicate the option's payoff. Then, at all times prior to the option's expiry, that portfolio will have to have the same price as the option itself, for otherwise there will be an arbitrage opportunity (which is not supposed to exist in an efficient market: should such an opportunity present itself, its immediate exploitation by smart market agents would immediately correct the prices to exclude it). The remarkable feature of that approach to options pricing is that it enables — at least, in theory — the option seller to hedge his risks by managing a portfolio that would replicate the option. Indeed, that makes it possible for the seller to fulfil his contractual obligations in case the option will be exercised! We will discuss the concepts of arbitrage and replication in more detail below.

In conclusion we will make the following observation. Throughout this chapter, we will be dealing with at least two different probability measures on a common measurable space $(\Omega, \mathcal{F})$ specifying what (of relevance) can happen in the world during the lifetime of a given option. One (denoted, as usual, by $\mathbf{P}$) will be the usual "real-world" (a.k.a. statistical, or actuarial) probability. One can think about its values as representing long-run relative frequencies of the respective events. The other one (well, there may be more than one such object, but there is only one in "nice cases"; we will denote it by $\mathbf{P}^*$) will be what is called the "risk-neutral" (a.k.a. arbitrage-free, or fair) probability, which is used to compute the prices of financial derivatives (as the expectations, under that probability measure, of their discounted payoffs).

That second probability $\mathbf{P}^*$ has little to do with the statistical probability and has no natural interpretation in terms of relative frequencies. The characteristic properties that $\mathbf{P}^*$ must have will be that (i) it is *equivalent* to $\mathbf{P}$ in the sense that, for any $A \in \mathcal{F}$, one has $\mathbf{P}(A) > 0$ iff $\mathbf{P}^*(A) > 0$

(that is, the same events can occur with positive probability under both $\mathbf{P}$ and $\mathbf{P}^*$), and that (ii) the discounted underlying asset price must be a martingale under $\mathbf{P}^*$.

For that reason, the probability measure $\mathbf{P}^*$ is often referred to as the *equivalent martingale measure* (EMM). It actually appears when solving the above-mentioned derivative replication problem, i.e., the problem of constructing financial portfolios that will replicate — not on average, not with high probability, but with certainty! — the payoffs of financial derivatives. It is somewhat paradoxical that the solution to the problem of uncertainty extermination is given in terms of an artificially constructed new probability measure.

13.2 Binomial Markets

These are very simple models[7] that can yet be quite helpful for introducing and understanding important notions and ideas of arbitrage free pricing.

Assume that time is discrete: $t = 0, 1, \ldots, T$ (one says that the model has T periods). There are two assets (or underlyings) in the market:

• a bond (or bank account) yielding a riskless rate r of return in each time period: the bond price at time t is

$$B_t = (1+r)^t, \quad t = 0, 1, \ldots, T$$

(so that the interest rate is assumed to be constant), and

• a risky asset (stock), with price S_t at time t that has the dynamics we have already encountered in Example 11.1: for fixed positive values $d < u$, at each time step, the previous price value is multiplied by d (a "down movement") or by u (an "up movement"),

$$S_{t-1} \begin{array}{l} \nearrow S_t = uS_{t-1}, \\ \searrow S_t = dS_{t-1}, \end{array} \qquad t = 1, 2, \ldots, T. \tag{13.1}$$

Note that we said *nothing* neither about the probabilities with which the up- and down-movements occur, nor about the character of dependence between those movements for different t values. It turns out that none of that matters from the viewpoint of arbitrage free pricing! All what will matter is that the only possible values of S_t are uS_{t-1} and dS_{t-1}.

[7]First introduced in: Cox, J.C., Ross, S.A. and Rubinstein, M. (1979) Option pricing: A simplified approach. *J. Fin. Econ.* 7, 229–263. The models are a.k.a. CRR (Cox–Ross–Rubinstein) models.

As we will see below, the continuous time Black–Scholes model can be obtained as a limit of a special sequence of scaled multi-period binomial models as $T \to \infty$. For those models, the assumption of independence of up- and down-movements in different time periods and a special choice of their probabilities will be essential, the emerging in the limit process being the already familiar to us geometric BM (11.17).

It only makes sense to consider that model when

$$d < 1 + r < u. \tag{13.2}$$

Indeed, otherwise the market would be trivial! If $1 + r \le d$, then the stock will always be at least as good as the bank account, so the only meaningful investor's action would be to borrow money and invest in stock. If $u \le 1+r$, then the stock is a risk-free loss: all the money will go to the bank account! Later, we will review the meaning of assumption (13.2) from the viewpoint of the general theory.

The natural choice of the sample space to represent all possible outcomes in our T-period binomial market is

$$\Omega := \{\omega = (v_1, \ldots, v_T) : v_t = u \text{ or } d, \; t = 1, \ldots, T\}. \tag{13.3}$$

To visualise possible trajectories of the price process, one often uses the so-called *recombinant tree* of which the vertices represent possible stock price values at the respective times, and the edges show possible movements of the prices from one time period to another (see Fig. 13.1).

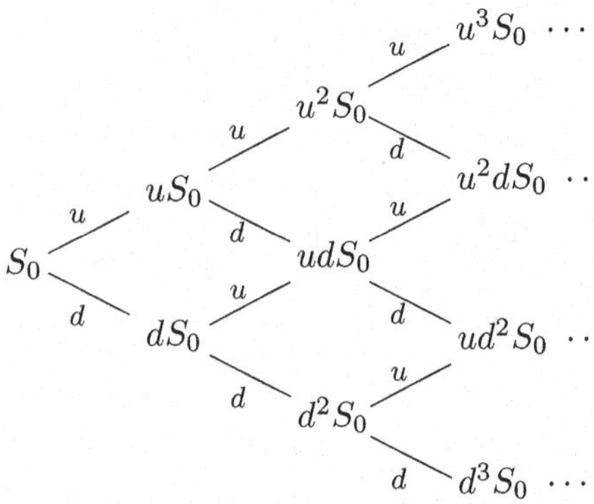

Fig. 13.1 The recombinant tree diagram representing the first three time periods in the binomial market model.

It is important to note that the RV $S_t = S_t(\omega) = v_1 \cdots v_t S_0$ actually depends on the first t components of $\omega = (v_1, \ldots, v_T)$ only. So the price process $\{S_t\}$ is adapted to the filtration $\{\mathcal{F}_t\}$ introduced in Example 11.1.

The next assumption of the model is that, at any time t, both assets can be traded: one can sell or buy as many shares or bonds (or parts thereof) as one wishes, at their current prices (and no transaction will affect the prices!), paying no fees or charges. This is somewhat unrealistic, of course, but it will do as a crude approximation allowing one to better understand the key ideas.

The so-called *short selling* of stock is also allowed. Short selling (or "going short") is the selling of a stock that the seller does not own. Basically, going short means that one borrows[8] the stock (usually, from a broker) to sell it to the third party (at the current market price). To close a short position in the future, the seller will then have to purchase the stock to return it to the lender.

Unsurprisingly, the opposite of "going short" is called "going long": a long position in a security means that the holder of the position owns the security.

Mathematically, a derivative security (or contingent claim) with maturity (expiry) date T in our market is a function

$$X = X(\omega) = g(S_T(\omega)) \geq 0 \tag{13.4}$$

of the underlying asset price S_T at time T. The financial interpretation of X is that the contract will pay its owner the amount X at time T.

As we pointed out on p. 386, this is actually what is called a "vanilla option". Exotic (or path-dependent) options depend on the history of the price process as well, so for them $X(\omega) = g(S_1(\omega), \ldots, S_T(\omega))$.

Moreover, the above definition specifies what is referred to as a *European option*, which means that it may be exercised only at the expiration date of the option. More popular are *American options* that may be exercised at any time before the expiration date (see Example 11.7; we will briefly discuss them later).

Example 13.2. *European call and put.* A European call option gives the owner the right, but not the obligation, to buy a share of the stock at expiry time T at the specified strike price K, regardless of the market stock price on that day.[9] What payoff function g will correspond to that contract in the general formula (13.4)? We already saw in Example 13.1 that the

[8] Likewise, short selling the risk free asset simply means borrowing money (at the interest rate r in our market).

[9] We will always be looking at options on one share. It is clear that, for an option of the same kind on N shares of the stock, the price will simply be N times the price of the option on one share.

expiry value of the call is $S_T - K$ if $S_T \geq K$, and zero otherwise, i.e.,

$$X = g(S_T) \quad \text{with} \quad g(s) := (s - K)^+, \tag{13.5}$$

(recall that $x^\pm := 0 \vee (\pm x)$ denotes the positive/negative part of x).

A European put option gives the owner the right to *sell* the share at the strike price K. Hence the owner will only exercise it when $K > S_T$. To do that, she will first buy a share of the stock at the market price S_T—and then will get the strike price K for that share from the option seller. Hence the expiry claim value for the put is zero if $S_T \geq K$, and is equal to $K - S_T$ if $S_T < K$. That is,

$$X = g(S_T) \quad \text{with} \quad g(s) := (s - K)^-. \tag{13.6}$$

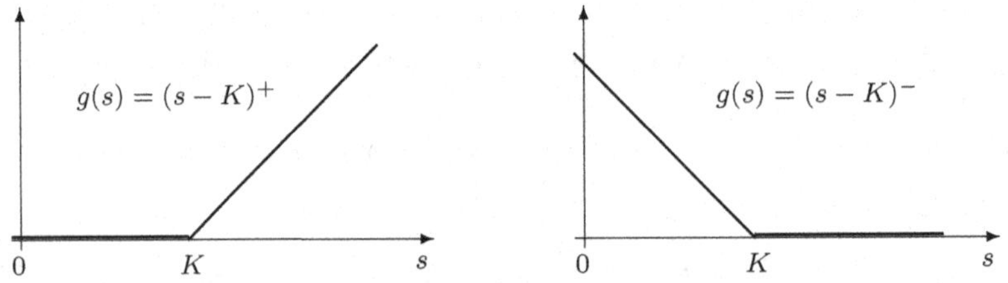

Fig. 13.2 The plots of the payoff functions $g(s)$ for Europeans call (left) and put (right) with strike K.

In what follows, simply "call" ("put") will always refer to a European call (put) option.

In the general case, the owner of a claim X receives the payment $X \geq 0$ at time $t = T$. At that time, the value $X = g(S_T)$ will already be known. But how much does the claim cost at time $t = 0$ (or, more generally, at a time $t < T$)?

The key idea is that the time $t = 0$ value (and hence the fair price) of the claim should be the *minimum* initial amount of money that allows the seller to fulfil the claim obligations with certainty.

All the instruments one has to achieve that are the stock and bond that one can trade. Could one form a portfolio of these two assets of which the time $t = T$ value will be $\geq X(\omega)$ *for all possible states of the world* $\omega \in \Omega$? Any portfolio with that property is called a *hedge* for the claim. A portfolio with time $t = T$ value $X(\omega)$ for all $\omega \in \Omega$ is called a *perfect hedge.*

Now both the claim X and its perfect hedge generate one and the same cash flow at time $t = T$. That implies that their time $t = 0$ prices must be equal as well. If it were not so, short selling the dearer one and using the money to buy the other one would create an arbitrage opportunity! To get a better understanding of how it works, we will have a close look at the simplest situation where $T = 1$.

13.3 The Single-Period Binomial Market

Now it is time for us to introduce the following concept: a *trading strategy* (in a single-period binomial market) is a pair (Δ, b), where Δ is the number of shares and b is the number of bond units bought (or held) at time $t = 0$.

Any of these two values can be negative: $\Delta < 0$ means that $|\Delta|$ shares were sold short, $b < 0$ means that the amount $|b|B_0 = |b|$ was borrowed from the bank (or, equivalently, that many bond units were short sold).

The time t value of the portfolio governed by that strategy is:

$$V_t = V_t(\omega) = \begin{cases} \Delta S_0 + bB_0 = \Delta S_0 + b & \text{at time } t = 0, \\ \Delta S_1 + bB_1 = \Delta S_1 + b(1+r) & \text{at time } t = 1. \end{cases}$$

This portfolio will be a hedge given that $V_1(\omega) \geq X(\omega)$, $\omega \in \Omega$. There are only two ω's in our Ω in the case $T = 1$: $\omega = u$ and $\omega = d$. Denoting the respective values of $X(\omega)$ by X_u and X_d, the hedge condition is equivalent to the following system of linear inequities:

$$\begin{cases} \Delta u S_0 + b(1+r) \geq X_u, \\ \Delta d S_0 + b(1+r) \geq X_d, \end{cases}$$

or, equivalently,

$$b \geq -\frac{uS_0}{1+r}\Delta + \frac{X_u}{1+r}, \tag{13.7}$$

$$b \geq -\frac{dS_0}{1+r}\Delta + \frac{X_d}{1+r}. \tag{13.8}$$

Solution to this system of inequalities is represented in Fig. 13.3: the points (Δ, b) satisfying the inequalities all lie above the respective thick lines (labelled by the respective inequality numbers, (13.7) and (13.8)), so that the points satisfying both relations lie in the sector above both thick lines. Therefore, all such points represent hedges for our claim X. Now which one of them is the cheapest, i.e., has the smallest time $t = 0$ value? That is, we want to find the point (Δ, b) at which the minimum

$$\min_{\text{all hedges } (\Delta, b)} (\Delta S_0 + b)$$

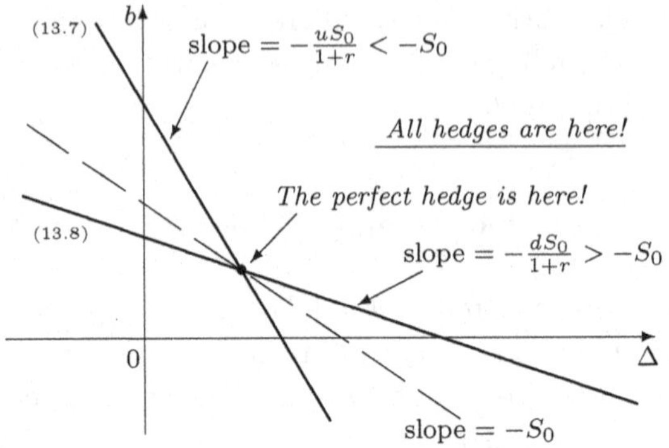

Fig. 13.3 The region in the (Δ, b)-plane *above* both thick lines (which are the boundaries of the half-planes given by the respective inequalities, (13.7) and (13.8)) contains all hedges. The vertex represents the perfect hedge. All the portfolios represented by the points on the dashed line have the same time $t = 0$ value: $\Delta S_0 + b = \text{const}$.

is attained. It is easy to see that that point is just the vertex of the sector of hedges, i.e., the intersection of the thick lines, which is given by the solution to the linear system

$$\begin{cases} \Delta u S_0 + b(1+r) = X_u, \\ \Delta d S_0 + b(1+r) = X_d. \end{cases}$$

Indeed, all the points (Δ, b) lying on the dashed line passing through the vertex clearly have one and the same initial value, as the equation of that line is $\Delta S_0 + b = \text{const}$. All the imperfect hedges are represented by point that lie *strictly above* the dashed line, and hence must have greater values of $\Delta S_0 + b$.

Solving the linear system for the perfect hedge yields

$$\Delta = \frac{X_u - X_d}{(u-d)S_0} = \frac{g(uS_0) - g(dS_0)}{uS_0 - dS_0}, \tag{13.9}$$

which is a discrete-time version of the"delta-hedging rule" for derivative securities, where the number of shares in the hedging portfolio equals the derivative of the option value w.r.t. the price of the underlying (cf. (13.38); note that (13.9) is a discrete analog of the derivative of g), and

$$b = \frac{X_u - \Delta u S_0}{1+r} = \frac{uX_d - dX_u}{(1+r)(u-d)}. \tag{13.10}$$

Portfolio (13.9), (13.10) is a perfect hedge for X, and its time $t = 0$ value is

$$V_0 = \Delta S_0 + b = \frac{X_u - X_d}{u-d} + \frac{uX_d - dX_u}{(1+r)(u-d)}$$
$$= \frac{1}{1+r}\Bigg[\underbrace{\frac{1+r-d}{u-d}}_{=:p^*}X_u + \underbrace{\frac{u-(1+r)}{u-d}}_{=1-p^*}X_d\Bigg]$$
$$= \mathbf{E}^*\frac{X}{1+r} =: X^*, \tag{13.11}$$

where $\mathbf{E}^*$ denotes expectation under the probability $\mathbf{P}^*$ on $\Omega = \{u,d\}$ given by $\mathbf{P}^*(\{u\}) = 1 - \mathbf{P}^*(\{d\}) = p^*$, where $p^* = \frac{1+r-d}{u-d} \in (0,1)$ owing to condition (13.2).

What happened was that the initial value of the replicating portfolio proved to have the form of the expectation, under some artificially introduced probability measure $\mathbf{P}^*$, of the claim value that we discounted by dividing it by $1+r$, thus expressing it in time $t=0$ dollars.

We maintain that the thus calculated X^* is the *fair price* of the claim X at time $t=0$. Indeed, if the claim was sold at a different price $X_0 \neq X^*$ at that time, that would have created an arbitrage opportunity: making a net investment of \$0 at time $t=0$, we would have at time $t=1$ a non-negative amount which is actually positive for some states of the world $\omega \in \Omega$. This is best explained with the help of the following table showing actions to be taken at times $t=0$ and $t=1$ to make use of the arbitrage opportunity:

If $X_0 > X^*$:	
$t=0$	$t=1$
Sell the claim: X_0 Form the replicating portfolio: $-X^*$ Buy bonds: $-(X_0 - X^*)$	Cover the claim: $-X$ Replicating portfolio: X Sell bonds: $(1+r)(X_0 - X^*)$
Balance: 0	Balance: $(1+r)(X_0 - X^*) > 0$

If $X_0 < X^*$:	
$t=0$	$t=1$
Sell short Δ shares: ΔS_0 Sell short b bonds: b Buy the claim: $-X_0$ Buy bonds: $-(X^* - X_0)$	Deliver stock and bonds: $-X$ Payoff of the claim: X Sell bonds: $(1+r)(X^* - X_0)$
Balance: 0	Balance: $(1+r)(X^* - X_0) > 0$

So whether $X_0 > X^*$ or $X_0 < X^*$, one can still have riskless profit! It is only when the time $t = 0$ claim price is X^* that there is no such arbitrage opportunities.

This is actually the main principle of derivative pricing: the prices should *exclude* any arbitrage opportunities, as the latter are not supposed to be present in efficient markets. Replicating portfolios themselves are means to make all that work, and their construction and management is perhaps even more important than the pricing itself.

Now return to the claim price (13.11), which has the form of the expectation under some probability measure $\mathbf{P}^*$. There is something very special about that measure, namely, that

$$\mathbf{E}^* \frac{S_1}{1+r} = \frac{1}{1+r}\left[\frac{1+r-d}{u-d}uS_0 + \frac{u-(1+r)}{u-d}dS_0\right] = S_0. \tag{13.12}$$

That is, the discounted price process $\{(1+r)^{-t}S_t\}_{t=0,1}$ is a *martingale* under that measure $\mathbf{P}^*$ (note that $\mathcal{F}_0$ is trivial). It is obvious that there is a *unique* probability measure on Ω for which that martingale property of $\{(1+r)^{-t}S_t\}$ holds.

This is not just a nice coincidence, but the main result of the arbitrage-free pricing theory.

Example 13.3. *Pricing a European call.* From the call payoff representation (13.5) and the general pricing formula (13.11) we obtain that the price C of the call with strike K is given by

$$C = \frac{1}{1+r}\left[p^*(uS_0 - K)^+ + (1-p^*)(dS_0 - K)^+\right], \quad p^* = \frac{1+r-d}{u-d} \in (0,1).$$

It is easy to see that, as the function of strike K, the call price is a decreasing function (as to be expected: the higher the strike, the smaller the payoff). More precisely, it is piece-wise linear, starting at S_0 for $K = 0$ (buying a call with zero strike is the same as buying the share itself, right?), changing its slope at the point $K = dS_0$ (do you see why?) and then turning into zero for $K = uS_0$ (indeed, as the share price cannot be greater than that value, the right to buy it at that price is worthless).

Let us price and replicate the call assuming that $r = 0.25$, $u = 1.75$, $d = 0.5$ (so that (13.2) holds), $S_0 = 1$ and $K = 1$. As $p^* = \frac{1.25-0.5}{1.75-0.5} = 0.6$, one has

$$C = \frac{1}{1.25}[0.6 \times (1.75-1)^+ + 0.4 \times (0.5-1)^+] = \frac{0.75 \times 0.6}{1.25} = 0.36.$$

The replicating portfolio has the form:

$$\Delta = \frac{(uS_0 - K)^+ - (dS_0 - K)^+}{uS_0 - dS_0} = \frac{0.75 - 0}{1.75 - 0.5} = 0.6,$$

$$b = \frac{u(dS_0 - K)^+ - d(uS_0 - K)^+}{(1+r)(u-d)} = \frac{-0.5 \times 0.75}{1.25 \times 1.25} = -0.24.$$

Now check if the time $t = 0$ value of the replicating portfolio coincides with the call price we computed above, and that its time $t = 1$ value replicates the call payoff. One has

$$V_0 = \Delta S_0 + b = 0.6 \times 1 + (-0.24) \times 1 = 0.36 = C$$

indeed, and, if the stock goes up,

$$V_1(u) = \Delta u S_0 + b(1+r) = 0.6 \times 1.75 + (-0.24) \times 1.25 = 0.75,$$

while if it goes down,

$$V_1(d) = \Delta d S_0 + b(1+r) = 0.6 \times 0.5 + (-0.24) \times 1.25 = 0.$$

In both cases, the values coincide with $(S_1 - K)^+$, a perfect replication!

It may also be helpful to use the following simple graphical representation (which will be employed in the multi-period case as well). The left picture in Figure 13.4 depicts the stock prices at times $t = 0$ (the oval on the left, with the initial price) and $t = 1$ (two possible states of the world, represented by the two ovals on the right, with the values that we computed from the initial price and the known values of u and d). The right picture shows the respective call prices (in boxes), where we actually first knew the payoff prices on the right and then computed the time $t = 0$ price shown on the left.

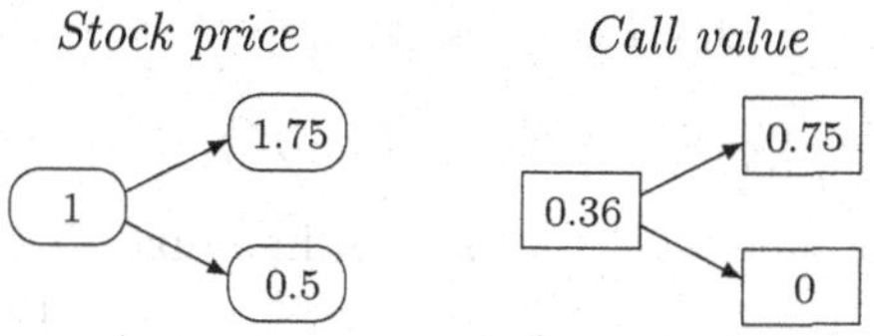

Fig. 13.4 Representation of possible movements of the stock and call prices in the single-period binomial market from Example 13.3.

The term "arbitrage" actually refers to the market of underlyings. It means that there exists a trading strategy with zero net investments, never making a loss and sometimes making profit. The absence (or presence) of such opportunities is one of the key properties for any market model.

When are there no arbitrage opportunities in our simple market? Zero net investment means that $0 = V_0 = \Delta S_0 + b$, so that $b = -\Delta S_0$ and

$$V_1 = \Delta S_1 + b(1+r) = \Delta S_1 - \Delta S_0(1+r) = \Delta(1+r)\left[\frac{S_1}{1+r} - S_0\right].$$

If condition (13.2) holds then $[\cdots] > 0$ when $S_1 = uS_0$, and $[\cdots] < 0$ when $S_1 = dS_0$, which means there is no arbitrage. However, when (13.2) fails one does have an arbitrage opportunity: if $u \leq 1 + r$, one should sell the stock short and buy bonds ($\Delta < 0$), and if $1 + r \leq d$, one should borrow money and buy the stock ($\Delta > 0$). Then $[\cdots] \geq 0$ whatever happens in the world, and $[\cdots] > 0$ for some outcomes, which means arbitrage.

Thus (13.2) is actually a necessary and sufficient condition of no-arbitrage (NA) for our simple market. Note that in the NA case one has

$$S_0 \in \left(\frac{S_1(d)}{1+r}, \frac{S_1(u)}{1+r} \right). \tag{13.13}$$

This means that the point S_0 can be represented as a convex combination of the interval's end points: for a $p^* \in (0, 1)$,

$$S_0 = p^* \frac{S_1(u)}{1+r} + (1 - p^*) \frac{S_1(d)}{1+r} = \mathbf{E}^* \frac{S_1}{1+r}, \tag{13.14}$$

which is the geometric meaning of the martingale property that we established earlier, in (13.12).

Example 13.4. *Pricing a European put.* There are two ways we can do that.

1) Similarly to what we did in Example 13.3, from the put payoff representation (13.6) and the general pricing formula (13.11) we obtain that the price P of the put with strike K is given by

$$P = \frac{1}{1+r} \left[p^* (uS_0 - K)^- + (1 - p^*)(dS_0 - K)^- \right].$$

In particular, for the numerical values from Example 13.3, we obtain

$$P = \frac{1}{1.25} [0.6 \times (1.75 - 1)^- + 0.4 \times (0.5 - 1)^-] = \frac{0.4 \times 0.5}{1.25} = 0.16.$$

2) Alternatively, one can use the so-called "put-call parity", which is a fundamental relation that can be derived using an arbitrage based argument.

Suppose that C and P are European call and put prices, respectively, on one share of a common stock, with a common strike K, in a general arbitrage-free market. Form a portfolio that consists of:

- one share of the stock,
- one put on one share of the stock,
- a short position in one call on one share of the stock.

The time $t = 0, 1$ values of the portfolio are then, respectively,

$$\begin{aligned} V_0 &= S_0 + P - C, \\ V_1 &= S_1 + (S_1 - K)^- - (S_1 - K)^+ \\ &= S_1 - ((S_1 - K)^+ - (S_1 - K)^-) = S_1 - (S_1 - K) = K, \end{aligned}$$

as one always has $x^+ - x^- = x$. Thus, at time $t = 1$ our portfolio generates one and the same cash flow K, regardless of the state of the world. That means that its time $t = 0$ price *must* be $V_0 = K/(1+r)$, as any other price would create an arbitrage opportunity. Indeed, if, say, $V_0 < K/(1+r)$, then at time $t = 0$ borrow V_0 dollars from the bank and buy the portfolio. Selling it at time $t = 1$ and returning the borrowed money to the bank (with interest) will leave you with $K - (1+r)V_0 > 0$ dollars *etc.*

Thus we established that

$$S_0 + P - C = \frac{K}{1+r}. \tag{13.15}$$

Hence one you know the price of one of the options, you can immediately price the other one as well. For our numerical values, we will have

$$P = \frac{K}{1+r} - S_0 + C = \frac{1}{1.25} - 1 + 0.36 = 0.16,$$

in perfect agreement with the directly computed put price.

Of course, in real life markets there are things like transaction costs, stocks pay dividends, most options are American *etc.*, so the parity does not hold exactly. However, its agreement with reality is usually reasonable. Here is an example[10]: on 23.06.1997, the *Deutsche Bank AG* stock price (traded at FWB[11]) was DM[12]97.70, while the call and put prices (for strike $K = 80$ with expiry on 18.06.1998) were DM 23.30 and DM 4.16, respectively. The relevant simple interest rate for that time period till expiry was 3.15%. One has $S_0 + P - C = 78.66$, while $K/(1+r) = 77.56$, which is not too large a difference, taking into account the above-mentioned deviations of reality from the model assumptions.

Note that the argument used to obtain the put-call parity is equally applicable to continuous time models as well. Thus, for European call and put on one share of a common stock, with common strike K and expiry T, assuming a constant compounding interest rate r, their time t prices C_t and P_t, respectively, will satisfy (in theory) a parity of the form

$$S_t + P_t - C_t = Ke^{-r(T-t)}, \quad t \in [0, T]. \tag{13.16}$$

13.4 Finite Single-Period Markets

Rather than proceeding to multi-period markets now, we will first have a look at NA conditions in single-period markets with more than two possible

[10]From p. 15 of the first edition of Bingham and Kiesel (2004).

[11]*Frankfurter Wertpapierbörse,* the Frankfurt Stock Exchange.

[12]DM stands for the *Deutsche Mark,* which was the official currency of the Federal Republic of Germany from 1948 to 2002, when it was replaced with the Euro.

states of the world, as this will give us additional insight into the meaning of the general NA condition expressed in terms of equivalent martingale measures.

Recall that, in the single-period binomial market, the NA condition had the form (13.13), which was equivalent to existence of a probability measure $\mathbf{P}^*$ on $\Omega = \{u, d\}$ with property (13.14). What if there were more than two possible states of the world and/or more than two assets in the market?

Here are our assumptions about the market:

• we are looking at a single period only: $t = 0$ and $t = 1$;

• there are finitely many possible states of the world at time $t = 1$: our outcome space has $N < \infty$ elements, $\Omega = \{\omega_1, \dots, \omega_N\}$;

• there are $n + 1$ traded assets in the market: n stocks, with time t prices $\boldsymbol{S}_t := (S_t^1, \dots, S_t^n)$, and a bond (or bank account) $B_t = (1 + r)^t$, $t = 0, 1$.

Next we recall a few basic concepts and facts from convex geometry[13]. A set $D \subset \mathbf{R}^m$ is called *convex* if, for any $\boldsymbol{x}_1, \boldsymbol{x}_2 \in D$, all the points on the straight line segment connecting the $\boldsymbol{x}_i$'s also lie in D:

$$\beta \boldsymbol{x}_1 + (1 - \beta)\boldsymbol{x}_2 \in D, \quad \beta \in (0, 1).$$

Further, the *convex hull* of a set $B \subset \mathbf{R}^m$, denoted by $\operatorname{conv}(B)$, is the smallest convex set D such that $B \subset D$ (it is not hard to see that such a set always exists, noticing that the intersection of convex sets containing B is also a convex set containing B). For a finite set $B := \{\boldsymbol{y}_1, \dots, \boldsymbol{y}_N\} \subset \mathbf{R}^m$, one has

$$D := \operatorname{conv}(B) = \left\{ \boldsymbol{x} \in \mathbf{R}^m : \boldsymbol{x} = \sum_{j=1}^{N} \alpha^j \boldsymbol{y}_j, \quad \alpha^j \ge 0, \ \sum_{j=1}^{N} \alpha^j = 1 \right\}.$$

Indeed, all $\boldsymbol{y}_j \in D$, $j \le N$, and D is convex: if $\boldsymbol{x}_i = \sum_{j=1}^N \alpha_i^j \boldsymbol{y}_j$ for some $\alpha_i^j \ge 0$, $\sum_{j=1}^N \alpha_i^j = 1$, $i = 1, 2$, then, for any $\beta \in (0, 1)$,

$$\beta \boldsymbol{x}_1 + (1 - \beta)\boldsymbol{x}_2 = \sum_{j=1}^{N} \underbrace{(\beta \alpha_1^j + (1 - \beta)\alpha_2^j)}_{=:\alpha^j} \boldsymbol{y}_j = \sum_{j=1}^{N} \alpha^j \boldsymbol{y}_j \in D,$$

as $\sum_{j=1}^N \alpha^j = \beta \sum_{j=1}^N \alpha_1^j + (1 - \beta) \sum_{j=1}^N \alpha_2^j = \beta + (1 - \beta) = 1$, $\alpha^j \ge 0$. It is not hard to see that D is the smallest convex set containing all the $\boldsymbol{y}_j$'s.

---

[13]Which can be found, e.g., in: Lay, S.R., *Convex Sets and Their Applications*, Dover, New York, 2007 (a republication of the original book published by Wiley, New York, 1982).

Denote by D_0 the *relative interior*[14] of the convex hull D:

$$D_0 = \left\{ \boldsymbol{x} \in \mathbf{R}^m : \boldsymbol{x} = \sum_{j=1}^{N} \alpha^j \boldsymbol{y}_j, \ \ \alpha^j > 0, \ \sum_{j=1}^{N} \alpha^j = 1 \right\}$$

(i.e., D_0 consists of all *strict convex combinations* of the $\boldsymbol{y}_j$'s).

By a *hyperplane* in $\mathbf{R}^m$ one means a linear manifold of dimensionality $m - 1$. Analytically, a hyperplane can be specified by one linear equation:

$$L := \{\boldsymbol{x} = (x^1, \cdots, x^m) \in \mathbf{R}^m : \boldsymbol{a} \cdot \boldsymbol{x} = c\}, \tag{13.17}$$

where $\boldsymbol{a} \cdot \boldsymbol{x} = \sum_{j=1}^{m} x^j a^j$ is the inner product of the two vectors, $c \in \mathbf{R}$ and $\boldsymbol{a} = (a^1, \ldots, a^m) \neq \mathbf{0}$ are fixed. It is easy to see that the vector $\boldsymbol{a}$ is orthogonal to L: for any $\boldsymbol{x}_1, \boldsymbol{x}_2 \in L$, one has $\boldsymbol{a} \cdot (\boldsymbol{x}_1 - \boldsymbol{x}_2) = 0$, and that, provided that $\boldsymbol{x}_0 \in L$, the equation for L can be re-written as

$$\boldsymbol{a} \cdot (\boldsymbol{x} - \boldsymbol{x}_0) = 0.$$

Any hyperplane divides the space into two half-spaces. The sign of c in (13.17) tells one on which side of L the origin lies: if $c < 0$ then $\mathbf{0}$ is on the side of L at which the vector $\boldsymbol{a}$ points, and $c > 0$ when $\mathbf{0}$ is on the opposite side, as illustrated in the case $m = 2$ in the picture below:

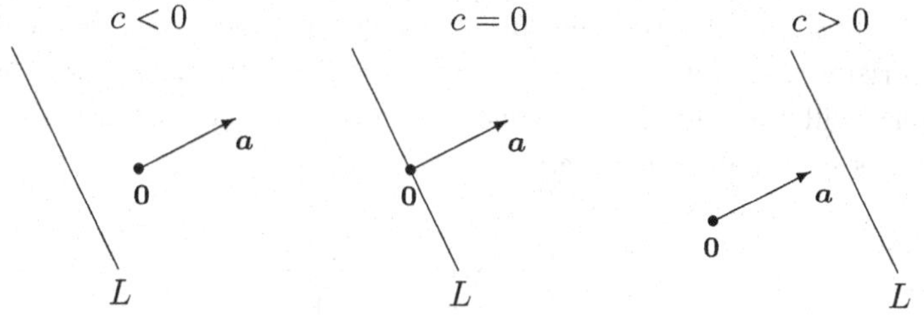

Analytically, the *open* half-space on the side of L at which $\boldsymbol{a}$ is pointing can be given as

$$H_L(\boldsymbol{a}) := \{\boldsymbol{x} \in \mathbf{R}^m : \boldsymbol{a} \cdot (\boldsymbol{x} - \boldsymbol{x}_0) > 0\},$$

where $\boldsymbol{x}_0 \in L$ is arbitrary. The following *separation theorem* is one of the most useful basic results of convex analysis:

Theorem 13.1. *Assume that $D \in \mathbf{R}^m$ is a convex set, D_0 its relative interior. Then, for any point $\boldsymbol{x}_0 \notin D_0$, there exists a hyperplane L with a normal vector $\boldsymbol{a}$ such that $\boldsymbol{x}_0 \in L$ and $D_0 \subset H_L(\boldsymbol{a})$. In other words, there exists a vector $\boldsymbol{a} \in \mathbf{R}^m$ such that $\boldsymbol{a} \cdot (\boldsymbol{x} - \boldsymbol{x}_0) > 0$ for all $\boldsymbol{x} \in D_0$.*

[14] The adjective "relative" indicates that the interior is taken in the linear manifold spanned by the set D. In the general case, the dimensionality of that manifold can be less than m, and then the "true interior" of D (in $\mathbf{R}^m$) will always be empty!

Now we are ready to state the main result of this section.

Theorem 13.2. *A finite single-period financial market is arbitrage-free iff there exists a probability* $\mathbf{P}^*$ *on* $(\Omega, 2^\Omega)$ *such that* $\mathbf{P}^*(\{\omega\}) > 0$ *for every* $\omega \in \Omega$ *and*

$$\mathbf{E}^* \frac{\boldsymbol{S}_1}{1+r} = \boldsymbol{S}_0, \tag{13.18}$$

i.e., the process $\{\boldsymbol{S}_t(1+r)^{-t}\}_{t=0,1}$ *is a (vector-valued) MG under* $\mathbf{P}^*$.

Thus, the NA condition (13.12) that we established for single-period binomial markets extends to the general case as well. Note that the necessary and sufficient condition of NA from the theorem is nothing else but the existence of the EMM (we will denote that condition by $\exists$EMM for brevity) that we mentioned on p. 389. We did not actually specify any "statistical" probability measure for our market, as that was not needed; all what we said was that all $\omega \in \Omega$ were *possible*, meaning that they all have positive probabilities. As $\mathbf{P}^*(\{\omega\}) > 0$ for any $\omega \in \Omega$, one can see that $\mathbf{P}^*$ will be equivalent to *any* probability $\mathbf{P}$ on $(\Omega, 2^\Omega)$ with the property $\mathbf{P}(\{\omega\}) > 0$ for all $\omega \in \Omega$.

Proof For simplicity, we will only consider the case $n = 2$ (so that there are two risky assets, with prices $\boldsymbol{S}_t = (S_t^1, S_t^2)$, $t = 0, 1$). Exactly the same argument will work for $n \geq 3$ (one just has to replace lines with planes, half-planes with half-space *etc.*).

Set

$$D := \operatorname{conv}\left\{\frac{\boldsymbol{S}_1(\omega_1)}{1+r}, \ldots, \frac{\boldsymbol{S}_1(\omega_N)}{1+r}\right\}.$$

There are two possibilities: either (I) $\boldsymbol{S}_0 \in D_0$ (which is an extension of (13.13) to the case of more than two states of the world) or (II) $\boldsymbol{S}_0 \notin D_0$.

We will prove that NA $\iff$ $\exists$EMM in three steps: Step 1 shows that $\exists$EMM $\iff$ (I), Step 2 proves that $\exists$EMM$\Longrightarrow$NA, and Step 3 demonstrates that NA $\Longrightarrow$ (I).

Step 1. From the definition of D_0, case (I) takes place iff the vector $\boldsymbol{S}_0$ is a strict convex combination of the points $\boldsymbol{S}_1(\omega_k)/(1+r)$, $k = 1, \ldots, N$: for some $p_k^* > 0$, $\sum_{k=1}^N p_k^* = 1$, one has

$$\boldsymbol{S}_0 = \sum_{k=1}^N p_k^* \frac{\boldsymbol{S}_1(\omega_k)}{1+r} =: \mathbf{E}^* \frac{\boldsymbol{S}_1}{1+r},$$

thus establishing equivalence of (I) and $\exists$EMM.

Step 2. Proof by contradiction: suppose that ∃EMM holds, but NA does not. That means there is an arbitrage opportunity, i.e., a trading strategy $(\boldsymbol{\Delta}, b)$ such that its value process $V_t := \boldsymbol{\Delta} \cdot \boldsymbol{S}_t + B_t$ has the property that $V_0 \equiv \boldsymbol{\Delta} \cdot \boldsymbol{S}_0 + b = 0$ (so that $b = -\boldsymbol{\Delta} \cdot \boldsymbol{S}_0$), but

$$\begin{aligned} V_1(\omega) &\equiv \boldsymbol{\Delta} \cdot \boldsymbol{S}_1(\omega) + b(1+r) \\ &= \boldsymbol{\Delta} \cdot \boldsymbol{S}_1(\omega) - \boldsymbol{\Delta} \cdot \boldsymbol{S}_0(1+r) \begin{cases} \geq 0 & \text{for all } \omega, \\ > 0 & \text{for at least one } \omega. \end{cases} \end{aligned} \tag{13.19}$$

However, that means that

$$\begin{aligned} \boldsymbol{\Delta} \cdot \boldsymbol{S}_0 &\overset{\text{Step 1}}{=} \boldsymbol{\Delta} \cdot \mathbf{E}^* \frac{\boldsymbol{S}_1}{1+r} = \mathbf{E}^* \frac{\boldsymbol{\Delta} \cdot \boldsymbol{S}_1}{1+r} = \sum_{k=1}^{N} p_k^* \frac{\boldsymbol{\Delta} \cdot \boldsymbol{S}_1(\omega_k)}{1+r} \\ &\overset{(13.19)}{>} \sum_{k=1}^{N} p_k^* \boldsymbol{\Delta} \cdot \boldsymbol{S}_0 = \boldsymbol{\Delta} \cdot \boldsymbol{S}_0, \end{aligned}$$

a contradiction! So our assumption was wrong, implying ∃EMM⟹NA.

Step 3. This will also be a proof by contradiction. Suppose that NA holds, but (I) does not. The latter means that (II) holds, and then by the separation Theorem 13.1 there exists a straight line L separating $\boldsymbol{S}_0$ from D_0. Denoting by $\boldsymbol{\Delta} := (\Delta^1, \Delta^2)$ a normal to L vector "pointing" in the direction of D_0, we have

$$\boldsymbol{\Delta} \cdot (\boldsymbol{x} - \boldsymbol{S}_0) > 0, \quad \boldsymbol{x} \in D_0.$$

Recalling how D_0 came about, we see that the above means that

$$\boldsymbol{\Delta} \cdot \left(\frac{\boldsymbol{S}_1(\omega)}{1+r} - \boldsymbol{S}_0 \right) \begin{cases} \geq 0 & \text{for all } \omega, \\ > 0 & \text{for at least one } \omega. \end{cases}$$

But this is an arbitrage opportunity, cf. (13.19)! This contradicts to the NA assumption we made, thus showing that NA ⟹ (I). The theorem is proved.

□

The EMMs (the measure $\mathbf{P}^*$ satisfying the condition in Theorem 13.2 does not need to be unique!) provide one with simple expressions for the prices of *attainable* contingent claims, i.e., such claims X that can be replicated (meaning that there exists a trading strategy $(\boldsymbol{\Delta}, b)$ of which the value process V_t has the property that $X(\omega) = V_1(\omega) := \boldsymbol{\Delta} \cdot \boldsymbol{S}_1(\omega) + b(1+r)$ for any state $\omega \in \Omega$).

Theorem 13.3. (The arbitrage pricing theorem.) *In a single-period arbitrage-free market, the discounted value process* $\{V_t(1+r)^{-t}\}_{t=0,1}$ *of any trading strategy* $(\mathbf{\Delta}, b)$ *is an MG under any EMM* $\mathbf{P}^*$:

$$\mathbf{E}^* \frac{V_1}{1+r} = V_0. \tag{13.20}$$

In particular, if X *is an attanable claim in the market, then its time* $t = 0$ *value is given by*

$$X^* := \mathbf{E}^* \frac{X}{1+r}, \tag{13.21}$$

where the right-hand side does not depend on the choice of the EMM $\mathbf{P}^*$.

Remark 13.1. The assertion of the theorem remains true for multi-period markets as well, provided that the trading strategy is *self-financing*, in the sense to be discussed on p. 406 in the context of the binomial model.

Proof of Theorem 13.3. For any portfolio $(\mathbf{\Delta}, b)$,

$$\mathbf{E}^* \frac{V_1}{1+r} = \mathbf{E}^* \frac{\mathbf{\Delta} \cdot \boldsymbol{S}_1}{1+r} + b = \mathbf{\Delta} \cdot \frac{\mathbf{E}^* \boldsymbol{S}_1}{1+r} + b \overset{(13.18)}{=} \mathbf{\Delta} \cdot \boldsymbol{S}_0 + b = V_0.$$

Now suppose that X is an attainable claim, i.e., for some portfolio $(\mathbf{\Delta}, b)$, one has $V_1(\omega) = X(\omega)$, $\omega \in \Omega$. Together with (13.20), that implies that

$$V_0 = \mathbf{E}^* \frac{V_1}{1+r} = \mathbf{E}^* \frac{X}{1+r} = X^*, \tag{13.22}$$

which proves (13.21) since, as our arbitrage argument on p. 395 shows, the fair time $t = 0$ price of the claim is given by the initial value V_0 of the replicating portfolio. Note that the above means that *any* portfolio replicating X will have the same initial value $V_0 = X^*$.

That the value of X^* does not depend on the choice of the EMM $\mathbf{P}^*$ follows from (13.22), as the left-hand side in that relation is one and the same for any EMM. □

Thus, Theorem 13.3 provides one with the arbitrage-free price for *any attainable* claim. It turns out that, in the general case, not all claims are attainable in a financial market. A market with the property that *any* claim is attainable in it is said to be *complete*. As we saw in Section 13.3, the single-period NA binomial market is complete. The next theorem gives a necessary and sufficient condition for a market to be complete, which is also stated in terms of the EMMs.

Theorem 13.4. (The completeness theorem.) *An arbitrage-free market is complete iff there exists a unique EMM.*

Like with the previous theorem, this result holds for any market, but we will only prove it in the context of the present section, i.e., for finite single-period markets.

Proof First observe that condition (13.18) can be re-written component-wise as the following system of $n+1$ linear equations for N unknowns $p_1^*, \ldots, p_N^*$:

$$\begin{aligned}
&p_1^* S_1^1(\omega_1) + \cdots + p_N^* S_1^1(\omega_N) = (1+r)S_0^1, \\
&\ldots \quad \ldots \quad \ldots \quad \ldots \quad \ldots \quad \ldots \\
&p_1^* S_1^n(\omega_1) + \cdots + p_N^* S_1^n(\omega_N) = (1+r)S_0^n, \\
&p_1^* + \cdots + p_N^* = 1.
\end{aligned}$$

So if the EMM is unique, the above system has a unique solution, which means that $n+1 = N$ and the system matrix

$$A := \begin{pmatrix} S_1^1(\omega_1) & S_1^1(\omega_2) & \cdots & S_1^1(\omega_N) \\ S_1^2(\omega_1) & S_1^2(\omega_2) & \cdots & S_1^2(\omega_N) \\ \cdots & \cdots & \cdots & \cdots \\ S_1^n(\omega_1) & S_1^n(\omega_2) & \cdots & S_1^n(\omega_N) \\ 1 & 1 & \cdots & 1 \end{pmatrix}$$

is non-singular: $\det A \neq 0$, and hence there exists the inverse A^{-1}.

Second, we observe also that a claim $X = X(\omega)$ can be replicated when there exists a portfolio $(\boldsymbol{\Delta}, b)$, $\boldsymbol{\Delta} = (\Delta^1, \ldots, \Delta^n)$, such that

$$\begin{aligned}
&\Delta^1 S_1^1(\omega_1) + \cdots + \Delta^n S_1^n(\omega_1) + b(1+r) = X(\omega_1), \\
&\ldots \quad \ldots \quad \ldots \quad \ldots \quad \ldots \quad \ldots \\
&\Delta^1 S_1^1(\omega_N) + \cdots + \Delta^n S_1^n(\omega_N) + b(1+r) = X(\omega_N),
\end{aligned} \tag{13.23}$$

which is a system of N linear equations for $n+1$ unknowns $\Delta^1, \ldots, \Delta^n, b(1+r)$. Note that the system's matrix is actually the transpose A^T.

Having made these observations, we proceed to proving the asserted equivalence.

$\Leftarrow$) Suppose that there exists a unique EMM. From our first observation, $n+1 = N$ and A is non-singular, and hence so is A^T (recall that $\det A^T = \det A$). Therefore, the linear system (13.23) has a unique solution for *any* right-hand side, which means that, for any claim X, there is a unique replicating strategy in the market, which proves that the market is complete.

$\Rightarrow$) As we know from Theorem 13.2, for an NA market there always exists an EMM. So we just have to show that if the market is complete then the EMM is unique. We will prove that by contradiction.

Suppose there are two EMMs, $\mathbf{P}^*$ and $\widetilde{\mathbf{P}}^*$ (with all $p_k^*, \widetilde{p}_k^* > 0$). Due to completeness, by Theorem 13.3 we have that, for *any* claim X,

$$\mathbf{E}^* \frac{X}{1+r} = \widetilde{\mathbf{E}}^* \frac{X}{1+r},$$

or, equivalently, $\sum_{k=1}^N p_k^* X(\omega_k) = \sum_{k=1}^N \widetilde{p}_k^* X(\omega_k)$. In particular, for the so-called "arrow-claims"

$$X_i(\omega_k) := \delta_{ik} = \begin{cases} 1 & \text{if } k = i, \\ 0 & \text{otherwise}, \end{cases} \qquad i = 1, \ldots, N,$$

one has

$$p_i^* = \sum_{k=1}^N p_k^* \delta_{ik} = \sum_{k=1}^N p_k^* X_i(\omega_k) = \sum_{k=1}^N \widetilde{p}_k^* X_i(\omega_k) = \sum_{k=1}^N \widetilde{p}_k^* \delta_{ik} = \widetilde{p}_i^*,$$

so that $\mathbf{P}^* = \widetilde{\mathbf{P}}^*$, meaning that the EMM is unique. The theorem is proved. □

Remark 13.2. We saw in the proof of Theorem 13.4 that, for a finite market to be complete, the total number of securities $n+1$ in it should match the number N of different states of the world allowed by the model. The morale is that the market should be *rich enough* for us to be able to replicate an arbitrary claim.

13.5 The Multi-Period Binomial Market

Now we return to the simplest case of only one risky asset with price S_t with dynamics (13.1) (we still do not need to assume anything about the probabilities and dependence of the price chances at different time steps) and bond with price $B_t = (1+r)^t$, but consider the case of the general time horizon $T \ge 1$.

By a *trading strategy* we will now mean a sequence of pairs (Δ_t, b_t), $t = 1, \dots, T$, with the property that

$$\Delta_t = \Delta_t(S_0, \dots, S_{t-1}), \quad b_t = b_t(S_0, \dots, S_{t-1}), \quad t = 1, \dots, T.$$

That is, as a process, the trading strategy $\{(\Delta_t, b_t)\}$ is predictable w.r.t. the natural filtration for $\{S_t\}$.

The interpretation of the process is that the portfolio following this strategy will have Δ_t shares and b_t bond units held during the time period $(t-1, t]$. As in the case of the single-period market, the quantities Δ_t and b_t can be negative, meaning short positions in the respective assets.

A trading strategy $\{(\Delta_t, b_t)\}$ is called *self-financing* if

$$\Delta_t S_t + b_t(1+r)^t = \Delta_{t+1} S_t + b_{t+1}(1+r)^t, \quad t = 1, \dots, T-1. \tag{13.24}$$

What does it mean that a portfolio follows a self-financing strategy? At time $t = 0$ one forms a portfolio consisting of Δ_1 shares and b_1 bonds, with the initial value $V_0 = \Delta_1 S_0 + b_1$. At time $t = 1$, the value of the portfolio becomes $V_1 = \Delta_1 S_1 + b_1(1+r)$, and one is now allowed to re-shuffle the portfolio, converting the available funds into Δ_2 shares and b_2 bonds, to be kept for the next time period (we assume that no costs are incurred by that). Self-financing means that the only funds that one can use to update the portfolio are the moneys one already has in the portfolio, so that the time $t = 1$ value of the updated portfolio must coincide with the time $t = 1$ value of the old portfolio:

$$\Delta_1 S_1 + b_1(1+r) = \Delta_2 S_1 + b_2(1+r),$$

which is (13.24) with $t = 1$. The same applies to each time step.

To find the fair price of a given claim $X = g(S_T)$, we want to exploit the same replication idea as was used in the previous sections: construct a self-financing portfolio $\{(\Delta_t, b_t)\}$ with value process $V_t := \Delta_t S_t + b_t(1+r)^t$, $t = 1, \ldots, T$ (and $V_0 = \Delta_1 S_0 + b_1$, in agreement with (13.24)) whose time $t = T$ value would *exactly replicate* the claim, i.e.,

$$V_T(\omega) \equiv \Delta_T(\omega) S_t(\omega) + b_T(\omega)(1+r)^T = X(\omega), \quad \omega \in \Omega.$$

Thus the problem is whether such a replicating portfolio exists for an arbitrary claim $X = g(S_T)$ in our market. The remarkable thing is that we can split the multi-period problem into T single-period ones, and we already know how to solve those!

First we observe that there is NA in the multi-period model iff there is NA in *each* of the time periods $t = 1, \ldots, T$. In one direction this is obvious. Indeed, NA in the tth time period corresponds to the absence of arbitrage strategies of this form: $\Delta_k = b_k = 0$ for all $k < t$, then one trades at the beginning of the tth period and cashes one's portfolio at the end of that period, doing nothing afterwards. As these are just special strategies for the whole multi-period model, the absence of arbitrage in the latter implies that single-period arbitrage is also impossible. The opposite implication is slightly harder to explain; in two words, the idea is that if there is arbitrage in the multi-period model, then there *must* be at least one period in which arbitrage is possible.

That observation means that we have the same NA condition (13.2) for the multi-period binomial market as in the single-period case. Next we will try to construct a replicating portfolio for a given claim of the form $X = g(S_T)$, extending the approach from Section 13.3.

We know the *terminal values* of the claim, so we will start at the end of the time horizon and work backwards. Suppose $t = T - 1$ and $S_{T-1} = s$. Then the only possible time $t = T$ stock prices are us and ds. What amount do we need to have in the replicating portfolio at time $t = T - 1$, *when the stock price is* s, to be able to replicate the two possible claim values, $g(us)$ and $g(ds)$, at time $t = T$ (see Fig. 13.5)?

This, however, is exactly the single-period claim replication problem that we solved in Section 13.3. Using the answers (13.9) and (13.10) we obtained there (just replace X_u with $g(us)$, X_d with $g(ds)$, and $1 + r$ with $(1+r)^T$, for the latter will be the bond price at the end of the time period) yields the following values for the numbers of shares and bonds to be held in the replicating portfolio during the Tth time period given that the stock

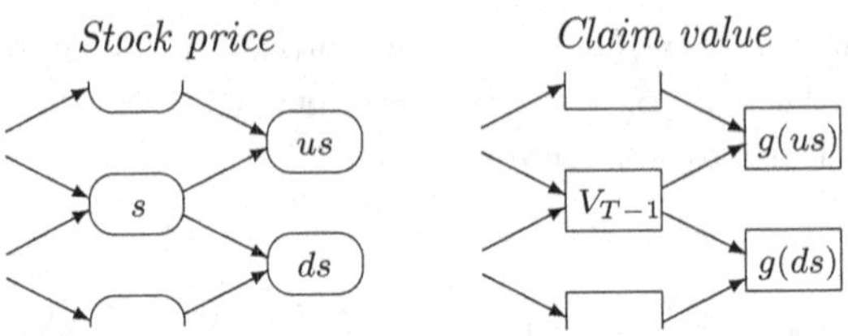

Fig. 13.5 A fragment of the recombinant tree for the stock price, corresponding to the time $t = T-1$ price $S_{T-1} = s$ and the only two values that the price can take after that at time T (left), and the respective fragment of the tree for the replicating portfolio values for the claim $g(S_T)$ (right).

price at the beginning of the period was s:

$$\Delta_T[s] = \frac{g(us) - g(ds)}{us - ds} = \frac{V_T[us] - V_T[ds]}{us - ds},$$
$$b_T[s] = \frac{ug(ds) - dg(us)}{(1+r)^T(u-d)} = \frac{uV_T[ds] - dV_T[us]}{(1+r)^T(u-d)},$$

where $\Delta_t[x]$, $b_t[x]$ and $V_t[x]$ stand for the tth period numbers of shares and bonds in the replicating portfolio and the portfolio time t value, respectively, when the stock price at the beginning of the period is x. The time $t = T-1$ value of that portfolio is

$$V_{T-1}[s] = \frac{1}{1+r}\left[p^* g(us) - (1-p^*)g(ds)\right] = \mathbf{E}^*\left(\frac{g(S_T)}{1+r}\,\middle|\, S_{T-1} = s\right)$$

with the same value of p^* as in (13.11).

Next we move one time step back and repeat the exercise to find the $\Delta_{T-1}[s]$, $b_{T-1}[s]$ and $V_{T-2}[s]$, assuming now that the time $t = T-2$ stock price value is s (note that the end of period bond price is now $(1+r)^{T-1}$):

$$\Delta_{T-1}[s] = \frac{V_{T-1}[us] - V_{T-1}[ds]}{us - ds}, \quad b_{T-1}[s] = \frac{uV_{T-1}[ds] - dV_{T-1}[us]}{(1+r)^{T-1}(u-d)},$$

and

$$\begin{aligned} V_{T-2}[s] &= \mathbf{E}^*\left(\frac{V_{T-1}[S_{T-1}]}{1+r}\,\middle|\, S_{T-2} = s\right) \\ &= \frac{1}{1+r}\left[p^* V_{T-1}[us] - (1-p^*)V_{T-1}[ds]\right] \\ &= \frac{1}{(1+r)^2}\sum_{j=0}^{2}\binom{2}{j}(p^*)^j(1-p^*)^{2-j} g(u^j d^{2-j} s), \end{aligned}$$

and so on. So one can see that the general formulae for $t = 1, \ldots, T$ are:

$$\Delta_t[s] = \frac{V_t[us] - V_t[ds]}{us - ds}, \quad b_t[s] = \frac{uV_t[ds] - dV_t[us]}{(1+r)^t(u-d)}, \tag{13.25}$$

and

$$V_t[s] = \frac{1}{(1+r)^{T-t}} \sum_{j=0}^{T-t} \binom{T-t}{j} (p^*)^j (1-p^*)^{T-t-j} g(u^j d^{T-t-j} s). \quad (13.26)$$

Thus, starting with the time T (i.e., expiry) claim values that are known for different possible states of the world from the claim description, one can compute recursively the replicating portfolio composition and its values at all times, at all possible states the world can be in at those times.

It is important to observe that, for each time period, one has the same "probabilities" p^* (for u) and $1-p^*$ (for d), regardless of the stock prices values, which means that different time step price movements are *independent under the probability measure* $\mathbf{P}^*$. Moreover, the model inherits the single-period market property that, given $S_t = s_t$, one has

$$s_t = \mathbf{E}^*\left(\frac{S_{t+1}}{1+r}\middle| S_t = s_t\right) = \mathbf{E}^*\left(\frac{S_{t+1}}{1+r}\middle| S_0 = s_0, S_1 = s_1, \ldots, S_t = s_t\right)$$

for any previous stock prices values $s_0, s_1, \ldots, s_{t-1}$ (due to the Markov property of the RW with independent jumps; it would be appropriate to assume that the s-values are consistent in the sense that $s_k = us_{k-1}$ or $= ds_{k-1}$ for all $k = 1, \ldots, t$). That is, under $\mathbf{P}^*$, the discounted stock price process $S_t^* := S_t(1+r)^{-t}$ is a martingale w.r.t. its natural filtration:

$$S_t^* = \mathbf{E}^*\left(S_{t+1}^* \middle| S_0^*, S_1^*, \ldots, S_t^*\right), \quad t = 0, 1, \ldots, T-1.$$

It is easy to see that, like in the single-period case, $\mathbf{P}^*$ is the *only probability measure* on the sample space (13.3) under which $\{S_t^*\}$ is an MG.

Example 13.5. Price and replicate a call with strike $K = 80$ on the stock in a two-period binomial market with $u = 1.5$, $d = 0.5$, $r = 0$ (meaning that all the prices are already discounted) and $S_0 = 120$.

First of all note that $d < 1 + r < u$, so this is an NA market and we can use the techniques we developed above. Next we draw the stock price diagram, starting with the left-most oval, corresponding to the time $t = 0$ price $S_0 = 120$, and then working to the right, multiplying the already computed values by u and d each step (Fig. 13.6).

Now we compute $p^* = \frac{1+r-d}{u-d} = 0.5$, $1 - p^* = 0.5$, and then proceed to computing the call prices, starting from the last column of the claim values diagram, as shown in Fig. 13.7.

First find the time $t = 1$ replicating portfolio composition at the node labelled by b in Fig. 13.7. Using (13.25) with $t = 2$, we get

$$\Delta_2[180] = \frac{190 - 10}{270 - 90} = 1, \quad b_2[180] = \frac{1.5 \times 10 - 0.5 \times 190}{1^2(1.5 - 0.5)} = -80.$$

For the node labelled by c, we have

$$\Delta_2[60] = \frac{10 - 0}{90 - 30} = \frac{1}{6}, \quad b_2[60] = \frac{1.5 \times 0 - 0.5 \times 10}{1^2(1.5 - 0.5)} = -5.$$

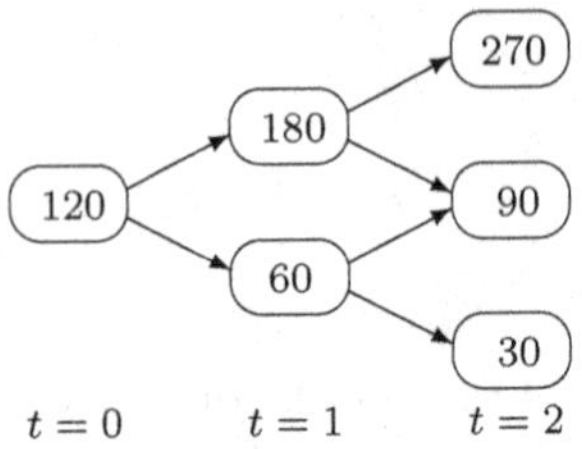

Fig. 13.6 The stock price diagram for Example 13.5.

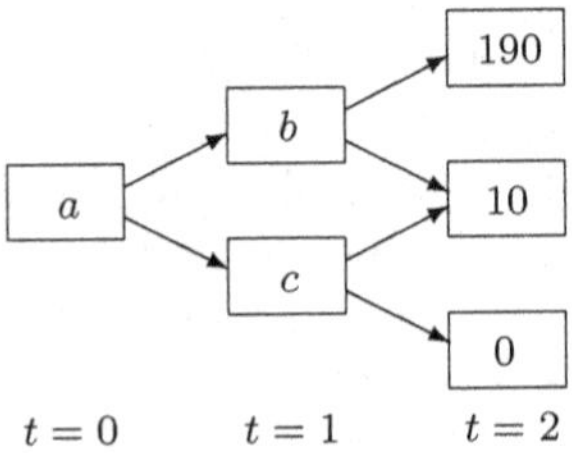

Fig. 13.7 The initial call price diagram for Example 13.5. Only the terminal claim prices are shown: $190 = (270 - K)^+$, $10 = (90 - K)^+$ and $0 = (30 - K)^+$, corresponding to the respective stock prices in the last column in Fig. 13.6.

Now we can compute the time $t = 1$ claim values:

$$V_1[180] = \Delta_2[180] \times 180 + b_2[180](1 + r) = 1 \times 180 - 80 \times 1 = 100,$$

$$V_1[60] = \Delta_2[60] \times 60 + b_2[60](1 + r) = \frac{1}{6} \times 60 - 5 \times 1 = 5,$$

and then write them in the boxes labelled by b and c, respectively, in the diagram in Fig. 13.7. These are the amounts of money one needs at that time, in those two different states of the world, to achieve (within the model framework) perfect replication of the call at time $t = 2$.

It remains to repeat the procedure, now for the last remaining blank node (labelled by a) in the call price diagram, to find

$$\Delta_1[120] = \frac{100 - 5}{180 - 60} = \frac{19}{24}, \quad b_1[120] = \frac{1.5 \times 5 - 0.5 \times 100}{1^2(1.5 - 0.5)} = -42\tfrac{1}{2}$$

which yields the time $t = 0$ claim value

$$V_0[120] = \Delta_1[120] \times 120 + b_1[120] = \frac{19}{24} \times 120 - 42\tfrac{1}{2} = 52\tfrac{1}{2}.$$

This is the investment one needs to make at time $t = 0$ to construct the following self-financing replicating portfolio for our call:

$$(\Delta_1, b_1) = (\tfrac{19}{24}, -42\tfrac{1}{2}) \longrightarrow \begin{cases} (\Delta_2, b_2) = (1, -80) & \text{if } S_1 = uS_0 = 180, \\ (\Delta_2, b_2) = (\tfrac{1}{6}, -5) & \text{if } S_1 = dS_0 = 60. \end{cases}$$

Note that all the b_t-values are negative, i.e., the bond position is always short in the replicating portfolio, meaning the we need to borrow money from the bank at all times. Also, note that $\Delta_2[180] = 1$, i.e., we put one share of the stock in the portfolio at the beginning of the second time period, when the current stock price is 180. This is because, according to the model, whatever happens during that time period, the call *will be exercised* at time $t = 2$ (the stock price will be above the strike anyway!). Hence the issuer of the call (who is running the replicating portfolio) *must* have that share to deliver it to the owner of the call at the end of the period.

Now return to the general problem of claim pricing and observe that the time $t = 0$ claim price is given by (13.26) as

$$\begin{aligned} V_0[S_0] &= \frac{1}{(1+r)^T}\sum_{j=0}^{T}\binom{T}{j}(p^*)^j(1-p^*)^{T-j}g(u^j d^{T-j}S_0) \\ &= \mathbf{E}^*\frac{g(S_0e^{Z_t})}{(1+r)^T}, \end{aligned} \tag{13.27}$$

where $Z_n := Y_1 + \cdots + Y_n$, $n \geq 0$, is an RW with i.i.d. jumps

$$Y_k = \begin{cases} \ln u & \text{w.p. } p^*, \\ \ln d & \text{w.p. } 1-p^* \end{cases}$$

on the probability space $(\Omega, 2^\Omega, \mathbf{P}^*)$, where Ω is specified by (13.3).

Example 13.6. *Pricing European calls in binomial setting.* In particular, the time $t = 0$ price of the European call with the pay-off function (13.5) is

$$\begin{aligned} C &= \frac{1}{(1+r)^T}\sum_{j=0}^{T}\binom{T}{j}(p^*)^j(1-p^*)^{T-j}\underbrace{(S_0u^jd^{T-j}-K)^+}_{=0 \text{ if } S_0u^jd^{T-j}\leq K} \\ &= \frac{1}{(1+r)^T}\sum_{j=M}^{T}\binom{T}{j}(p^*)^j(1-p^*)^{T-j}(S_0u^jd^{T-j}-K) \\ &= S_0\sum_{j=M}^{T}\binom{T}{j}\Big(\underbrace{\frac{up^*}{1+r}}_{=:p^\dagger}\Big)^j\Big(\underbrace{\frac{d(1-p^*)}{1+r}}_{=1-p^\dagger}\Big)^{T-j} \\ &\qquad - \frac{K}{(1+r)^T}\sum_{j=M}^{T}\binom{T}{j}(p^*)^j(1-p^*)^{T-j} \\ &= S_0\mathbf{P}(U^\dagger \geq M) - \frac{K}{(1+r)^T}\mathbf{P}(U^* \geq M), \end{aligned} \tag{13.28}$$

where we set $M := \lfloor \frac{\ln K - \ln S_0 - T\ln d}{\ln u - \ln d}\rfloor + 1$ (so that $(\cdots)^+ = 0$ for $j < M$) and $U^\dagger \sim B_{n,p^\dagger}$ and $U^* \sim B_{n,p^*}$ are binomial RVs.

The binomial model is rather crude and, in particular, quite unrealistic in its assumptions about one period price movements. One can hope, however, to get a better model by splitting the given (continuous) time interval $[0, T]$ into a *large* number n of time periods of length $\delta_n := T/n$ and then using the binomial model with n periods, with appropriately chosen values of $u = u_n$, $d = d_n$, and $r = r_n$. We will now see what happens then, as $n \to \infty$.

Consider a fixed maturity time $T > 0$, and suppose that trading of the stock with the time t price can only occur at discrete time epochs $t = 0, \delta_n, 2\delta_n, \ldots, n\delta_n = T$. Assume, further, that during each time period $((j-1)\delta_n, j\delta_n]$, $j = 1, 2, \ldots, n$, the stock price S_t (note that the subscript t is not integer-valued anymore, but is rather a value from the interval $[0, T]$) can move to only one of the two future values:

$$S_{(j-1)\delta_n} \begin{array}{l} \nearrow S_{j\delta_n} = u_n S_{(j-1)\delta_n}, \\ \searrow S_{j\delta_n} = d_n S_{(j-1)\delta_n}, \end{array} \qquad j = 1, 2, \ldots, n, \qquad \begin{array}{l} u_n = e^{\sigma\sqrt{\delta_n}}, \\ d_n = e^{-\sigma\sqrt{\delta_n}}, \end{array}$$

where the parameter $\sigma > 0$ is called "volatility". Thus, each (small) time step of length T/n, the price can experience also a small change, but of much higher magnitude ($1/\sqrt{n}$ rather than $1/n$) as $n \to \infty$:

$$u_n = 1 + \sigma\sqrt{\delta_n} + O(n^{-1}), \quad d_n = 1 - \sigma\sqrt{\delta_n} + O(n^{-1}).$$

Assume also the (continuously compounding) constant force of interest r, so that

$$B_{j\delta_n} = e^{r\delta_n} B_{(j-1)\delta_n} = (1 + r_n) B_{(j-1)\delta_n}, \quad r_n = e^{r\delta_n} - 1 = r\delta_n + O(n^{-2}).$$

Now observe that, under the respective risk-neutral measures $\mathbf{P}_n^*$ (note that to different n there will correspond different Ω's, that will be sets of strings of length n, so our measures will be on different spaces), the RWs that appear in the general pricing formula (13.27) will have i.i.d. jumps with values

$$Y_{n,k} = \begin{cases} \ln u_n = +\sigma\sqrt{T/n} & \text{w.p. } p_n^*, \\ \ln d_n = -\sigma\sqrt{T/n} & \text{w.p. } 1 - p_n^*, \end{cases} \qquad p_n^* = \frac{e^{r\delta_n} - e^{-\sigma\sqrt{\delta_n}}}{e^{\sigma\sqrt{\delta_n}} - e^{-\sigma\sqrt{\delta_n}}},$$

so that

$$Z_{n,n} := Y_{n,1} + \cdots + Y_{n,n} = \sigma\sqrt{T} \times \frac{1}{\sqrt{n}} (Y_{n,1}^0 + \cdots + Y_{n,n}^0),$$

where $Y_{n,j}^0 = \pm 1$ w.p. p_n^* and $1 - p_n^*$, respectively. Observe that although the possible values of $Y_{n,j}^0$ no longer depend on n, their probabilities — that tend to $\frac{1}{2}$ as $n \to 0$ — still do, so we need to keep the first subscript n in the notation for them.

Taking into account the above representation for $Z_{n,n}$, the reader will not be surprised to learn that the CLT will be applicable to the distribution of that RV. In the special case of pricing a European call with maturity T and strike K, that will be displayed in normal DF approximations to the binomial probabilities $\mathbf{P}(U^\dagger \geq M)$ and $\mathbf{P}(U^* \geq M)$ that appear in the call price (13.28). If one *carefully*[15] apples the CLT in the above setting (which is a good exercise), one will find that the limiting as $n \to \infty$ time $t = 0$ price of the European call is equal to

$$S_0\Phi(h) - Ke^{-rT}\Phi(h - \sigma\sqrt{T}), \quad h := \frac{\ln(S_0/K) + (r + \frac{1}{2}\sigma^2)T}{\sigma\sqrt{T}}. \tag{13.29}$$

This is the famous *Black–Scholes formula* giving the price of the European call with maturity T and strike K under the assumption of a continuous time financial market with the stock price modelled by the geometric BM as $S_t = S_0 \exp\{\mu t + \sigma W_t\}$ and the bond price given by $B_t = e^{rt}$, $t \in [0, T]$. It can be derived using NA pricing principles in the latter framework, as we will see below in Section 13.7. Once can view the convergence of the call prices in binomial markets to the one in the Black–Scholes market as a display of the functional CLT that we briefly mentioned in Section 2.10 (although here we need a somewhat more general setting for it).

13.6 Martingales and Claim Pricing

The results that we obtained and discussed in the previous sections can be extended to general (discrete and continuous time) markets. As these extensions would require mathematical techniques far beyond the level we assume for the present text, we will not discuss them in detail, but rather summarise the key facts in a relatively simple setup, sometimes omitting some technicalities. For a more rigourous treatment of the subject, the interested reader is referred to the literature listed in Section 13.9.

Assume that the price process $\{\boldsymbol{S}_t := (S_t^1, \ldots, S_t^n)\}_{t \geq 0}$, its components being the time t prices of n different (risky) underlying assets, is an SP on a probability space $(\Omega, \mathcal{F}, \mathbf{P})$, and let $\mathbf{F} = \{\mathcal{F}_t\}_{t \geq 0}$ be its natural filtration. Suppose also that we have a risk-free security (bond, or bank account) of which the price is given by

$$B_t = \begin{cases} (1+r)^t & \text{in discrete time: } t = 0, 1, 2, \ldots, \\ e^{rt} & \text{in continuous time: } t \geq 0, \end{cases}$$

[15] As we said, the distribution of $Y_{n,j}^0$ still depends (although slightly!) on n, and that in fact matters.

for a constant[16] $r \geq 0$.

The main question remains how to price a European contingent claim with a given maturity T, i.e., an $\mathcal{F}_T$-measurable RV $X \geq 0$ (that is, a non-negative RV which is a function of the price values $\boldsymbol{S}_t$ with $t \leq T$). The "risk-neutral", or "arbitrage-free" price of the claim is obtained as the value of the self-financing[17] portfolio that replicates the claim X at its maturity time T. Here are the three key facts of the no-arbitrage pricing theory:

- If there exists an EMM $\mathbf{P}^*$ on $(\Omega, \mathcal{F})$, i.e., a probability distribution on that measurable space such that (i) $\mathbf{P}^*(A) > 0$ iff $\mathbf{P}(A) > 0$, $A \in \mathcal{F}$, and (ii) the SP

$$\boldsymbol{S}_t^* := \begin{cases} (1+r)^{-t}\boldsymbol{S}_t & \text{(in discrete time)} \\ e^{-rt}\boldsymbol{S}_t & \text{(in continuous time)} \end{cases}, \quad t \geq 0,$$

is an MG on $(\Omega, \mathcal{F}, \mathbf{P}^*)$, then the *market is arbitrage-free.*

- For an *attainable claim* X with maturity T, its time t arbitrage-free price is given by

$$\mathcal{P}_t(X) = \begin{cases} \mathbf{E}^*\left((1+r)^{t-T}X \middle| \mathcal{F}_t\right) & \text{(in discrete time)} \\ \mathbf{E}^*\left(e^{-r(T-t)}X \middle| \mathcal{F}_t\right) & \text{(in continuous time)} \end{cases}, \qquad t \in [0, T].$$

In particular, since $\mathcal{F}_0$ is trivial (as the price vector $\boldsymbol{S}_0$ is non-random), one has $\mathcal{P}_0(X) = \mathbf{E}^*(1+r)^{-T}X$ (or $= \mathbf{E}^* e^{-rT}X$).

- If the EMM $\mathbf{P}^*$ is *unique*, then the market is *complete*, i.e., *any claim is attainable* and so can always be priced.

Remark 13.3. Observe that the discounted claim value price process $M_t := (1+r)^{-t}\mathcal{P}_t(X)$, $t = 0, 1, \ldots, T$ (in discrete time; in continuous time, the remark will apply to $M_t := e^{-rt}\mathcal{P}_t(X)$, $t \in [0, T]$), will, like the discounted stock and bond prices, also be an MG under $\mathbf{P}^*$. Indeed, the SP $M_t = (1+r)^{-T}\mathbf{E}^*(X | \mathcal{F}_t)$ will just be a Lévy MG, see Problem 4 on p. 345.

Now what about American derivative securities (ADS), i.e., claims that can be exercised by their owners at any time prior to their expiry? Recall

---

[16]Extension to variable deterministic interest rates is relatively straightforward, but there are also more sophisticated random interest rate models.

[17]The definition of a self-financing trading strategy in continuous time is somewhat more technical than in discrete time, but the meaning remains the same: one is only allowed to "reshuffle" the composition of the portfolio, without adding or withdrawing any funds from it. See (13.37) for a formal definition in the Black–Scholes framework.

that we have already briefly discussed ADS in Example 4.2 (p. 134) and Example 11.7 (p. 310).

Assume for simplicity that time is discrete. Denote by τ the ST at which the owner of an ADS X with expiry T decides to terminate the contract: if $\tau = t < T$ then that means exercising the ADS at time t, while $\tau = T$ both when the owner exercises the ADS at time T and when she lets it expire. Then, at that time τ, she receives the payoff

$$X_\tau = \sum_{t=0}^{T} \underbrace{X_t \mathbf{1}_{\{\tau=t\}}}_{=:Y_t},$$

where Y_t is clearly $\mathcal{F}_t$-measurable (as τ is an ST!). Thus, the ADS X is equivalent to a "compound European claim", a portfolio consisting of $T+1$ European claims Y_t, $t = 0, 1, \ldots, T$, with different expiries t. Note that if one can replicate each of those Y_t's, then the combination of the replicating portfolios will give a perfect hedge for the ADS X, provided that its owner uses the ST τ to terminate it. Hence the time $t = 0$ price (similarly for other times as well) $\mathcal{P}_0(X|\tau)$ of the ADS, when one uses the ST τ, will be given by

$$\begin{aligned}\mathcal{P}_0(X|\tau) &= \sum_{t=0}^{T} \mathbf{E}^* \frac{X_t \mathbf{1}_{\{\tau=t\}}}{(1+r)^t} = \sum_{t=0}^{T} \mathbf{E}^* \frac{X_\tau \mathbf{1}_{\{\tau=t\}}}{(1+r)^\tau} \\ &= \mathbf{E}^* \left[\frac{X_\tau}{(1+r)^\tau} \underbrace{\sum_{t=0}^{T} \mathbf{1}_{\{\tau=t\}}}_{=1, \text{ as } \tau \le T} \right] = \mathbf{E}^* \frac{X_\tau}{(1+r)^\tau}.\end{aligned}$$

But what ST τ should one use? To avoid arbitrage, one should actually take the one that maximises the price of X, leading to the claim price

$$\mathcal{P}_0(X) = \sup_{\text{all STs } \tau \le T} \mathbf{E}^* \frac{X_\tau}{(1+r)^\tau}. \tag{13.30}$$

The optimal ST will usually have the form of a special "hitting time".

As a rule, explicit formulae are not available for ADSs, with the one exception: *American calls are equivalent to European calls* (in that case, one always has $\tau \equiv T$; note that we only deal with the case where our stocks do not pay dividends during the options lifetimes). There is no such equivalence for puts!

Theorem 13.5. *Let $C_A(t)$ be the time t price of an American call with maturity T and strike K on one share of a stock, and let $C_E(t)$ be the time t price of a European call with the same maturity and strike, on one share of the same stock. Then $C_A(t) = C_E(t)$ for all $t \in [0, T]$.*

Proof We will only consider the discrete time case (the argument in continuous time is no different).

That $C_A(t) \geq C_E(t)$ is obvious, as the American option has additional features compared to the European version. One can also use the time t version of the pricing formula (13.30) to show that.

To establish the (less obvious) opposite inequality, we will first establish that, assuming that our calls have maturity T and strike price K (so that, given that the American call is exercised at time t, the cash flow generated thereby at that time is $X_t := (S_t - K)^+$), one has

$$C_E(t) \geq \left(S_t - \frac{K}{(1+r)^{T-t}}\right)^+. \tag{13.31}$$

Indeed, at time t form a portfolio consisting of a long position on one share of the stock and short position on $K(1+r)^{t-T}$ bond units, so that its time t value is $V_t = S_t - \frac{K}{(1+r)^{T-t}}$. Now at time T the value of the portfolio will become

$$V_T = S_T - K \leq (S_T - K)^+ = C_E(T).$$

Therefore, to avoid arbitrage, one must also have $V_t \leq C_E(t)$. Since always $C_E(t) \geq 0$, the inequality (13.31) follows.

It remains to observe that, for all $t \leq T$,

$$C_E(t) \overset{(13.31)}{\geq} \left(S_t - \frac{K}{(1+r)^{T-t}}\right)^+ \geq (S_t - K)^+ = X_t,$$

where the second inequality is obvious. That is, there is no point in exercising the call at time $t < T$ as that would only result in the cash flow X_t which does not exceed the price of the European call $C_E(t) \leq C_A(t)$ (as we saw above): just selling the option will result in at least the same cash flow. So one can just wait till the expiry, which means that the premature exercise feature of the American call is useless, so that $C_A(t) \equiv C_E(t)$. □

13.7 The Black–Scholes Framework

In this section we will discuss the most popular continuous time model dealt with in mathematical finance. In its basic form, the market has only two assets: the riskless bank account (bond) with a constant interest rate r, having the price dynamics

$$B_t = e^{rt}, \quad t \in [0, T]$$

(or, equivalently, $dB_t = rB_t\,dt$, $B_0 = 1$), and a risky asset (a stock) of which the price S_t satisfies the Black–Scholes SDE

$$dS_t = \mu S_t dt + \sigma S_t dW_t, \quad t \in [0, T], \tag{13.32}$$

with $\mu \in \mathbf{R}$, $\sigma > 0$ and $S_0 > 0$ constant, $\{W_t\}$ being the standard BM. One motivation for using such a model is that the SDE can be re-written as

$$\frac{dS_t}{S_t} = \mu\,dt + \sigma dW_t,$$

meaning that the short-term return on the stock consists of two components: a "systematic return" μdt with a constant rate, which is perturbed by the normal white noise dW_t, the *volatility* coefficient σ specifying the effect of the noise on the return.

We saw in Example 11.13 that the process $Z_t = Z_0 \exp\{\lambda t + \sigma W_t\}$ satisfies the SDE

$$dZ_t = (\lambda + \sigma^2/2) Z_t dt + \sigma Z_t dW_t,$$

which has exactly the same form as (13.32) provided that $\mu = \lambda + \sigma^2/2$. Therefore, the SDE (13.32) has the following solution:

$$S_t = S_0 \exp\{(\mu - \sigma^2/2)t + \sigma W_t\}, \quad t \in [0, T]. \tag{13.33}$$

That is, the stock price in the Black-Scholes market is given by a geometric BM process.

There is a lot of criticism of the above model, including the fact that the market data do not support the assumption that the stock log-prices are normally distributed (they usually follow skew distributions with tails much "heavier" than the normal ones) and the observation that the option prices computed basing on the model do not have some important properties observed in the market data. Nevertheless, the Black–Scholes model still "remains in service" (although by no means it is used by everybody, under any circumstances!). Here is one possible explanation[18] of that:

> All models sweep dirt under the rug. A good model makes the absence of the dirt visible. In this regard, we believe that the Black-Scholes model of options valuation, now often unjustly maligned, is a model for models; it is clear and robust. Clear, because it is based on true engineering; it tells you how to manufacture an option out of stocks and bonds and what that will cost you, under ideal dirt-free circumstances that it defines. Its method of valuation is analogous to figuring out the price of a can of fruit salad from the cost of fruit, sugar, labor and transportation. The world of markets doesn't exactly match the ideal circumstances Black-Scholes requires, but the model is robust because it allows an intelligent trader to qualitatively adjust for those mismatches. You know what you are assuming when you use the model, and you know exactly what has been swept out of view.

So we will have a close look at this classical model, starting with answering the following two fundamental questions:

[18] Given in the *Financial Modelers' Manifesto* by E. Derman and P. Wilmott (published on 08.01.2009; to find the whole text, start at `http://www.wilmott.com/`).

Q1. Is the Black–Scholes market *arbitrage-free*? If yes, then one can price attainable claims.

Q2. Is the Black–Scholes market *complete*? Of course, asking this question only makes sense if the answer to Q1 is positive. If the answer to Q2 is also positive, one can price *any claim* in the market.

To answer the above questions, we will turn to the key facts stated in Section 13.6. The market will be NA is there exists an EMM. So we first need to specify the underlying filtered probability space $(\Omega, \mathcal{F}, \mathbf{F}, \mathbf{P})$.

The natural choice for Ω is the sample space for the price process, i.e., the space $C[0,T]$ of continuous functions on $[0,T]$: our "points" $\omega \in \Omega$ are continuous functions $[0,T] \mapsto \mathbf{R}$. The usual choice of $\mathcal{F}$ is the σ-field of Borel subsets $\mathcal{B}(C[0,T])$. This is the smallest σ-field of subsets of $\Omega = C[0,T]$ containing all open sets[19] in the uniform norm $\|f\| := \max_{t\in[0,T]} |f(t)|$ on $C[0,T]$ (as the space is separable[20], this is the same as the smallest σ-field on Ω containing all *open balls*).

It can be shown that $\mathcal{B}(C[0,T])$ coincides with the so-called cylindrical σ-field, which is generated by cylinders, i.e., sets of the form $\mathcal{C}_s(B) := \{f \in C[0,T] : f(s) \in B\}$ for some $s \in [0,T]$ and $B \in \mathcal{B}$. The filtration $\{\mathcal{F}_t\}_{t\in[0,T]}$ is specified by taking $\mathcal{F}_t$ to be the σ-field generated by cylinders $\mathcal{C}_s(B)$ with $s \in [0,t]$.

Finally, $\mathbf{P}$ is the distribution of the geometric BM process (13.33) on the so chosen $(\Omega, \mathcal{F})$.

Alternatively, one could take $\mathbf{P}$ to be the *Wiener measure* on $(\Omega, \mathcal{F})$, i.e., the distribution of the BM process $\{W_t\}_{t\in[0,T]}$ on that measurable space, the interpretation of ω becoming not the trajectory of the price process S_t itself, but, rather, of the driving it BM process W_t. In fact, we already referred to such a construction in Chapter 2 (p. 66).

Since the answers to our questions Q1 and Q2 are given in terms of EMMs, we first need to know what probability measures on our $(\Omega, \mathcal{F})$ will be equivalent to the distribution of (13.33). The answer to that has a simple form: they will all be distributions of processes of the form

$$Z_t := S_0 \exp\left\{\int_0^t a_s ds + \sigma W_t\right\}, \quad t \in [0,T], \tag{13.34}$$

where $\{a_t\}_{t\in[0,T]}$ is an adapted process on $(\Omega, \mathcal{F}, \mathbf{F}, \mathbf{P})$ (satisfying some integrability conditions).

[19] A subset $A \subset S$ of a normed space $(S, \|\cdot\|)$ is called open if, for any $x_0 \in A$, there exists an $r > 0$ such that the *open ball* $\{x \in S : \|x - x_0\| < r\}$ is a subset of A.

[20] A normed space $(S, \|\cdot\|)$ is called separable if it contains an everywhere dense countable subset. That is, there exists a sequence of points $x_1, x_2, \ldots \in S$ such that any open ball in S will contain at least one point x_k from that sequence.

To understand why this is so, recall that equivalence of probability distributions $\mathbf{P}$ and $\mathbf{P}^*$ on $(\Omega, \mathcal{F})$ means that they assign positive probabilities to the same elements of $\mathcal{F}$. Equivalently, $\mathbf{P}$ and $\mathbf{P}^*$ assign zero probabilities to the same elements of $\mathcal{F}$, which, in turn, is equivalent to $\mathbf{P}$ and $\mathbf{P}^*$ assigning probability one to the same sets $A \in \mathcal{F}$.

Since, under the distribution $\mathbf{P}$ of (13.33), w.p. 1 the initial value of the process was the constant S_0, this should also be the case under any equivalent to $\mathbf{P}$ probability measure $\mathbf{P}^*$. Hence the appearance of S_0 in the process (13.34).

Further, $\sigma^{-1} \ln S_t = \sigma^{-1}(\mu - \sigma^2/2)t + W_t$, and so, w.p. 1, for any $s \in [0, T]$, the quadratic variation of the trajectory $\sigma^{-1} \ln S_t$ on $[0, s]$ should be equal to s (see p. 322; note that the presence of a smooth drift can have no affect on the quadratic variation of the process). Hence that should be the case under any equivalent to $\mathbf{P}$ probability $\mathbf{P}^*$ as well. It turns out, however, that the only SP whose trajectories have quadratic variation on $[0, s]$ given by s, $s \in [0, T]$, is the BM process, modulo an additive nice enough drift. Hence the expression in the exponential in the representation (13.34). Of course, we did not prove anything here: that was just an attempt to clarify what is happening.

Now it remains to find out for what SPs $\{a_t\}_{t\in[0,T]}$ the process $\{e^{-rt}Z_t\}$ with $\{Z_t\}$ specified by (13.34) will be an MG. But Remark 11.3 already contains the answer to that question: as $\{e^{-rt}Z_t\}$ is clearly an Itô process (as a smooth function of such a process), it will be an MG iff the coefficient of the dt-term in its stochastic differential is zero. Using the product rule of Itô calculus (Theorem 11.10) and the result of Example 11.13, one has

$$\begin{aligned} d(e^{-rt}Z_t) &= d\Bigg(\underbrace{\exp\left\{\int_0^t a_s ds\right\}}_{=:X_t} \times \underbrace{\exp\{-rt + \sigma W_t\}}_{=:Y_t}\Bigg) \\ &= Y_t dX_t + X_t dY_t \\ &= Y_t a_t X_t dt + X_t\left(\left(-r + \frac{\sigma^2}{2}\right)Y_t dt + \sigma Y_t dW_t\right)\right) \\ &= X_t Y_t\left(a_t - r + \frac{\sigma^2}{2}\right)dt + \sigma X_t Y_t dW_t. \end{aligned}$$

As $X_t Y_t > 0$, the coefficient of dt is zero iff $a_t = r - \sigma^2/2$, $t \in [0, T]$. That is, there is a unique solution to that problem, and we have the following key result.

Theorem 13.6. *There exists a unique EMM in the Black–Scholes market. Under that probability* $\mathbf{P}^*$, *the price process has the geometric BM process' dynamics specified by*

$$S_t = S_0 \exp\{(r - \sigma^2/2)t + \sigma\widetilde{W}_t\}, \quad t \in [0, T],$$

or, equivalently, by

$$dS_t = rS_t dt + \sigma S_t d\widetilde{W}_t, \quad t \in [0, T], \tag{13.35}$$

where $\{\widetilde{W}_t\}$ *is a standard BM process.*[21]

Therefore, according to what we said in Section 13.6, the Black–Scholes market is *arbitrage free and complete*, and so one can *price any contingent claim* X *in the market,* using the valuation formula

$$\mathcal{P}_t(X) = \mathbf{E}^*\big(e^{-r(T-t)}X\big|\mathcal{F}_t\big).$$

There are straightforward extensions of the above result to more general Black–Scholes models, involving variable spot interest rates, multiple stocks and dividend-paying stocks. The interested reader will find them (and much more) in the books listed in Section 13.9 below.

Remark 13.4. It seems counterintuitive that the "systematic return" μ of the stock completely disappeared from our calculations by now! Arbitrage free pricing formulae only involve the volatility σ and interest rate r of the model. Moreover, by switching to discounted prices (expressing everything in the time $t = 0$ dollars), one removes the interest rate r as well, the volatility σ remaining the only relevant parameter of the model. Why is that so?

Recall that the arbitrage free price of a claim is the smallest capital required to perfectly hedge the claim *under any circumstances*, whatever happens in the world. You may also remember that, when discussing discrete markets, we said that it only mattered what values the future prices can take, their probabilities being irrelevant: the constructed replicating strategy will always work! Likewise here, although the situation in continuous time is much more subtle: roughly speaking, the set of all possible trajectories of the price process is specified by the value of σ (via a very special behaviour of the quadratic variation of the process trajectories), the drift values being irrelevant.

Example 13.7. *Pricing European calls in the Black–Scholes market.* For the call with payoff function (13.5), arguing similarly to our derivation in Example 13.6 and noting that $\{S_T > K\} = \{T^{-1/2}W_T > -h_0\}$, where $T^{-1/2}W_T \sim N(0,1)$ and

$$h_0 := \frac{\ln(S_0/K) + (r - \frac{1}{2}\sigma^2)T}{\sigma\sqrt{T}} \equiv h - \sigma\sqrt{T}$$

[21]This $\{\widetilde{W}_t\}$ is, generally speaking, different from the BM process $\{W_t\}$ that appeared in the market description. Note that $\{W_t\}$ was a BM on $(\Omega, \mathcal{F}, \mathbf{P})$, while $\{\widetilde{W}_t\}$ is a BM on $(\Omega, \mathcal{F}, \mathbf{P}^*)$.

(cf. (13.29)), we obtain that

$$\begin{aligned}
\mathcal{P}_0(X) &= e^{-rT}\mathbf{E}^*(S_T-K)^+ = e^{-rT}\mathbf{E}^*(S_T-K)^+\mathbf{1}_{\{S_T>K\}}\\
&= e^{-rT}\int_{-h_0}^{\infty}\Big(S_0e^{(r-\sigma^2/2)T+\sigma\sqrt{T}x}-K\Big)\frac{e^{-x^2/2}}{\sqrt{2\pi}}\,dx\\
&= S_0\int_{-h_0}^{\infty}\frac{e^{-(x-\sigma\sqrt{T})^2/2}}{\sqrt{2\pi}}\,dx - Ke^{-rT}\int_{-h_0}^{\infty}\frac{e^{-x^2/2}}{\sqrt{2\pi}}\,dx\\
&= S_0\Phi\big(h_0+\sigma\sqrt{T}\big) - Ke^{-rT}\Phi(h_0)\\
&= S_0\Phi(h) - Ke^{-rT}\Phi\big(h-\sigma\sqrt{T}\big),
\end{aligned} \tag{13.36}$$

since

$$e^{-\sigma^2T/2+\sigma\sqrt{T}x}\frac{e^{-x^2/2}}{\sqrt{2\pi}} = \frac{e^{-(x-\sigma\sqrt{T})^2/2}}{\sqrt{2\pi}}$$

is the density of $N(\sigma\sqrt{T},1)$ and

$$\int_{-u}^{\infty}\frac{e^{-x^2/2}}{\sqrt{2\pi}}\,dx = 1-\Phi(-u) = \Phi(u)$$

due to the symmetry of $N(0,1)$. This, of course, is the already familiar to us Black–Scholes formula (13.29) that appeared earlier in our course as the limit of call prices in a sequence of binomial markets. The price of the European put can be computed in a similar fashion (or immediately obtained from the put-call parity).

Theorem 13.6 provides us with arbitrage free (or risk neutral) price for any contingent claim in the Black–Scholes market. Recall that that price is nothing else but the smallest capital needed to replicate the cash flow provided by the claim. If the issuer of the claim does not know how to realise such a replication via a suitable self-financing trading strategy, then knowing the price itself will be of little help for him. So how does one replicate claims in the Black–Scholes market?

First of all, we need to define the concept of self-financing in continuous time. In discrete time, a trading strategy (Δ_t, b_t), $t = 1,\dots,T$, was called self-financing provided that the following relation would hold for the portfolio time t value V_t (cf. (13.24)):

$$V_t := \Delta_tS_t + b_t(1+r)^t = \Delta_{t+1}S_t + b_{t+1}(1+r)^t, \quad t<T.$$

This is equivalent to the following condition on the increments of the value process:

$$\begin{aligned}
V_{t+1}-V_t &= \Delta_{t+1}S_{t+1} + b_{t+1}(1+r)^{t+1} - \Delta_{t+1}S_t - b_{t+1}(1+r)^t\\
&= \Delta_{t+1}(S_{t+1}-S_t) + b_{t+1}(B_{t+1}-B_t), \quad t<T.
\end{aligned}$$

The latter proves to be the most suitable form for extension to the continuous time case: a bivariate process (Δ_t, b_t), $t \in [0, T]$, is said to be a *trading strategy* if it is predictable[22] and such that the integrals $\int_0^u \Delta_t dS_t$ are well-defined. The time t value of the portfolio following the strategy is given by

$$V_t := \Delta_t S_t + b_t B_t.$$

The trading strategy (Δ_t, b_t) is called *self-financing* if

$$dV_t = \Delta_t dS_t + b_t dB_t, \quad t < T. \tag{13.37}$$

So, in the continuous time case, the absence of arbitrage means that there is no self-financing (in the sense of (13.37)) trading strategy that has zero initial value and a.s. non-negative value V_T at the terminal time T, with $\mathbf{P}(V_T > 0) > 0$. Replicating (or hedging) a claim X means constructing a self-financing trading strategy that would generate the same cash flow as X at expiry.

It is actually more complicated than that. Formally, one needs to restrict the class of admissible trading strategies, to avoid "doubling strategies" akin to the classical martingale betting scheme (see Problem 9(ii) on p. 345). In particular, one can restrict oneself to only considering the so-called "tame strategies" that are characterised by the property that, for any time $t \in [0, T]$, the value of the respective portfolio satisfies $V_t \geq -c$ for some fixed $c < \infty$. In practical terms, that means that the trader does not have access to unlimited credit, which is not unreasonable.

Without such restrictions, situation becomes nasty. Assume for simplicity that the stock price process has dynamics $dS_t = dW_t$ (one can present a similar argument in the Black–Scholes setting as well) and that $r = 0$, $T = 1$. Consider a self-financing strategy with $V_0 = 0$ and $\Delta_t := (1 - t)^{-1/2}$, $t \in [0, 1)$ (that is, we borrow the amount S_0 to buy one share at time $t = 0$). We do not need to specify b_t here since, by the product rule of Itô's calculus (Theorem 11.10),

$$dV_t = d(\Delta_t S_t + b_t B_t) = S_t d\Delta_t + \Delta_t dS_t + db_t,$$

and so it follows from (13.37) (that takes here the form $dV_t = \Delta_t dS_t$ as $dB_t = 0$ owing to $r = 0$) that $db_t = -S_t d\Delta_t$.

As Δ_t is non-random and $\int_0^t \Delta_s^2 ds < \infty$ for $t \in [0, 1)$, we know from Remark 11.1 that $V_t = \int_0^t \Delta_s dS_s = \int_0^t \Delta_s dW_s$, $t \in [0, 1)$, is just a time-changed BM process. It is easy to see that it has the same distribution as $\{W_{a(t)}\}_{t\in[0,1)}$, where $a(t) := -\ln(1 - t)$ increases from 0 to ∞ as t runs from 0 to 1. Hence $\{W_{a(t)}\}_{t\in[0,1)}$ will a.s. hit level one at some time $\tau < 1$ (and we can actually derive the density of the first hitting time τ from Theorem 11.5(ii)). Now modify our trading strategy by cashing the stock component of the portfolio at that ST τ. The resulting portfolio has initial value $V_0 = 0$ and the terminal value $V_1 = V_\tau = 1$, which means arbitrage!

[22] We did not formally introduce the concept of a predictable continuous time process. This is because such concepts are more subtle in the continuous time case, which places them well outside the scope of the present book. The reader may still wish to figure out what they could mean (or, even better, to find them in the recommended literature).

For a self-financing strategy that replicates a given claim $X = g(S_T)$ (so that $V_T(\omega) = X(\omega)$ a.s.), the respective time t portfolio value should be equal to the that time claim value: $\mathcal{P}_t(X) = V_t$, $t \le T$, for otherwise there will be an arbitrage opportunity (cf. p. 395). However, as we know from the general theory (Section 13.6), one also has

$$\begin{aligned}\mathcal{P}_t(X) &= \mathbf{E}^*\big(e^{-r(T-t)}X\big|\,\mathcal{F}_t\big) = e^{-r(T-t)}\mathbf{E}^*\big(g(S_T)\big|\,\mathcal{F}_t\big)\\ &= e^{-r(T-t)}\mathbf{E}^*\big(g(S_T)\big|\,S_t\big) =: f(t, S_t),\end{aligned}$$

where the third equality holds as the price process is Markovian, and the last relation follows from the definition of CE. Now assuming that the function $f(t, s)$ is smooth enough, one can apply Itô's formula (11.41), representation (13.35) and (11.39) to get

$$\begin{aligned}dV_t &= d\mathcal{P}_t(X) = df(t, S_t)\\ &= \partial_t f(t, S_t)dt + \partial_s f(t, S_t)dS_t + \frac{1}{2}\,\partial_{ss} f(t, S_t)(dS_t)^2\\ &= \left(\partial_t f(t, S_t) + \frac{1}{2}\,\partial_{ss} f(t, S_t)\sigma^2 S_t^2\right)dt + \partial_s f(t, S_t)dS_t.\end{aligned}$$

Equating this to (13.37), we see that the only terms in the equality that contain dW_t are the ones with dS_t, and hence their coefficients must coincide. Hence we showed that the replicating strategy must be of the form

$$\Delta_t = \partial_s f(t, S_t), \quad b_t = (f(t, S_t) - \Delta_t S_t)/B_t, \quad t \in [0, T]. \tag{13.38}$$

Thus the number of shares in the hedge is given by the partial derivative of the time t claim value (cf. the discrete version of that rule (13.9) for the binomial model).

This partial derivative is often referred to as just *the delta* and is one of the so-called *sensitivites* of the claim price with respect to the elements of the model, which are called the *Greeks*. Along with delta, they include the *gamma* $\Gamma := \partial_{ss} f(t, S_t)$ (measuring the sensitivity of the hedging portfolio for the claim to the change of the stock price), the *vega*[23] $\mathcal{V} := \partial_\sigma \mathcal{P}_t(X)$, the *theta* $\Theta := \partial_T \mathcal{P}_t(X)$ and the *rho* $\rho := \partial_r \mathcal{P}_t(X)$.

Example 13.8. *Sensitivities for a European call and put.* Assume w.l.o.g. that $t = 0$ and find the delta, gamma and vega for the European call $C = C(S_0, T, K, r, \sigma)$ in the

[23]This is not a Greek letter. The vega is an exception: for some unclear reasons, it was named after the brightest star in the constellation Lyra (this sensitivity is also denoted occasionally by the Greek letters epsilon, kappa, tau, lambda and zeta). The use of *vega* has probably persisted due to its starting with a v (like the *volatility* whose influence on the prices it refers to).

Black–Scholes market. Using (13.36) and observing that $\partial_{S_0} h = (\sigma\sqrt{T}\, S_0)^{-1}$, $\Phi'(h) = \phi(h) := e^{-h^2/2}/\sqrt{2\pi}$, we have

$$\begin{aligned}\Delta = \partial_{S_0} C &= \Phi(h) + S_0\Phi'(h)\partial_{S_0} h - Ke^{-rT}\Phi'(h-\sigma\sqrt{T})\partial_{S_0} h \\ &= \Phi(h) + \frac{1}{\sigma T\sqrt{2\pi}}\left[e^{-h^2/2} - \frac{K}{S_0 e^{rT}}\, e^{-(h-\sigma\sqrt{T})^2/2}\right] \\ &= \Phi(h) + \frac{e^{-h^2/2}}{\sigma T\sqrt{2\pi}}\left[1 - \frac{K}{S_0 e^{rT}}\, e^{h\sigma\sqrt{T}} e^{-\sigma^2 T/2}\right] = \Phi(h)\end{aligned}$$

since $e^{h\sigma\sqrt{T}} = \exp\{\ln(S_0/K) + rT + \frac{1}{2}\sigma^2 T\}$. Therefore, $\Gamma = \partial_{S_0 S_0} C = \partial_{S_0}\Phi(h) = \Phi'(h)\partial_{S_0} h = \phi(h)(\sigma\sqrt{T}\, S_0)^{-1}$. Further calculations show that $\mathcal{V} = \phi(h) S_0\sqrt{T}$.

To compute the Greeks for the put, one can use the put-call parity (13.16), immediately yielding $\partial_{S_0} P = \partial_{S_0} C - \partial_{S_0} S_0 = \Phi(h) - 1$ *etc.* Note that since $\Phi(h) \in (0,1)$, the delta for the call is always positive, while the delta for the put is always negative, which makes sense: the issuer of the call should be able to *deliver* the stock in case the owner decides to exercise the option, while issuer of the put should be able to "absorb" purchasing the stock.

Note that, like the price process $\{S_t\}$, both components of the replicating portfolio (13.38) are diffusion SPs, meaning that the portfolio has to be continuously managed and reshuffled. In real life, that would not be feasible, of course, because of transaction costs, and so more advanced modelling is needed to reflect that factor.

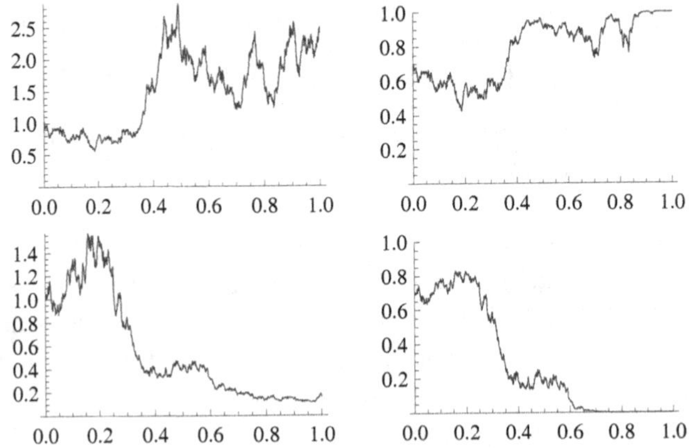

Fig. 13.8 Simulated trajectories of the stock prices ($\mu = 0$, $S_0 = \sigma = 1$; left panes) and the deltas for the respective replicating portfolios for the European call with $K = 1$ and $T = 1$ (right panes).

Figure 13.8 shows two realisations of the simulated stock price following (13.32) with $\mu = 0$, $S_0 = \sigma = 1$, on $[0, T] = [0, 1]$ (left plots). The right plots depict the respective trajectories of $\{\Delta_t\}$, the numbers of shares in the replicating portfolio for the call with strike $K = 1$ and maturity $T = 1$. Note that, in the upper row, the option is "in the money": the time T stock price is above the strike, so that the option will be exercised. Accordingly, the respective Δ_t-values are pretty close to unity where the stock price is high, and the closer to the maturity time, the closer the value of Δ_t to 1: the issuer will need that share at time T! In the lower row, the option is "out of the money" at time T and, accordingly, the value of Δ_t vanishes for times close to expiry.

13.8 Pricing Barrier Options

The so-called *barrier options* are rather popular versions of path-depended derivatives in which the payoff of a usual call (or put) is modified so that it is nullified if the stock price hits (for "kick-out" options) or does not hit (for "kick-in" options) a pre-specified "barrier". In the one-sided case, the barrier is given by a function H_t, $t \in [0, T]$ and so, in the case of an upper barrier, the payoffs of the kick-out and kick-in modifications of the vanilla call $X = (S_T - K)^+$ with maturity T and strike K have the form:

$$X_{\text{out}} := (S_T - K)^+ \mathbf{1}_{\{\sup_{t\in[0,T]}(S_t - H_t)\le 0\}},$$
$$X_{\text{in}} := (S_T - K)^+ \mathbf{1}_{\{\sup_{t\in[0,T]}(S_t - H_t)> 0\}}.$$

Clearly, always $X = X_{\text{out}} + X_{\text{in}}$, so that pricing one of the two barrier options will immediately yield the price of the other. There exist lower barrier and two-sided versions of barrier options. It is obvious that the barrier options are cheaper than their vanilla versions, and so are more attractive in case the buyer is interested in hedging against specific risks the option is paying attention to.

Barriers H_t are usually flat and partial, i.e., cover only part of the option's lifetime: $H_t = H\mathbf{1}_{t\in[t_1,t_2]}$ for some fixed $H > 0$, $0 \le t_1 < t_2 \le T$.

How to price barrier options in the general case? In fact, we already have all the tools for that prepared in Chapters 11 and 12. Consider, for example, the case of a kick-out call option with a barrier given by a closed set $B \subset (0, T] \times [0, \infty)$ of which the complement D in $[0, T] \times [0, \infty)$ is an open connected set containing the point $(0, S_0)$, and of which the boundary ∂B includes a part of the "expiry line" $\{(t, x) : t = T,\ x \in \mathbf{R}\}$. Then the barrier call payoff can be specified by setting the "boundary value function"

ψ on ∂B as follows:

$$\psi(t,x) = \begin{cases} 0, & t < T, \\ (x-K)^+, & t = T, \end{cases} \qquad (t,x) \in \partial B,$$

introducing $\tau := \inf\{t > 0 : S_t \in \partial B\}$ and letting our claim to be

$$X := \psi(\tau, S_\tau).$$

Thus, should the price process "touch" the barrier prior to maturity, the payoff will be zero.

In the case of a flat partial one-sided upper barrier call with barrier height H and "barrier window" $[t_1, t_2] \subset [0, T]$ (so that there is no barrier outside that time interval), one has (see also Fig. 13.9):

$$B = \{(t,x) : t \in [t_1, t_2],\ x \geq H\} \cup \{(t,x) : t = T\}. \tag{13.39}$$

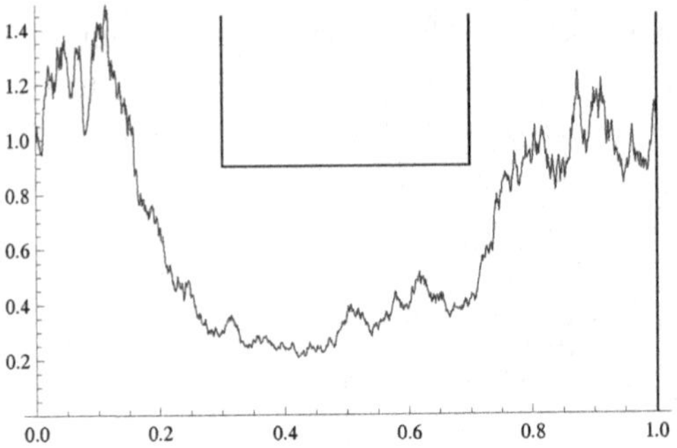

Fig. 13.9 The boundary set ∂B in the case of a partial upper barrier option with $t_1 = 0.3$, $t_2 = 0.7$ and $H = 0.9$ is shown (see (13.39)), together with a simulated price trajectory (with the same parameters as in Fig. 13.8) that does not hit the barrier prior to expiry (no "kicking-out").

As we know from the general theory, the time $s < T$ arbitrage free price of the option will be given by

$$\begin{aligned}\mathcal{P}_s(X) &= \mathbf{E}^*\big(e^{-r(T-s)}\psi(\tau, S_\tau)\big|\, \mathcal{F}_s\big) = \mathbf{E}^*\big(e^{-r(T-s)}\psi(\tau, S_\tau)\big|\, S_s\big) \\ &= C(s, S_s), \quad \text{where} \quad C(s,x) := \mathbf{E}^*_{s,x} e^{-r(T-s)}\psi(\tau, S_\tau),\end{aligned}$$

where the second equality is due to the fact that, under $\mathbf{P}^*$, the price process is a Markov process (namely, the diffusion driven by (13.35)). We have already encountered expectations of that kind in Chapter 12 and know that they satisfy the Feynman–Kac formula (see Remark 12.6 on p. 366).

According to (13.35), we have $\mu(s,x) = rx$, $\sigma(s,x) = \sigma x$, and letting $f(s,x) \equiv r$, $g(s,x) \equiv 0$, the formula yields that $C(s,x)$ satisfies the PDE

$$\partial_s C + rx\partial_x C + \frac{1}{2}\sigma^2 x^2 \partial_{xx} C = rC, \quad (s,x) \in D, \tag{13.40}$$

with the boundary conditions $\psi(s,x)$ on ∂B. In mathematical finance, the equation is known as the *Black–Scholes PDE.*

Thus one can price barrier options numerically by solving (13.40), which may not always be convenient, especially when the pricing has to be done for different time epochs, in a changing interest rates environment. A popular alternative approach consists in using the Monte Carlo method (often in its crude, "brute force" version), simulating a large number of trajectories of the price process following (13.35) to evaluate the expectation (under $\mathbf{E}^*$) giving the desired price (see Section 10.4). That may be quite time consuming, as the whole trajectory of the price process has to be simulated (to see if it ever hit the barrier prior to expiry). There is, however, a smarter way combining (if necessary) the Monte Carlo method with analytical results for diffusion SPs.

Consider the barrier kick-out call with the upper barrier (13.39) and start with re-writing the barrier non-hitting condition in terms of the underlying BM $\{W_t\}$ related to the price process via the representation[24]

$$S_t = S_0 \exp\{(r - \sigma^2/2)t + \sigma W_t\}, \quad t \in [0,T].$$

Clearly,

$$\{S_t < H,\ t \in [t_1,t_2]\} = \{W_t < \sigma^{-1}(\ln(H/S_0) - (r - \sigma^2/2)t),\ t \in [t_1,t_2]\},$$

which is the event that the trajectory of the BM stays under the linear boundary

$$l(t) := x_1 \frac{t_2 - t}{t_2 - t_1} + x_2 \frac{t - t_1}{t_2 - t_1}, \quad t \in [t_1,t_2],$$

which is the straight line segment connecting the points (t_1,x_1) and (t_2,x_2), where $x_j = \sigma^{-1}(\ln(H/S_0) - (r - \sigma^2/2)t_j)$, $j = 1,2$. So the expectation giving the time $t = 0$ price of the call (we omit the discounting factor for brevity) can be re-written, conditioning on the values of W_{t_1} and W_{t_2}, as

$$\begin{aligned}\mathbf{E}^*_{0,S_0} X &= \mathbf{E}^*(S_T - K)^+ \mathbf{1}_{\{W_t < l(t),\ t \in [t_1,t_2]\}} \\ &= \int_{-\infty}^{x_2}\int_{-\infty}^{x_1} \mathbf{E}^*\big[\cdots \big| W_{t_1} = z_1, W_{t_2} = z_2\big] f_{t_1,t_2}(z_1,z_2)\,dz_1 dz_2, \end{aligned} \tag{13.41}$$

[24] We had $\widetilde{W}_t$ in (13.35), to stress that that was a BM under $\mathbf{E}^*$, which was different from the original BM from (13.32). Here we will omit the tilde for brevity.

where $f_{t_1,t_2}(z_1, z_2)$ is the density of (W_{t_1}, W_{t_2}) (see (11.22)). Now note that, relative to the "present" time t_2, the first factor in the product

$$(S_T - K)^+ \mathbf{1}_{\{W_t < l(t),\ t\in[t_1,t_2]\}}$$

refers to the "future", while the second one is "from the past", so that by the Markov property they will be conditionally independent given the present (cf. Section 6.1). Hence the conditional expectation in the integrand is equal to the product

$$\begin{aligned}
&\mathbf{E}^*\big[(S_T - K)^+ \big| W_{t_1} = z_1, W_{t_2} = z_2\big] \\
&\qquad \times \mathbf{E}^*\big[\mathbf{1}_{\{W_t < l(t),\ t\in[t_1,t_2]\}} \big| W_{t_1} = z_1, W_{t_2} = z_2\big] \\
&= \mathbf{E}^*\big[(S_T - K)^+ \big| W_{t_2} = z_2\big] \mathbf{E}^*\big[\mathbf{1}_{\{\cdots\}} \big| W_{t_1} = z_1, W_{t_2} = z_2\big],
\end{aligned}$$

where we again used the Markov property to remove the condition $W_{t_1} = z_1$ from the first factor. Observe that that factor is now nothing else but the time t_2 price (modulo the discounting factor $e^{-r(T-t_2)}$) of the European call with maturity T and strike K given that the current stock price is $S_0 \exp\{(r - \sigma^2/2)t_2 + \sigma z_2\}$, and so admits a closed-form expression given by the Black–Scholes formula (13.29). The second factor, being the expectation of an indicator function, is just the probability of the respective event:

$$\mathbf{P}^*\big(W_t < l(t),\ t \in [t_1, t_2] \big| W_{t_1} = z_1, W_{t_2} = z_2\big). \tag{13.42}$$

This is clearly the probability that the Brownian bridge with "end points" fixed at (t_1, z_1) and (t_2, z_2) stays below the linear boundary $l(t)$ on the time interval $[t_1, t_2]$. Using the representation for that process in terms of the standard Brownian bridge $\{W_t^0\}$ given in Remark 12.7:

$$\frac{t_2 - t}{t_2 - t_1} z_1 + \sqrt{t_2 - t_1}\, W^0_{(t-t_1)/(t_2-t_1)} + \frac{t - t_1}{t_2 - t_1} z_2, \quad t \in [t_1, t_2],$$

and the time change $u := \frac{t-t_1}{t_2-t_1}$, we see that the probability (13.42) equals

$$\begin{aligned}
&\mathbf{P}\Bigg(W_u^0 < \underbrace{\frac{x_1 - z_1}{\sqrt{t_2 - t_1}}}_{=:b_1}(1 - u) + \underbrace{\frac{x_2 - z_2}{\sqrt{t_2 - t_1}}}_{=:b_2} u,\ u \in [0, 1]\Bigg) \\
&\quad = 1 - \mathbf{P}\Big(\sup_{u\in[0,1]} [W_u^0 - (b_1 + (b_2 - b_1)u)] \geq 0\Big) \\
&\quad = 1 - e^{-2b_1 b_2} = 1 - \exp\Big\{-\frac{2(x_1 - z_1)(x_2 - z_2)}{t_2 - t_1}\Big\}
\end{aligned}$$

by (12.39). Thus we obtained a closed-form expression for the integrand on the right-hand side of (13.41), and it remains to compute that double integral now.

It is not hard to see that, arguing similarly to our derivation of the Black-Scholes formula in Example 13.7, one can obtain an expression for the barrier call price in terms of the values of trivariate normal DFs. These, however, are actually not very easy to evaluate, so it would perhaps be better to use the Monte Carlo method to estimate the integral (13.41). That, however, will only require simulating realisations of the bivariate normal vector (W_{t_1}, W_{t_2}) rather than those of the whole trajectory of the BM process, which is a huge economy in computer time!

13.9 Recommended Literature

BINGHAM, N.H. AND KIESEL, R. *Risk-Neutral Valuation: Pricing and Hedging of Financial Derivatives.* 2nd edn. Springer, New York, 2004. [*A mathematically rigorous book, written clearly and with precision, but with a practical, reader-oriented focus. The reader will need an undergraduate level background in probability theory.*]

HULL, J.C. *Options, Futures, and Other Derivatives.* 6th edn. Pearson Education, Saddle River, NJ, 2006. [*A well-written (voluminous) introductory text on the futures and options markets, for those with a limited background in mathematics.*]

SHIRYAEV, A.N. *Essentials of Stochastic Finance: Facts, Models, Theory.* World Scientific, Singapore, 1999. [*An encyclopedic book on results and methods for financial analysis, presenting not only the essentials of probability as it is applied to finance, but also covering empirical facts and contemporary developments.*]

SHREVE, S. *Stochastic Calculus for Finance I: The Binomial Asset Pricing Model.* Springer, New York, 2004. [From the book cover: *The first volume presents the binomial asset-pricing model primarily as a vehicle for introducing in the simple setting the concepts needed for the continuous-time theory in the second volume.*]

SHREVE, S. *Stochastic Calculus for Finance II: Continuous-Time Models.* Springer, New York, 2008. [From the book cover: *This second volume develops stochastic calculus, martingales, risk-neutral pricing, exotic options and term structure models, all in continuous time.*]

13.10 Problems

1. For a single-period binomial market, sketch the graph of the call price $C = C(K)$ as a function of strike K. What features of this function do you think will be common to the call prices in more general markets as well?

2. A *straddle* is a portfolio consisting of a call and a put on the same underlying stock, with the same strike and same expiry. Plot the graph of the payoff function (of the terminal stock price) of the straddle (on one share of the stock).

3. A *bull spread* is a portfolio formed by buying a call with strike price K_1 and selling a call with strike price $K_2 > K_1$. Both calls are on the same underlying stock and have the same expiry date T. A bull spread is obviously a contingent claim.

   (i) Plot the payoff function of a bull spread.

   (ii) What is cheaper: a bull spread or a call with the strike K_1 (on the same underlying stock, with the same expiry date)? Explain.

   (iii) Consider a single-period binomial market. Assume that the current stock price is $S_0 = 5$, the possible values of S_1 are 4 and 6, $r = 0.1$, $K_1 = 3$ and $K_2 = 5$. Price the bull spread using the risk-neutral valuation ("fair price") formula (13.11).

   (iv) Under the assumptions from part (iii), construct a replicating portfolio for the bull spread and verify that the portfolio does replicate the claim.

4. Consider a single-period binomial financial market with $\Omega = \{\omega_1, \omega_2\}$, with the current (time $t = 0$) asset prices $S_0 = 5$ and $B_0 = 1$, and the terminal (time $t = 1$) stock prices $S_1(\omega_1) = 20/3$ and $S_1(\omega_2) = 40/9$. Assume the interest rate $r = 1/9$.

   (i) Show that the market is arbitrage-free.

   (ii) Consider a contingent claim X with $X(\omega_1) = 7$ and $X(\omega_2) = 2$. Find the arbitrage-free value of this claim at time $t = 0$.

   (iii) Construct a replicating portfolio for the claim X and verify that its values at times $t = 0$ and $t = 1$ coincide with the claim price you found in part (ii) and the claim value X, respectively.

5. For a single-period binomial market:

   (i) sketch the graph of the put price $P = P(K)$ as a function of strike K;

   (ii) price a put with strike $K = 3.8$ on one share of the stock with current price $S_0 = 4$, if $r = 5\%$ and the possible values of the time $t = 1$ stock price are $S_1 = 3.6$ and $S_1 = 4.6$;

   (iii) find the time $t = 0$ value of the call with the same strike.

6. Consider a single period trinomial market: $\Omega = \{d, m, u\}$ (i.e., at time $t = 1$ there are three possible states of the world now), with $S_1(d) = dS_0$, $S_1(m) = mS_0$ and $S_1(u) = uS_0$, where $0 < d < m < u$. The time $t = 0, 1$ bond prices are $B_0 = 1$ and $B_1 = 1 + r$, respectively, where $d < 1 + r < u$.

   (i) Use Theorem 13.2 to verify (or disprove) that the market is arbitrage free.

   (ii) Let X be a contingent claim, with some values $X_d = X(d)$, $X_m = X(m)$ and $X_u = X(u)$ for the possible states of the world in our model. Assuming that $X_d < X_m < X_u$, draw a diagram showing the set of all hedges (Δ, b) for the claim (similar to Fig. 13.3 for the binomial market). Does there exist a perfect hedge (i.e., a replicating portfolio) for X?

7. Consider a single period trinomial market with two stocks (stock 1 and stock 2) and one bond with $r = 0$. Assume that $\Omega = \{\omega_1, \omega_2, \omega_3\}$ (i.e., at time $t = 1$ there are three possible states of the world), with the initial (time $t = 0$) stock prices $S_0^1 = 4$ and $S_0^2 = 5$ for stock 1 and stock 2, respectively, and the time $t = 1$ stock prices

$$\begin{array}{ll} S_1^1(\omega_1) = 6, & S_1^2(\omega_1) = 6; \\ S_1^1(\omega_2) = 4, & S_1^2(\omega_2) = 4; \\ S_1^1(\omega_3) = 2, & S_1^2(\omega_3) = 7. \end{array}$$

   (i) Depict the time $t = 0$ and $t = 1$ stock prices by points on the (S_1^1, S_2^1)-plane.

   (ii) Use Theorem 13.2 to prove or disprove that the market is arbitrage free.

   (iii) Find the EMM $\mathbf{P}^*$ and use the Completeness Theorem 13.4 to prove or disprove that the market is complete.

   (iv) Price the call on (one share of) stock 1 with expiry $T = 1$ and strike $K = 5$.

   (v) Find the strategy replicating the call from part (iv). Do we really need stock 2 to replicate the call on stock 1?

8. Consider a $T = 2$ period binomial market with $u = 1.75$, $d = 0.5$ and $r = 0.25$. Assuming the time $t = 0$ stock price $S_0 = 400$,

   (i) use the diagram method (see Section 13.5) to price a put option with maturity $T = 2$ and strike $K = 450$;

   (ii) construct a replicating portfolio for this option;

   (iii) directly verify that the replicating strategy you found in part (ii) is self-financing;

   (iv) price a call with the same maturity and strike price.

9. Consider a four-period binomial financial market with $u = 4/3$, $d = 2/3$ and $r = 0$ (i.e., the prices have already been discounted). Let the current (time $t = 0$) stock price be $S_0 = 81$. Use the diagram method to price a call with maturity $T = 4$ and strike $K = 120$.

10. On 28 July 2000 ($t = 0$), the BHP[25] stock price at the ASX[26] was $S_0 = 18.50$.

    (i) Assuming the (continuously compounded) interest rate $r = 0.062$ and volatility $\sigma = 0.2$, use the Black-Scholes formula to find the price of the call option with maturity one month later ($T = 1/12$ year) and strike $K = 18.50$.

    (ii) The real-life buy/sell prices of the option were 0.57/0.63, respectively. Taking the mean of these two values as the market price of the option, find the volatility value such that the Black-Scholes formula gives the closest value to the market price. This value is called the *implied volatility*.

11. Suppose that the current stock price is $S_0 = 100$. Assume volatility 40% and (continuously compounded) interest rate $r = 4\%$ (the time unit is one year). Use the Black-Scholes formula and the put-call parity to find the prices of the call and put options with strike $K = 110$ and time to maturity (i) 18 months; (ii) one month. Comment on your findings (in particular, on the prices you found in (ii)).

[25]BHP Billiton is a multinational resources company.
[26]Australian Securities Exchange.

Answers to Problems

Chapter 2

1. (i) 0.35. (ii) 0.15. (iii) 0.5.

2. (i), (ii) $1/10! \approx 2.756 \times 10^{-7}$. (iii) 0.1. (iv) $1/90 \approx 0.011$. (v) 0.2.

3. 1/2 (the problem tacitly assumes that children's genders are independent and that the probability of having a boy is 0.5—which is not quite true).

4. 1/3.

5. 0.4909; 0.3818; 0.1273.

6. 0.4615; 0.3846; 0.1538.

7. (i) 0.99. (ii) 0.09.

8. (i) 0.0498. (ii) 0.0446; 0.9826. (iii) 0.2798.

9. 0.2257; 0.0141.

10. (i) $F(h(x))$. (ii) $1-F(h(x)-0)$ ($F(y-0) := \lim_{x\nearrow y} F(x)$; this is $= F(y)$ if F is continuous at y). (iii) For part (i): $F_Y(x) = 0$ if $x < 0$; $= \sqrt{x}/2$ if $x \in (0,4)$; $= 1$ if $x \geq 4$. For part (ii): $F_Y(x) = 0$ if $x < 1/2$; $= 1 - 1/2x$ if $x \geq 1/2$.

11. (i) $\Phi(\min\{x_1, x_2\})$. (ii) $\Phi(x_1)\Phi(x_2)$.

12. $F(x) = 0$ if $x < 0$; $= 1/5 + 2\pi x^3/1875$ if $0 \leq x < 5$; $= 1$ if $x \geq 5$; the expected level is $4 - \pi/6 \approx 3.4764$.

13. (i) $p_k = (b/k)p_{k-1} = (b^2/k(k-1))p_{k-2} = \cdots = (b^k/k!)p_0$. (ii) The factor $(a + b/n) < 0$ for all large enough n, which leads to a contradiction if $b \neq -a(n+1)$ (as all $p_n \geq 0$). So must have $b = -a(n+1)$ for some n. Then $p_k = |a|((n-k+1)/k)p_{k-1} = \cdots = \binom{n}{k}|a|^k p_0$. (iii) $a+b > 0$ since $p_1 = (a+b)p_0$. Setting $n := 1 + b/a$ we get $p_k = a((n+k-1)/k)p_{k-1} = \cdots = a^k(\Gamma(n+k)/\Gamma(n)k!)p_0$.

14. (i) T is the sum of six independent geometrically distributed RVs T_j with parameters $q_j = (j-1)/6$, $j = 1, \ldots, 6$ (i.e., $\mathbf{P}(T_j = k) = (1-q_j)q_j^{k-1}$, $k \geq 1$); its GF is $g_T(z) = 120z^6/\prod_{j=1}^{5}(6 - jz)$. (ii) $\mathbf{E}\,T = 14.7$; $\mathrm{Var}\,(T) = 38.99$.

15. (i) Alternatively to the hint, note that $|\varphi_X(i(t+h)) - \varphi_X(it)| = |\mathbf{E}\, e^{itX}(e^{ihX} - 1)| \leq \mathbf{E}\,|e^{ihX} - 1| \to 0$ by the dominated convergence theorem as $|e^{ihX}-1| \le 2$ and $|e^{ihX}-1| \to 0$ as $h \to 0$. (ii) $0 \le \mathbf{E}\left|\sum_j a_j e^{it_jX}\right|^2 = \mathbf{E}\left(\sum_j \overline{\sum_k}\right) = \mathbf{E}\left(\sum_{j,k} a_j\overline{a_k}e^{it_jX}e^{-it_kX}\right)$. (iii) Real: $\overline{\varphi_X(it)} = \mathbf{E}\,\overline{e^{itX}} = \mathbf{E}\,e^{-itX} = \varphi_{-X}(it) = \varphi_X(it)$ as $-X$ has the same distribution as X; even: $\varphi_X(-it) = \varphi_{-X}(it) = \varphi_X(it)$.

16. $[a,b] = \bigcap_{n\ge1}(a-1/n, b+1/n)$.

17. $\cap$: If $A, A_1, A_2, \dots \in \mathcal{F}_i$ for all i, then $A^c, \bigcup_n A_n \in \mathcal{F}_i$ for all i, and hence they both $\in \bigcap_i \mathcal{F}_i$. $\cup$: A counterexample: let $\Omega = \{1,2,3\}$ with $\mathcal{F}_1 = \{\emptyset, \{1\}, \{2,3\}, \Omega\}$, $\mathcal{F}_2 = \{\emptyset, \{2\}, \{1,3\}, \Omega\}$. For $A_1 = \{1\} \in \mathcal{F}_1$, $A_2 = \{2\} \in \mathcal{F}_2$, the union $A_1 \cup A_2 \notin \mathcal{F}_1 \cup \mathcal{F}_2$.

18. Any open ball can be represented as a countably infinite union of small cubes, which are elements of the product σ-field. And vice versa, any open parallelepiped (generating the product σ-field) is a union of countably many open balls.

19. Can be directly verified: $(a,b]^c = (-\infty, a] \cup (b, +\infty]$ *etc.*

20. Follows from Problem 18.

21. Can take $B_j = \mathbf{R}$ for some indices j. This is equivalent to selecting a subcollection of indices in (2.34) as $\{X_j \in \mathbf{R}\} = \Omega$.

22. More convenient to deal with unions and increasing sequences $B_n \subset B_{n+1}$. Take $B_n = \bigcup_{j=1}^n A_j$ for disjoint A_j's. Then $\bigcup_{n=1}^\infty A_n = \bigcup_{n=1}^\infty B_n$, and finite additivity implies that $\mathbf{P}(B_n) = \sum_{j=1}^n \mathbf{P}(A_j)$.

23. For independent simple RVs $X = \sum_j x_j \mathbf{1}_{A_j}$ and $Y = \sum_k y_k \mathbf{1}_{B_k}$ we have $\mathbf{P}(X = x_j, Y = y_k) = \mathbf{P}(X = x_j)\mathbf{P}(Y = y_k)$, i.e., A_j and B_k are independent for any pair j,k. Hence $\mathbf{E}\,XY = \sum_{j,k} x_j y_k \mathbf{E}\,\mathbf{1}_{A_j}\mathbf{1}_{B_k} = \sum_{j,k} x_j y_k \mathbf{P}(A_jB_k) = \sum_{j,k} x_j y_k \mathbf{P}(A_j)\mathbf{P}(B_k) = (\sum_j) \times (\sum_k) = \mathbf{E}\,X \times \mathbf{E}\,Y$.

24. For $A := \{X > Z\}$, the RV $Y := (X - Z)\mathbf{1}_A \ge 0$, but $\mathbf{E}\,Y = \mathbf{E}(X;\,A) - \mathbf{E}(Z;\,A) = 0$. So must have $\mathbf{P}(A) = \mathbf{P}(Y > 0) = 0$. Similarly $\mathbf{P}(X < Z) = 0$.

25. Find the minimum of the function $f(a) := \mathbf{E}(X-a)^2 = \mathbf{E}\,X^2 - 2a\mathbf{E}\,X + a^2$, $a \in \mathbf{R}$. Or use the representation $\mathbf{E}(X-a)^2 = \mathbf{E}[(X - \mathbf{E}\,X) + (\mathbf{E}\,X - a)]^2 = \mathrm{Var}(X) + (\mathbf{E}\,X - a)^2$, where the first term does not depend on a, whereas the second one is non-negative and $= 0$ when $a = \mathbf{E}\,X$.

26. Using the hint, if $F(x)$ is continuous at $x = a$, then $\frac{d}{da}\mathbf{E}|X - a| = F(a) - (1 - F(a)) = 2F(a) - 1 \le 0$ if $a \le$ the median of f and $a \ge 0$ if $a \ge$ the median. The minimum is attained at the median of F.

27. Similarly to the second approach in solution to Problem 25, consider $\mathbf{E}(Y - g(X))^2 = \mathbf{E}[(Y - \mathbf{E}(Y|X)) + (\mathbf{E}(Y|X) - g(X))]^2 = \mathbf{E}(Y - \mathbf{E}(Y|X))^2 + \mathbf{E}(\mathbf{E}(Y|X) - g(X))^2$ (we used properties (iv) and (v) of conditional expectations to see that $\mathbf{E}[(Y - \mathbf{E}(Y|X))(\mathbf{E}(Y|X) - g(X))] = \mathbf{E}\,\mathbf{E}[\cdots|X] = \mathbf{E}\{(\mathbf{E}(Y|X) - g(X))\mathbf{E}[Y - \mathbf{E}(Y|X)|\,X]\} = 0$ as $\mathbf{E}[Y - \mathbf{E}(Y|X)|\,X] = 0$ by properties (i) and (v)). Again, the first term on the RHS does not depend on $g(\cdot)$, whereas the second one is non-negative and $= 0$ when $g(x) = \mathbf{E}(Y|\,X = x)$.

28. Fix an arbitrary small $\varepsilon > 0$. Events $B_n := \{\sup_{m>n}|X_n - X_0| > \varepsilon\}$ are decreasing, and from (2.38) and (2.9) one has $\mathbf{P}\left(\bigcap_{n\ge1} B_n\right) = 0$. So $\mathbf{P}(A) = 1$

for the complementary event $A := \left(\bigcap_{n\geq 1} B_n\right)^c = \bigcup_{n\geq 1}\bigcap_{m>n}\{|X_m - X_0| \leq \varepsilon\}$ (using the hint). In words, the event A means that "there exists an n such that for any $m > n$, $|X_m - X_0| \leq \varepsilon$"—convergence to X_0!

29. Let (2.40) hold for any bounded continuous g. Fix an $\varepsilon > 0$. Let $t \in \mathbf{R}$ be a continuity point of F_{X_0}. Take $g_t(x) := \mathbf{1}_{\{x\leq t\}} + (1-(x-t)/\varepsilon)\mathbf{1}_{\{t<x\leq t+\varepsilon\}}$ (plot this function!). Then $F_{X_n}(t-\varepsilon) \leq \mathbf{E}\, g_{t-\varepsilon}(X_n) \leq F_{X_n}(t) \leq \mathbf{E}\, g_t(X_n) \leq F_{X_n}(t+\varepsilon)$, $n \geq 0$. Therefore for all sufficiently large n, $F_{X_0}(t-\varepsilon) - \varepsilon \leq \mathbf{E}\, g_t(X_0) - \varepsilon \leq F_{X_n}(t) \leq \mathbf{E}\, g_t(X_0) + \varepsilon \leq F_{X_0}(t+\varepsilon)+\varepsilon$. It remains to take ε small enough: by continuity at t, both $F_{X_0}(t\pm\varepsilon)$ will be arbitrary close to $F_{X_0}(t)$.

Suppose $F_{X_n}(t) \to F_{X_0}(t)$ as $n \to \infty$ at any continuity point t of F_{X_0}. Let $g(x)$ be a bounded ($|g(x)| \leq C < \infty$) continuous function. Fix an $\varepsilon > 0$. Choose an $N < \infty$ such that $\pm N$ are continuity points of F_{X_0} and $\mathbf{P}(|X_0| > N) < \varepsilon$. Then $\mathbf{E}\, g(X_n) = \mathbf{E}\,(g(X_n);\, |X_n| \leq N) + \mathbf{E}\,(g(X_n);\, |X_n| > N)$, where the last term's absolute value $\leq C\mathbf{P}(|X_n| > N) \leq C(F_{X_n}(-N)+1-F_{X_n}(N)) \to C(F_{X_0}(-N)+1-F_{X_0}(N)) \leq C\varepsilon$ by assumption. As any function continuous on a closed interval is *uniformly continuous* on the interval, we can choose $-N = x_1 < \cdots < x_m = N$ such that x_j are continuity points of F_{X_0} (not a problem as F_{X_0} can have at most countably many jumps as the sum of the jumps ≤ 1) and $|g(x) - g(x_j)| < \varepsilon$, $x \in [x_j, x_{j+1}]$, $j = 1, \ldots, m-1$. It remains to use $|\mathbf{E}\,(g(X_n);\, |X_n| \leq N) - \sum_j g(x_j)(F_{X_n}(x_{j+1}) - F_{X_n}(x_j))| \leq \max_j \max_{x_j\leq x\leq x_{j+1}} |g(x) - g(x_j)| \leq \varepsilon$ and the fact of convergence of $F_{X_n}(x_j) \to F_{X_0}(x_j)$.

30. Suffices to show that, for any bounded continuous $h(x)$, $\mathbf{E}\, h(G(X_n)) \to \mathbf{E}\, h(G(X_0))$ as $n \to \infty$. As $g(x) := h(G(x))$ is a bounded continuous function, this follows from (2.40).

31. Just use $\frac{d^k}{dz^k} z^j = 0$ if $j < k$, $= k!$ if $j = k$, and $j! z^{j-k}/(j-k)!$ if $j > k$.

32. For $0 < s < t$, $\mathbf{E}\, W_s W_t = \mathbf{E}\, W_s^2 + \mathbf{E}\, W_s(W_t - W_s) = \mathbf{E}\, W_s^2 = s$ as by independence of increments $\mathbf{E}\, W_s(W_t - W_s) = \mathbf{E}\, W_s \mathbf{E}\,(W_t - W_s) = 0$.

Similarly, $\mathbf{E}\, N_s N_t = \mathbf{E}\, N_s^2 + \mathbf{E}\, N_s(N_t - N_s) = \mathbf{E}\, N_s^2 + \mathbf{E}\, N_s \mathbf{E}\,(N_t - N_s) = (\lambda s + \lambda^2 s^2) + \lambda s \times \lambda(t-s) = \lambda s + \lambda^2 st$ (as for an RV $Y \sim Po(\mu)$, $\mathbf{E}\, Y = \mathrm{Var}\,(Y) = \mu$). So $\mathrm{Cov}\,(N_s, N_t) = \lambda s + \lambda^2 st - \lambda s \times \lambda t = \lambda s$ (i.e., same as for $\{\lambda^{1/2} W_t\}$!).

33. Firstly, $(X_{t_1}, \ldots, X_{t_m}) = (t_1^{-1} W_{t_1}, \ldots, t_m^{-1} W_{t_m})$ is a normal random vector as a linear transformation of the normal random vector $(W_{t_1}, \ldots, W_{t_m})$. As we know (see (2.61)), the distribution of a normal random vector is determined by its mean vector and covariance matrix. So we just need to verify that (i) $\mathbf{E}\, X_t = \mathbf{E}\, W_t = 0$ (obvious) and (ii) $\mathbf{E}\, X_s X_t = \mathbf{E}\, W_s W_t \equiv s$, $s < t$ (from Problem 32). But $\mathbf{E}\, X_s X_t = st\mathbf{E}\, W_{1/s} W_{1/t} = st \times \min\{1/s, 1/t\} = s$.

34. For any fixed $n \geq 1$, Y has the same distribution as $\sum_{j=1}^n X_j$ with i.i.d. RVs $X_j \sim Po(\lambda/n)$ having $\mathbf{E}\, X_j = \sigma^2 := \mathrm{Var}\,(X_j) = \lambda/n$ and $\beta := \mathbf{E}\,|X_j - \mathbf{E}\, X_j|^3 = \mathbf{E}\,(\cdots;\, X_j = 0) + \mathbf{E}\,(\cdots;\, X_j > 0) \leq (\lambda/n)^3 + \mathbf{E}\, X_j^3 = (\lambda/n)^3 + ((\lambda/n) + 3(\lambda/n)^2 + (\lambda/n)^3) = (\lambda/n)(1 + o(1))$ [as $n \to \infty$]. Since $\mathbf{E}\, Y = \mathrm{Var}\,(Y) = \lambda$, $\Phi(\lambda + x\sqrt{\lambda})$ is a normal DF with the same first two moments as Y. By the Berry–Esseen theorem, the quantity in question will be bounded by $8\beta\sigma^{-3} n^{-1/2} \approx 8\lambda^{-1/2}$.

35. For simplicity's sake assume that F is continuous. Fix an arbitrary small

$\varepsilon > 0$ and choose an $N < \infty$ and $-\infty = t_0 < t_1 < \cdots < t_N < t_{N+1} = \infty$ such that $F(t_{j+1}) - F(t_j) < \varepsilon$, $j = 0, 1, \ldots, N$. As we said, by the LLN, for any j, $F_n^*(t_j) - F(t_j) \to 0$ a.s. as $n \to \infty$. So for all large enough n, $\max_j |F_n^*(t_j) - F(t_j)| < \varepsilon$. Now for any $t \in (t_j, t_{j+1})$, $-\varepsilon + F_n^*(t_j) - F(t_j) \le F_n^*(t_j) - F(t_{j+1}) \le F_n^*(t) - F(t) \le F_n^*(t_{j+1}) - F(t_j) \le F_n^*(t_{j+1}) - F(t_{j+1}) + \varepsilon$, so for all large enough n, $|F_n^*(t) - F(t)| < 2\varepsilon$. [In the general case, the proof is almost identical, one just has to consider separately finitely many points t where F has relatively large jumps.]

36. The density of W_{t_1} is equal to $f_{t_1}(x) = (2\pi t_1)^{-1/2} e^{-x^2/2t_1}$. As $W_{t_2} - W_{t_1} \sim N(0, t_2 - t_1)$ and is independent of W_{t_1}, the conditional density $f_{t_2|t_1}(y|x)$ of W_{t_2} given $W_{t_1} = x$ is equal to $(2\pi(t_2 - t_1))^{-1/2} e^{-(y-x)^2/2(t_2-t_1)}$. From (2.80) the unconditional density of (W_{t_2}, W_{t_1}) is then given by $f_{t_1,t_2}(x, y) = f_{t_1}(x) f_{t_2|t_1}(y|x) = (2\pi)^{-1}(t_1(t_2 - t_1))^{-1/2} \exp\{-x^2/2t_1 - (y - x)^2/2(t_2 - t_1)\}$.

In the general case, use the recurrence $f_{t_1,\ldots,t_{n-1},t_n}(x_1, \ldots, x_{n-1}, x_n) = f_{t_1,\ldots,t_{n-1}}(x_1, \ldots, x_{n-1}) \times f_{t_n|t_{n-1}}(x_n|x_{n-1})$ (a special case of (2.80)) to get

$$\prod_{j=1}^{n} \frac{1}{\sqrt{2\pi(t_j - t_{j-1})}} \exp\left\{-\frac{(x_j - x_{j-1})^2}{2(t_j - t_{j-1})}\right\}, \qquad t_0 = x_0 = 0.$$

37. $N\left(\frac{t_2-s}{t_2-t_1}x_1 + \frac{s-t_1}{t_2-t_1}x_2, \frac{(s-t_1)(t_2-s)}{t_2-t_1}\right)$ (from (2.80) the density of the desired conditional distribution is equal to $f_{t_1,s,t_2}(x_1, y, x_2)/f_{t_1,t_2}(x_1, x_2)$).

38. (i) $\mathbf{P}(nU_{1:n} > t) = \mathbf{P}\left(\bigcap_{j\le n}\{U_j > t/n\}\right) = (1 - t/n)^n \to e^{-t}$, so the limiting distribution is $Exp(1)$. (ii) As $N_t^{(n)} = \sum_{j=1}^n \mathbf{1}_{\{U_j \le t/n\}} \sim B_{n,t/n}$ (binomial with "success probability" $p = t/n$), we have $\mathbf{P}(nU_{2:n} > t) = \mathbf{P}(N_t^{(n)} \le 1) = (1-t/n)^n + n(t/n)(1-t/n)^{n-1} \to (1+t)e^{-t}$. The corresponding density is given by $-d(\cdots)/dt = te^{-t}$, $t > 0$ (gamma distribution $\Gamma(2, 1)$). (iii) As $n \to \infty$,

$$\mathbf{P}(nU_{k:n} > t) = \mathbf{P}(N_t^{(n)} \le k-1) = \sum_{j=0}^{k-1} \binom{n}{j} p^j (1-p)^{n-j} \to e^{-t} \sum_{j=0}^{k-1} \frac{t^j}{j!}, \quad (13.43)$$

so that the limiting distribution for $N_t^{(n)}$ is $Po(1)$. You can easily verify that the increments of the limiting process will be independent, so it is the Poisson process with rate 1.

Differentiating the right-hand side of (13.43), we see that the density of the limiting for $nU_{k:n}$ distribution is $-d(\cdots)/dt = \frac{t^{k-1}}{(k-1)!} e^{-t}$, $t > 0$ ($\gamma(k, 1)$, which agrees with the fact the kth jump in the Poisson process with rate 1 has the same distribution, cf. (5.6)).

Chapter 3

1. If $A = S$, the stated relation becomes $\mathbf{P}(X_{n+1} \in B | X_{n-1} \in C) =$

$\mathbf{P}(X_{n+1} \in B)$, i.e., X_{n-1} and X_{n+1} are independent. Not true in the general case.

2. (3.2): For any $B_j, A_i \subset \mathcal{S}$, $x \in S$,

$$\begin{aligned}\mathbf{P}(X_{n+1} \in B_1, X_{n+2} \in B_2, \dots \mid X_n = x, X_{n-1} \in A_1, X_{n-2} \in A_2, \dots) \\ = \mathbf{P}(X_{n+1} \in B_1, X_{n+2} \in B_2, \dots \mid X_n = x). \quad (13.44)\end{aligned}$$

(3.3):

$$\begin{aligned}\mathbf{P}(X_{n+1} \in B_1, X_{n+2} \in B_2, \dots ; X_{n-1} \in A_1, X_{n-2} \in A_2, \dots \mid X_n = x) \\ = \mathbf{P}(X_{n+1} \in B_1, X_{n+2} \in B_2, \dots \mid X_n = x) \\ \times \mathbf{P}(X_{n-1} \in A_1, X_{n-2} \in A_2, \dots \mid X_n = x).\end{aligned}$$

As the first relation can be obtained by dividing both sides of the second one by $\mathbf{P}(X_{n-1} \in A_1, X_{n-2} \in A_2, \dots \mid X_n = x)$, they are equivalent to each other. Clearly (3.1) is a special case of (13.44). To derive (13.44) from (3.1), use

$$\begin{aligned}&\mathbf{P}(X_{n+1} \in B_1, X_{n+2} \in B_2 \mid X_n = x, X_{n-1} \in A_1, X_{n-2} \in A_2, \dots) \\ &= \sum_{j \in B_1} \mathbf{P}(X_{n+1} = j, X_{n+2} \in B_2 \mid X_n = x, X_{n-1} \in A_1, X_{n-2} \in A_2, \dots) \\ &\overset{(3.10)}{=} \sum_{j \in B_1} \mathbf{P}(X_{n+2} \in B_2 \mid X_{n+1} = j, X_n = x, X_{n-1} \in A_1, X_{n-2} \in A_2, \dots) \\ &\qquad \times \mathbf{P}(X_{n+1} = j \mid X_n = x, X_{n-1} \in A_1, X_{n-2} \in A_2, \dots) \\ &\overset{(3.1)}{=} \sum_{j \in B_1} \mathbf{P}(X_{n+2} \in B_2 \mid X_{n+1} = j, X_n = x)\mathbf{P}(X_{n+1} = j \mid X_n = x) \\ &\overset{(3.10)}{=} \sum_{j \in B_1} \mathbf{P}(X_{n+1} = j, X_{n+2} \in B_2 \mid X_n = x) \\ &= \mathbf{P}(X_{n+1} \in B_1, X_{n+2} \in B_2 \mid X_n = x),\end{aligned}$$

and so on.

3. The LHS of (3.1) is equal to

$$\frac{\mathbf{P}(X_{n+1} \in B, X_n = x, X_{n-1} \in A_1, X_{n-2} \in A_2, \dots)}{\mathbf{P}(X_n = x, X_{n-1} \in A_1, X_{n-2} \in A_2, \dots)},$$

where the numerator equals

$$\begin{aligned}
&\sum_{j\in B, i_1\in A_1,\dots} \mathbf{P}(X_{n+1}=j, X_n=x, X_{n-1}=i_1,\dots)\\
&= \sum_{j\in B, i_1\in A_1,\dots} \mathbf{P}(X_{n+1}=j|\, X_n=x, X_{n-1}=i_1,\dots)\,\mathbf{P}(X_n=x, X_{n-1}=i_1,\dots)\\
&\overset{(3.6)}{=} \sum_{j\in B, i_1\in A_1,\dots} \mathbf{P}(X_{n+1}=j|\, X_n=x)\,\mathbf{P}(X_n=x, X_{n-1}=i_1,\dots)\\
&= \mathbf{P}(X_{n+1}\in B|\, X_n=x)\,\mathbf{P}(X_n=x, X_{n-1}\in A_1,\dots).
\end{aligned}$$

4. Let $j,k \in S_r$ have periods $d_j, d_k \geq 1$, respectively. Since $p_{jj}^{(u)} > 0$, $p_{kk}^{(u)} > 0$ for $u = s+t$ from the proof of Corollary 3.1, both d_j and d_k divide u. Hence from (3.18), for all n such that $p_{jj}^{(n)} > 0$, d_k divides n, and for all n such that $p_{kk}^{(n)} > 0$, d_j divides n. It only remains to recall the definition of the period of a state. DIY.

5. DIY.

6. DIY.

7. (i) $S = \{1, 2, \dots, a-1\}$, X_n = capital of player A after nth play. Transition diagram: DIY (possible transitions are to neighbouring states, from 1 and $a-1$ can move to the same state). Transition matrix:

$$P = \begin{pmatrix} q\,p\,0\,0 \dots 0\,0\,0 \\ q\,0\,p\,0 \dots 0\,0\,0 \\ 0\,q\,0\,p \dots 0\,0\,0 \\ 0\,0\,q\,0 \dots 0\,0\,0 \\ \cdot\;\cdot\;\cdot\;\cdot \dots \cdot\;\cdot\;\cdot \\ 0\,0\,0\,0 \dots 0\,p\,0 \\ 0\,0\,0\,0 \dots q\,0\,p \\ 0\,0\,0\,0 \dots 0\,q\,p \end{pmatrix}.$$

All states are essential, one class (irreducible MC), aperiodic.

(ii) $S = \{0, 1, 2, \dots, a-1, a\}$. Transition diagram: DIY (possible transitions are to neighbouring states, but from 0 and a can only move to the same state). Transition matrix:

$$P = \begin{pmatrix} 1\,0\,0\,0 \dots 0\,0\,0 \\ q\,0\,p\,0 \dots 0\,0\,0 \\ 0\,q\,0\,p \dots 0\,0\,0 \\ 0\,0\,q\,0 \dots 0\,0\,0 \\ \cdot\;\cdot\;\cdot\;\cdot \dots \cdot\;\cdot\;\cdot \\ 0\,0\,0\,0 \dots 0\,p\,0 \\ 0\,0\,0\,0 \dots q\,0\,p \\ 0\,0\,0\,0 \dots 0\,0\,1 \end{pmatrix}.$$

States 1 and a are absorbing, all the rest are non-essential.

8. An MC since given the current position, transition probabilities do not depend on the past history.

(i) The distance between corners 1 and 3 is 5, so, for some $c > 0$, $p_{11} = 0$, $p_{12} = c/4$, $p_{13} = c/5$, $p_{14} = c/3$; $1 = \sum_j p_{1j} = \frac{47}{60}c$ yields $c = 60/47$. Similarly for other states:

$$P = \frac{1}{47}\begin{pmatrix} 0 & 15 & 12 & 20 \\ 15 & 0 & 20 & 12 \\ 12 & 20 & 0 & 15 \\ 20 & 12 & 15 & 0 \end{pmatrix}.$$

(ii) Irreducible and aperiodic (all states communicate, and, say, from 1 you can get back to 1 both in two and three steps, so 1 is aperiodic, and by solidarity all other states are aperiodic, too). Since P is doubly stochastic, $\pi = (1/4, 1/4, 1/4, 1/4)$.

(iii) $\mathbf{P}(X_1 = 1, X_2 = 4, X_4 = 2) = \sum_{j=1}^4 p_j p_{j1} p_{14} p_{42}^{(2)} = p_{14} p_{42}^{(2)} \sum_{j=1}^4 p_j p_{j1} \approx 0.029$ (as $p_{42}^{(2)} = 600/47^2$).

(iv) Not an MC as $\mathbf{P}(X_3 = 1 | X_2 = 2, X_1 = 1) = 0$ (cannot get back to 1), but $\mathbf{P}(X_3 = 1 | X_2 = 2, X_1 = 3) > 0$ (for an MC, both would be equal to the common value $\mathbf{P}(X_3 = 1 | X_2 = 2)$). To get an MC, consider $Y_n = (X_{n-1}, X_n)$ (cf. Example 3.15).

9. Transition diagram: DIY. State 1 is non-essential, $\{2, 3, 4\}$ is a closed class of essential aperiodic states.

(ii)

$$P = \begin{pmatrix} 0 & 1 & 0 & 0 \\ 0 & 0 & 1 & 0 \\ 0 & 0 & 0 & 1 \\ 0 & 1/2 & 1/2 & 0 \end{pmatrix}.$$

(ii)

$$P^2 = \begin{pmatrix} 0 & 0 & 1 & 0 \\ 0 & 0 & 0 & 1 \\ 0 & 1/2 & 1/2 & 0 \\ 0 & 0 & 1/2 & 1/2 \end{pmatrix}, \qquad P^{16} = \frac{1}{256}\begin{pmatrix} 0 & 20 & 102 & 104 \\ 0 & 52 & 102 & 102 \\ 0 & 51 & 103 & 102 \\ 0 & 51 & 102 & 103 \end{pmatrix}.$$

10. (i) DIY. (ii) $\{2, 8, 10\}$ are non-essential states; $\{1, 4, 5\}$ is a periodic ($d = 2$) closed class, $\{6, 7, 9\}$ is an aperiodic closed class, 3 is an absorbing state ($\{3\}$ a single state closed class). (iii) (a) No changes. (b) States 1, 4 and 5 become non-essential.

11. (i) Note: $\mathbf{P}(X_n = \pm 1) = 1/4$, $\mathbf{P}(X_n = 0) = 1/2$. If $n > m + 1$, X_n and X_m are independent. For $n = m + 1$:

$$\begin{aligned} p_{jk}(m, m+1) &\overset{TPF}{=} \mathbf{P}(Y_{m+1} + Y_{m+2} = 2k | Y_m + Y_{m+1} = 2j) \\ &= \sum_{r \in \{-1,1\}} \mathbf{P}(Y_{m+2} = 2k - r)\, \mathbf{P}(Y_{m+1} = r | Y_m + Y_{m+1} = 2j). \end{aligned}$$

For $j = \pm 1$ only $r = \pm 1$, respectively, is possible. For $j = 0$, the last conditional probability is 1/2. From this infer that the transition matrix $(p_{jk}(m,n))_{j,k=-1,0,1}$ is:

$$\begin{pmatrix} 1/4 & 1/2 & 1/4 \\ 1/4 & 1/2 & 1/4 \\ 1/4 & 1/2 & 1/4 \end{pmatrix}, \quad n > m+1, \quad \begin{pmatrix} 1/2 & 1/2 & 0 \\ 1/4 & 1/2 & 1/4 \\ 0 & 1/2 & 1/2 \end{pmatrix}, \quad n = m+1.$$

(ii) See the hint: we get different values for $k = -1$, $j = 0$, $m = \pm 1$. Alternatively, if it were a (homogeneous) MC, we would have

$$(p_{jk}(m, m+2))_{j,k=-1,0,1} = (p_{jk}(m, m+1))^2_{j,k=-1,0,1},$$

which is obviously not the case.

12. (i) $p_{12}p_{22}p_{21} = 1/16$ (by (3.15)). (ii) $p_{22}p_{21}\sum_{j=1}^{3} p_j p_{j2} = 17/320$. (iii) $p_{22}^{(3)}p_{22}^{(2)}\sum_{j=1}^{3} p_j p_{j2} = \frac{35}{64} \times \frac{31}{48} \times \frac{17}{60} \approx 0.1$.

13. (i) $(\boldsymbol{p}P)_2 = 1/2$. (ii) $(\boldsymbol{p}P^2)_2 = 1/3$. (iii) $p_{12}^{(3)} = (P^3)_{12} = 1/4$. (iv) All states are essential and communicate with each other (irreducible MC), aperiodic (as $p_{11}^{(2)} > 0$, $p_{11}^{(3)} > 0$). (v) From (iv), the MC is ergodic, so the limit exists and is equal to $\pi_2 = 0.4$ (from (vi)). (vi) $\pi_1 = \pi_2 = 0.4$, $\pi_3 = 0.2$ (solving (3.29)).

14. (i) Irreducible and aperiodic, hence ergodic. The limits exist and are given by the stationary probability $\pi_1 = 1/3$ ($\boldsymbol{\pi} = (1/3, 2/3)$ is a unique solution to (3.29)).

(ii) State 1 is non-essential, 2 is absorbing. The limit exists and is equal to zero.

(iii) Both states are essential and periodic with $d = 2$. The MC is not ergodic and does not "forget" about its initial distribution! Clearly,

$$P^{2m} = \begin{pmatrix} 1 & 0 \\ 0 & 1 \end{pmatrix}, \quad P^{2m+1} = \begin{pmatrix} 0 & 1 \\ 1 & 0 \end{pmatrix},$$

so the limit does not exist:

$$\mathbf{P}(X_n = 1) = (\boldsymbol{p}P^n)_1 = \begin{cases} p_1 = 1/4 \text{ for } n = 2m, \\ p_2 = 3/4 \text{ for } n = 2m+1. \end{cases}$$

15. (i) DIY ($S = \{0, 1, 2, \ldots\}$ with possible transitions to neighbouring states and also to the current state as well).

(ii)

$$P = \begin{pmatrix} 1-p^2 & p^2 & 0 & 0 & 0 & \ldots \\ (1-p)^2 & 2p(1-p) & p^2 & 0 & 0 & \ldots \\ 0 & (1-p)^2 & 2p(1-p) & p^2 & 0 & \ldots \\ 0 & 0 & (1-p)^2 & 2p(1-p) & p^2 & \ldots \\ \ldots & \ldots & \ldots & \ldots & \ldots & \ldots \end{pmatrix}.$$

(iii) A special case of the general RW with jumps $0, \pm 1$ (see Example 3.18). Need $\sum_{j\geq 0} K_j < \infty$, where

$$K_j = \frac{p_{01}p_{12}\cdots p_{j-1,j}}{p_{j,j-1}\cdots p_{21}p_{10}} = r^j, \quad r = \frac{p^2}{(1-p)^2}.$$

So the MC is ergodic iff $r < 1$, i.e., $p < 1/2$. The stationary distribution:

$$\pi_j = K_j / \sum_{i\geq 0} K_i = (1-r)r^j, \quad j = 0, 1, 2, \ldots.$$

(iv) Total costs on day n:

$$d\mathbf{1}_{\{X_n=0\}}\mathbf{1}_{\{\text{both produced items defective}\}} + cX_{n+1},$$

so the expected value (in the stationary regime) is

$$d\pi_0(1-p)^2 + c\sum_{j=0}^{\infty} j\pi_j = d(1-r)(1-p)^2 + \frac{cr}{1-r} = \frac{d(1-2p)}{(1-p)^2} + \frac{cp^2}{1-2p}.$$

16. Gambler's ruin; if $X_n =$ the capital of B at time n, $p = 0.6$. For $N = 5$, $M = 10$: $u_5 \approx 0.870$; $N = 10$, $M = 20$: $u_10 \approx 0.983$.

17. (i) Transition diagram: DIY;

$$P = \begin{pmatrix} 0.6 & 0.2 & 0.2 \\ 0.2 & 0.3 & 0.5 \\ 0.5 & 0 & 0.5 \end{pmatrix}.$$

(ii) $\mathbf{P}(X_1 = 1, X_3 = 1, X_6 = 1 | X_0 = 2) = p_{21}p_{11}^{(2)}p_{11}^{(3)} = 0.035$. (iii) $\pi = (35/69, 10/69, 24/69)$.

18. 1/6 in both cases.

19. Case $p = q = 1/2$: equation

$$w_i = 1 + pw_{i+1} + qw_{i-1} \tag{13.45}$$

becomes $\nabla^2 w_i = 2$, where $\nabla w_i = w_i - w_{i-1}$ is the backward difference operator. General solution: $w_i = i^2 + ai + b$ (verify! Cf. the differential equation $d^2w(t)/dt^2 = 2$). From the boundary conditions $w_0 = w_{M+N} = 0$ find the constants a, b.

Case $p \neq q$: The general solution to (13.45) has the form $w_i = w_i^* + u_i$, where w_i^* is a partial solution to that equation and u_i is the general solution to the homogeneous equation $u_i = pu_{i+1} + qu_{i-1}$ (verify!). As (13.45) is equivalent to $py_{i+1} - qy_{i-1} = -1$ for $y_i = \nabla w_i$, which is solved by $y_i = c =$ const when $pc - qc = -1$, i.e., $c = 1/(q-p)$, we get $w_i = y_1 + \cdots + y_i = i/(q-p)$ (cf. (9.19)).

So $w_i = i/(q-p) + u_i$, and from here $u_i = 0$. Now use (3.43) and the second boundary condition ($w_{N+M} = 0$).

20. DIY (in (v) and (vi), use the CLT).

Chapter 4

1. By definition (4.1),

$$V_n(i) = \max_{\{a_{T-n+1},\dots,a_T\}} \mathbf{E}\left[\sum_{t=T-n+1}^{T} R(X_t,a_t)\middle| X_{T-n+1}=i\right]$$
$$= \max_{\{\cdots\}}\left\{R(X_{T-n+1},a_{T-n+1}) + \mathbf{E}\left[\sum_{t=T-n+2}^{T} R(X_t,a_t)\middle| X_{T-n+1}=i\right]\right\},$$

where by the TPF the last expectation can be computed as

$$\sum_j \mathbf{E}\left[\cdots \mid X_{T-n+1}=i, X_{T-n+2}=j\right]\mathbf{P}\left(X_{T-n+2}=j \mid X_{T-n+1}=i\right)$$
$$= \sum_j \mathbf{E}\left[\cdots \mid X_{T-n+2}=j\right] p_{ij}(a_{T-n+1})$$

by Markov property. Complete the argument!

2. (i) $X_t = Z_t$ if has not bought yet, $X_t = 0$ otherwise, $t = 1,2,3,4$ (with $Z_4 = \infty$ to make the person purchase the land!). Actions: $a = 1$ is "buy"; $a = 0$ is "do nothing". Transition probabilities for $a = 0$ at different times t are specified by the table, while $a = 1$ always means transition to 0. Reward function: $R(x,0) = 0$, $R(x,1) = -x$.

(ii) As the process in non-homogeneous (in time), we have

$$V_n(i) = \max_a\left[R(i,a) + \mathbf{E}_a(V_{n-1}(X_{5-n}) \mid X_{4-n}=i)\right], \quad n = 1,2,3$$

(since $T = 3$, $T-n+1 = 4-n$ now) with $V_0(x) = -x$ (inflicting a huge penalty in case the person has not bought land during the three days).

(iii) Decision tree: DIY. Optimal policy: in week one, buy if the price is 2.2 else wait; in week two, buy if the price is 2.2 or 2.3, else wait; in week three buy if you haven't yet. The minimum expected price is 2.2574 (i.e., \$225,740).

3. The optimality equation becomes

$$V_n(x) = \max_{a\in[0,1]} \mathbf{E}\, V_{n-1}(x + axZ) = \max_{a\in[0,1]}\Big(pV_{n-1}(x(1+a)) + qV_{n-1}(x(1-a))\Big),$$

so that, since $V_0(x) = \log x$,

$$V_1(x) = \log x + \max_{a\in[0,1]}(p\log(1+a) + q\log(1-a)).$$

(i) If $p \le 1/2$, the function on the right-hand side is decreasing in $a \in [0,1]$, so the optimal $a = 0$, $V_1(x) = \log x$. Repeating the argument, we derive that all $V_n(x) = \cdots = V_1(x) = V_0(x) = \log x$, and the optimal action is always $a = 0$.

(ii) If $p > 1/2$, the maximum is attained at $a^* = 2p - 1 = p - q$. As $V_1(x) = \log x + c$, $c = \mathbf{E}\,(1 + a^* Z) = \text{const}$, repeating the argument yields that a^* is optimal at each step.

4. (i) $X_t = Z_t$ if has not sold yet; $X_t = 0$ otherwise. Actions: $a = 1$ is "sell"; $a = 0$ is "do nothing". The evolution of $\{X_t\}$: given $X_t = 0$, $X_{t+1} = 0$ for any a; given $X_t = x > 0$, $X_{t+1} = 0$ if $a = 1$ and $X_{t+1} = Z_{t+1}$ if $a = 0$. Reward function: $R(x,0) = 0$, $R(x,1) = x$. The sum of one-step reward equals the only term ($\neq 0$) giving the selling price.

(ii) $V_n(x) = \max_{a=0,1}\big[R(x,a) + \mathbf{E}_a(V_{n-1}(X_1)|\, X_0 = x)\big] = \max\big\{\mathbf{E}_0(V_{n-1}(X_1)|\, X_0 = x),\, x\big\}$, where the subscript $_a$ indicates that the expectation is taken under action a. If $x = 0$, then $V_n(x) = 0$, so can only consider case $x > 0$, and then

$$V_n(x) = \max\big\{\mathbf{E}\, V_{n-1}(Z),\, x\big\}, \quad Z \sim U(0,1),$$

with the initial condition $V_0(x) = 0$ (as nothing can be gained after time $T = 4$). Solution:

$n = 1$: For $x > 0$, $V_1(x) = x$ and the optimal action (for which the maximum is attained) is always $a = 1$.

$n = 2$: $V_2(x) = \max\{\mathbf{E}\, V_1(Z),\, x\} = \max\{\mathbf{E}\, Z,\, x\} = \max\{1/2,\, x\}$. Optimal action: $a = 1$ iff $x > 1/2$.

$n = 3$: Here

$$V_3(x) = \max\big\{\mathbf{E}\, V_2(Z),\, x\big\} = \max\big\{\mathbf{E}\,\max\{1/2,\, Z\},\, x\big\} = \max\{5/8,\, x\}$$

since (using the hint)

$$\mathbf{E}\,\max\{c,\, Z\} = c\,\mathbf{P}\,(Z \le c) + \int_c^1 x\,dx = \frac{1}{2}(1 + c^2).$$

The optimal action: $a = 1$ iff $x > 5/8$.

$n = 4$: Now

$$V_4(x) = \max\big\{\mathbf{E}\, V_3(Z),\, x\big\} = \max\big\{\mathbf{E}\,\max\{5/8,\, Z\},\, x\big\} = \max\{89/128,\, x\}.$$

The optimal action: $a = 1$ iff $x > 89/128$.

(iii) Day 1: sell if $Z_1 > 89/128 \approx 0.695$. Day 2: sell if $Z_2 > 5/8 = 0.625$. Day 3: sell if $Z_1 > 1/2 = 0.5$. Day 4: sell. Maximum expected price: $\mathbf{E}\, V_4(X_1) = \mathbf{E}\,\max\{89/128,\, Z\} \approx 0.742$.

5. As we proved that $\{s_n\}$ is non-decreasing, it suffices to show that $s_2 = \infty$. Since $V_1 \ge s - c$, we get $\mathbf{E}\, V_1(s + Y_1) \ge \mathbf{E}\,(s + Y_1 - c) = s - c + \mu > s - c$, so from (4.4) with $n = 2$ we see that (4.5) holds for all s, i.e., $s_2 = \infty$.

6. (i) $\mathbf{E}\,Y = \sum_{j\le n}\lambda_j \mathbf{E}\,X_j = \mu$; $\mathrm{Var}\,(Y) = \sigma^2\sum_{j\le n}\lambda_j^2 = \sigma^2\big(\lambda_1^2+\cdots+\lambda_{n-1}^2+ (1-\lambda_1-\cdots-\lambda_{n-1})^2\big)$. Solve $\partial(\cdots)/\partial\lambda_j = 0$, $j=1,\dots,n-1$, to get $\lambda_j = 1/n$ for all j (and note that this is a minimum indeed!).

(ii) First show that $f(\boldsymbol{\lambda}) := \mathbf{E}\,u\Big(\sum_{j=1}^n \lambda_j X_j\Big)$ is a strictly concave function of $\boldsymbol{\lambda} = (\lambda_1,\dots,\lambda_n)$: for any $\boldsymbol{\lambda}'$, $\boldsymbol{\lambda}''$ and $\alpha\in(0,1)$,

$$f(\boldsymbol{\lambda}) > \alpha f(\boldsymbol{\lambda}') + (1-\alpha) f(\boldsymbol{\lambda}''). \tag{13.46}$$

Next assume that the maximum of $f(\boldsymbol{\lambda})$ is attained at a point $\boldsymbol{\lambda}'$ such that $\lambda_i' \ne \lambda_j'$ for some $i\ne j$. Define $\boldsymbol{\lambda}''$ by setting $\lambda_j'' := \lambda_i'$, $\lambda_i'' := \lambda_j'$, and $\lambda_k'' := \lambda_k'$ for all $k\ne i,j$. As $\sum\lambda_k'' X_k$ has the same distribution as $\sum \lambda_k' X_k$ (since $X_1,\dots,X_n$ are exchangeable), $f(\boldsymbol{\lambda}'') = f(\boldsymbol{\lambda}')$ is also a minimum, and by taking $\alpha = 1/2$ we see from (13.46) that for the midpoint $\boldsymbol{\lambda} = (\boldsymbol{\lambda}'+\boldsymbol{\lambda}'')/2$, $f(\boldsymbol{\lambda}) > (f(\boldsymbol{\lambda}')+f(\boldsymbol{\lambda}''))/2 = f(\boldsymbol{\lambda}')$, a contradiction! So must have $\lambda_i' = \lambda_j'$ for all i,j at the maximum point.

7. Since $u_n(x) = \max\{x, \mu_{n-1}\}$, assuming $\mu_{n-1}\le \mu_n$, we get

$$\begin{aligned}\mu_{n+1}-\mu_n &= \alpha\mathbf{E}\,(u_{n+1}(Z) - u_n(Z))\\ &= \alpha(\mu_n-\mu_{n-1})\,\mathbf{P}\,(Z\le\mu_{n-1}) + \alpha\mathbf{E}\,((Z-\mu_{n-1});\ \mu_{n-1}< Z\le \mu_n).\end{aligned}$$

Clearly, $0\le$ (right-hand side) $\le \alpha|\mu_n-\mu_{n-1}|$. Similar argument if $\mu_{n-1}>\mu_n$.

Chapter 5

1. Putting $f(s) := \mathbf{P}\,(X>s)$, condition (5.4) is equivalent to $f(t+s) = f(t)f(s)$. Setting $t = s, 2s, \dots$, this yields $f(ms) = f(s)^m$, $m = 1,2,\dots$. This first implies that $f(1/n) = (f(1))^{1/n} = e^{-\lambda/n}$ with $\lambda := -\ln f(1)$, and then that $f(m/n) = e^{-\lambda m/n}$ for any $m,n = 1,2,\dots$ So for all *rational* $s = m/n$, the tail $f(s) = e^{-\lambda s}$. For *irrational* values s this now holds by the monotonicity of $f(s)$.

2. $\varphi_{\tau_j}(it) = \int_0^\infty e^{itx}\lambda e^{-\lambda x}dx = (1-it/\lambda)^{-1}$, $\varphi_{T_k}(it) = (\varphi_{\tau_1}(it))^k = (1-it/\lambda)^{-k}$. For $k>1$, integrating by parts we obtain

$$\begin{aligned}f_{T_k}(x) &= \frac{1}{2\pi}\int_{-\infty}^{\infty}\frac{e^{-itx}dt}{(1-it/\lambda)^k} = \left[\frac{\lambda e^{-itx}}{2\pi i(k-1)(1-it/\lambda)^{k-1}}\right]_{-\infty}^{\infty}\\ &\quad + \frac{1}{2\pi}\int_{-\infty}^{\infty}\frac{\lambda x e^{-itx}dt}{(k-1)(1-it/\lambda)^{k-1}} = \frac{\lambda x}{k-1}\times f_{T_{k-1}}(x).\end{aligned}$$

As $f_{T_1}(x) = \lambda e^{-\lambda x}$, $x>0$, we get $f_{T_k}(x) = \lambda^k x^{k-1}e^{-\lambda x}/(k-1)!$, $x>0$ (i.e., the density of the $\gamma(k,\lambda)$-distribution).

3. The random index $j(t)$ is not independent of $\{\tau_j\}$: loosely speaking, the larger the interval $[T_{j-1}, T_j)$, the more likely it will cover the point t. So the distribution of $\tau_{j(t)} = T_{j(t)} - T_{j(t)-1}$ differs from $Exp(\lambda)$. In fact, we can easily find that its density

$$f_{\tau_{j(t)}} = \begin{cases} \lambda^2 x e^{-\lambda x}, & 0 < x \le t, \\ \lambda(1+\lambda t)e^{-\lambda x}, & x > t. \end{cases}$$

(consider the two cases $x \le t$ and $x > t$ separately but similarly, using the TPF—partitioning the sample space into events $\{j(t) = n\}$—and our knowledge of the distribution of T_k).

4. The RVs $X_j := N_{t_j} - N_{t_{j-1}} \sim Po(\lambda(t_j - t_{j-1}))$, $j = 1, 2, \ldots, m+1$, where $t_0 = 0$, $t_{m+1} = t$, are independent. So, for any integer $k_j \ge 0$, $k_1 + \cdots + k_{m+1} = k$,

$$\begin{aligned} &\mathbf{P}(N_{t_1} = k_1, N_{t_2} = k_1 + k_2, \ldots, N_{t_m} = k_1 + \cdots + k_m \mid N_t = k) \\ &\quad = \mathbf{P}(X_1 = k_1, \ldots, X_m = k_m \mid X_1 + \cdots + X_{m+1} = k) \\ &\quad = \mathbf{P}(X_1 = k_1, \ldots, X_m = k_m, X_{m+1} = k_{m+1}) / \mathbf{P}(N_t = k) \\ &\quad = e^{\lambda t} \frac{k!}{(\lambda t)^k} \prod_{j=1}^{m+1} \frac{(\lambda(t_j - t_{j-1}))^{k_j}}{k_j!} e^{-\lambda(t_j - t_{j-1})} = \frac{k!}{k_1! \cdots k_{m+1}!} p_1^{k_1} \cdots p_{m+1}^{k_{m+1}} \end{aligned}$$

which is the *multinomial* distribution with "success" probabilities $p_j = (t_j - t_{j-1})/t$, $j = 1, \ldots, m+1$. But, for k independent $U(0,t)$-distributed points Y_j, the event $\{kF_k^*(t_1) = k_1, \ldots, kF_k^*(t_m) = k_1 + \cdots + k_m\} = \{k_1 \text{ points hit } (0, t_1), \ldots, k_m \text{ points hit } (t_{m-1}, t_m)\}$ clearly has the same multinomial probability.

5. (i) $e^{-1} \approx 0.3679$. (ii) $\mathbf{P}(N_4 - N_2 = 2) = \frac{2^2}{2!} e^{-2} \approx 0.2707$. (iii) 4. (iv) $\mathbf{P}(N_3 = 2) \frac{3^2}{2!} e^{-3} \approx 0.2240$. (v) The RVs $X_1 := N_{(3,4]}$, $X_2 := N_{(4,6]}$ and $X_3 := N_{(6,7]}$ are *independent*, and

$$\begin{aligned} &\mathbf{P}(N_{(4,7]} = 2, N_{(3,6]} = 1) = \mathbf{P}(X_1 + X_2 = 1, X_2 + X_3 = 2) \\ &\quad = \mathbf{P}(\cdots, X_2 = 0) + \mathbf{P}(\cdots, X_2 = 1) = \mathbf{P}(X_1 = 1, X_2 = 0, X_3 = 2) \\ &\qquad + \mathbf{P}(X_1 = 0, X_2 = 1, X_3 = 1) = 2.5e^{-4} \approx 0.0458. \end{aligned}$$

(vi) $\frac{31}{16} e^{-2} \approx 0.2622$. Similarly, $X_1 := N_{(1,4]}$, $X_2 := N_{(4,5]}$ and $X_3 := N_{(5,7]}$ are independent and

$$\begin{aligned} &\mathbf{P}(N_{(4,7]} = 2 \mid N_{(1,5]} = 2) = \mathbf{P}(X_2 + X_3 = 2 \mid X_1 + X_2 = 2) \\ &\quad = \mathbf{P}(X_1 + X_2 = 2, X_2 + X_3 = 2) / \mathbf{P}(X_1 + X_2 = 2), \end{aligned}$$

where the numerator is equal to $\sum_{j=0}^{2} \mathbf{P}(\cdots, X_2 = j) = \mathbf{P}(X_1 = 2, X_2 = 0, X_3 = 2) + \mathbf{P}(X_1 = 1, X_2 = 1, X_3 = 1) + \mathbf{P}(X_1 = 0, X_2 = 2, X_3 = 0)$.

6. (i) The failures of the machines occur at the times T_j, $j \le 4$, of the first four jumps in the Poisson process $\{N_t\}$ with mean $\lambda = 1/10$, so the probability in question is equal to

$$\begin{aligned}\mathbf{P}(N_{40} < 4 | T_1 = 3, T_2 = 13) &= \mathbf{P}(N_{40} < 4 | N_{13} = 2) = \mathbf{P}(N_{40} - N_{13} < 2) \\ &= \mathbf{P}(N_{27} = 0) + \mathbf{P}(N_{27} = 1) \approx 0.2487\end{aligned}$$

by the lack of memory property and independent (stationary) increments.

(ii) $\mathbf{P}(N_7 \geq 4) = 1 - \sum_{j=0}^{3} \mathbf{P}(N_7 = j) \approx 0.0058$.

7. The "tanker process" $\{M_t\}$ is obtained by independent thinning (with probability 0.2) of the Poisson process with parameter 1 (hour^{-1}), hence itself is the Poisson process with parameter 0.2. (i) $\mathbf{P}(M_{24} > 0) = 1 - e^{-4.8} \approx 0.9918$. (ii) Since each of the passed ships is a tanker w.p. $p = 0.2$, the number X of tankers follows the binomial distribution:

$$\mathbf{P}(X = 6) = \binom{30}{6} p^6 (1-p)^{24} \approx 0.1795.$$

8. (i) DIY (you will get a pure step function). (ii) By independence,

$$\mathbf{E}\, Y_t = \mathbf{E}\, S_{N_t} = \sum_{n=0}^{\infty} \mathbf{E}\, S_n \mathbf{P}(N_t = n) = \frac{1}{2} \sum_{n=0}^{\infty} n\, \mathbf{P}(N_t = n) = \frac{1}{2} \mathbf{E}\, N_t = 35$$

as $\mathbf{E}\, S_n = n\mathbf{E}\, X_1 = n/2$. (iii) Similar to that of an RW with a positive trend of 35 (units/day) [note that the process Y_t has independent increments]. The LLN will hold, as well as the CLT (verify that $\mathrm{Var}(Y_t) < \infty$).

9. Calls to phones $j = 1, \ldots, 6$ follow independent Poisson processes $\{N_t^{(j)}\}$ with rates $\lambda_j = 5$, $j \leq 4$, and $\lambda_j = 10$, $j = 5, 6$ [rates are in hour^{-1}]. (i) The probability $= \mathbf{P}(N_{0.5}^{(1)} = 0) = e^{-2.5} \approx 0.0821$. (ii) The total number of calls by time t is $N_t = \sum_{j=1}^{6} N_t^{(j)}$, the Poisson process with rate $\sum_{j=1}^{6} \lambda_j = 40$. So the time of the first call is an $Exp(40)$-RV, with the density $f(t) = 40e^{-40t}$, $t > 0$ (hours). (iii) $\lambda_5 / \sum_{j=1}^{6} \lambda_j = 10/40 = 0.25$. (vi) Independent thinning. The process $\{M_t^{(1)}\}$ of calls answered by the first clerk is Poissonian with rate $0.5 \times 10 = 5$, so the probability is equal to $\mathbf{P}(M_1^{(1)} = 1) = 5e^{-5} \approx 0.0337$.

Chapter 6

1. Both implications follow from the fact that, for any $\varepsilon \in (0,1)$, one has $\mathbf{P}(|X_{t+h} - X_t| > \varepsilon) = \mathbf{P}(X_{t+h} \neq X_t) = \sum_j \mathbf{P}(X_t = j)(1 - p_{jj}^{(h)}) \to 0$ if $p_{jj}^{(h)} \to 1$, $j \in S$, as $h \to 0$. The convergence of the sum to 0 implies that all the terms in it must tend to zero, too, as they are non-negative. So if all $\mathbf{P}(X_t = j) > 0$, then must have $p_{jj}^{(h)} \to 1$, $j \in S$, and hence $p_{jk}^{(h)} \to 0$, $j \neq k$, as the row sums in $P^{(h)}$ are all ones.

2. $\boldsymbol{\pi} P^{(t)} = \sum_{k \geq 0} (t^k/k!) \boldsymbol{\pi} A^k = \boldsymbol{\pi} + \sum_{k \geq 1} (t^k/k!) \boldsymbol{\pi} A \cdot A^{k-1} = \boldsymbol{\pi}$.

3. From (6.28), for the function $f(t) := \mathbf{P}(X_t = 0)$ we have:

$$f'(t) = \lambda(q - f(t) + pf^2(t)). \tag{13.47}$$

As $f(t) \le 1$ and $f(t) \nearrow$ as $t \to \infty$, there exists $f := \lim_{t\to\infty} f(t)$, and hence from (13.47) it follows that there also exists $\lim_{t\to\infty} f'(t) = \lambda(q - f + pf^2) =: c \ge 0$. Cannot have $c > 0$, as that would imply $f(t) > 1$ for large enough t, so $c = 0$. Solving $q - f + pf^2 = 0$ we get $f = q/p$, $f = 1$. As $f(0) = 0$, $f(t)$ will converge to the minimum of the two numbers. So $p_{10}^{(\infty)} = \min(q/p, 1)$. In the general case $X_0 = k$, as extinction occurs when all k independent branching processes starting with one of the initial k particles become extinct, we have $p_{k0}^{(\infty)} = (p_{10}^{(\infty)})^k$.

4. (i) $X_t =$ the number of alive particles at time t. Rates: $\lambda_j = \lambda$, $j \ge 0$, $\mu_j = j/\alpha$, $j \ge 1$. (ii) $K_j = \rho^j/j!$, $\rho = \lambda\alpha$, so $\sum_{j\ge 0} K_j = e^\rho < 0$ and hence the process is always ergodic; stationary distribution: $Po(\rho)$. (iii) ρ.

5. Given the value $X_t = x$, the next transition in the process will occur after a random time $\sim Exp(\lambda + \mu)$ (the minimum of two independent exponential RVs— the times till the first after time t jumps in the independent customers' and bus' arrival Poisson processes), and w.p. $\lambda/(\lambda+\mu)$ the new value will be $x+1$ (a new customer arrives first), and with the complementary probability $\mu/(\lambda+\mu)$ it will be $x - \min\{x, N\} \ge 0$ (a bus arrives first). After the first jump, the process will continue evolving according to the same rules. So given the state of the process at time t, its evolution after time t does not depend on its past.

The only non-zero values of the generator entries are: $a_{00} = -\lambda$, $a_{jj} = -(\lambda+\mu)$, $j \ge 1$, $a_{j,j+1} = \lambda$ for all $j \ge 0$, $a_{j0} = \mu$, $1 \le j \le N$, $a_{j,j-N} = \mu$, $j > N$. Sketch the transition diagram accordingly.

6. (i) $Exp(n\lambda)$. (ii) $T_k - T_{k-1} \sim Exp((n-k+1)\lambda)$ (at time T_{k-1}, due to the memoryless property of the exponential distribution, the residual lifetimes of the remaining $n-(k-1)$ atoms are independent $Exp(\lambda)$-RVs). (iii) $X_t = \sum_{j\le n} \mathbf{1}_{\{\tau_j > t\}}$, so $\mathbf{E}\, X_t = n\mathbf{P}(\tau_1 > t) = ne^{-\lambda t}$. The half-life $t = t_{0.5}$ solves (for large n) $X_t/n = 1/2$. By the LLN, $X_t/n \approx \mathbf{E}\,\mathbf{1}_{\{\tau_j > t\}} = e^{-\lambda t}$, so that $t_{0.5} = \lambda^{-1}\ln 2 \approx 0.693/\lambda$.

7. The only non-zero entries of the generator are: $a_{j,j-1} = 0.2j$, $a_{jj} = -j$, $a_{j,j+1} = 0.4j$, $a_{j,j+2} = 0.4j$, $j = 1, 2, \dots$. Sketch the diagram accordingly.

Since $n_t := \mathbf{E}(X_t | X_0 = 1) = \partial\varphi(t,z)/\partial z|_{z=1}$, $\varphi(t,z) = \mathbf{E}(z^{X_t} | X_0 = 1)$, differentiating (6.28) w.r.t. z at $z = 1$ yields $n_t' = \sum_{m=0}^{\infty} a_{1m} m n_t = n_t$ with the initial condition $n_0 = 1$, so that $n_t = e^t$. So $\mathbf{E}(X_t | X_0 = 66) = 66 n_t = 66e^t$.

8. (i) $S = \{1, \dots, 5\}$ with "1"$= uu$, "2"$= ud$, "3"$= du$, "4"$= dd1$, "5"$= dd2$, where uu means that both machines are up, ud that the first machine is up and the second one is down etc, ddi means that both machines are down and the repairman works on the ith machine, $i = 1, 2$ (he can only work on one machine at a time). The generator:

$$A = \begin{pmatrix} -(\lambda_1+\lambda_2) & \lambda_2 & \lambda_1 & 0 & 0 \\ \mu_2 & -(\lambda_1+\mu_2) & 0 & 0 & \lambda_1 \\ \mu_1 & 0 & -(\lambda_2+\mu_1) & \lambda_2 & 0 \\ 0 & \mu_1 & 0 & -\mu_1 & 0 \\ 0 & 0 & \mu_2 & 0 & -\mu_2 \end{pmatrix}.$$

Sketch the diagram accordingly (in fact, it is easier to first sketch the diagram and then derive the generator).

(ii) $\boldsymbol{\pi} = (7/34, 8/34, 5/34, 10/34, 4/34)$.

9. (i) $A = \alpha D$, where

$$D = \begin{pmatrix} -3 & 1 & 1 & 1 \\ 1 & -3 & 1 & 1 \\ 1 & 1 & -3 & 1 \\ 1 & 1 & 1 & -3 \end{pmatrix}.$$

(ii) $\boldsymbol{\pi} = (1/4, 1/4, 1/4, 1/4)$. (iii) As $D^2 = -4D$, we have $A^k = (-4)^{k-1}\alpha^k D$, $k \geq 1$, so that by (6.7)

$$P^{(t)} = I - \frac{1}{4}\sum_{k\geq 1} \frac{(-4\alpha)^k}{k!} D = I + \frac{1}{4}(1 - e^{-4\alpha t})D,$$

i.e., $p_{jj}^{(t)} = (1 + 3e^{-4\alpha t})/4$, $p_{jk}^{(t)} = (1 - e^{-4\alpha t})/4$, $j \neq k$.

10. For $X_t =$ the number of accidents up to time t, the only non-zero entries of the generator are $-a_{jj} = a_{j,j+1} = a + jb$, $j = 0, 1, 2, \ldots$. So for $\boldsymbol{p}_0^{(t)}$ we have $(\boldsymbol{p}_0^{(t)})' = \boldsymbol{p}(P^{(t)})' = \boldsymbol{p}P^{(t)}A = \boldsymbol{p}_0^{(t)}A$, which is equivalent to $(p_{00}^{(t)})' = -ap_{00}^{(t)}$,

$$(p_{0k}^{(t)})' = (a + (k-1)b)p_{0,k-1}^{(t)} - (a + kb)p_{0k}^{(t)}, \quad k \geq 1.$$

Substitute the expressions for $p_{0k}^{(t)}$ from conditions into the above equations.

Chapter 7

1. (i) Let ST stand for "service time", WT for "waiting time", and DT=ST+WT for "delay time". For FIFO:

Jobs:	1	2	3	4	5	Total
ST	1.3	0.7	4.1	2.9	3.1	
WT	0	1.3	2.0	6.1	9.0	
DT	1.3	2.0	6.1	9.0	12.1	30.5

The average delay time is 6.1 (hours).

(ii) For SIFO:

Jobs:	1	2	3	4	5	Total
ST	0.7	1.3	2.9	3.1	4.1	
WT	0	0.7	2.0	4.9	8.0	
DT	0.7	2.0	4.9	8.0	12.1	27.7

The average delay time is 5.54 (hours).

(iii) \$305 vs \$277.

2. The number $= 1$ w.p. $1 - \pi_0$ and $= 0$ w.p. π_0, so its expected value equals $1 - \pi_0 = \lambda/\mu$.

3. The time from the epoch the server becomes idle till the next arrival $\sim Exp(\lambda)$ due to the memoryless property of the exponential distribution (as arrivals follow the Poisson process). So $\mathbf{E}$ (IP)$= 1/\lambda$ (IP stands for "idle period", BP for "busy period").

The long-run fraction of the time when the server is idle equals $\pi_0 = 1-\rho$. As to each idle period there corresponds a single busy period (following immediately after the idle period), we must have

$$\frac{\mathbf{E}\,(\text{IP})}{\mathbf{E}\,(\text{IP}) + \mathbf{E}\,(\text{BP})} = 1 - \rho, \quad \mathbf{E}\,(\text{BP}) = \frac{\rho}{1-\rho} \times \frac{1}{\lambda} = \frac{1}{\mu - \lambda}.$$

4. $M/M/1$ with arrival rate $\lambda = 3$ (day^{-1}) and service rate $\mu = 24/7 > \lambda$, so that the stationary regime exists, in which the expected number of customers in the QS is $L = \lambda/(\mu - \lambda) = 7$. The total cost (in \$/day) without making the change: $150 + 10L = 220$.

With the new service rate $\mu' = 4$ we get $L' = \lambda/(\mu' - \lambda) = 3$, the total cost (denoting by C the operating cost) is $C + 10L' = C + 30$, so the change is economically attractive if $C <$ \$190.

5. $M/M/\infty$ (i) See Problem 4 on p. 191. (ii) $L = \lambda/\mu$ (λ being the arrival and μ the service rates), $D = 1/\mu$ (= the mean service time, as there is no queue in the system).

6. This is a B+DP with $S = \{0, 1, \ldots, N\}$ and transition rates $\lambda_j = \lambda$, $0 \le j < N$, $\mu_j = j\mu$, $1 \le j \le N$. So $K_j = \lambda^j/\mu^j j! = \rho^j/j!$, $j \le N$.

7. The second distribution is much more skewed to the right, with a "strong" mode at 0, whereas the first one is "bell-shaped". To cope with the load in the first case, all three repairmen are working 55% of the time (with only 13% of the time all three being idle). The superworker is idle for 36% of the time, for when the demand arises, he satisfies it very quickly.

8. A finite source queue with $a = 1$ server (bulldozer) and M customers (dumpers). [Service provided is loading a "customer", and after the completion of the service, the "customer" is back in an exponentially distributed random time.] For any M, $\rho = \lambda/\mu = 12/5$ as $\lambda = 1/(\text{mean delivery time}) = 12$ and $\mu = 5$ (hour^{-1}).

If A is the average number of dumpers loaded in an hour, than the first factor in (7.21) is equal to $10^7/(10^3 A)$, whereas

$$A = \text{loading rate} \times \text{fraction of time the bulldozer is busy} = 5(1-\pi_0).$$

So the expected cost is

$$10^4 \times \frac{100-40M}{5(1-\pi_0)} = 8 \times 10^4 \times \frac{2.5+M}{1-\pi_0},$$

and we just have to minimise in M the last factor (denote it by f_M) using

$$\pi_0^{-1} = 1 + \sum_{k=1}^{M} M(M-1)\cdots(M-k+1)\rho^k.$$

As $f_1 \approx 4.958$, $f_2 \approx 4.777$, and $f_M > 2.5 + M \geq 5.5$ for $M \geq 3$, the optimum value $M = 2$.

9. DIY!

10. B+DP with birth rates $\lambda_k = \lambda$, $j = 0, 1, \ldots$, and death rates

$$\mu_j = \begin{cases} \mu & \text{if } j = 1,2,3; \\ 2\mu & \text{if } j = 4,\ldots,9; \\ 3\mu & \text{if } j > 9 \end{cases}$$

(variable service rates). So for $\rho = \lambda/\mu$,

$$K_j = \begin{cases} \rho^j & \text{if } j = 1,2,3; \\ \rho^3 \dfrac{\lambda^{j-3}}{(2\mu)^{j-3}} = 2^{-(j-3)}\rho^j & \text{if } j = 4,\ldots,9; \\ 2^{-6}\rho^9 \dfrac{\lambda^{j-9}}{(3\mu)^{j-9}} = 2^{-6}3^{-(j-9)}\rho^j & \text{if } j > 9. \end{cases}$$

Since $\sum_{j\geq 0} K_j < \infty$ iff $\sum_{j\geq 10} K_j = c\sum_{j\geq 10}(\rho/3)^j < \infty$, the system is stable iff $\rho < 3$. Steady-state distribution: $\pi_j = K_j\pi_0$, $j \geq 1$, where

$$\pi_0^{-1} = \sum_{j\geq 0} K_j = \frac{1-\rho^4}{1-\rho} + \frac{\rho^4(1-(\rho/2)^6)}{2-\rho} + \frac{\rho^{10}}{2^6(3-\rho)}.$$

11. A finite source queue (note that the number of burnt out lamps is large enough to influence the arrival rates) with $M = 10^4$, $\lambda = 0.01$ (day^{-1}). As we know that $L = 10^3$, from (7.17) the average delay $D \approx 11.1$, i.e., 159% of the contract time of 7 days. No good.

12. By Bayes' formula, in the steady state,

$$\begin{aligned}\tilde{\pi}_0 &:= \mathbf{P}\left(X_t = k \,|\, \text{arrival during}(t, t+dt)\right) \\ &= \frac{\mathbf{P}\left(\text{arrival during}(t, t+dt)|\, X_t = k\right)\mathbf{P}\left(X_t = k\right)}{\sum_{j=0}^{M}\mathbf{P}\left(\text{arrival during}(t, t+dt)|\, X_t = j\right)\mathbf{P}\left(X_t = j\right)} \\ &= \frac{(M-k)\lambda dt \times \pi_k}{\sum_{j=0}^{M}(M-j)\lambda dt \times \pi_j} = \frac{M-k}{M-L}\pi_k, \qquad k = 0, 1, \ldots, M.\end{aligned}$$

The probability $\tilde{\pi}_0$ gives the fraction of the times when a newly arriving customer sees k other customers in the QS. For k close to M, the arrival flow is *thin* (as most of the customers would already be in the queue), so only a few of new arrivals will see k customers in the system, so $\tilde{\pi}_k \ll \pi_k$. For small k, the arrival flow is *dense*, so $\tilde{\pi}_k$ will be relatively large.

13. (i) X_t = the number of broken trucks at time t; $S = \{0, 1, 2\}$; rates: $\lambda_0 = 2\lambda$, $\lambda_1 = \lambda$ with $\lambda = 1/40$; $\mu_1 = \mu$, $\mu_2 = 2\mu$ with $\mu = 1/4$ (days^{-1}). A finite source queue (machine repair problem) with $a = 2$, $M = 2$. (ii) As $K_1 = 1/5$ and $K_2 = 1/100$, $\pi_0 = 100/121 \approx 0.826$. (iii) $\pi_2 = K_2\pi_0 = 1/121 \approx 0.008$.

Chapter 8

1. Assuming for simplicity's sake that $n := t/\mu + x\sigma\sqrt{t/\mu^3}$ is integer, we have, using (8.2),

$$\{\xi_t \geq x\} = \left\{N_t \geq \frac{t}{\mu} + x\sigma\sqrt{\frac{t}{\mu^3}}\right\} = \{T_n \leq t\} = \left\{\frac{T_n - \mu n}{\sigma\sqrt{n}} \leq -x\sqrt{\frac{t}{\mu n}}\right\}.$$

By the CLT, the probability of the last event $\approx \Phi(-x\sqrt{t/\mu n}) \approx 1 - \Phi(x)$ since $t/\mu n \to 1$ as $t \to \infty$.

2. Let RVs $X_A \sim f_A$, $X_B \sim f_B$. Clearly, $1/4 - X_A$ also follows density f_B, so that $C_A = C_B$, $m_A (= \mathbf{E}\, X_A) = 1/4 - m_B$, $\sigma_A^2 (= \mathrm{Var}\,(X_A)) = \sigma_B^2$. So need to do all the calculations for f_B only.

(i) As $1 = \int f_B(x)\, dx = C_B \int_0^{1/4} x\, dx = C_B/32$, $C_A = C_B = 32$.

(ii) Case B: $F_B(x) := \int_0^x f_B(t)\, dt = 16x^2$, $0 \leq x \leq 1/4$, $m_B = \int x f_B(x)\, dx = 1/6$, so the density of the stationary residual lifetime distribution $g_B(x) = 6(1 - 16x^2)$, $0 \leq x \leq 1/4$. Mean value $3/32$.

Case A: as $1 - F_A(x) = F_B(1/4 - x)$ and $m_A = 1/4 - m_B = 1/12$, we get $g_A(x) = 12(1 - 4x)^2$, $0 \leq x \leq 1/4$, with mean $1/16$.

(iii) Case A: $N_8 \approx 8/m_A = 96$. Case B: $N_8 \approx 8/m_B = 48$.

(iv) By Theorem 8.2, $(N_8 - 8/m)/\sigma\sqrt{8/m^3}$ is approx. $N(0,1)$-distributed (by Theorem 8.2. Hence, as for $\xi \sim N(0,1)$ one has $\mathbf{P}\left(|\xi| < 1.645\right) \approx 0.9$, we can take the interval with the end points $8/m \pm 1.645\sigma\sqrt{8/m^3}$. Only need to find

$$\sigma_A^2 = \sigma_B^2 = \mathbf{E}\, X_B^2 - m_B^2 = 32\int_0^{1/4} x^2 \times x\, dx - \frac{1}{6^2} = \frac{1}{288}.$$

So the desired intervals are: $(84.60, 107.40)$ in case A, $(43.97, 52.03)$ in case B.

As the number of renewals till time $t = 8$ is *smaller* in case B (with the same variance of the times between renewal epoches), the number of RVs "involved in the uncertainty in N_8" is also *smaller*—hence the uncertainty itself is smaller, too.

3. $F + H * F = F + \left(\sum_{n=1}^{\infty} F^{*n}\right) * F = F + \sum_{n=2}^{\infty} F^{*n} = H.$

4. Substituting $M = D + D * H$:

$$M - [D + M * F] = D + D * H - [D + (D + D * H) * F] = D * (H - F - H * F) = 0$$

from the renewal equation.

5. Since $\mathbf{1}_{\{T_n \le t\}}\mathbf{1}_{\{T_k \le t\}} = \mathbf{1}_{\{T_k \le t\}}$ for $n \le k$, we get

$$N_t^2 = \left(\sum_{n\ge1} \mathbf{1}_{\{T_n \le t\}}\right)\left(\sum_{k\ge1} \mathbf{1}_{\{T_k \le t\}}\right) = \sum_{n\ge1} \mathbf{1}_{\{T_n \le t\}} + 2\sum_{k\ge2} (k-1)\mathbf{1}_{\{T_k \le t\}}.$$

Taking the expectations of the both sides: $H_2(t) = H(t) + 2\sum_{k\ge2}(k-1)F^{*k}(t)$. It remains to notice that

$$H * H = \left(\sum_{n\ge1} F^{*n}(t)\right) * \left(\sum_{k\ge1} F^{*k}(t)\right) = 2\sum_{k\ge2}(k-1)F^{*k}(t).$$

6. Using the hint:

$$((I - F) * J)(x) = \int_{-\infty}^{\infty} (I(x-y) - F(x-y))\, dJ(y)$$
$$= \int_0^{\infty} (\mathbf{1}_{\{x-y\ge0\}} - F(x-y))\, dy = \int_0^{x} (1 - F(x-y))\, dy = \mu F_1(x),$$

so

$$H_S = \frac{1}{\mu} J * (I - F) * \left(\sum_{n=0}^{\infty} F^{*n}\right) = \frac{1}{\mu} J * \left(\sum_{n=0}^{\infty} F^{*n} - \sum_{n=1}^{\infty} F^{*n}\right) = \frac{1}{\mu} J.$$

7. Here we will use the alternative approach. For $p_k := \mathbf{P}(\tau_2 = k)$, $k = 1, 2, \ldots$, set $g(z) := \mathbf{E}\, z^{\tau_2} = \sum_{k\ge1} p_k z^k$. Then

$$g_1(z) := \mathbf{E}\, z^{\tau_1} = \frac{1}{\mu}\sum_{k=1}^{\infty} z^k \sum_{j=k}^{\infty} p_j = \frac{1}{\mu}\sum_{j=1}^{\infty} p_j \sum_{k=1}^{j} z^k = \frac{z}{\mu}\sum_{j=1}^{\infty} p_j \frac{1 - z^j}{1 - z} = \frac{z(1 - g(z))}{\mu(1 - z)},$$

so that the GF of the delayed renewal function (we write $H_S(\{k\}) = H_S(k) -$

$H_S(k-1)$, as for the DFs for distributions on the integers) equals

$$\sum_{k=1}^{\infty} z^k H_S(\{k\}) = \sum_{n=1}^{\infty} g_1(z)g^{n-1}(z) = \frac{g_1(z)}{1-g(z)} = \frac{z}{\mu(1-z)} = \frac{1}{\mu}\sum_{k\geq 1} z^k,$$

which coincides with the GF corresponding to the "DF" k/μ, $k \geq 0$ (assigning the weights $1/\mu$ to each of the points $k = 1, 2, \ldots$).

Chapter 9

1. (i) Setting $f(a,b) := \mathbf{E}\,(X_0-(aX_1+b))^2 = \sigma_0^2+m_0^2+a^2(\sigma_1^2+m_1^2)+2abm_1+b^2-2a(C+m_0m_1)-2bm_1$ and solving $\partial f/\partial a = \partial f/\partial b = 0$ for the stationary point, we get $\widehat{X}_0 = m_0+C\sigma_1^{-2}(X_1-m_1) \equiv m_0+\rho\sigma_0\sigma_1^{-1}(X_1-m_1)$, where $\rho := C/\sigma_0\sigma_1$ is the correlation between X_0 and X_1, with the error $\mathbf{E}\,(X_0-\widehat{X}_0)^2 = \sigma_0^2(1-\rho^2)$.

(ii) Set $\boldsymbol{X} := (X_1,\ldots,X_n)$, $\boldsymbol{a} := (a_1,\ldots,a_n)$, $\boldsymbol{m} := (m_1,\ldots,m_n)$, $\boldsymbol{c} := (C_{01},\ldots,C_{0n}) \in \mathbf{R}^n$ and $C := (C_{jk})_{j,k=1,\ldots,n}$. A linear predictor for X_0 has now the form $\boldsymbol{a}\boldsymbol{X}^T + b$, where ^{T} stands for transposition. Since

$$\mathbf{E}\,(X_0 - \boldsymbol{a}\boldsymbol{X}^T - b)^2 = \mathbf{E}\,((X_0-m_0) - \boldsymbol{a}(\boldsymbol{X}-\boldsymbol{m})^T)^2 + (b - m_0 + \boldsymbol{a}\boldsymbol{m}^T)^2,$$

we just need to minimise in $\boldsymbol{a}$ the expectation on the right-hand side and then put $b := m_0 - \boldsymbol{a}\boldsymbol{m}^T$.

So can now assume that all $m_j = 0$. The system

$$\frac{\partial}{\partial a_j}\mathbf{E}\,(X_0 - \boldsymbol{a}\boldsymbol{X}^T)^2 = 0, \quad j = 1,\ldots,n,$$

for the stationary point is clearly equivalent to $\boldsymbol{c} = \boldsymbol{a}C$ (compute the derivatives!). So when the covariance matrix C is non-degenerate, the best linear prediction is given by $\widehat{X}_0 = m_0 + \boldsymbol{c}C^{-1}(\boldsymbol{X}-\boldsymbol{m})^T$. [What if C is degenerate? Cf. Section 9.4.] Prediction error: DIY.

2. (i) Since $(-X_1, X_2)$ has the same distribution as (X_1,X_2), $\mathrm{Cov}\,(X_1,X_2) = \mathrm{Cov}\,(-X_1,X_2) = -\mathrm{Cov}\,(X_1,X_2) = 0$. So from Problem 1, the best linear predictor for X_2 from X_1 is $\mathbf{E}\,X_2 = 1/3$ (noting that the density of X_2 is $2(1-x)$, $x \in [0,1]$—why?). (ii) Note that the conditional distribution of X_2 given $X_1 = x \in [-1,1]$ is $U(0, 1-|x|)$ (cf. Theorem 10.2). Hence the best predictor is given by the conditional expectation $\mathbf{E}\,(X_2|X_1) = (1-|X_1|)/2$.

3. Putting $z := e^{i\lambda}$ we have

$$\left(\frac{\sin(n\lambda/2)}{\sin(\lambda/2)}\right)^2 = \left(\frac{z^{n/2}-z^{-n/2}}{z^{1/2}-z^{-1/2}}\right)^2 = \frac{(1-z^n)(1-z^{-n})}{(1-z)(1-z^{-1})}$$
$$= \sum_{j=0}^{n-1} z^j \sum_{k=0}^{n-1} z^{-k} = \sum_{k=-n}^{n} (n-|k|)z^k.$$

Use (9.2).

4. Since $(a-b)^2 \geq 0$, we get $2ab \leq a^2 + b^2$ and hence $(\xi_1 + \xi_2)^2 = \xi_1^2 + \xi_2^2 + 2\xi_1\xi_2 \leq 2(\xi_1^2 + \xi_2^2)$. Take the expectations of both sides and note that one can assume w.l.o.g. that $\mathbf{E}\,\xi_j = 0$, $j = 1, 2$.

5. By independence, $\mathbf{E}\,X_t = \mathbf{E}\,Y \times \mathbf{E}\,\cos(\lambda_0 t + \varphi) = 0$. So

$$\mathrm{Cov}\,(X_{t+h}, X_t) = \mathbf{E}\,(X_{t+h}X_t) = \mathbf{E}\,Y^2 \times \mathbf{E}\left[\cos(\lambda_0(t+h)+\varphi)\cos(\lambda_0 t + \varphi)\right]$$

$$= \sigma^2 \int_0^{2\pi} \cos(\lambda_0(t+h)+v)\cos(\lambda_0 t + v)\frac{dv}{2\pi}$$

$$= \frac{\sigma^2}{4\pi}\int_0^{2\pi}\left[\cos(\lambda_0(t+h)+v)\cos(\lambda_0 t+v) + \sin(\lambda_0(t+h)+v)\sin(\lambda_0 t + v)\right]dv$$

$$= \frac{\sigma^2}{4\pi}\int_0^{2\pi}\cos(\lambda_0 h)\,dv = \frac{\sigma^2}{2}\cos(\lambda_0 h) =: \gamma(h)$$

using the statement from the hint (which holds since the integrand is a 2π-periodic function) and a formula for $\cos(x-y)$. As the mean function of $\{X_t\}$ is constant and the ACVF depends on the lag h only, the process is (weakly) stationary.

6. (i) $1 - 0.7z + 0.1z^2 = 0$; $z_1 = 5$, $z_2 = 2$. (ii) $1 - 1.4z + 0.45z^2 = 0$; $z_1 = 2$, $z_2 = 10/9$. (iii) $1 + 0.4z - 0.45z^2 = 0$; $z_1 = 2$, $z_2 = -10/9$. (iv) $1 - \frac{3}{4}z + \frac{9}{16}z^2 = 0$; $z_{1,2} = 2/3 \pm i\sqrt{4/3}$.

Plotting the ACFVs: DIY using Example 9.10.

7. (i) Using $f(\lambda) = (5 + 2e^{i\lambda} + 2e^{-i\lambda})/2\pi = (2 + e^{i\lambda})(2 + e^{-i\lambda})/2\pi = |2 + e^{-i\lambda}|^2/2\pi$, which corresponds to an MA(1) process with $a_0 = 2$, $a_1 = 1$: $X_t = 2Y_t - Y_{t-1}$, $\{Y_t\} \sim \mathrm{WN}(0,1)$. (But also $f(\lambda) = |2 + e^{i\lambda}|^2/2\pi = |2e^{-i\lambda} + e^{-2i\lambda}|^2/2\pi$ etc! So there are other processes having the same ACVF!) (ii) Direct computation (using (9.16)): $\gamma(0) = \sum_{k\geq 0} a_k^2 = 2^2 + 1^2 = 5$, $\gamma(\pm 1) = \sum_{k \geq 0} a_k a_{k+1} = 2$, $\gamma(\pm 2) = \gamma(\pm 3) = \cdots = 0$. Alternatively, you could use (9.10).

8. (i) As $(1 - \beta B)X_t = Y_t$, we get $X_t = (1-\beta B)^{-1}Y_t = \sum_{k=0}^{\infty}\beta^k Y_{t-k}$. Hence, for $h > 0$, $\gamma(h) = \mathbf{E}\,(X_t X_{t+h}) = \sigma^2\beta^h\sum_{k=0}^{\infty}\beta^{2k}$ (using $\mathbf{E}\,(Y_tY_s) = \delta_{st}$), and as $\gamma(h)$ is an even function of h, for an arbitrary h,

$$\gamma(h) = \frac{\sigma^2\beta^{|h|}}{1-\beta^2} = 6.25 \times 0.8^{|h|}.$$

(ii)

$$f(\lambda) = \frac{\sigma^2}{2\pi}\left|\sum_{k=0}^{\infty}\beta^k(e^{-\lambda h})^k\right|^2 = \frac{\sigma^2}{2\pi|1-\beta e^{-\lambda h}|^2} = \frac{2.8125}{\pi(4.1 - 4\cos\lambda)}.$$

Plot: DIY.

(iii) A sharp peak at $\lambda = 0$ and fast decay as λ goes away from zero means that only very slowly oscillating components will pass the LF, whereas all rapid oscillations will effectively be annihilated.

(iv) $\gamma(h) = 6.25 \times (-0.8)^{|h|}$; $f(\lambda) = 2.8125/\pi(4.1 + 4\cos\lambda)$; maxima are at the points $\lambda = \pm\pi$, with very small values in vicinity of $\lambda = 0$, which means that slowly varied components will (nearly) be removed by the LF, while the rapidly oscillating ones will pass.

(v) $\widehat{X}_{t+s} = \beta^s X_t$.

9. Since

$$(1-\beta B)^{-1}(1+\alpha B) = \sum_{k=0}^{\infty} \beta^k B^k (1+\alpha B) = 1 + \sum_{k=1}^{\infty} (\alpha+\beta)\beta^{k-1} B^k,$$

we get $X_t = \sum_{k=0}^{\infty} a_k Y_{t-k}$ with $a_0 = 1$, $a_k = (\alpha+\beta)\beta^{k-1}$, $k > 1$. Hence from (9.16)

$$\gamma(0) = \sigma^2 \sum_{k=0}^{\infty} a_k^2 = \sigma^2 \left[1 + (\alpha+\beta)^2 \sum_{k=0}^{\infty} \beta^{2k}\right] = \sigma^2 \frac{1 + 2\alpha\beta + \alpha^2}{1-\beta^2},$$

and similarly for $\gamma(h)$, $h \neq 0$.

10. (i) $b(z) := 1 + z + 0.5z^2 = 0$; $z_{1,2} = -1 \pm i$, so that $|z_{1,2}| = \sqrt{2} > 1$, which means that $\{X_t\}$ is stable/causal. (ii) No change as the ARMA process has the same characteristic polynomial $b(z)$.

(iii)

$$f(\lambda) = \frac{4}{2\pi|b(e^{-i\lambda})|^2} = \frac{2}{\pi(2.25 + 3\cos\lambda + \cos(2\lambda))}.$$

The plot (DIY!) will have two rather sharp peaks at $\lambda^* \approx \pm 2.4$ and almost vanish at $\lambda = 0$. This means that there will be "almost no" slowly oscillating components in $\{X_t\}$, and the "strongest" (i.e., having the maximum amplitude variance) oscillations will be at frequencies around λ^*.

(iv) As the LF has the form $B - 2 + B^{-1}$, its transfer function is $A(\lambda) = e^{i\lambda} - 2 + e^{-i\lambda} = 2(\cos\lambda - 1)$. The new spectral density:

$$g(\lambda) = |A(\lambda)|^2 f(\lambda) = \frac{8(\cos\lambda - 1)^2}{\pi(2.25 + 3\cos\lambda + \cos(2\lambda))}.$$

(v) As $A(0) = 0$, the output will be identically zero (since the input is an "oscillation at zero frequency").

(vi) As the LF has the form $B - 2 + B^{-1} = B\nabla^2$ and $\nabla^2 1 = \nabla^2 t = 0$, $\nabla^2 t^2 = 2$, we have $B\nabla^2(c_0 + c_1 t + c_2 t^2) = 2c_2$. Verification: DIY.

Chapter 10

1. See Example 3.4.

2. For continuous F, $F(t_u) = u$ for $t_u := \sup\{t : F(t) \le u\}$, $u \in (0,1)$. Hence $\mathbf{P}(F(X) \le u) = \mathbf{P}(X \le t_u) = F(t_u) = u$. For discontinuous F, if there is a jump at point t such that $F(t-0) \le u < F(t)$ for a given u, then $t_u = t$ and

$\mathbf{P}(F(X) \le u) = F(t_u) > u$. [For example: if $X \sim B_p$, then $F(X) = 1 - p$ w.p. $1 - p$, $F(X) = 1$ w.p. p.]

3. Observe that $F(t_u - 0) \le u \le F(t_u)$, $t_u := \sup\{t : F(t) \le u\}$, $u \in (0, 1)$. Since $\{X > t_u\} \subset \{F(X) > u\}$ for any $u \in (0, 1)$, by the TPF one has

$$\begin{aligned}\mathbf{P}(V \le u) &= \underbrace{\mathbf{P}(V \le u | X < t_u)}_{=1} \mathbf{P}(X < t_u) + \mathbf{P}(V \le u | X = t_u) \underbrace{\mathbf{P}(X = t_u)}_{=F(t_u) - F(t_u - 0)} \\ &= F(t_u - 0) + \mathbf{P}(V \le u | X = t_u)(F(t_u) - F(t_u - 0)). \qquad (13.48)\end{aligned}$$

If F is continuous at t_u, the RHS equals $F(t_u - 0) = F(t_u) = u$. If F has a jump $F(t_u) - F(t_u - 0) > 0$ at t_u then, given $X = t_u$, one has $V = (1 - U)F(t_u - 0) + UF(t_u)$, so that

$$\begin{aligned}\mathbf{P}(V \le u | X = t_u) &= \mathbf{P}((1 - U)F(t_u - 0) + UF(t_u) \le u) \\ &= \mathbf{P}(U(F(t_u) - F(t_u - 0)) \le u - F(t_u - 0)) \\ &= \frac{u - F(t_u - 0)}{F(t_u) - F(t_u - 0)}.\end{aligned}$$

Substituting that into the RHS of (13.48) again yields $\mathbf{P}(V \le u) = u$.

4. Show analytically that the ChF of the density (2.17) is $\varphi(it) = e^{it\mu - \sigma^2 t^2/2}$; in particular, we get $\varphi_X(it) = e^{-t^2/2}$ for $X \sim N(0, 1)$. On the other hand, the ChF of Y is $\mathbf{E}\, e^{it(\mu + \sigma X)} = e^{it\mu}\varphi_X(it\sigma) = \varphi(it)$.

Alternatively, for $g(x) = \mu + \sigma x$, the inverse function is $h(y) = (x - \mu)/\sigma$ with $h'(y) = 1/\sigma$, and so by (2.27) we get the density (2.17) from (2.18).

5. (i) As $1 = \int f_{a,b}(x)\, dx = C_{a,b} \int_b^\infty x^{-a-1} dx = C_{a,b}/ab^a$, we get $C_{a,b} = ab^a$.
(ii)

$$F_{a,b}(x) = \begin{cases} 0 & \text{if } x \le b, \\ 1 - (b/x)^a & \text{otherwise.} \end{cases}$$

Sketches here and in (i): DIY.

(iii) As $\mathbf{P}(x \le X \le y) = F(y) - F(x - 0)$: a) $1 - (b/y)^a$; b) $(b/x)^a - (b/y)^a$.

(iv) First find $Q(u) = b(1 - u)^{-1/a}$ (solving $u = F_{a,b}(x)$ for x). Algorithm: 1. Simulate an RV $U \sim U(0, 1)$. 2. Return $X := b(1 - U)^{-1/a}$. 3. Stop. For a sample of n independent copies of X, repeat n times steps 1 and 2 (using independent copies of U).

(v) As $\tau_1 := T_1$ and $\tau_2 := T_2 - T_1$ are independent $Exp(1)$-RVs, can use Problem 2 to first get $U_j := 1 - e^{-\tau_j} \sim U(0, 1)$, $j = 1, 2$, and then use part (iv) to get $X_j = 3(1 - U_j)^{-1/2} = 3e^{\tau_j/2}$ $(= 4.1683, 5.9095)$.

(vi) Wrong, since the inverse (quantile) function is readily available. The rejection method is less efficient as part of simulated (auxiliary) RVs is lost.

6. (i) $F(x) = 1/2 + \arctan(x)/\pi$, $-\infty < x < \infty$.

(ii) First find $Q(u) = \tan(\pi(u - 1/2))$. Algorithm: 1. Simulate an RV $U \sim U(0, 1)$. 2. Return $X := \tan(\pi(U - 1/2))$. 3. Stop. For a sample of n independent copies of X, repeat n times steps 1 and 2 (using independent copies of U).

7. Taking logarithms, we see that X is defined by:

$$\sum_{j=1}^{X}(-\ln U_j) \le \lambda < \sum_{j=1}^{X+1}(-\ln U_j).$$

As $-\ln U_j$, $j = 1, 2, \ldots$, are i.i.d. $Exp(1)$-RVs, $X =$ the number of jumps in the Poisson process with unit rate prior to time t, i.e., $X \sim Po(\lambda)$. Algorithm: 1. Set $k := 0$, $A := 1$. 2. Set $k := k + 1$. 3. Simulate an independent $U \sim U(0,1)$. 4. Set $A := A \times U$. 5. If $A \ge e^{-\lambda}$, go to step 2. 6. Return $X := k - 1$. 7. Stop. (You may wish to rewrite the algorithm using, say, "do while"—in case you have problems with "go to".)

8. Algorithm: 1. Set $t := t_1$, $k := 0$, $A := 1$. 2. Set $k := k + 1$. 3. Simulate an independent $U \sim U(0,1)$. 4. Set $A := A \times U$. 5. If $A \ge e^{-\lambda t}$, go to step 2. 6. If $t = t_1$ then {Return $N_{t_1} := k - 1$. Set $t := t_2$. Go to step 2.} End If. 7. Return $N_{t_2} := k - 1$. 8. Stop.

9. From Example 2.5, the conditional distribution of $N_s - N_{t_1}$ given $N_{t_2} - N_{t_1} = m$ is binomial $B_{m,p}$ with "success" probability $p = (s - t_1)/(t_2 - t_1)$. So you may wish to set $N_s := N_{t_1} + X_1 + \cdots + X_m$, where $X_j \sim B_p$ are independent Bernoulli RVs. Or use the inverse function method as following:

Algorithm: 1. Generate a $U \sim U(0,1)$ 2. Set $j := 0$, $p_0 := (1-p)^m$, $u_0 := p_0$. 3. If $U > u_j$ then $\{p_{j+1} := p_j \times (m - j)p/(j+1)(1-p)$, $u_{j+1} := u_j + p_{j+1}$, $j := j + 1$, go to step 3.$\}$ End If. 4. Return $N_s := N_{t_1} + j$. 5. Stop.

10. (i) Recall that the increments $W_{t_j} - W_{t_{j-1}} \sim N(0, t_j - t_{j-1})$, $j = 1, \ldots, m$, are independent RVs. Algorithm: 1. Simualte m independent $N(0,1)$-RVs $X_1, \ldots, X_m$ (using, say, the Box-Muller algorithm). 2. Set $W_0 := 0$. 3. For $j = 1$ to m step 1 $\{$Set $W_{t_j} := W_{t_{j-1}} + (t_j - t_{j-1})^{1/2} X_j\}$ End For. 4. Return $(W_0, W_{t_1}, \ldots, W_{t_m})$. 5. Stop.

(ii) If $s > t_m$, can take $W_s := W_{t_m} + (s - t_j)^{1/2} X_{m+1}$, where $X_{m+1} \sim N(0,1)$ is independent of $X_1, \ldots, X_m$ (i.e., we just add the point $t_{m+1} = s$ to the grid). If $t_{j-1} < s < t_j$ for some $j \le m$, use Problem 37. We get

$$W_s := W_{t_{j-1}} + \frac{s - t_{j-1}}{t_j - t_{j-1}}(W_{t_j} - W_{t_{j-1}}) + \left(\frac{(s - t_{j-1})(t_j - s)}{t_j - t_{j-1}}\right)^{1/2} X_{m+1}.$$

11. Note that $f(t)$ is the density of the RV $T = A + (B - A)X$, where the density of X is $g(x) = 12x^2(1-x)$, $x \in [0,1]$ (apply (2.27)). As $g(x) = 1$ outside $[0,1]$ and is bounded by $16/9$ ($= \max_x g(x) = g(2/3)$) inside it, and the inverse of the DF is not readily available (have to solve a fourth order algebraic equation), can use the rejection method taking $B = [0,1] \times [0, 16/9]$. Algorithm: 1. Simulate independent $U_1, U_2 \sim U(0,1)$. 2. Set $(V, W) := (U_1, 16U_2/9)$ [which is uniformly distributed in B]. 3. If $W > g(V)$ go to step 1. 4. Return $T := A + (B - A)V$. 5. Stop.

12. Same idea, but now with A being the three-dimensional volume under the density surface and B the enclosing cylinder:

$$A := \{(x_1, x_2, x_3) : (x_1, x_2) \in D, 0 \le x_3 \le f(x_1, x_2)\}, \quad B := D \times [0, C].$$

If (V_1, V_2) is uniformly distributed over D, then for its polar coordinates (R, Θ) we have $\Theta \sim U(0, 2\pi)$, $\mathbf{P}(R \le x) = (x/r)^2$, $0 \le x \le r$. So using the inverse function method for R ($Q(u) = ru^{1/2}$), we can simulate the vector as $(V_1, V_2) := rU_2^{1/2}(\cos(2\pi U_1), \sin(2\pi U_1))$, where $U_1, U_2 \sim U(0,1)$ arè independent. (Alternatively, one could use the "hit-and-run" method sampling points $(Y_1, Y_2) := (2U_1 - 1, 2U_2 - 1)$ uniformly distributed in the square $[-1,1]^2$, discarding those for which $Y_1^2 + Y_2^2 > 1$ and returning $(V_1, V_2) := r(Y_1, Y_2)$ for the retained points.)

Algorithm: 1. Simulate independent (V_1, V_2) uniformly distributed over D and $W \sim U(0, C)$. 2. If $W > f(V_1, V_2)$, go to step 1. 3. Return $(X_1, X_2) := (V_1, V_2)$. 4. Stop.

13. Solving $F(x) = u$ for x we get $Q(u) = x_0 + a(-\ln(1-u))^{1/c}$. So can use the inverse function method.

Algorithm: 1. Simulate $U \sim U(0,1)$. 2. Return $X := x_0 + a(-\ln U)^{1/c}$. 3. Stop. [As usual, we used the fact that $1 - U \sim U(0,1)$ if $U \sim U(0,1)$.]

14. (i) The DF for f_1 is

$$F(x) = \int_{-\infty}^{x} f_1(y)\, dy = \int_0^x \frac{dy}{(1+y)^2} = \frac{x}{1+x}, \qquad x > 0.$$

So the inverse function is $Q(u) = u/(1-u)$, $u \in (0,1)$. Algorithm: 1. Generate $U \sim U(0,1)$. 2. Return $X := U/(1-U)$.

Note that f_2 is a "symmetrisation" of f_1: if $Y = \pm 1$ w.p. 1/2 and $X \sim f_1$ are independent RVs, then $YX \sim f_2$. Hence the algorithm: 1. Generate independent $U_1, U_2 \sim U(0,1)$. 2. Return $X := (1 - 2\mathbf{1}_{\{U_1 < 1/2\}}) \times U_2/(1 - U_2)$.

(ii) The desired relation

$$\frac{\theta}{\pi(x^2 + \theta^2)} \le \frac{2}{\pi}\left(\theta + \frac{1}{\theta}\right) \times \frac{1}{2(1+x)^2},$$

is equivalent to $(1+x)^2 \le (x^2 + \theta^2)(\theta^2 + 1)\theta^{-2}$. The latter follows from the obvious inequality $0 \le (x/\theta - \theta)^2$.

15. (i) Using the representation $X = \mu + \sigma Y$, where $Y \sim N(0,1)$, we get:

$$\begin{aligned} X_1 &= 1 + 2(-2\ln U_2)^{1/2}\cos(2\pi U_1) = -1.9575, \\ X_2 &= 1 + 2(-2\ln U_2)^{1/2}\sin(2\pi U_1) = 1.8855, \\ X_3 &= 1 + 2(-2\ln U_4)^{1/2}\cos(2\pi U_3) = -0.8364. \end{aligned}$$

(ii) Using the inverse function method (cf. Example 10.8), we will get $(X_0, \ldots, X_6) = (2,1,3,2,3,3,2)$. Indeed, as $\boldsymbol{p}_0 = (0.2, 0.7, 0.1)$ and $0.2 < U_1 < 0.2 + 0.7$, we set $X_0 = 2$. Next we use the second row of P, and as $0 < U_2 < 0.4$, we get $X_1 = 1$ *etc.*

(iii) You could proceed by using the facts we established in Section 6.3: given $X_0 = k$, the process stays at k for an $Exp(\lambda_k + \mu_k)$-RV τ_1 and then jumps to $k+1$ (w.p. $\lambda_k/(\lambda_k + \mu_k)$) or to $k-1$. From Example 10.5, we get $\tau_1 = -(\lambda_4 + \mu_4)^{-1} \ln U_1 = 0.4391$, and as $U_2 < 5/9$, the process jumps to $k = 3$, and so on. So the process has jumps at times $T_1 = \tau_1 = 0.4391$, $T_2 = T_1 + \tau_2 = 0.7298$, $T_3 = T_2 + \tau_3 = 1.0294$ and $T_4 = T_3 + \tau_4 = 1.1862$, and the values change as follows: $4 \to 3 \to 4 \to 5 \to ?$ (we need one more U to simulate the value the process takes at time T_4).

16. [a] (i) $F_a(x) = 1 - (1-x)^2$, $x \in [0,1]$, so $Q(u) = 1 - (1-u)^{1/2}$ and we can use $T := 1 - U^{1/2}$ to get 0.6682, 0.0768, 0.2623, 0.5369. (ii) Using $B = [0,1] \times [0,2]$ and $(V_1, W_1) := (U_1, 2U_2)$, we get $f_a(V_1) > W_1$ (acceptance), $T_1 = 0.1101$. Similarly we accept $T_2 = 0.5442$. [b] (i) $F_b(x) = (1 - \cos(\pi x))/2$, $x \in [0,1]$, so $Q(u) = \pi^{-1} \arccos(1-2u)$ leading to 0.2153, 0.7490, 0.5282, 0.3066. (ii) Using $B = [0,1] \times [0, \pi/2]$ and $(V_1, W_1) := (U_1, 0.5\pi U_2)$, we get $f_b(V_1) < W_1$ (rejection). Using the second pair $(U_3, 0.5\pi U_4)$, we accept $T_1 = 0.5442$.

17. We have to show that

$$\mathbf{E}\,[\mathbf{E}\,(Z^2|Y) - (\mathbf{E}\,(Z|Y))^2] \le \mathbf{E}\,Z^2 - (\mathbf{E}\,Z)^2.$$

As $\mathbf{E}\,[\mathbf{E}\,(Z^2|Y)] = \mathbf{E}\,Z^2$ by property (iv) of conditional expectations (see p. 56), this is equivalent (setting $X := \mathbf{E}\,(Z|Y)$) to $\mathbf{E}\,X^2 \ge (\mathbf{E}\,X)^2$. The last relation is obvious from $0 \le \mathbf{E}\,(X - \mathbf{E}\,X)^2 = \mathbf{E}\,X^2 - (\mathbf{E}\,X)^2$.

18. Differentiating $e^{-ta}\varphi_X^m(t)$ w.r.t. t we get the following equation for the stationary point: $m\varphi'_X(t)/\varphi_X(t) = a$. Now use

$$\varphi'_X(t) = (\mathbf{E}\,e^{tX})' = \mathbf{E}\,Xe^{tX} = \int xe^{tx}p(x)\,dx = \varphi_X(t)\int xp_t(x)\,dx.$$

Chapter 11

1. (i) $\mathbf{E}\,(S_{n+m}|\,S_n) = \mathbf{E}\,(S_n + X_{n+1} + \cdots + X_{n+m}|\,S_n) \overset{\text{CE1}}{=} \mathbf{E}\,(S_n|\,S_n) + \mathbf{E}\,(X_{n+1} + \cdots + X_{n+m}|\,S_n) \overset{\text{CE2,CE3}}{=} S_n + \mathbf{E}\,(X_{n+1} + \cdots + X_{n+m}) = S_n + m\mu$.

(ii) Using CE2 and the hint, $S_n = \mathbf{E}\,(X_1 + \cdots + X_n|\,S_n) = \sum_{k\le n} \mathbf{E}\,(X_k|\,S_n) = n\mathbf{E}\,(X_1|\,S_n)$. Hence $\mathbf{E}\,(X_1|\,S_n) = S_n/n$.

(iii) $\mathbf{E}\,(S_{n+m}^2|\,S_n) = \mathbf{E}\,((S_n + X_{n+1} + \cdots + X_{n+m})^2|\,S_n) = \mathbf{E}\,(S_n^2|\,S_n) + 2\mathbf{E}\,(S_n(X_{n+1} + \cdots + X_{n+m})|\,S_n) + \mathbf{E}\,((X_{n+1} + \cdots + X_{n+m})^2|\,S_n) = S_n^2 + 2S_n\mathbf{E}\,(X_{n+1} + \cdots + X_{n+m}) + \mathbf{E}\,(X_{n+1} + \cdots + X_{n+m})^2 = S_n^2 + 2S_n m\mu + m\sigma^2 + m^2\mu^2 = (S_n + m\mu)^2 + m\sigma^2$.

(iv) $\mathbf{E}\,(S_m|\,S_n) = \sum_{k\le m} \mathbf{E}\,(X_k|\,S_n) = \frac{m}{n} S_n$ from (ii).

2. (i) $\mathbf{E}\,(N_{t+s}|\,\mathcal{F}_t) = \mathbf{E}\,(N_t + N_{t+s} - N_s|\,\mathcal{F}_t) \overset{\text{CE1}}{=} \mathbf{E}\,(N_t|\,\mathcal{F}_t) + \mathbf{E}\,(N_{t+s} - N_s|\,\mathcal{F}_t) \overset{\text{CE2,CE3}}{=} N_t + \mathbf{E}\,(N_{t+s} - N_s) = N_t + \lambda s$.

(ii) $\mathbf{E}(N_{t+s}^2|\mathcal{F}_t) = (N_t + \lambda s)^2 + \lambda s$ (the argument is basically the same as in solution to part (iii) in the previous problem, just recall that $\mathbf{E}\, N_s = \mathrm{Var}\,(N_s) = \lambda s$).

(iii) $\mathbf{E}(N_s|\mathcal{F}_t) = N_s$, $\mathbf{E}(N_s^2|\mathcal{F}_t) = N_s^2$ by CE2, as N_s (and hence N_s^2 as well) is $\mathcal{F}_t$-measurable.

(iv) We know that $\mathbf{E}(N_s|N_t) =: \eta(N_s)$ is a function of N_t, with $\eta(n) = \mathbf{E}(N_s|N_t = n)$. From Example 2.5 we know that $(N_s|N_t = n) \sim B_{n,s/t}$ (see also the proof of Theorem 5.2). Hence $\eta(n) = sn/t$, and so $\mathbf{E}(N_s|N_t) = \frac{s}{t}N_t$. Similarly, one can find that $\mathbf{E}(N_s^2|N_t) = \frac{s(t-s)}{t^2}N_t + \frac{s^2}{t^2}N_t^2$.

3. (i) $\{\tau \le t\} = \Omega$ if $m \le t$ and $= \emptyset$ if $m > t$. In either case, the set is an element of $\mathcal{F}_t$. This is an ST.

(ii) $\{\tau_1 \wedge \tau_2 \le t\} = \{\tau_1 \le t\} \cup \{\tau_2 \le t\} \in \mathcal{F}_t$ as each of the events in the union belongs to $\mathcal{F}_t$. Likewise, $\{\tau_1 \vee \tau_2 \le t\} = \{\tau_1 \le t\} \cap \{\tau_2 \le t\} \in \mathcal{F}_t$. This is an ST.

(iii) $\{\tau > t\} = \bigcap_{k=0}^{t}\{X_{k+1}/X_k \le 1\}$, where the last event in the intersection does not need to belong to $\mathcal{F}_t$ (as X_{t+1} is, generally speaking, not $\mathcal{F}_t$-measurable). So $\{\tau \le t\} = \{\tau > t\}^c$ does not need to belong to $\mathcal{F}_t$ as well, and hence τ may not be an ST.

(iv) $\{\tau > t\} = \bigcap_{j=0}^{t}\{\sum_{k=0}^{t} X_k \le X_t^2\} \in \mathcal{F}_t$ as each of the events in the intersection $\in \mathcal{F}_t$. So $\{\tau \le t\} = \{\tau > t\}^c \in \mathcal{F}_t$ as well. This is an ST.

(v) $\{\tau = t\} = \{X_t > 10\} \cap \left[\bigcap_{k=t+1}^{10}\{X_k \le 10\}\right]$. Here, generally speaking, $[\cdots] \notin \mathcal{F}_t$, so τ does not need to be an ST.

4. That $\{X_t\}$ is adapted follows from the definition of CE. Integrability is obvious from CE4. The latter also implies the MG property, as for $0 \le s < t$, one has $\mathbf{E}(X_t|\mathcal{F}_s) = \mathbf{E}(\mathbf{E}(Y|\mathcal{F}_t)|\mathcal{F}_s) = \mathbf{E}(Y|\mathcal{F}_s) = X_s$.

5. As X_{t_1}, X_{t_2} are $\mathcal{F}_{t_3}$-measurable, $\mathbf{E}(X_{t_2} - X_{t_1})(X_{t_4} - X_{t_3}) = \mathbf{E}((\cdots)|\mathcal{F}_{t_3}) = \mathbf{E}\left[(X_{t_2} - X_{t_1})\mathbf{E}(X_{t_4} - X_{t_3}|\mathcal{F}_{t_3})\right] = 0$ since $\mathbf{E}(X_{t_4} - X_{t_3}|\mathcal{F}_{t_3}) = \mathbf{E}(X_{t_4}|\mathcal{F}_{t_3}) - X_{t_3} = 0$, $\{X_t\}_{t\ge 0}$ being an MG. Square integrability is needed to ensure that the expectation of the product exists (by Cauchy–Bunyakovskii's inequality).

6. The SP is clearly adapted. Integrability: $\mathbf{E}|X_n| = \mathbf{E}|S_n^2 - n\sigma^2| \le \mathbf{E}\, S_n^2 + n\sigma^2 = \mathrm{Var}\,(S_n) + (\mathbf{E}\, S_n)^2 + n\sigma^2 = 2n\sigma^2 < \infty$.

(i) $\mathbf{E}(X_{n+1}|\mathcal{F}_n) = \mathbf{E}((S_n + Y_{n+1})^2 - (n+1)\sigma^2|\mathcal{F}_n) = \mathbf{E}(S_n^2 + 2S_nY_{n+1} + Y_{n+1}^2|\mathcal{F}_n) - (n+1)\sigma^2$, where $\mathbf{E}(S_n^2|\mathcal{F}_n) = S_n^2$, $\mathbf{E}(S_nY_{n+1}|\mathcal{F}_n) = S_n\mathbf{E}(Y_{n+1}|\mathcal{F}_n) = S_n\mathbf{E}\, Y_{n+1} = 0$, $\mathbf{E}(Y_{n+1}^2|\mathcal{F}_n) = \mathbf{E}\, Y_{n+1}^2 = \sigma^2$. So $\mathbf{E}(X_n|\mathcal{F}_n) = S_n^2 - n\sigma^2 = X_n$, that's an MG!

(ii) As all $X_k = S_k^2 - k\sigma^2$, $k = 1, \ldots, n$, are functions of $Y_1, \ldots, Y_n$, one has $\mathcal{F}_n' \subset \mathcal{F}_n$. Hence, using CE4, $\mathbf{E}(X_n|\mathcal{F}_n') = \mathbf{E}[\mathbf{E}(X_n|\mathcal{F}_n)|\mathcal{F}_n'] = \mathbf{E}(X_n|\mathcal{F}_n') = X_n$ from (i). Done.

7. DIY. This is no different from the proof of Theorem 11.6 (use the fact the this is a process with independent increments and, for part (iii), the explicit form of the Poisson process' MGF).

8. (i) We know from Example 11.2 that $\{S_n\}$ is an MG. By Theorem 11.2, $0 = \mathbf{E}\, S_0 = \mathbf{E}\, S_\tau = a\mathbf{P}(S_\tau = a) + b\mathbf{P}(S_\tau = b) = (a-b)\mathbf{P}(S_\tau = a) + b$, so that $\mathbf{P}(S_\tau = a) = \frac{b}{b-a}$, $\mathbf{P}(S_\tau = b) = \frac{-a}{b-a}$.

(ii) By Theorem 11.2 applied to the MG $X_n := S_n^2 - n\sigma^2 \equiv S_n^2 - n$, and the result of (i), one has $0 = \mathbf{E}\, X_0 = \mathbf{E}\, X_\tau = \mathbf{E}\,(S_\tau^2 - \tau) = \frac{a^2 b}{b-a} + \frac{b^2(-a)}{b-a} - \mathbf{E}\,\tau = -ab - \mathbf{E}\,\tau$. Therefore, $\mathbf{E}\,\tau = -ab$.

9. (i) The SP $\{Z_n\}$ is clearly adapted to $\mathbf{F}$. Integrability follows from the bound $|Z_n| \le \max_{j\le n} |X_j - X_{j-1}| \sum_{k\le n} |Y_k| \le \sum_{k\le n} C_n$. The MG property: using CE2, $\mathbf{E}\,(Z_{n+1} - Z_n|\, \mathcal{F}_n) = \mathbf{E}\,(Y_{n+1}(X_{n+1} - X_n)|\, \mathcal{F}_n) = Y_{n+1}\mathbf{E}\,(X_{n+1} - X_n|\, \mathcal{F}_n) = 0$ as $\{X_n\}$ is an MG.

(ii) $Y_1 = 1$, and for $n > 1$, one has $Y_n = 2Y_{n-1}$ if $X_{n-1} - X_{n-2} = -1$ (loss) and $Y_n = 1$ if $X_{n-1} - X_{n-2} = 1$ (win).

(iii) The RVs $\xi_n := X_n - X_{n-1}$ are i.i.d., $\mathbf{P}\,(\xi_1 = \pm 1) = \frac{1}{2}$. Hence one has $\mathbf{P}\,(\tau = n) = \mathbf{P}\,(\xi_1 = \xi_2 = \cdots = \xi_{n-1} = -1, \xi_n = 1) = 2^{-n}$, $n = 1, 2, \ldots$ (the geometric distribution). To find the expectation, either compute $\sum_{n=1}^{\infty} n2^{-n}$ or, equivalently, first compute the GF $g_\tau(z) = z/(2-z)$ of τ (it's just the sum of a geometric series) and then take $\mathbf{E}\,\tau = \gamma_\tau'(1) = 2$. The statement of Theorem 11.2 does not hold as $0 = Z_0$, but as your fortune at the time of your first win is always $Z_\tau = 1$ (verify that!), one has $\mathbf{E}\, Z_\tau = 1$.

(iv) $\mathbf{E}\,(Z_n; \tau > n) = \mathbf{E}\,(Z_n; \xi_1 = \xi_2 = \cdots = \xi_n = -1) = -\sum_{j=1}^{n-1} 2^j \times 2^{-n} = -\frac{2^n-1}{2-1} 2^{-n} = 2^{-n} - 1 \not\to 0$ as $n \to \infty$. Condition (11.11) is not satisfied.

10. (i) As no filtration is specified, we consider the natural one. Integrability follows from boundedness ($0 < X_n \le 2^{2n}$). Setting $Y_n := S_n - S_{n-1}$, $n \ge 1$, we have $\mathbf{E}\,(X_{n+1}|\, \mathcal{F}_n) = \mathbf{E}\,(2^{-S_n - Y_{n+1}}|\, \mathcal{F}_n) = 2^{-S_n}\mathbf{E}\,(2^{-Y_{n+1}}|\, \mathcal{F}_n) = X_n \mathbf{E}\, 2^{-Y_{n+1}} = X_n$ since $\mathbf{E}\, 2^{-Y_{n+1}} = 2^{-1} \times \frac{6}{7} + 2^2 \times \frac{1}{7} = 1$.

(ii) Note that $\{X_n \le 0.1\} = \{S_n \ge \log_2 10\} = \{S_n \ge 4\}$ (as S_n is integer-valued). Hence $\tau = \min\{n \ge 0 : S_n \ge 4\} = \min\{n \ge 0 : S_n = 4\}$ as the RW $\{S_n\}$ is skip-free, and so $\mathbf{P}\,(S_\tau = 4) = 1$ (the RW $\{S_n\}$ crosses any positive level w.p. 1 as it has positive trend: $\mathbf{E}\, Y_1 = 1 \times \frac{6}{7} - 2 \times \frac{1}{7} = \frac{4}{7}$.) By Theorem 11.2, $0 = \mathbf{E}\, Z_0 = \mathbf{E}\, Z_\tau = \mathbf{E}\,(S_\tau - \frac{4}{7}\tau) = 4 - \frac{4}{7}\mathbf{E}\,\tau$. Hence $\mathbf{E}\,\tau = 7$.

11. (i) Use the natural filtration. Integrability is obvious from boundedness. Set $Y_n := S_n - S_{n-1}$, $n \ge 1$. We have $\mathbf{E}\,(X_{n+1}|\, \mathcal{F}_n) = \mathbf{E}\,((q/p)^{S_n + Y_{n+1}}|\, \mathcal{F}_n) = (q/p)^{S_n}\mathbf{E}\,((q/p)^{Y_{n+1}}|\, \mathcal{F}_n) = X_n \mathbf{E}\,(q/p)^{Y_{n+1}} = X_n((q/p)^1 p + (q/p)^{-1} q) = X_n$.

(ii) Clearly, either $S_\tau = a$ or $S_\tau = b$. Set $r := q/p$. By Theorem 11.2, $1 = \mathbf{E}\, X_0 = \mathbf{E}\, X_\tau = r^a \mathbf{P}\,(S_\tau = a) + r^b \mathbf{P}\,(S_\tau = b) = (r^a - r^b)\mathbf{P}\,(S_\tau = a) + r^b$, so that $\mathbf{P}\,(S_\tau = a) = 1 - \mathbf{P}\,(S_\tau = b) = (r^b - 1)/(r^b - r^a)$.

Apply Theorem 11.2 to $\{Z_n\}$: $0 = \mathbf{E}\, Z_0 = \mathbf{E}\, Z_\tau = \mathbf{E}\,(S_\tau - (p-q)\tau) = \mathbf{E}\, S_\tau - (2p-1)\mathbf{E}\,\tau = a\mathbf{P}\,(S_\tau = a) + b(1 - \mathbf{P}\,(S_\tau = a)) - (2p-1)\mathbf{E}\,\tau$. Hence $\mathbf{E}\,\tau = (b(1-r^a) + a(r^b - 1))/[(2p-1)(r^b - r^a)]$.

12. Suppose $\omega \in \Omega$ is such that $\{U_1(\omega), U_2(\omega), \ldots\}$ is not everywhere dense in [0,1]. That means that there exist points $t_1 < t_2$ in $[0,1]$ s.t. $U_j(\omega) \notin (t_1, t_2)$, $j = 1, 2, \ldots$ Hence, for the EDFs F_n^* corresponding to the samples consisting of the first n RVs U_j, one has $F_n^*(t_1) = F_n^*(t_2)$, $n = 1, 2, \ldots$ But that clearly contradicts (2.87) as the limiting DF is $F(t) = t$, $t \in [0,1]$. Therefore the set of such ω's must have zero probability.

13. The RW X is normal as a linear transformation of a Gaussian vector, so one just needs to compute its mean and variance. Clearly, $\mathbf{E}\,(2W_{t_1} - W_{t_2}) =$

$2\mathbf{E}\,W_{t_1} - \mathbf{E}\,W_{t_2} = 0$, $\mathrm{Var}\,(X) = \mathbf{E}\,X^2 = \mathbf{E}\,(2W_{t_1} - W_{t_2})^2 = 4\mathbf{E}\,W_{t_1}^2 - 4\mathbf{E}\,W_{t_1}W_{t_2} + \mathbf{E}\,W_{t_2}^2 = 4t_1 - 4(t_1 \wedge t_2) + t_2 = t_2$. So $2W_{t_1} - W_{t_2} \sim N(0, t_2)$.

Alternatively, one may wish to use independence of the increments of the BM: $X = 2W_{t_1} - W_{t_2} = W_{t_1} - (W_{t_2} - W_{t_1})$, the summands on the right-hand side being independent with distributions $N(0, t_2)$ and $N(0, t_2 - t_1)$, resp. Hence the same answer as above.

14. As in Problem 13, X will be normal with zero mean. One can compute $\mathrm{Var}\,(X)$ from the covariance function of the BM (cf. solution to Problem 13; DIY!) or using independence of increments and the following alternative representation for the RV: $X = 2W_2 + (W_3 - W_2) + 2(W_4 - W_3) \sim N(0, 13)$.

15. In the notation of (2.33), we have $\boldsymbol{X} = (W_{t_1} - W_{t_0}, \ldots, W_{t_n} - W_{t_{n-1}})$, $\boldsymbol{Y} = (W_{t_1}, \ldots, W_{t_n})$, and $\boldsymbol{g}(\boldsymbol{x}) = \boldsymbol{x}A$, where

$$A = \begin{pmatrix} 1 & 1 & 1 & \cdots & 1 \\ 0 & 1 & 1 & \cdots & 1 \\ 0 & 0 & 1 & \cdots & 1 \\ \cdot & \cdot & \cdot & \cdots & \cdot \\ 0 & 0 & 0 & \cdots & 1 \end{pmatrix}.$$

Hence $\boldsymbol{g}^{-1}$ is also a linear transform, $|J(\boldsymbol{g}^{-1}(\boldsymbol{y}))| = |\det A^{-1}| = |\det A|^{-1} = 1$, and so

$$f_{\boldsymbol{Y}}(\boldsymbol{y}) = f_{\boldsymbol{X}}(\boldsymbol{g}^{-1}(\boldsymbol{y})) = \prod_{j=1}^{n} f_{W_{t_j} - W_{t_{j-1}}}(x_j - x_{j-1}),$$

which is clearly the same as $f_{t_1,\ldots,t_k}(x_1, \ldots, x_n)$ (see (11.22)).

16. Cf. solution to the previous problem. The transformation matrix is

$$A = \begin{pmatrix} \sqrt{t_1 - t_0} & \sqrt{t_1 - t_0} & \sqrt{t_1 - t_0} & \cdots & \sqrt{t_1 - t_0} \\ 0 & \sqrt{t_2 - t_1} & \sqrt{t_2 - t_1} & \cdots & \sqrt{t_2 - t_1} \\ 0 & 0 & \sqrt{t_3 - t_2} & \cdots & \sqrt{t_3 - t_2} \\ \cdots & \cdots & \cdots & \cdots & \cdots \\ 0 & 0 & 0 & \cdots & \sqrt{t_n - t_{n-1}} \end{pmatrix}.$$

For the arithmetic BM, multiply the matrix by σ (and don't forget that there can be a non-zero drift!).

17. (i) Observing that $2W_3 \sim N(0, 12)$ and $(W_5|\,2W_3 = x) \sim N(x/2, 2)$ by independent increments, by the formula (2.80) for conditional densities, $f_{(2W_3, W_5)}(x, y) = f_{W_5|2W_3}(y|x) f_{2W_3}(x) = \frac{1}{\sqrt{2\pi \cdot 2}} e^{-(y - x/2)^2/4} \frac{1}{\sqrt{2\pi \cdot 12}} e^{-x^2/24} = \frac{1}{4\pi\sqrt{6}} \exp\{-\frac{y^2}{4} + \frac{xy}{4} - \frac{5x^2}{48}\}$.

(ii) Arguing as in (i), compute $f_{(W_2, 2W_3)}(x, y) = \frac{1}{4\pi\sqrt{2}} \exp\{-\frac{y^2}{8} + \frac{xy}{2} - \frac{3x^2}{4}\}$. Next note that, due to independent increments, $(W_5|\,(W_2, 2W_3)) = (x, y)) \stackrel{d}{=} (W_5|\,2W_3 = y) \sim N(y/2, 2)$, as in (i). Now use $f_{(W_2, 2W_3, W_5)}(x, y, z) = f_{W_5|(W_2, 2W_3)}(z|(x, y)) f_{(W_2, 2W_3)}(x, y) = \frac{1}{\sqrt{2\pi \cdot 2}} e^{-(z - y/2)^2/4} f_{(W_2, 2W_3)}(x, y) =$

$\frac{1}{8\pi\sqrt{2\pi}}\exp\{-\frac{3x^2}{4}+\frac{xy}{2}-\frac{3y^2}{16}+\frac{yz}{4}-\frac{z^2}{4}\}$.

18. Clearly, $\mathbf{E}\,\widetilde{W}_t = 0$. For $0 < s < t$, one has $\mathbf{E}\,\widetilde{W}_s\widetilde{W}_t = st\mathbf{E}\,W_{1/s}W_{1/t} = s$ since $\frac{1}{s}\wedge\frac{1}{t}=\frac{1}{t}$. Since $\{\widetilde{W}_t\}$ is a Gaussian process (indeed, $(\widetilde{W}_{t_1},\dots,\widetilde{W}_{t_n}) = (t_1W_{1/t_1},\dots,t_nW_{1/t_n})$ is just a linear transform of the Gaussian vector $(W_{1/t_1},\dots,W_{1/t_n})$, and so is Gaussian itself), Theorem 11.4 yields the desired assertion.

19. Using the MG $\{Y_t := W_t^2 - t\}_{t\ge 0}: 0 = \mathbf{E}\,Y_0 = \mathbf{E}\,Y_\tau = \mathbf{E}\,(W_\tau^2-\tau) = \mathbf{E}\,(a+b\tau-\tau) = a-(1-b)\mathbf{E}\,\tau$. Hence $\mathbf{E}\,\tau = \frac{a}{1-b}$.

20. (i) Using the first MG: $0 = \mathbf{E}\,W_0 = \mathbf{E}\,W_\tau = \mathbf{E}\,v_\tau = \mathbf{E}\,(2\tau-4) = 2\mathbf{E}\,\tau-4$, so that $\mathbf{E}\,\tau = 2$.

(ii) Using the second MG $\{Y_t := W_t^2-t\}_{t\ge0}: 0 = \mathbf{E}\,Y_0 = \mathbf{E}\,Y_\tau = \mathbf{E}\,(W_\tau^2-\tau) = \mathbf{E}\,((2\tau-4)^2-\tau) = 4\mathbf{E}\,\tau^2 - 17\mathbf{E}\,\tau + 16 = 4\mathbf{E}\,\tau^2 - 18$, so that $\mathbf{E}\,\tau^2 = 4.5$.

(iii) Using the third MG $\{Z_t := e^{uW_t-u^2t/2}\}_{t\ge0} : 1 = \mathbf{E}\,Z_0 = \mathbf{E}\,Z_\tau = \mathbf{E}\,e^{uW_\tau-u^2\tau/2} = \mathbf{E}\,e^{u(2\tau-4)-u^2\tau/2} = \mathbf{E}\,e^{-4u-(u^2/2-2u)\tau} = e^{-4u}\mathbf{E}\,e^{-s\tau}$, so that $\mathbf{E}\,e^{-s\tau} = e^{4u}$ for $s := u^2/2-2u$. Solving for u: $u = 2\pm\sqrt{2(s+2)}$; choosing "$-$" (why?), we obtain $l_\tau(s) = \exp\{8-4\sqrt{2(s+2)}\}$.

(iv) By Theorem 11.2, $0 = \mathbf{E}\,W_0 = \mathbf{E}\,W_\tau$, $0 = \mathbf{E}\,Y_0 = \mathbf{E}\,Y_\tau = \mathbf{E}\,W_\tau^2 - \mathbf{E}\,\tau$, so that $\mathbf{E}\,W_\tau = 0$, $\mathbf{E}\,W_\tau^2 = \mathbf{E}\,\tau = 2$.

21. (i) Clearly, $\mathbf{P}\,(W_\tau = a) = 1-\mathbf{P}\,(W_\tau = b)$, and, by Theorem 11.2 for the MG $\{W_t\}$, one has $0 = \mathbf{E}\,W_0 = \mathbf{E}\,W_\tau = a\mathbf{P}\,(W_\tau = a) + b\mathbf{P}\,(W_\tau = b) = (a-b)\mathbf{P}\,(W_\tau = a) + b$, so that $\mathbf{P}\,(W_\tau = a) = \frac{b}{b-a}$.

(ii) $\mathbf{E}\,\tau = -ab$ (see (12.31)).

(iii) Using the third MG and Theorem 11.2, for any $u\in\mathbf{R}$, one has $1 = \mathbf{E}\,Z_0 = \mathbf{E}\,Z_\tau = \mathbf{E}\,e^{uW_\tau-u^2\tau/2} = \mathbf{E}\,\mathbf{E}\,\big(e^{uW_\tau-u^2\tau/2}\big|\,\tau\big) = \mathbf{E}\,\big(e^{-u^2\tau/2}\mathbf{E}\,\big(e^{uW_\tau}\,\big|\,\tau\big)\big)$. By symmetry, $W_\tau = \pm a$ independently of τ, with equal probabilities, so we obtain $1 = \frac{1}{2}(e^{au}+e^{-au})\mathbf{E}\,e^{-u^2\tau/2}$. Therefore $l_\tau(s) = \mathbf{E}\,e^{-s\tau} = 1/\cosh(\sqrt{2s}a)$, $s\ge0$.

(iv) By the chain rule, $l_\tau(s)' = \frac{\sinh(\sqrt{2s}a)}{\cosh^2(\sqrt{2s}a)}\frac{a}{\sqrt{2s}} \to a^2$ as $s\to0$, which is the same answer as in (ii) in the case $a=-b$.

(v) DIY. Just keep differentiating.

22. One can assume w.l.o.g. that the simple processes have a common time partition $0 = t_0 < t_1 < \dots < t_n = T$: $f_t = X_k$ and $g_t = Y_k$ for $t\in[t_{k-1},t_k)$, for some $\mathcal{F}_{t_{k-1}}$-measurable RVs X_k and Y_k, $k = 1,2,\dots,n$. Then

$$\mathbf{E}\,I_t(f)I_t(g) = \mathbf{E}\sum_{k=1}^n X_k(W_{t_k}-W_{t_{k-1}})\sum_{j=1}^n Y_j(W_{t_j}-W_{t_{j-1}})$$

$$= \sum_{k=1}^n \underbrace{\mathbf{E}\,\big(X_kY_k(W_{t_k}-W_{t_{k-1}})^2\big)}_{=:A_k} + \sum_{j\ne k}\underbrace{\mathbf{E}\,X_kY_j(W_{t_k}-W_{t_{k-1}})(W_{t_j}-W_{t_{j-1}})}_{=:B_{jk}},$$

where, by CE4, $A_k = \mathbf{E}\,\mathbf{E}\,\big(X_kY_k(W_{t_k}-W_{t_{k-1}})^2\big|\,\mathcal{F}_{t_{k-1}}\big) = \mathbf{E}\,\big(X_kY_k\mathbf{E}\,\big((W_{t_k}-W_{t_{k-1}})^2\big|\,\mathcal{F}_{t_{k-1}}\big)\big) = \mathbf{E}\,X_kY_k\mathbf{E}\,(W_{t_k}-W_{t_{k-1}})^2 = (t_k-t_{k-1})\mathbf{E}\,X_kY_k$, whereas, for $j<k$, $B_{jk} = \mathbf{E}\,\mathbf{E}\,\big(X_kY_j(W_{t_k}-W_{t_{k-1}})(W_{t_j}-W_{t_{j-1}})\big|\,\mathcal{F}_{t_{k-1}}\big) = \mathbf{E}\,\big(X_kY_j(W_{t_j}-$

$W_{t_{j-1}})\mathbf{E}\,(W_{t_k}-W_{t_{k-1}}|\,\mathcal{F}_{t_{k-1}})) = 0$ as the last CE is zero. Thus, $\mathbf{E}\,I_t(f)I_t(g) = \sum_{k=1}^n (t_k - t_{k-1})\mathbf{E}\,X_k Y_k = \int_0^t \mathbf{E}\,f_s g_s\,ds$.

The identity is actually an immediate consequence of Itô's isometry (Theorem 11.7(iii)) and the following simple general fact: suppose $L : H_1 \to H_2$ is a linear mapping of one inner product space H_1 into another (H_2) such that, for any $u \in H_1$ one has $\|Lu\| = \|u\|$. Then the inner products are also preserved: for any $u, v \in H_1$, one has $(Lu, Lv) = (u, v)$. To verify that, just compare the first expression with the last one in the following chain of equalities: $\|u\|^2 + \|v\|^2 + 2(u,v) = \|u+v\|^2 = \|L(u+v)\|^2 = \|Lu\|^2 + \|Lv\|^2 + 2(Lu, Lv) = \|u\|^2 + \|v\|^2 + 2(Lu, Lv)$.

23. One has $Y_t = f(X_t)$ with $f(x) = e^x$ (so that $f'(x) = f''(x) = f(x)$) and $dX_t = g_t dW_t - \frac{1}{2}g_t^2 dt$. By Itô's formula, $dY_t = f'(X_t)dX_t + \frac{1}{2}f''(X_t)(dX_t)^2 = f(X_t)(g_t dW_t - \frac{1}{2}g_t^2 dt) + \frac{1}{2}f(X_t)g_t^2 dt = f(X_t)g_t dW_t$, showing that $\{Y_t\}$ is an MG (Remark 11.3).

24. Here $X_t = f(W_t)$ with $f(x) = \cos x$, $f'(x) = -\sin x$, $f''(x) = -\cos x$, and so Itô's formula yields $dX_t = -\sin(W_t)dW_t - \frac{1}{2}\cos(W_t)dt$.

25. (i) For $f(x) = e^{-2x}$ one has $f'(x) = -2f(x)$, $f''(x) = 4f(x)$, so by Itô's formula $df(X_t) = -2f(X_t)(dt + dW_t) + \frac{1}{2}\,4f(X_t)(dt + dW_t)^2 = -2f(X_t)dt - 2f(X_t)dW_t + 2f(X_t)dt = -2f(X_t)dW_t$.

(ii) It is an MG indeed, using Remark 11.3.

26. (i) Clearly, $c = S_0 = 5$ and $S_t = f(t, W_t)$ with $f(t,x) = 5e^{at+bx}$, so that $\partial_t f = af$, $\partial_x f = bf$, $\partial_{xx} f = b^2 f$. So, by Itô's formula, $dS_t = af(t,W_t)dt + \partial_x f(t,W_t)dW_t + \frac{1}{2}b^2 f(t,W_t)(dW_t)^2 = (a + \frac{b^2}{2})S_t dt + bS_t dW_t$. Comparing that with the assumed SDE, we obtain that $a + \frac{b^2}{2} = 0.2$, $b = 1$. So yes, the suspected form of S_t is correct, with $a = -0.3$, $b = 1$, $c = 5$.

(ii) Either you apply Itô's formula to the verified representation $X_t := 1/S_t = (5e^{-0.3t+W_t})^{-1} = 0.5e^{0.3t-W_t}$, or you apply it directly to $f(S_t) = 1/S_t$ using the assumed SDE. Using the latter approach, one has $f'(x) = -x^{-2}$, $f''(x) = 2x^{-3}$, so that $df(S_t) = -S_t^{-2}(0.2S_t dt + S_t dW_t) + \frac{1}{2}\,2S_t^{-3}(dS_t)^2 = -0.2S_t^{-1}dt - S_t^{-1}dW_t + S_t^{-3}S_t^2 dt = 0.8X_t dt - X_t dW_t$, QED.

27. (i) Here $f(x) = \sqrt{x}$, $f'(x) = \frac{1}{2}\,x^{-1/2}$, $f''(x) = -\frac{1}{4}\,x^{-3/2}$, $x > 0$. So, by Itô's formula, $dZ_t = \frac{1}{2}\,V_t^{-1/2}((1 - V_t)dt + 2\sqrt{V_t}\,dW_t) + \frac{1}{2}(-\frac{1}{4}\,V_t^{-3/2})(2\sqrt{V_t})^2 dt = -\frac{1}{2}\,V^{1/2}(t)dt + dW_t = -\frac{1}{2}\,Z dt + dW_t$, which is a special case of (11.46) (with $\alpha = \sigma = 1$). Clearly, $Z_0 = \sqrt{V_0} = 1$.

(ii) Use the product rule of Itô's calculus (Theorem 11.10) with $X_t = e^{-t/2}$, $Y_t = 1 + \int_0^t e^{s/2}dW_s$. As one the factors is smooth ($dX_t = -\frac{1}{2}\,e^{-t/2}dt = -\frac{1}{2}\,X_t dt$), by Remark 11.2 this will be just the usual product rule: $dZ_t = dX_t \cdot Y_t + X_t \cdot dY_t = -\frac{1}{2}\,X_t Y_t dt + X_t e^{-t/2}dW_t = -\frac{1}{2}\,Z_t dt + dW_t$. Obviously, $Z_0 = e^0\left(1 + \int_0^0\right) = 1$.

Chapter 12

1. (i) As $X_t = X_s + (t - s) + 2(W_t - W_s)$, the conditional distribution of X_t

given $X_s = x$ is $N(x+t-s, 4(t-s))$, so that

$$p(s,x;t,y) = \frac{1}{2\sqrt{2\pi(t-s)}} \exp\left\{-\frac{(y-x-t+s)^2}{8(t-s)}\right\}.$$

(ii) As $\mu \equiv 1$, $\sigma \equiv 2$, the BKE for $v = v(s,x)$ is: $\partial_s v = -\partial_x v - 2\partial_{xx} v$. The FKE for $u = u(t,y)$: $\partial_t u = -\partial_y u + 2\partial_{yy} u$.

(iii) One has $u(t,y) = \varphi(8t, y - x_0 - t)$, $\varphi(s,z) := (\pi s)^{-1/2} e^{-z^2/s}$. Compute

$$\partial_s \varphi = -\frac{\varphi}{2s} + \frac{z^2\varphi}{s^2}, \quad \partial_z \varphi = -\frac{2z\varphi}{s}, \quad \partial_{zz}\varphi = -\frac{2\varphi}{s} + \frac{4z^2\varphi}{s^2},$$

and then use the chain rule to compute $\partial_t u = \partial_s \varphi \partial_t s + \partial_z \varphi \partial_t z = \left(-\frac{\varphi}{2s} + \frac{z^2\varphi}{s^2}\right) \times 8 + \left(-\frac{2z\varphi}{s}\right) \times (-1) = \left(-\frac{4}{s} + \frac{8z^2}{s^2} + \frac{2z}{s}\right)\varphi$ *etc.* to verify the claim. As $t \to 0$, $u(t,y) \to 0$ for all $y \neq x_0$, while $\lim_{t\to 0} u(t,x_0) = \infty$. As $\int u(t,y)dy = 1$ for all $t > 0$, one can say that the limit is the Dirac delta function.

2. (i) As $\mu(s,x) = -\alpha x$, $\sigma(s,x) \equiv \sigma = \text{const}$, the BKE for $v = v(s,x)$ is: $\partial_s v = \alpha x \partial_x v - \frac{1}{2}\sigma^2 \partial_{xx} v$. The FKE for $u = u(t,y)$: $\partial_t u = -\partial_y(-\alpha y u) + \frac{1}{2}\sigma^2 \partial_{yy} u = \alpha u + \alpha y \partial_y u + \frac{1}{2}\sigma^2 \partial_{yy} u$.

(ii) As we saw in Remark 11.1, if f_t is a non-random function and $a(t) := \int_0^t f_s^2 ds$, then $dW_{a(t)} = f_t d\widetilde{W}_t$. Here we have $f_t = \sigma e^{\alpha t}$ yielding $\int_0^t f_s^2 ds = \sigma^2(e^{2\alpha t} - 1)/2\alpha$ and, since $dZ_{a(t)} de^{-\alpha t} = 0$, by the product rule one has $dY_t = Z_{a(t)} de^{-\alpha t} + e^{-\alpha t} dZ_{a(t)} = -\alpha e^{-\alpha t} Z_{a(t)} dt + e^{-\alpha t}\sigma e^{\alpha t} d\widetilde{W}_t = -\alpha Y_t dt + \sigma d\widetilde{W}_t$.

(iii) The distribution of $e^{-\alpha t} Z_{a(t)} = e^{-\alpha t}(x + W_{a(t)})$ is $N(e^{-\alpha t}x, e^{-2\alpha t}a(t)) = N(e^{-\alpha t}x, \sigma^2(1-e^{-2\alpha t})/(2\alpha))$, hence the density of $Y(t)$ is

$$\frac{1}{\sqrt{\pi\sigma^2(1-e^{-2\alpha t})/\alpha}} \exp\left\{-\frac{(y - xe^{-\alpha t})^2}{\sigma^2(1-e^{-2\alpha t})/\alpha}\right\} \to \frac{1}{\sqrt{2\pi\sigma_\infty^2}} \exp\left\{-\frac{y^2}{2\sigma_\infty^2}\right\}$$

as $t \to \infty$, $\sigma_\infty^2 := \frac{\sigma^2}{2\alpha}$. The existence of the limit means the existence of a stationary distribution independent of the initial values of the process, i.e., the ergodicity of the process. (Cf. Example 11.15.)

3. (i) As $\mu(s,x) = 1 - x$, $\sigma(s,x) = \sqrt{2x}$, the BKE for $v = v(s,x)$ is: $\partial_s v = (x-1)\partial_x v - x\partial_{xx} v$. The FKE for $u = u(t,y)$: $\partial_t u = -\partial_y((1-y)u) + \partial_{yy}(yu) = u + (y+1)\partial_y u + y\partial_{yy} u$.

(ii) We will use (12.23): in this case, for $y_0 > 0$,

$$g_{y_0}(y) = \exp\left\{-\int_{y_0}^{y} \frac{2(1-z)}{2z}\, dz\right\} = \frac{y_0}{y} e^{y-y_0}, \quad y > 0,$$

so that, for $y > 0$,

$$\pi(y) = \frac{C_1 \int_{y_0}^{y} g_{y_0}(u)\, du + C_2}{\sigma^2(y) g_{y_0}(y)} = \frac{C_1 y_0 e^{-y_0} \int_{y_0}^{y} u^{-1} e^{u} du + C_2}{2y(y_0/y)e^{y-y_0}} = C_3 e^{-y},$$

as we must put $C_1 := 0$ to ensure that $\pi(y) \geq 0$ for all $y > 0$. Clearly, $\int_0^\infty \pi(y)\,dy = 1$ implies that $C_3 = 1$, so that the stationary distribution is $Exp(1)$.

4. (i) As $\mu(s,x) = \frac{1}{2}x$, $\sigma(s,x) \equiv 1$, the BKE for $v = v(s,x)$ has the form: $\partial_s v = -\frac{1}{2}x\partial_x v - \frac{1}{2}\partial_{xx}v$. The FKE for $u = u(t,y)$: $\partial_t u = -\partial_y(\frac{1}{2}yu) + \partial_{yy}(\frac{1}{2}u) = -\frac{1}{2}u - \frac{1}{2}y\partial_y u + \frac{1}{2}\partial_{yy}u$.

(ii) For the derivation, see Case 1 in Section 12.4. The differential equation is: $0 = -\frac{1}{2}xV'(x) - \frac{1}{2}V''(x)$, with $V(a) = 0$, $V(b) = 1$. Solving: $(\ln V')' = V''/V' = -x$, so $\ln V' = -\frac{1}{2}x^2 + C_0$, or $V' = C_1 e^{-x^2/2}$. Hence $V(x) = C_2\Phi(x) + C_3$, Φ being the standard normal DF. Using the boundary conditions yields $V(x) = (\Phi(x) - \Phi(a))/(\Phi(b) - \Phi(a))$, $x \in [a,b]$.

(iii) Plot sketching: DIY. Differences in shape can be explained referring to the effect of the drift coefficient that can be thought of as representing a "force" pushing the "walking particle" away from the origin.

(iv) $\mathbf{P}\left(\max_{t\geq 0} X_t > b \middle|\, X_0 = x\right) = \lim_{a\to-\infty} V(x) = \Phi(x)/\Phi(b)$, $x \leq b$ (of course, $\mathbf{P}(\cdots) = 1$ for $x > b$).

5. (i) The BKE for $v = v(s,x)$ has the form: $\partial_s v = \alpha x\partial_x v - \frac{1}{2}(1-x^2)\partial_{xx}v$. The FKE for $u = u(t,y)$: $\partial_t u = (\alpha-1)u + (\alpha-2)y\partial_y u + \frac{1}{2}(1-y^2)\partial_{yy}u$.

(ii) One has $\pi(y) = C(1-y^2)^{\alpha-1}$, $y \in (0,1)$. Note that stationary distributions are actually scaled (to the interval $(-1,1)$) *beta distributions* with parameters α, α (cf. p. 301). Sketching the plots: DIY. Note that the graphs are U-shaped for $\alpha < 1$, the stationary distribution is $U(-1,1)$ for $\alpha = 1$, and the stationary densities are unimodal and tend to zero as $t \to \mp 1$ when $\alpha > 1$, all the densities being even functions. An explanation of the differences can be given referring to the meaning of the parameter α: the bigger its value, the stronger the force pushing the "walking particle" towards the origin. Expand!

(iii) [a] For $\alpha = 1/2$,

$$V(x) = \frac{\arcsin x - \arcsin a}{\arcsin b - \arcsin a}, \qquad a \leq x \leq b.$$

[a] For $\alpha = 1$,

$$V(x) = \frac{\ln\frac{1+x}{1-x} - \ln\frac{1+a}{1-a}}{\ln\frac{1+b}{1-b} - \ln\frac{1+a}{1-a}}, \qquad a \leq x \leq b.$$

(iv) Use the method presented in Case 2 in Section 12.4. The answer is: $U(x) = 2\ln 2 - (1-x)\ln(1-x) - (1+x)\ln(1+x)$, $x \in (-1,1)$.

6. SDE (12.32) means this: $X_t = x_0 + \int_0^t \alpha X_u du + s\int_0^t \sqrt{X_u}\,dW_u$. Taking the expectations: $m_X(t) = x_0 + \int_0^t m_X(u)\,du$. Differentiating: $m_X'(t) = \alpha m_X(t)$, so $m_X(t) = x_0 e^{\alpha t}$. Use Itô's formula for $f(x) = x^2$ to get

$$dX_t^2 = 2X_t(\alpha X_t dt + s\sqrt{X_t}dW_t) + s^2 X_t dt,$$

i.e., $X_t^2 - x_0^2 = 2\alpha\int_0^t X_u^2 du + 2s\int_0^t X_u^{3/2} dW_u + s^2\int_0^t X_u du$. Taking expectations: $m_{X^2}(t) - x_0^2 = 2\alpha\int_0^t m_{X^2}(u)du + s^2\int_0^t m_X(u)\,du$, leading to the differential equation $m_{X^2}'(t) = 2\alpha m_{X^2}(t) + s^2 m_X(t) = 2\alpha m_{X^2}(t) + s^2 x_0 e^{\alpha t}$, $m_{X^2}(0) = x_0^2$. Please solve it yourself.

7. (i) The BKE for $v = v(s,x)$ is: $\partial_s v = -x(1-x)\partial_x v - \frac{1}{2}x(1-x)\partial_{xx}v$. The FKE for $u = u(t,y)$: $\partial_t u = 2(y-1)u + (1-3y+y^2)\partial_y u + \frac{1}{2}y(1-y)\partial_{yy}u$.

(ii) For the derivation, see Case 1 in Section 12.4. The equation for $V(x) := \mathbf{P}(X_\tau = 1 | X_0 = x)$, where $\tau := \min\{t \geq 0 : X_t = 0 \text{ or } X_t = 1\}$, $x \in (0,1)$, has the form $0 = -x(1-x)V'(x) - \frac{1}{2}x(1-x)V''(x)$, i.e., $0 = V' + \frac{1}{2}V''$, with boundary conditions $V(0) = 0$, $V(1) = 1$. This is a linear differential equation with constant coefficients. Solution: $V(x) = (1-e^{-2x})/(1-e^{-2})$, $x \in (0,1)$.

8. (i) The BKE for $v = v(s,x)$ has the form: $\partial_s v = (\gamma_1 x - \gamma_2(1-x))\partial_x v - \frac{1}{2}x(1-x)\partial_{xx}v$. The FKE for $u = u(t,y)$: $\partial_t u = (\gamma_1 + \gamma_2 - 1)u + (1 - \gamma_2 + (\gamma_1 + \gamma_2 - 2)y)\partial_y u + \frac{1}{2}y(1-y)\partial_{yy}u$.

(ii) The stationary density has the form $\pi(y) = 6y(1-y)$, $y \in (0,1)$. (Note that this a beta distribution.)

9. This one is easy: $\mathbf{E}\,X_t = \mathbf{E}\,W_t - t\mathbf{E}\,W_1 = 0$ and, for $0 \leq s \leq t \leq 1$, one has $\mathbf{E}\,X_s X_t = \mathbf{E}\,W_s W_t - s\mathbf{E}\,W_1 W_t - t\mathbf{E}\,W_s W_1 + st\mathbf{E}\,W_1^2 = s(1-t)$, as desired.

Chapter 13

1. The call price $C(K)$ is given in Example 13.3. It is obvious that $C(K)$ is a piece-wise linear function of K with the following features: $C(0) = S_0$, the slope is equal to $-1/(1+r)$ for $K < dS_0$ (in that interval, both $(\cdots)^+ > 0$), to $-p^*/(1+r) > -1/(1+r)$ for $dS_0 < K < uS_0$ (in that interval, only the first of $(\cdots)^+$ is positive), to 0 for $K > uS_0$, where $C(K) = 0$ (plot it yourself!).

The features common to all calls: (i) $C(0) = S_0$ (zero strike price means one just receives the share at maturity, so the current price of such call should be equal to the current share price); (ii) $C(K)$ is a decreasing function of K (the larger K, the smaller the payoff, so it should cost less); (iii) $C(K) \to 0$ as $K \to \infty$ (as the payoff then vanishes).

2. The payoff function is equal to the sum of those of the call and put, so, from (13.5) and (13.6), it is given by $(s-K)^+ + (s-K)^- = |s-K|$. You must be able to plot that function yourself!

3. (i) The payoff function is $g(s) = (s-K_1)^+ - (s-K_2)^+$. This is a piece-wise linear function with zero slope outside the interval $[K_1, K_2]$, such that $g(0) = 0$ and $g(\infty) = K_2 - K_1$. Plot it.

(ii) The spread must be cheaper as its payoff is equal to that of the call for $s \leq K_2$ and is strictly smaller than the call's for $s > K_2$.

(iii) Since $S_0 = 5$, $dS_0 = 4$ and $uS_0 = 6$, one has $d = \frac{4}{5}$, $u = \frac{6}{5}$ (NA as $d < 1 + r = 1.1 < u$) and $p^* = \frac{1+r-d}{u-d} = \frac{3}{4}$. So, from the pricing formula (13.11), the spread price equals $\frac{1}{1+r}\left[p^* g(6) + (1-p^*)g(4)\right] = \frac{1}{1.1}\left[\frac{3}{4}\times 2 + \frac{1}{4}\times 1\right] = \frac{35}{22} \approx 1.59$.

(iv) One has $\Delta = \frac{g(6)-g(4)}{6-4} = \frac{1}{2}$, $b = \frac{1.2g(4)-0.8g(6)}{1.1\times 0.4} = -\frac{10}{11}$. Verifying replication: $V_1(u) = \frac{1}{2}\times 6 - \frac{10}{11}\times 1.1 = 3 - 1 = 2$, $V_1(d) = \frac{1}{2}\times 4 - \frac{10}{11}\times 1.1 = 2 - 1 = 1$, which coincide with the values of $g(6)$ and $g(4)$, respectively. OK!

4. (i) One has to verify (13.2). Here $d = S_1(\omega_2)/S_0 = \frac{8}{9}$, $u = S_1(\omega_1)/S_0 = \frac{4}{3} = \frac{12}{9}$, so indeed $d < 1 + r = \frac{10}{9} < u$.

(ii) One has $p^* = \frac{10/9-8/9}{12/9-8/9} = \frac{1}{2}$, so that the claim value is $X^* = \frac{1}{1+r}\mathbf{E}^*X = \frac{9}{10}\left[\frac{1}{2}\times 7 + \frac{1}{2}\times 2\right] = \frac{81}{20} = 4.05$.

(iii) One has $\Delta = \frac{7-2}{20/3-40/9} = \frac{9}{4} = 2.25$, $b = \frac{(12/9)\times 2-(8/9)\times 7}{(10/9)\times(4/9)} - \frac{36}{5} = -7.2$, so that the replicating portfolio is $(\Delta, b) = \left(\frac{9}{4}, -\frac{36}{5}\right)$. Its time $t = 0$ value is $V_0 = \Delta S_0 + b = \frac{9}{4}\times 5 - \frac{36}{5} = \frac{81}{20}$, which agrees with the result of part (ii). Its time $t = 1$ values are:

$$V_1 = \Delta S_1 + b(1+r) = \begin{cases} \frac{9}{4}\times\frac{20}{3} - \frac{36}{5}\times\frac{10}{9} = 7 & \text{if } \omega = \omega_1, \\ \frac{9}{4}\times\frac{40}{9} - \frac{36}{5}\times\frac{10}{9} = 2 & \text{if } \omega = \omega_2, \end{cases} \qquad \text{OK!}$$

5. (i) We computed the price in Example 13.4. Like the call price discussed in Problem 1, it is a piece-wise linear function, changing its slope value at the points dS_0 and uS_0. Plot it. Alternatively, one can express the put price in terms of the call price using the put-call parity (13.15) and use the plot from Problem 1 (flip it upside down!).

(ii) Here $d = \frac{3.6}{4} = 0.9$, $u = \frac{4.6}{4} = 1.15$, so that $d < 1 + r = 1.05 < u$, this is an NA market, one can use the pricing formula with $p^* = \frac{1.05-0.9}{1.15-0.9} = 0.6$: $P = \frac{1}{1.05}\left[0.6\times(4.6-3.8)^- + 0.4\times(3.6-3.8)^-\right] = \frac{8}{105} \approx 0.076$.

(iii) From the put-call parity (13.15), $C = S_0 + P - \frac{K}{1+r} = \frac{16}{35} \approx 0.457$.

6. (i) NA is equivalent to existence of $p_j^* > 0$, $\sum_{j=1}^3 p_j^* = 1$, such that $S_0 = \frac{1}{1+r}(p_1^* dS_0 + p_2^* mS_0 + p_3^* uS_0)$ or, which is the same, $p_1^* d + p_2^* m + p_3^* u = 1 + r$. Dividing both sides by $p_1^* + p_3^* = 1 - p_2^*$, we get

$$\underbrace{\frac{p_1^*}{p_1^* + p_3^*}}_{=:p} d + \underbrace{\frac{p_3^*}{p_1^* + p_3^*}}_{=1-p} u = \frac{1+r}{1-p_2^*} - \frac{p_2^*}{1-p_2^*} m.$$

Note that, for $p_2^* > 0$ small enough, the right-hand side will be arbitrary close to $1+r$, and so will still lie in the interval (d, u). Hence, according to our argument in the binomial case, there will exist a $p \in (0,1)$ such that $pd + (1-p)u =$ the right-hand side of the displayed formula. This proves existence of the EMM (p_j^*).

(ii) DIY. The set of all hedges is the intersection of the three half-planes in the (Δ, b)-plane that are bounded (from below) by the straight lines given by the equations $\Delta kS_0 + b(1+r) = X_k$, $k = d, m, u$. The perfect hedge would be the point at the intersection of these three lines, but they do not intersect at a common point (why?), so no perfect hedge exists.

7. (i) DIY.

(ii) NA holds iff there exists an EMM $\mathbf{P}^*$. That, in turn, is equivalent to having the point $\boldsymbol{S}_0 = (S_0^1, S_0^2)$ inside the triangle conv$\{\boldsymbol{S}_1(\omega_1), \boldsymbol{S}_1(\omega_2), \boldsymbol{S}_1(\omega_3)\}$, which is the case (see the plot you made in part (i)).

(iii) Need to find p_j^*, $j = 1, 2, 3$, such that $\boldsymbol{S}_0 = \sum_{j=1}^3 p_j^* \boldsymbol{S}_1(\omega_j)$, or, component-wise,

$$\begin{cases} 4 = 6p_1^* + 4p_2^* + 2p_3^* \\ 5 = 6p_1^* + 4p_2^* + 7p_3^* \\ 1 = p_1^* + p_2^* + p_3^* \end{cases}$$

the last equation just meaning that the p_j^*'s form a probability distribution. The system has unique solution $p_1^* = 0.2$, $p_2^* = 0.6$, $p_3^* = 0.2$. Thus, the EMM is unique, so that the market is complete.

(iv) For the claim $X := (S_1^1 - K)^+ = (S_1^1 - 5)^+$, its arbitrage free price is given by $X^* = \frac{1}{1+r}\mathbf{E}^*X = p_1^*(S_1^1(\omega_1) - 5)^+ + p_2^*(S_1^1(\omega_2) - 5)^+ + p_3^*(S_1^1(\omega_3) - 5)^+ = 0.2 \times (6-5)^+ + 0.6 \times (4-5)^+ + 0.2 \times (2-5)^+ = 0.2$.

(v) Need to find (Δ^1, Δ^2, b) such that $\Delta^1 S_1^1(\omega) + \Delta^2 S_1^2(\omega) + b(1+r) = X(\omega)$ for all $\omega \in \Omega$. That is, we have three equations (corresponding to the three possible states of the world):

$$\begin{cases} 6\Delta^1 + 6\Delta^2 + b = 1 \\ 4\Delta^1 + 4\Delta^2 + b = 0 \\ 2\Delta^1 + 7\Delta^2 + b = 0. \end{cases}$$

The system has unique solution $(\Delta^1, \Delta^2, b) = (0.3, 0.2, -2)$.

Yes, we do need stock 2 for replication, as the above strategy is the unique solution of the linear system equivalent to the replication condition.

8. (i) Diagrams: please DIY. The time $t = 0$ put price is $P = 66.56$.

(ii) The replicating portfolio has the following form: at time $t = 0$, use $(\Delta_1, b_1) = (-0.256, 168.96)$; at time $t = 1$, if $S = 200$ then use $(\Delta_2, b_2) = (-1, 288)$, while if $S = 700$ then use $(\Delta_2, b_2) = (-\frac{4}{35}, 89.6)$.

(iii) We simply have to verify that at time $t = 1$ one has $\Delta_1 S_1 + b_1(1+r) = \Delta_2 S_1 + b_2(1+r)$, whatever the state of the world. This is so indeed, the common value being 160 if $S_1 = 200$ and 32 if $S_1 = 700$.

(iv) Use the put-call parity.

9. Diagrams: please DIY. Here $p^* = \frac{1-2/3}{4/3-2/3} = 0.5$, the call payoff $X = (S_4 - 120)^+$. Working backwards, from the five possible payoff values (after four periods) $(16-120)^+ = 0$, $(32-120)^+ = 0$, $(64-120)^+ = 0$, $(128-120)^+ = 8$ and $(256-120)^+ = 136$, we find that the time $t = 0$ call price is 10.5.

10. (i) The Black-Scholes formula gives $C \approx 0.1184$. (ii) The implied volatility is approx. 0.2594.

11. (i) $C \approx 17.95$, $P \approx 21.54$. (ii) $C \approx 1.46$, $P \approx 11.09$. Observe that both prices are close to 20 in (i), while the call is much cheaper in (ii), the put price being close to the value $(S_0 - K)^- = 10$. One can explain that by noting that, in the latter case, the time to maturity is small and so it is unlikely that the stock price will change much. If S_T is close to 100, the call is next to worthless (its price is positive as it is still possible that $S_T > 110$), while the put value $(S_T - K)^-$ is close to $(100-110)^- = 10$.

Greek Alphabet

A	α	alpha
B	β	beta
Γ	γ	gamma
Δ	δ	delta
E	ε, ϵ	epsilon
Z	ζ	zeta
H	η	eta
Θ	θ, ϑ	theta
I	ι	iota
K	κ	kappa
Λ	λ	lambda
M	μ	mu
N	ν	nu
Ξ	ξ	xi
O	o	omicron
Π	π	pi
P	ρ	rho
Σ	σ, ς	sigma
T	τ	tau
Υ	υ	upsilon
Φ	ϕ, φ	phi
X	χ	chi
Ψ	ψ	psi
Ω	ω	omega

The letters υ and ς are (almost) never used in mathematical formulae.[1]

[1]Why do you think this is so?

Notations

$:=$	defining equality ($a := b$ assigns the value b to a)
$\emptyset$	the empty set
$\|S\|$	the number of elements in the set S (the cardinality of S)
$\#\{S : C\}$	the number of elements in S such that C holds for them
$\lfloor x \rfloor$	the integer part of x
$\nearrow$	$t \nearrow s$ if t increases to s
$\searrow$	$t \searrow s$ if t decreases to s
$\xrightarrow{\text{distr}}$	convergence in distribution
$\leftrightarrow$	states $j \leftrightarrow k$ communicate with each other
$\vee$	$t \vee s := \max\{t, s\}$
$\wedge$	$t \wedge s := \min\{t, s\}$
x^+	$= 0 \vee x$, the positive part of x
x^-	$= 0 \vee (-x)$, the negative part of x
∇	the backward difference operator
$\mathbf{1}_A$	the indicator of the set (event) A
2^S	the power set (consists of all subsets of the set S)
B	the backward shift operator (for time series)
B_p	the Bernoulli distribution with parameter p
$B_{n,p}$	the binomial distribution with parameters n, p
$\mathcal{B}$	the σ-algebra of Borel sets on $\mathbf{R}$
$C_{\boldsymbol{X}}$	the covariance matrix of the random vector $\boldsymbol{X}$
$\mathrm{Cov}\,(X, Y)$	the covariance between the RVs X and Y
^{c}	the set complement: $A^c = \{\omega \in \Omega : \omega \notin A\}$
$\mathbf{E}$	the expectation symbol
$\mathbf{E}\,(X; A)$	$= \mathbf{E}\,(X\mathbf{1}_A)$
$\mathbf{E}\,(X\| A)$	the conditional expectation of X given the event A
$\mathbf{E}\,(X\| Y)$	the conditional expectation of X given the RV Y
$Exp(\lambda)$	the exponential distribution with parameter λ
$\mathcal{F}$	σ-algebra (σ-field) of events on Ω
F_X	the (cumulative) distribution function of the RV X
$g_X(z)$	the generating function of the RV X
I	the unity matrix/operator

I_a	the degenerate at the point a distribution
inf	infimum (the exact lower bound)
$\mathbf{N}$	the set of all natural numbers: $\{1, 2, \dots\}$
$N(\mu, \sigma^2)$	the normal distribution with mean μ and variance σ^2
$o(h)$	any function vanishing faster than h (as $h \to 0$)
$\mathbf{P}$	the probability symbol
P	the transition matrix ($= (p_{jk})$) of a Markov chain
$P^{(n)}$	the n-step transition matrix (in an MC)
P_X	the distribution of the RV X
$Po(\lambda)$	the Poisson distribution with parameter λ
$\mathbf{R}$	the real line: $\{x : -\infty < x < \infty\}$
$\mathbf{R}^d$	the d-dimensional Euclidean space
$R(i, a)$	reward function (state i, action a)
sup	supremum (the exact upper bound)
^{T}	the operation of transposition
$U(a, b)$	the uniform distribution on (a, b)
$V_n(i)$	the optimal value function
Var (X)	the variance of the RV X
$\mathbf{Z}$	the set of all integers: $\{\dots, -2, -1, 0, 1, 2, \dots\}$
$\Gamma(\alpha)$	the gamma function
$\gamma(\lambda, \alpha)$	the gamma distribution with parameters λ, α
$\Phi(x)$	the standard normal distribution function
$\varphi_X(t)$	the moment generating function of the RV X
Ω	sample space
ω	an elementary outcome ("chance": a point $\omega \in \Omega$)

Abbreviations

ACF	autocorrelation function
ACVF	autocovariance function
a.s.	almost surely
B+DP	birth-and-death process
BB	Brownian bridge
BKE	backward Kolmogorov equation
BM	Brownian motion
CE	conditional expectation
ChF	characteristic function
CLT	central limit theorem
DF	distribution function (same as cdf)
DIY	do it yourself
EDF	empirical (or sample) distribution function
EMM	equivalent martingale measure
FDD	finite-dimensional distribution
FKE	forward Kolmogorov equation
GCD	greatest common divisor
GF	generating function
iff	if and only if
i.i.d.	independent and identically distributed
JMP	jump Markov process
LF	linear filter
LLN	law of large numbers
MC	Markov chain
MG	martingale
MGF	moment generating function
MP	Markov process
QS	queueing system
RV	random variable
RP	renewal process
RW	random walk
SDE	stochastic differential equation
SP	stochastic process
ST	stopping time
TPF	total probability formula
TS	time series
w.l.o.g.	without loss of generality
WN	white noise
w.r.t.	with respect to
w.p.	with probability

Index

Printed in the United States
By Bookmasters